Numerical Modelling of Material Deformation Processes

Research, Development and Applications

Peter Hartley, Ian Pillinger
and Clive Sturgess (Eds.)

Numerical Modelling of Material Deformation Processes

Research, Development and Applications

With 274 Figures

Springer-Verlag
London Berlin Heidelberg New York
Paris Tokyo Hong Kong
Barcelona Budapest

P. Hartley, PhD
I. Pillinger, PhD
C.E.N. Sturgess, PhD
School of Manufacturing and Mechanical Engineering, University of Birmingham, Edgbaston, Birmingham B15 2TT, UK

Cover illustrations: Ch. 15, Fig. 18. Square box: deformed mesh at 27 min. Ch. 10, Fig. 8. Closed die forging of a horizontal rectangular bar: generalized plastic strain.

British Library Cataloguing in Publication Data
Numerical modelling of material deformation processes:
research, development and applications.
I. Hartley, P. (Peter) II. Pillinger, Ian
III. Sturgess, C.E.N. (Clive E.N.)
620.112
ISBN-13: 978-1-4471-1747-6 e-ISBN-13: 978-1-4471-1745-2
DOI: 10.1007/978-1-4471-1745-2
Library of Congress Cataloging-in-Publication Data
Numerical modelling of material deformation processes:
research, development and applications/edited by P. Hartley, I. Pillinger, C.E.N. Sturgess.
p. cm.
Includes bibliographical references and index.
1. Metal-work–Mathematical models. 2. Deformations (Mechanics) – Mathematical models. 1. Hartley, Peter, 1951–. II. Pillinger, I. (Ian), 1955–/ III. Sturgess, C.E.N. (Clive Edward N.), 1942–.
TS213.N86 1992 91–30699
671.3–dc20 CIP

Softcover reprint of the hardcover 1st edition 1992

Typeset by Photo·graphics, Honiton, Devon
69/3830–543210 Printed on acid-free paper

To Our Families

Claire
David
Elle
Rebecca
Sally
Sheila
Tom

Preface

The principal aim of this text is to encourage the development and application of numerical modelling techniques as an aid to achieving greater efficiency and optimization of metal-forming processes. The contents of this book have therefore been carefully planned to provide both an introduction to the fundamental theory of material deformation simulation, and also a comprehensive survey of the "state-of-the-art" of deformation modelling techniques and their application to specific and industrially relevant processes. To this end, leading international figures in the field of material deformation research have been invited to contribute chapters on subjects on which they are acknowledged experts.

The information in this book has been arranged in four parts: Part I deals with plasticity theory, Part II with various numerical modelling techniques, Part III with specific process applications and material phenomena and Part IV with integrated computer systems.

The objective of Part I is to establish the underlying theory of material deformation on which the following chapters can build. It begins with a chapter which reviews the basic theories of classical plasticity and describes their analytical representations. The second chapter moves on to look at the theory of deforming materials and shows how these expressions may be used in numerical techniques. The last two chapters of Part I provide a review of isotropic plasticity and anisotropic plasticity.

The five chapters of Part II each take one particular numerical modelling technique, describe in detail the theoretical basis of the approach and examine some examples of the model. Chap. 5 considers the upper-bound method and shows how predictions of forging loads can be obtained very quickly, although predictions of flow may be approximate. Chap. 6 looks at the boundary-element method which has yet to find widespread use in forming process modelling. This chapter concentrates on non-Newtonian flow models of particular relevance for the deformation of polymers. Plane-strain extrusion of polymers with square and triangular cross-sections and the calendering of rolled polymer sheet are described. Consideration is given to how different numerical techniques can be combined to take advantage of the useful characteristics of each. The remaining chapters in Part II look at the finite-element method. Chap. 7 examines the rigid-plastic finite-element method and provides examples of its application to isothermal forging of aluminium and also to the forming of titanium cylinders. Chap. 8 introduces the viscoplastic finite-element method and shows how it can be applied to a range of processes such as continuous casting, rolling, forging and superplastic forming of thin sheet. Computational aspects such as thermal coupling, error estimates, explicit and implicit schemes and mesh optimization are discussed. Part

II concludes with a chapter on the elastic-plastic finite-element method and describes simulations of two very different industrial processes: cropping and rolling.

Part III consists of seven chapters, the first four of which concentrate on a specific process. Chap. 10 looks at the numerical modelling of forging; Chap. 11 deals with rolling processes; Chap. 12 describes extrusion and drawing; and Chap. 13 concentrates on sheet-forming processes. In each case, the process and numerical technique used is described and particular problems associated with modelling each process are discussed. Certain difficulties arising from interface friction are common to all the techniques and processes considered, and each chapter contains comments on how this is dealt with in each case.

The remaining three chapters of Part III deal with topics related to material behaviour. Chap. 14 introduces the complex subject of modelling the behaviour of metal–matrix composites. A rigid-plastic finite-element method is used to model the deformation behaviour of the matrix and an elastic-plastic technique used to assess the stress distribution at the interface between the matrix and the inclusions. Chap. 15 provides a thorough description of superplasticity and describes viscoplastic finite-element methods used to model superplastic sheet forming. The final chapter of Part III gives an overview of the criteria that have been proposed to predict ductile fracture. This chapter describes how these criteria have been used in conjunction with an elastic-plastic finite-element method in order to assess which of them has the potential for general process application.

The use of finite-element simulations is slowly becoming more common in a variety of industries, but the greatest benefit will only be achieved when used in conjunction with other CAD/CAM systems, for tool design for example. The two chapters of Part IV address this question. Chap. 17 considers the integration of numerical modelling techniques with CAD/CAM systems and puts particular emphasis on the interfaces required to make such integrated systems effective. Chap. 18 considers the use of artificial intelligence methodologies and describes some current developments in the integration of tool design and flow simulation methods using knowledge-based techniques.

It is hoped that this book will contribute to a greater understanding of numerical techniques for material deformation process modelling. The wealth of practical applications contained in the following chapters demonstrate the tremendous potential that exists for using numerical modelling methods to increase the efficiency of industrial material deformation processes and to improve product quality, structure and reliability.

Birmingham
July 1991

P.H.
I.P.
C.E.N.S.

Contents

Part IV. Computer Integrated Systems

Contributors

Dr C. Agelet de Saracibar
Escuela Tecnica Superior de Ingenerios de Caminos,
Canales y Puertos de Barcelona,
Universidad Politécnica de Cataluña,
08034 Barcelona,
Spain

Professor J.M. Alexander
Emeritus Professor, University of Wales,
Adjunct Professor, Ohio University,
Rowan Cottage,
Furze Hill Road,
Headley Down,
Hampshire GU3F 8NP,
UK

Professor T. Altan
Engineering Research Center for Net Shape Manufacturing,
Ohio State University,
1971 Neil Avenue,
Columbus, Ohio 43210,
USA

Dr-Ing. G.H. Arfmann
Institut für Bildsame Formgebung,
RWTH Aachen,
Intzestrasse 10,
510 Aachen,
Germany

Dr P. Bate
School of Metallurgy and Materials,
University of Birmingham,
Edgbaston,
Birmingham B15 2TT,
UK

Dr M. Bellet
Centre de Mise en Forme des Materiaux,
Ecole des Mines de Paris
et URA CNRS No 1374,
Sophia-Antipolis 06560,
Valbonne,
France

Dr J. Bonet
Department of Civil Engineering,
University College of Swansea,
University of Wales,
Singleton Park,
Swansea SA2 8PP,
Wales, UK

Professor A.N. Bramley
School of Mechanical Engineering,
University of Bath,
Claverton Down,
Bath BA2 7AY,
UK

Professor S. Cescotto
MSM – Institut du Génie Civil,
Quai Banning 6,
B-4000 Liège,
Belgium

Professor J.-L. Chenot
Centre de Mise en Forme des Materiaux,
Ecole des Mines de Paris
et URA CNRS No 1374,
Sophia-Antipolis 06560,
Valbonne,
France

Dr S.E. Clift
School of Mechanical Engineering,
University of Bath,
Claverton Down,
Bath BA2 7AY,
UK

Professor C.K.H. Dharan
Department of Mechanical Engineering,
University of California,
Berkeley,
California 94720,
USA

Dr F.R. Hall
Interdisciplinary Research Centre in Materials for High Performance Applications,
University of Birmingham,
Edgbaston,
Birmingham B15 2TT,
UK

Dr P. Hartley
School of Manufacturing and Mechanical Engineering,
University of Birmingham,
Edgbaston,
Birmingham B15 2TT,
UK

Professor J.E. Jackson, Jr
Department of Aerospace,
College of Engineering,
University of Alabama,
Tuscaloosa,
Alabama 35487,
USA

Professor S. Kobayashi
Department of Mechanical Engineering,
University of California,
Berkeley,
California 94720,
USA

Prof. Dr-Ing. R. Kopp
Institut für Bildsame Formgebung,
RWTH Aachen,
Intzestrasse 10,
510 Aachen,
Germany

Professor J.G. Lenard
Department of Mechanical Engineering,
University of Waterloo,
Waterloo, Ontario,
Canada, N2L 3G1

Dr R. Miller
Engineering Research Center for Net Shape Manufacturing,
Ohio State University,
1971 Neil Avenue,
Columbus,
Ohio 43210,
USA

Professor K. Mori
Department of Mechanical Engineering,
Kyoto Institute of Technology,
Matsugasaki,
Sakyo-ku,
Kyoto 606,
Japan

Dr F.H. Osman
School of Mechanical Engineering,
University of Bath,
Claverton Down,
Bath BA2 7AY,
UK

Professor E. Oñate
Escuela Tecnica Superior de Ingenerios de Caminos,
Canales y Puertos de Barcelona,
Universidad Politécnica de Cataluña,
08034 Barcelona
Spain

Dr N. Phan-Thien
Department of Mechanical Engineering,
University of Sydney,
Sydney,
NSW 2006,
Australia

Dr M. Pietrzyk
Instytut Metallurgii AGH,
Mickiewicza 30,
30–059 Krakow,
Poland

Dr I. Pillinger
School of Manufacturing and Mechanical Engineering,
University of Birmingham,
Edgbaston,
Birmingham B15 2TT,
UK

Mr M.S. Ramesh
College of Engineering,
Department of Mechanical Engineering,
Clemson University,
Clemson,
South Carolina 29634,
USA

Mr S.M. Roberts
School of Manufacturing and Mechanical Engineering,
University of Birmingham,
Edgbaston,
Birmingham B15 2TT,
UK

Professor R.I. Tanner
Department of Mechanical Engineering,
University of Sydney,
Sydney,
NSW 2006,
Australia

Professor P. Van Houtte
Department of Metallurgy and Materials Engineering,
Katholieke Universiteit Leuven,
de Croylaan 2,
3030 Heverlee,
Leuven,
Belgium

Dr R.D. Wood
Department of Civil Engineering,
University College of Swansea,
University of Wales,
Singleton Park,
Swansea SA2 8PP,
Wales, UK

Part I
Introduction to Plasticity Theory

1 Classical Plasticity

J.M. Alexander

1.1 General Theory

A complete treatment of the classical mathematical theory of plasticity is not possible within the confines of a single chapter; comprehensive early texts have been written by Nadai (1931), Hill (1950), and Prager and Hodge (1951). Later treatments of the theory of plasticity, discussed specifically with the needs of engineers in mind, are given by Johnson and Mellor (1973), Ford and Alexander (1977), Rowe (1979) and Alexander (1981). The present discussion is based on that given in Alexander, Brewer and Rowe (1987, Vol. 2, Chap. 2) and is intended simply to consider the basic fundamentals of the theory, to show how a better understanding of the behaviour of materials subjected to deformation processes can be achieved.

In material deformation processes the amount of straining or deformation is many times larger than that which occurs during simple elastic straining. The solution of a problem in elasticity can usually be achieved by making the simplifying assumption that the external shape is unaltered during elastic straining. This is justified for most materials of interest since their elastic range never exceeds about 1.0% strain. Processes of material deformation demand that strains of more than 50 times this value be imposed. The first result of having such large deformations is the necessity for specifying some better measure of strain than simply (change in length)/(original length). Another consequence of having large strains of this order is that, as far as determining the changes of shape and the forces required are concerned, it will be permissible to neglect elastic strains. This considerably simplifies the theoretical approach, but it should be remembered that elastic deformation is very important when considering residual stresses.

Bearing in mind the large deformations which are imposed in most processes for the deformation of materials, it may be queried why it is that such large deformations can be imposed when, in a tensile test, most materials will never withstand more than about 50% elongation without fracture or instability occurring. Consideration of the major processes reveals the answer – they are all predominantly compressive in nature. Any stress system can be regarded as the sum of an all-round *hydrostatic stress* (usually taken as being equal to the *mean stress*) and stresses equal to the differences between the actual stresses and this mean stress. It was shown experimentally by Bridgman (1952), and later again by Crossland (1954), that there is no plastic flow occasioned by this hydrostatic stress. The plastic flow is caused by the *deviatoric* stress system, so called because the stresses take the values by which the actual stresses "deviate" from this mean stress. Although the mean stress is not responsible for any plastic flow it has a profound influence on fracture, and the more compressive its value the more deformation can be imposed before failure occurs due to fracture. A corollary of the observation that the mean stress occasions no plastic flow is that there can be no permanent change of volume. Thus, if elastic strains are neglected, it may reasonably be assumed that there is no change of volume *at all* during plastic flow.

Now, all material deformation processes take place under conditions of complex stressing, as distinct from simple uniaxial tension or compression. Thus, in formulating a theory, a general three-dimensional system of stress must be introduced and it may be asked how the behaviour of material subjected to complex systems of stress can be correlated with its behaviour in simple tension or compression. To answer this question it is convenient to invoke the concepts of *equivalent* or *effective stress* and *strain*.

Having discussed the principles on which the theory rests, it is now possible to develop the equations of plastic flow. Initially, what may be called the *laws* of elasticity and the *laws* of plasticity can be set up and compared, as follows:

- *Laws of elasticity*
  Hooke's law
  Equations of equilibrium (of stresses and forces)
  Equations of compatibility (of strains and displacements)
  Boundary conditions for both stresses and displacements
- *Laws of plasticity*
  Stress–strain relations
  Equations of equilibrium
  Equations of compatibility
  Yield criterion (function of stresses required for plastic flow)
  Boundary conditions for both stresses and displacements

Because of the magnitude of the strains and the nonlinearity of the relationships involved in plastic deformation, it is necessary to formulate the equations in terms of incremental strains or strain rates. This difficulty does not arise for elastic straining, since the strains are always sufficiently small for the total strains to be used. Thus, Hooke's law for a three-dimensional system of stress, referred to x,y,z Cartesian coordinates, is as follows:

$$\epsilon_{xx} = \frac{1}{E}[\sigma_{xx} - \nu(\sigma_{yy}+\sigma_{zz})],\ \epsilon_{yy} = \frac{1}{E}[\sigma_{yy}-\nu(\sigma_{zz}+\sigma_{xx})],\ \epsilon_{zz} = \frac{1}{E}[\sigma_{zz}-\nu(\sigma_{xx}+\sigma_{yy})]$$

$$\gamma_{xy} = \frac{1}{G}\tau_{xy},\ \gamma_{yz} = \frac{1}{G}\tau_{yz},\ \gamma_{zx} = \frac{1}{G}\tau_{zx} \tag{1.1}$$

In these well-known equations of elasticity representing Hooke's law, τ_{xy} is the stress in the y direction acting on a plane normal to the x direction, γ_{xy} is the corresponding engineering shear strain, σ_{xx} is a positive (tensile) stress and ϵ_{xx} a positive strain in the x direction, etc., E is *Young's modulus*, G is the *shear modulus*, ν is *Poisson's ratio*, and $E = 2G(1+\nu)$. By adding together the first three equations of Eqs (1.1) the elastic volume change is determined as:

$$(\epsilon_{xx}+\epsilon_{yy}+\epsilon_{zz}) = \frac{1-2\nu}{E}(\sigma_{xx}+\sigma_{yy}+\sigma_{zz}) \tag{1.2}$$

The factor $E/(1-2\nu)$ is three times the *bulk modulus* K of the material, since $(\sigma_{xx}+\sigma_{yy}+\sigma_{zz})/3$ is the mean stress, or hydrostatic component of the stress system. Thus, if the volume change is to be zero in plastic flow, Poisson's ratio ν must be replaced by the factor $\frac{1}{2}$ in the stress–strain relations, since the right-hand side of Eq. (1.2) then becomes zero. Young's modulus E has no meaning for plastic flow in which the elastic strains are neglected, and must be replaced by an analogous parameter which will be considered later.

In any element of material subjected to a complex system of stress there are three mutually perpendicular directions in which the local direct stresses attain either maximum or minimum values. These direct stresses act on planes on which the shear stresses are zero, and these planes are known as "principal" planes, the direct stresses as "principal" stresses. If these principal directions are denoted by 1, 2 and 3, Eqs (1.1), for example, reduce to three equations:

$$\epsilon_1 = \frac{1}{E}[\sigma_1 - \nu(\sigma_2+\sigma_3)],\ \epsilon_2 = \frac{1}{E}[\sigma_2 - \nu(\sigma_3+\sigma_1)],\ \epsilon_3 = \frac{1}{E}[\sigma_3 - \nu(\sigma_1+\sigma_2)] \quad (1.3)$$

It will be seen later that the stress–strain equations of plastic flow will be similar to Eqs (1.3), but with ν replaced by $\frac{1}{2}$ and with strains such as ϵ_1 replaced either by the incremental strain $\delta\epsilon_1$, or strain rate $\dot{\epsilon}_1 = \partial\epsilon_1/\partial t$, where t is time or some quantity which increases in proportion to time. The parameter $1/E$ will disappear, however, being replaced by a term which will have the same value in each of the Eqs (1.3). The actual value of this term, determining the relationship existing between the plastic strain and the stresses causing plastic flow will depend both on the work-hardening characteristics of the material and/or the constraints of neighbouring elements or boundaries of the element of material under consideration.

1.1.1 The Principal Stress Cubic

To give an actual value to this term, consider in more detail the concepts of equivalent stress and equivalent strain already referred to. Firstly, considering the equivalent stress, if the material is isotropic and homogeneous, then this equivalent stress must be a function of the "invariants" of the stress system. Without considering the mathematical theory in detail, these invariants may be described as those functions of the stresses at a point in the material which do not vary when referred to different sets of Cartesian coordinate axes through the point. One of the easiest ways of deriving them is to obtain the cubic equation whose roots give the principal stresses at the point. In full, this cubic equation is as follows:

$$p^3 - p^2(\sigma_{xx}+\sigma_{yy}+\sigma_{zz}) - p(\tau_{xy}^2+\tau_{yz}^2+\tau_{zx}^2-\sigma_{xx}\sigma_{yy}-\sigma_{yy}\sigma_{zz}-\sigma_{zz}\sigma_{xx})$$
$$-(\sigma_{xx}\sigma_{yy}\sigma_{zz}+2\tau_{xy}\tau_{yz}\tau_{zx}-\sigma_{xx}\tau_{yz}^2-\sigma_{yy}\tau_{zx}^2-\sigma_{zz}\tau_{xy}^2) = 0 \quad (1.4)$$

1.1.1.1 Determination of the Principal Stresses and Their Directions

The problem of solving cubic equations was apparently first studied by the Italian mathematician Cardano in the sixteenth century, as pointed out by Johnson and Mellor (1973). Cardano's method consists essentially of replacing the variables of the cubic equation with trigonometrical functions, thus allowing an analytical solution to be found. A detailed discussion of his method, including some computer and programmable calculator programs has been given by Alexander (1981) and Alexander et al. (1987).

1.1.2 The Yield Criterion

The three roots of Eq. (1.4) (p being the unknown quantity) give the principal stresses $p_1 = \sigma_1$, $p_2 = \sigma_2$ and $p_3 = \sigma_3$ at the point considered. Now it should not matter how the original set of x,y,z axes is orientated with regard to evaluating σ_1, σ_2 and σ_3 from this equation. These principal stresses should always have the same value for the particular state of stress concerned. For this to be true the coefficients of p^3, p^2 and p should always be the same in Eq. (1.4). In other words, the functions of the stresses in the terms within the brackets in Eq. (1.4) are invariant with respect to rotation of the x,y,z axes in the element considered. This is a most important concept to understand, since it can then be said that any property of the idealized isotropic homogeneous material, such as the criterion of yielding, criterion of fracture, criterion of creep, or fatigue even, should be a function of the invariants of the stress system. The invariants are usually denoted as follows.

1.1.2.1 Invariants of the Stress Tensor

These are essentially the coefficients of the principal stress cubic (Eq. (1.4)), as follows:

$$
\begin{aligned}
J_1 &= \sigma_1+\sigma_2+\sigma_3 = \sigma_{xx}+\sigma_{yy}+\sigma_{zz}\\
J_2 &= -\sigma_1\sigma_2-\sigma_2\sigma_3-\sigma_3\sigma_1 = \tau_{xy}^2+\tau_{yz}^2+\tau_{zx}^2-\sigma_{xx}\sigma_{yy}-\sigma_{yy}\sigma_{zz}-\sigma_{zz}\sigma_{xx} \qquad (1.5)\\
J_3 &= \sigma_1\sigma_2\sigma_3 = \sigma_{xx}\sigma_{yy}\sigma_{zz}+2\tau_{xy}\tau_{yz}\tau_{zx}-\sigma_{xx}\tau_{yz}^2-\sigma_{yy}\tau_{zx}^2-\sigma_{zz}\tau_{xy}^2
\end{aligned}
$$

Thus, for this particular problem, the yield criterion, and consequently the effective stress, must be a function of the invariants J_1, J_2 and J_3. The invariant J_1 is seen to be three times the arithmetic mean stress $\sigma = (\sigma_1+\sigma_2+\sigma_3)/3$, and since plastic flow has been shown experimentally to be independent of the mean stress, the effective stress should not depend on J_1, at least as far as yielding is concerned. As stated before, J_1 would be important in establishing a criterion of fracture. J_2 and J_3, in contrast, involve the deviatoric stresses and are therefore directly related to the criterion of yielding.

1.1.2.2 Hydrostatic and Deviatoric Stresses

The easiest way of making the yield criterion independent of the mean stress is by making it a function of the shear stress components of the stress system, since these are unaffected by the magnitude of the mean stress in the system.

As mentioned before, the given system of stress represented by the principal stresses σ_1, σ_2 and σ_3 may be regarded as the sum of a system of hydrostatic stresses equal to the arithmetic mean stress $[\sigma_m = (\sigma_1+\sigma_2+\sigma_3)/3]$ plus the deviatoric stress σ_1', σ_2', σ_3'. The deviatoric stresses are so called because they represent the amount by which the actual stresses *deviate* from the mean stress. Thus

$$\sigma_1' = \sigma_1-\sigma_m, \quad \sigma_2' = \sigma_2-\sigma_m \quad \text{and} \quad \sigma_3' = \sigma_3-\sigma_m \qquad (1.6)$$

The stresses which are effective in causing plastic flow may be regarded either as functions of the deviatoric stresses (a concept preferred by mathematicians) or as functions of the shear stresses (a concept preferred by engineers).

1.1.2.3 Yield Criteria of Maxwell (Huber–von Mises) and Tresca (Guest)

There are two important yield criteria which are used in practice, namely that of Maxwell, often attributed to von Mises or Huber, and that of Tresca or Guest. A discussion of the criteria is given by Sopwith and Morrison (1956) and a list of references is to be found in the book by Hill (1950). It is possible to show that both of these criteria are functions of the invariants of the stresses and also to attach some physical significance to the functions involved.

The Maxwell criterion, in its simplest form, may be expressed in terms of the equivalent stress, given by the invariant function

$$\bar{\sigma} = \frac{1}{\sqrt{2}} \sqrt{(\sigma_1 - \sigma_2)^2 + (\sigma_2 - \sigma_3)^2 + (\sigma_3 - \sigma_1)^2} \tag{1.7}$$

This quantity is readily shown to be proportional to either the root mean square shear stress, or the shear stress which exists on the octahedral planes (the planes whose normals make equal angles with the three principal stress directions in the element); or the square root of the energy of distortion. The factor $1/\sqrt{2}$ has been chosen so that for uniaxial stressing ($\sigma_2 = \sigma_3 = 0$), $\bar{\sigma}$ is equal to the applied stress σ_1. When $\bar{\sigma}$ attains a certain value (the *yield* or *flow* stress, often denoted by Y or σ_f and usually derived experimentally), yielding or plastic flow will occur.

The Tresca criterion may be expressed in an analogous way, in terms of the invariant function

$$\bar{\sigma} = \sigma_{max} - \sigma_{min} \tag{1.8}$$

This quantity is seen to be proportional to the "maximum" maximum shear stress operating and can be seen to be inferior to the Maxwell criterion in that no "weight" is given to the intermediate stress. The difference between the two criteria never exceeds about 15% in terms of the stresses.

The equivalent or effective strain increment may again be expressed in terms of invariant functions of the strain increments, either as (for von Mises)

$$\delta\bar{\epsilon} = \sqrt{\tfrac{2}{9}\left[(\delta\epsilon_1 - \delta\epsilon_2)^2 + (\delta\epsilon_2 - \delta\epsilon_3)^2 + (\delta\epsilon_3 - \delta\epsilon_1)^2\right]} \tag{1.9}$$

or as (for Tresca)

$$\delta\bar{\epsilon} = \tfrac{2}{3}(\delta\epsilon_{max} - \delta\epsilon_{min}) \tag{1.10}$$

Again, the factors $\sqrt{\frac{2}{9}}$ and $\frac{2}{3}$ in these equations have been chosen so that for uniaxial stressing (i.e. $\delta\epsilon_2 = \delta\epsilon_3 = -\frac{1}{2}\delta\epsilon_1$, since $\delta\epsilon_1 + \delta\epsilon_2 + \delta\epsilon_3 = 0$), $\delta\bar{\epsilon}$ is equal to the applied strain increment $\delta\epsilon_1$. It may be noted in passing that the increment of *work done* in any deformation is given by the equation

$$\delta W = \sigma_1\delta\epsilon_1 + \sigma_2\delta\epsilon_2 + \sigma_3\delta\epsilon_3 \tag{1.11}$$

The Tresca criterion gives an incorrect prediction of the amount of work done in any deformation. In extensive experiments carried out long ago by Taylor and Quinney, Hohenemser, Morrison and Shepherd, and many others, it was found that the Maxwell criterion does, in general, represent the yielding behaviour of most ductile materials.

1.1.2.4 *Stress–Strain Relations for Plastic Flow*

To return to the question of determining the stress–strain relations in plastic flow, the following relations may be written, by analogy with the elastic equations (1.3)

$$\delta\epsilon_1 = \delta\lambda[\sigma_1-\tfrac{1}{2}(\sigma_2+\sigma_3)],\ \delta\epsilon_2 = \delta\lambda[\sigma_2-\tfrac{1}{2}(\sigma_3+\sigma_1)],\ \delta\epsilon_3 = \delta\lambda[\sigma_3-\tfrac{1}{2}(\sigma_1+\sigma_2)] \tag{1.12}$$

The multiplier $\delta\lambda$ is assumed to be a constant for the particular element and particular moment in time considered, and is analogous to $1/E$ in the elastic equations, the factor $\frac{1}{2}$ being analogous to ν. By substituting the values for the strain increments given by Eqs (1.12) in Eq. (1.9), which may be written in a simpler form (using the fact that $\delta\epsilon_1+\delta\epsilon_2+\delta\epsilon_3 = 0$) as

$$\delta\bar{\epsilon} = \sqrt{\tfrac{2}{3}[\delta\epsilon_1^2+\delta\epsilon_2^2+\delta\epsilon_3^2]} \tag{1.9a}$$

it is easily shown that

$$\delta\bar{\epsilon} = \delta\lambda\bar{\sigma} \tag{1.12a}$$

Thus $\delta\lambda = \delta\bar{\epsilon}/\bar{\sigma}$, and Eqs (1.12) may be written as follows

$$\delta\epsilon_1 = \frac{\delta\bar{\epsilon}}{\bar{\sigma}}[\sigma_1-\frac{1}{2}(\sigma_2+\sigma_3)],\ \delta\epsilon_2 = \frac{\delta\bar{\epsilon}}{\bar{\sigma}}[\sigma_2-\frac{1}{2}(\sigma_3+\sigma_1)],\ \delta\epsilon_3 = \frac{\delta\bar{\epsilon}}{\bar{\sigma}}[\sigma_3-\frac{1}{2}(\sigma_1+\sigma_2)] \tag{1.13}$$

It can be seen from this that the parameter in the stress–strain relations of plastic flow which is analogous to Young's modulus in elasticity is the quantity $\bar{\sigma}/\delta\bar{\epsilon}$. The analogy is now complete in that Poisson's ratio ν is replaced by the factor $\frac{1}{2}$ required for zero volume change and Young's modulus E is replaced by the ratio of the equivalent stress divided by the equivalent strain increment, namely $\bar{\sigma}/\delta\bar{\epsilon}$.

If the material work-hardens the effective stress and effective strain increment will be interdependent, through the basic stress–strain curve of the material. This may be written as $\bar{\sigma} = H(\int d\bar{\epsilon})$ where $\bar{\sigma}$ and $\int d\bar{\epsilon}$ are simply the true stress and integrated plastic strain increment measured in a tension or compression test. The slope of this plastic stress–strain curve will be $H' = d\bar{\sigma}/d\bar{\epsilon}$, so that $\delta\bar{\epsilon}$ in Eqs (1.13) may be replaced by the quantity $\delta\bar{\sigma}/H'$. Clearly in many cases it will not be possible to solve these equations analytically, and a step-by-step method of solution must be adopted. If the material does not work-harden the effective strain increment is not so important, and the factor $\delta\bar{\epsilon}/\bar{\sigma} = \delta\lambda$ is generally eliminated between Eqs (1.13) by division. Such problems can often be solved for work-hardening materials by using a constant mean value of the yield stress, $\bar{Y}$ or Y_m.

1.1.2.5 *Incremental and Large Strains*

It is now possible to consider what is meant by strain when considering the case of large deformations. To fix ideas, consider the uniaxial stressing of a solid round rod in tension. If the current length of the specimen is l, and there is imposed on it an increment of plastic strain $\delta\epsilon_1$ in the axial direction (still neglecting elastic strains), then $\delta\epsilon_1 = \delta l/l$, where δl is the small increase of length of the specimen. The total strain in a large extension of the specimen can be defined as

$$\epsilon_1 = \int_0^{\epsilon_1} d\epsilon_1$$

and if l_0 is the original length of the specimen, and l_1 the final length, then

$$\epsilon_1 = \int_0^{\epsilon_1} d\epsilon_1 = \int_{l_0}^{l_1} \frac{dl}{l} = \ln\left(\frac{l_1}{l_0}\right) \tag{1.14}$$

This is the well-known logarithmic or true strain. If, as in common engineering handbooks, the symbol $e_1 = (l_1 - l_0)/l_0$ is used, then $e_1 = (l_1/l_0) - 1$ and

$$\epsilon_1 = \ln(1 + e_1) \tag{1.15}$$

Similarly

$$\epsilon_2 = \int_0^{\epsilon_2} d\epsilon_2 = \int_{D_0}^{D_1} \frac{dD}{D}$$

where D is the current diameter of the specimen, etc. Thus

$$\epsilon_2 = \epsilon_3 = \ln(D_1/D_0) \tag{1.16}$$

Now since there is no volume change, $d\epsilon_2 = d\epsilon_3 = -\frac{1}{2}d\epsilon_1$, and by integration therefore, $\epsilon_2 = \epsilon_3 = -\frac{1}{2}\epsilon_1$. From Eqs (1.14) and (1.16) it is seen that

$$\ln\left(\frac{D_1}{D_0}\right) = -\frac{1}{2}\ln\left(\frac{l_1}{l_0}\right) \quad \text{or} \quad \frac{D_1}{D_0} = \sqrt{\frac{l_0}{l_1}}$$

or

$$l_1 D_1^2 = l_0 D_0^2 \tag{1.17}$$

as is required for constancy of volume. Thus, if true strains are used, the equation of constancy of volume for incremental strains, $\delta\epsilon_1 + \delta\epsilon_2 + \delta\epsilon_3 = 0$, may be extended to large strains, as $\epsilon_1 + \epsilon_2 + \epsilon_3 = 0$.

In any system of complex stress, to obtain a measure of the total amount of plastic straining suffered by any element of the material, it is necessary to add up all the increments of effective strain which the element has received, and the total effective strain may be written as

$$\bar{\epsilon} = \int_0^{\bar{\epsilon}} d\bar{\epsilon} \tag{1.18}$$

1.1.3 The Prandtl–Reuss Equations

These then, are the basic concepts underlying the theory of plasticity. The theory considered above applies only to a material in which elastic strains are neglected and there is zero volume change. This is permissible if it is desired only to estimate the deforming forces and displacements during plastic flow. If a more detailed picture of the deformation is required, as for example, an estimate of the residual stresses developed after some particular plastic deformation process, it would be necessary to include elastic strains. The problem then becomes more complicated, although the basic equations are easy enough to establish. In effect, the total strain increment is now simply the elastic strain increment plus the plastic strain increment and the basic stress–strain relations due to Prandtl (1924) and Reuss (1930) may be stated as follows

$$\left.\begin{aligned} \delta\epsilon_1 &= \frac{\delta\bar{\epsilon}}{\bar{\sigma}}[\sigma_1-\tfrac{1}{2}(\sigma_2+\sigma_3)] + \frac{1}{E}[\delta\sigma_1-\nu(\delta\sigma_2+\delta\sigma_3)] \\ \delta\epsilon_2 &= \frac{\delta\bar{\epsilon}}{\bar{\sigma}}[\sigma_2-\tfrac{1}{2}(\sigma_3+\sigma_1)] + \frac{1}{E}[\delta\sigma_2-\nu(\delta\sigma_3+\delta\sigma_1)] \\ \delta\epsilon_3 &= \frac{\delta\bar{\epsilon}}{\bar{\sigma}}[\sigma_3-\tfrac{1}{2}(\sigma_1+\sigma_2)] + \frac{1}{E}[\delta\sigma_3-\nu(\delta\sigma_1+\delta\sigma_2)] \end{aligned}\right\} \tag{1.19}$$

It should be noted that the plastic strain increments such as

$$\delta\epsilon_1^p = \frac{\delta\bar{\epsilon}}{\bar{\sigma}}[\sigma_1-\tfrac{1}{2}(\sigma_2+\sigma_3)]$$

are associated with the *total* stresses, whilst the elastic strain increments such as

$$\delta\epsilon_1^e = \frac{1}{E}[\delta\sigma_1-\nu(\delta\sigma_2+\delta\sigma_3)]$$

are associated with *incremental changes* of the stresses. These concepts are in accord with experimental observation, but contain the inherent assumption that elastic and plastic strains (or stresses) are separable. In practice, the yield stress is not only dependent on strain but also on strain rate and temperature, as discussed at length in Alexander et al. (1987, Vol. 1, Chap. 1).

1.2 Plane Strain – The Slip-Line Field

In all the preceding discussion the plastic deformation of an element of material has been assumed to occur as though it were a single fibre disconnected from the rest of the material. In any deformation process the geometrical configuration of the tools, frictional condition at the boundaries, and the constraint of neighbouring material will result in the plastically deforming material being subjected to a "field" of stress and strain which differs from point to point in the region of plastic deformation. The existence of this "field" of differing stress–strain states enormously complicates the task of obtaining a solution, and it is only by making drastic assumptions that any information can be obtained. Modern finite-element analysis can provide detailed predictions, as discussed later, but it requires large computing power which can be costly.

For the purpose of estimating the deforming loads it is often adequate in some processes such as strip rolling to neglect the variation in distortion over the deforming cross-section, and assume that "plane sections remain plane". To obtain information about the severity of the distortion, or of the effect of the process on the material, it is necessary to develop a "field" type of theory, which will give a picture of the deformation from point to point in the material. Up to the present the only rigorous means of developing such a "field" theory has been by making sweeping assumptions about the material and the way in which it is deformed. This has led to the development of *slip-line field theory*.

In addition to neglecting elastic strains (equivalent to assuming that the material has an infinite Young's modulus), work hardening is neglected, so that the material is assumed to flow under a constant yield stress. This hypothetical material is often referred to as a "rigid-plastic" material, since it is either being plastically deformed, or it is rigid in regions where the stresses are below those necessary to cause plastic flow. The constant yield strength is generally denoted by Y. In addition to these assumptions about the material, it is also assumed that deformation takes place under conditions such that the strain is zero in one direction (i.e. under "plane strain" conditions), because the theory can deal accurately only with two-dimensional flow.

Although at first sight these assumptions appear rather restrictive, in fact they are not, since many materials and processes approximate to these conditions. Since the strains are generally large it is permissible to neglect the small elastic strain, and the constant flow stress is closely representative of the behaviour of all metals during hot working and fairly well of work-hardened metals in cold working. Plane strain conditions are realized approximately in hot and cold strip rolling and machining and closely approximated in the extrusion of rectangular bars, ironing of cups, and other similar processes such as tube drawing, for example. The reason that the assumptions have to be made is to produce a workable "field" theory, so it is fortunate that these assumptions are not too unrealistic.

Considering Eqs (1.12) (the stress–strain relations for the rigid-plastic material), if the principal strain increment $\delta\epsilon_2$ is zero, the principal stress σ_2 in that direction is seen to be the mean of the other two principal stresses, and therefore equal to the mean stress $\sigma_m = (\sigma_1+\sigma_2+\sigma_3)/3$. For plane strain conditions, i.e. $\dot{\epsilon}_2 = \partial\epsilon_2/\partial t = 0$ (t being time), then $\dot{\epsilon}_3 = -\dot{\epsilon}_1$ in order that the condition of no volume change ($\dot{\epsilon}_1+\dot{\epsilon}_2+\dot{\epsilon}_3 = 0$) be satisfied.

The maximum shear stress in the plane of flow is $k = (\sigma_1-\sigma_3)/2$ and if σ_x, σ_y and τ_{xy} are the direct and shear stresses acting at a general point in the plane of flow, it can be seen (most easily by sketching the Mohr's circle) that

$$\left[\frac{(\sigma_x-\sigma_y)^2}{4} + \tau_{xy}^2\right]^{\frac{1}{2}} = k \tag{1.20}$$

Applying the Maxwell criterion of flow, we find from Eq. (1.7) that the effective stress is

$$\bar{\sigma} = \sqrt{\tfrac{3}{2}}\,(\sigma_1-\sigma_3) = \sqrt{3}k \tag{1.21}$$

Thus, if the stresses increase until the radius k of Mohr's circle attains $Y/\sqrt{3}$, yielding will then occur.

Applying the Tresca criterion of flow, Eq. (1.8) shows that the effective stress is

$$\bar{\sigma} = \sigma_1-\sigma_3 = 2k \tag{1.22}$$

and yielding will occur when the radius of Mohr's circle attains the value $Y/2$.

Put in another way, there is no difference between the functional relationship between the stresses representing either the Maxwell or the Tresca criterion under plane strain conditions because the deformation occurs in pure shear. This can be summarized by using Eq. (1.20) for the yield criterion, noting that yielding will occur when $k = Y/\sqrt{3}$ for Maxwell (von Mises), or $Y/2$ for Tresca.

By sketching Mohr's circles of stress and strain rate, it may be seen that the direct stress acting on the plane which has the maximum shear stress k acting on it is the mean stress σ_m. Moreover, this plane will have associated with it the maximum shearing strain and zero direct strain. In fact, in the deforming material the planes of maximum shear form an orthogonal curvilinear network. Each element is subjected to the maximum shear stress k and the mean normal stress σ on its faces. The directions of the principal stresses σ_1 and σ_3 at any point are at 45° to the lines of maximum shear stress (often referred to as α and β lines). The third principal stress σ_2 acts normal to the plane, in the direction of zero strain. Since the stress system at any point in the deforming material is made up of the sum of the two systems $(\sigma,\sigma,\sigma) + (k,0,-k)$ and the hydrostatic stress system causes no plastic flow, the deformation may be regarded as that of a simple pure shear. Thus the planes of maximum shear are the important unknowns in any problem; it can be shown mathematically that they are the characteristics of the equations governing the flow of the material. Furthermore, since there can be no volumetric change in the hypothetical rigid-plastic material, the hydrostatic component of the stress system can cause no dimensional change at all, and the distortion consists of a network of simple shear.

The orthogonal network of characteristics or lines of maximum shear has come to be called the "slip-line field". This is considered by some to be rather an undesirable terminology, since lines of slip have a special meaning for the metallurgist, and "shear-line field" is a better term, as often used by Prager and Hodge (1951), and Prager (1959). Nevertheless, metallurgical slip tends to occur in directions of maximum resolved shear stress. In a polycrystalline material the "slip lines" are reliably indicative of the deformation. It can be demonstrated by considering the equilibrium of the forces on any curvilinear element that

$$\left.\begin{aligned} \frac{\partial\sigma}{\partial\alpha} + \frac{\partial k}{\partial\beta} - \frac{2k}{r_\alpha} &= 0 \text{ (along an } \alpha \text{ line)} \\ \frac{\partial\sigma}{\partial\beta} + \frac{\partial k}{\partial\alpha} + \frac{2k}{r_\beta} &= 0 \text{ (along a } \beta \text{ line)} \end{aligned}\right\} \tag{1.23}$$

If φ is the angular displacement of a characteristic from some datum and is positive *anticlockwise*, then the respective curvatures are $1/r_\alpha = \partial\varphi/\partial\alpha$ and $1/r_\beta = \partial\varphi/\partial\beta$ and Eqs (1.23) become, after integration,

$$\left.\begin{aligned} \sigma - 2k\varphi + \int \frac{\partial k}{\partial\beta}\, d\alpha + \int 2\varphi \frac{\partial k}{\partial\alpha}\, d\alpha &= \text{Constant}_1, \text{ along an } \alpha \text{ line} \\ \sigma + 2k\varphi + \int \frac{\partial k}{\partial\alpha}\, d\beta + \int 2\varphi \frac{\partial k}{\partial\beta}\, d\beta &= \text{Constant}_2, \text{ along a } \beta \text{ line} \end{aligned}\right\} \tag{1.24}$$

These equations, without the integrals and with σ replaced by $-p$ (p being the pressure on a slip line), are known as *Hencky's equations*. The additional integral terms were first suggested by Christopherson et al. (1958), who found it necessary to allow the material to work harden, in order to obtain a solution to the machining problem.

Hencky's equations are as follows:

$$\left.\begin{aligned} \delta p + 2k\delta\varphi &= 0, \text{ along an } \alpha \text{ line} \\ \delta p - 2k\delta\varphi &= 0, \text{ along a } \beta \text{ line} \end{aligned}\right\} \tag{1.24a}$$

Analogous equations for the velocities were derived by Geiringer (see Hill 1950) from the condition of zero extension along the shear lines. If the components of the total velocity of any point are u along the α line and v along the β line, then the Geiringer equations are

$$\left.\begin{aligned} \frac{\partial u}{\partial \varphi} - v &= 0, \text{ along an } \alpha \text{ line} \\ \frac{\partial v}{\partial \varphi} + u &= 0, \text{ along a } \beta \text{ line} \end{aligned}\right\} \tag{1.25}$$

and are simply a statement of the fact that there is zero extension along the slip line.

The theory of the slip-line field has existed for many years, and major advances have been made, notably the realization by Prager (1953) and Green (1951), independently, that the distortion could be "mapped" as a velocity diagram, a geometrical method of construction developed by Prager (1953), and the inclusion of strain hardening.

Prager's geometrical method allows the solution of problems (involving mixed boundary conditions) hitherto intractable by numerical methods (Alexander 1955), and gives a convenient representation of the solution to any problem.

1.3 Velocity Discontinuities

One important consequence of the assumption of a rigid-plastic material is that the displacements need not be continuous through the material. In other words, it is possible for there to be "block slipping" between neighbouring sectors in the deforming material. Considering a curved line along which such a discontinuity occurs let it be assumed that at any point P on the curve, the material on one side of it has velocity with components v and u normal and tangential to the curve respectively, and on the other side of the curve has velocity with corresponding components $\hat{v}$ and $\hat{u}$.

It is easily seen that the material is neither to "pile up" nor form a cavity at the point P, $\hat{v} = v$. There is no such restriction in the tangential components, and it is possible for there to be a discontinuity of amount $\hat{u}-u$. Furthermore, since such a discontinuity must correspond with an infinitely large rate of shearing strain, *the line of velocity discontinuity must also be a slip line*, and Geiringer's appropriate equation for zero extension must be satisfied along both sides of it. From the first Eq. (1.25), $\delta u = v\delta\varphi$ on the first side of the curve and $\delta\hat{u} = \hat{v}\delta\varphi$ on the other, in moving along an elemental length of the slip line. Since $\hat{v} = v$, and $\delta\varphi$ is common to both sides of the curve, $\delta\hat{u} = \delta u$, and the discontinuity in tangential velocity $\hat{u}-u$ must be constant along the slip line. In a real material such discontinuities are not possible and the velocity discontinuity may be imagined as the limiting state of a band or layer of very intense shear strain.

1.4 Stress Discontinuities

Discontinuities in the stress components are also possible. Assume that such a line of stress discontinuity exists and then consider the equilibrium of two similar adjacent elements, one on each side of the line. If x and y axes are chosen parallel and normal to the line respectively, then clearly $\hat{\sigma}_y = \sigma_y$, and $\hat{\tau}_{xy} = \tau_{xy}$, but it is not necessary for $\hat{\sigma}_x$ to be equal to σ_x. It is not necessary for the material to be deforming plastically for a stress discontinuity to exist, although if it is plastic then the maximum shear stress will be k, on each side of the line. It is easily shown that a line of stress discontinuity can *never* coincide with a slip line. One of the most obvious examples of a stress discontinuity occurs at a free surface such as that of a bent beam.

It is also easy to show that slip lines are "reflected" at a stress discontinuity. Moreover, the change in mean stress between corresponding slip lines in the regions on each side of the stress discontinuity, viz. $\hat{\sigma}-\sigma$, is seen to be

$$\hat{\sigma}-\sigma = 2k \sin 2\theta \tag{1.26}$$

where θ is the angle between the slip line and the stress discontinuity.

Since slip lines are reflected at a stress discontinuity, it is easy to prove that if a stress discontinuity meets an axis of symmetry (which must be a principal plane) it is effectively "reflected" at 90°. Thus the point of incidence becomes the meeting point of four lines of stress discontinuity, since it lies on an axis of symmetry.

In the real material, the plastic stress discontinuity may be envisaged as the last vestige of an elastic layer coinciding with it.

1.5 Limit Analysis

Limit analysis is a powerful technique which has been developed to give a more realistic idea of the collapse loads of structures, as described by Baker et al. (1956). It consists in finding upper and lower bounds to the collapse load and these can be brought closer and closer together by assuming more accurate pictures of the deformations and loads, until the desired degree of precision is attained. In considering deformation processes, it may be of assistance in replacing the "plane sections remain plane" type of theory for estimating deforming loads. Particularly is this true for processes like extrusion in which plane sections certainly do not remain plane, and Johnson (1959) has pioneered the development of upper bound solutions to many forming problems.

The science of limit analysis applies generally to the rigid-plastic nonwork-hardening material only, and rests on two theorems, one relating to the lower bound, the other to the upper bound. These theorems are capable of mathematical proof (Hill 1950, Prager and Hodge 1951, Drucker 1954, Prager 1959); two concepts facilitate the development of the theorems, as follows:

1. *The statically admissible stress field.* This is a field of generalized stresses (which may be forces, bending moments, or torques) which is in statical equilibrium within itself and with the externally imposed stresses at the boundaries.
2. *The kinematically admissible velocity field.* This is a field of generalized

velocities (which may be strains, linear displacements, or angular rotation) which is kinematically compatible within itself and with the externally imposed displacements at the boundaries.

The theorems may be stated as follows:

1. *The lower bound theorem.* If a statically admissible stress field exists, such that the stresses are everywhere just below those necessary to cause yielding, then the loads associated with that field constitute a lower bound solution.
2. *The upper bound theorem.* If a kinematically admissible velocity field exists, the loads required to be applied to cause the velocity field to operate, constitute an upper bound solution.

The latter can be understood quite simply. If another kinematically admissible velocity field exists with a lower energy requirement, it will always come into play first as the load is increased. The postulated field can therefore never require less force than the real one.

Since the material is rigid, plastic deformation will only be possible if the yield criterion is exceeded at the joints (or along the slip lines) of the structure (or the material). Thus Theorem 2 really states that the loads required to cause the structure to deform as a mechanism will constitute an upper bound. It will be seen that many statically admissible stress fields and kinematically admissible velocity fields exist for any given problem. If one of each can be found to give the same answer, then the true collapse load has been found.

In general, slip-line fields are kinematically admissible velocity fields, but not necessarily statically admissible stress fields. This is because the velocity of the rigid material outside the deforming region is always considered, but not necessarily the stresses in the rigid material. Thus the solution is generally only a partial solution, giving an *upper* bound to the loads required, though in practice, recognizing the approximations involved, upper bound solutions are usually obtained from much more simplified fields.

The concept of velocity discontinuities introduced into the theory of the slip-line field gives a ready artifice for developing kinematically admissible velocity fields in plane strain problems. Similarly, stress discontinuities enable the development of statically admissible stress fields. Consequently, upper and lower bound solutions should be readily determined for plane-strain deformation. As a matter of fact, most research workers in this field have concentrated mainly on upper bound solutions to deformation problems, which are easier to visualize, and lower bound solutions have not been extensively determined.

These techniques are very powerful for finding the collapse load of structures, and have revolutionized design methods in civil engineering practice. They can be applied also to deformation processes such as metal working. A further practical advantage is that a machine allocated on the basis of an upper bound solution will be *at least* powerful enough to complete the process.

1.6 Instability

Instability during plastic deformation can occur in tension, compression or shear. Probably tensile plastic instability, as manifested in the simple tensile test by the "necking" phenomenon, is the most important, especially in relation to the forming of sheet metal under the action of tensile stresses in the plane of the sheet. Many engineering components such as automobile panels, refrigerators, washing machines etc., have to be made by deep drawing or pressing in biaxial tension, and tensile plastic instability can occur at relatively small strains. Therefore, this problem will be considered first. The manifestation of compressive plastic instability is found in large-deformation buckling which will be considered later, as will be the problem of plastic instability in adiabatic shear.

1.6.1 Tensile Plastic Instability

To fix ideas, consider the simple tension test. As the specimen extends under the action of the applied load, the material "work hardens" due to the deformation it is suffering. Thus it is necessary to impose an increasing stress on the cross-sectional area in order to maintain plastic flow. This can be accomplished by increasing the applied load, but there is another effect which must not be forgotten. The requirement for conservation of the plastic volume in the tensile test is such that as the specimen extends it also contracts, thus reducing the cross-sectional area on which the applied load is acting. This "automatically" increases the stress as deformation proceeds, so that it is not necessary to increase the load as much as it would have been if the effect were not present. There comes a time, in fact, when it is not necessary to increase the load at all and the specimen simply extends under a constant load, the steadily decreasing area of cross-section ensuring the rise of stress necessary to offset work hardening of the material. As deformation proceeds, the rise of stress due to the contraction of the specimen under the constant load "outweighs" that due to work hardening, and the specimen–load system becomes unstable. Fracture follows in a short while, the detailed behaviour during this phase depending to some extent on the characteristics of the testing machine used. "Necking" occurs in ductile metals, at some location along the specimen which must depend on the local imperfections of either manufacture or material.

It is easy to describe what happens in the simple tensile test to cause instability, but in sheet-metal stretch-forming processes in which the metal is subjected to a complex biaxial system of stress, the problem is more complicated. Swift (1952) considered the problem in some detail, and Mellor (1960) extended his analysis. An introductory approach to the problem may be made, as follows. Considering the tensile test as a first step, assume that the relationship between true stress and true strain is of the form $\bar{\sigma} = H(\bar{\epsilon})$ and that elastic strains may be neglected. If the current area of the specimen is A, and the current load is P, then the true stress is $\sigma_1 = P/A$, due to the applied load. This will be equal to the equivalent stress $\bar{\sigma}$, if the metal is deforming uniformly and plastically. Since elastic strains are being neglected, the equation Al = constant applies, where l is the current gauge length of the specimen, since there is no permanent volume change. Differentiating this equation it is found that

$$Adl + ldA = 0 \tag{1.27}$$

and hence the longitudinal strain increment

$$d\bar{\epsilon} = d\epsilon_1 = dl/l = -dA/A \tag{1.28}$$

An increase of load on the tension specimen of amount dP is now imposed. The increase of stress $d\sigma_1$ may be found by differentiating the equation $\sigma_1 = P/A$, to be

$$d\bar{\sigma} = d\sigma_1 = \frac{dP}{A} - \frac{PdA}{A^2} = \frac{dP}{A} - \bar{\sigma}\frac{dA}{A} \tag{1.29}$$

or

$$dP/dA = \bar{\sigma} - d\bar{\sigma}/d\bar{\epsilon} \tag{1.30}$$

Now, instability occurs when the load attains a stationary maximum value, i.e. $dP/d\bar{\epsilon} = 0$. The cross-sectional area will still be decreasing, however, so that dA has a finite negative value, and dP/dA will be zero. Thus, from Eq. (1.30), instability in the tension test occurs when

$$\bar{\sigma} = d\bar{\sigma}/d\bar{\epsilon} \tag{1.31}$$

This is a very important result, since it allows the determination of another quantity of interest from the true stress–strain curve, namely the point associated with the maximum load which a tensile specimen can withstand. This is the point giving the *ultimate tensile strength* (UTS) of the material, and it can now be seen that this may be determined from the true stress–true strain curve, by finding the point at which the tangent to the curve has the same magnitude as the true stress at that point.

When a complex state of stress exists in the material, a similar procedure may be adopted to find the conditions at which instability will occur. The method is essentially to find the equivalent stress and strain associated with the given stress system, and to find the states of stress and strain which will give rise to instability, or unrestrained deformation without increase of load. Since *biaxial* stress states are of importance in the forming of sheet, the problem of instability under biaxial tension was considered by Swift (1952) and Mellor (1960), who both introduced the stress ratio $x = \sigma_2/\sigma_1$, and assumed that the stress σ_3 normal to the sheet was zero.

To solve any given problem of this type, it is possibly simpler to consider what tensile plastic instability means in terms of the applied loads associated with that particular problem, rather than introducing the parameter x. Thus, for example, instability in a closed-end cylinder with thin walls subjected to an internal pressure p would occur when $dp/p = 0$. If the radius of the cylinder is r, and the wall thickness t, then the two principal stresses will be the circumferential stress σ_1, and the longitudinal or axial stress σ_2 given by the equations

$$\left.\begin{aligned} \sigma_1 &= \frac{pr}{t} \\ \sigma_2 &= \frac{pr}{2t} \end{aligned}\right\} \tag{1.32}$$

Differentiating these equations

$$\left.\begin{aligned} \frac{\delta\sigma_1}{\sigma_1} &= \frac{\delta p}{p} + \frac{\delta r}{r} - \frac{\delta t}{t} \\ \frac{\delta\sigma_2}{\sigma_2} &= \frac{\delta p}{p} + \frac{\delta r}{r} - \frac{\delta t}{t} \end{aligned}\right\} \tag{1.33}$$

Thus, instability of the vessel occurs when $\delta p/p = 0$, or

$$\frac{\delta\sigma_1}{\sigma_1} = \frac{\delta\sigma_2}{\sigma_2} = \frac{\delta r}{r} - \frac{\delta t}{t} \tag{1.34}$$

Substituting Eqs (1.32) into Eq. (1.7), recognizing that the radial stress σ_3 in a thin-walled cylinder is often assumed to be zero, it is found that $\bar{\sigma} = \sqrt{\frac{3}{2}}\sigma_1$, and from Eq. (1.9), $\delta\bar{\epsilon} = (2/\sqrt{3})\delta\epsilon_1$. Now $\delta r/r$ is the hoop strain $\delta\epsilon_1$ and $\delta t/t$ is the "through thickness" strain $\delta\epsilon_3$. From Eqs (1.13) it may be seen that $\delta\epsilon_2 = 0$ and $\delta\epsilon_3 = -\delta\epsilon_1$. Thus, considering Eq. (1.34), $\delta r/r - \delta t/t$ is equal to $2\delta\epsilon_1$, or $\sqrt{3}\delta\bar{\epsilon}$, and this equation may be rewritten in the form

$$\frac{\delta\bar{\sigma}}{\bar{\sigma}} = \sqrt{3}\delta\bar{\epsilon} \tag{1.34a}$$

since $\delta\bar{\sigma}/\bar{\sigma} = \delta\sigma_1/\sigma_1$ for a constant stress ratio σ_2/σ_1. In the limit, this condition of instability may be expressed as

$$\frac{d\bar{\sigma}}{d\bar{\epsilon}} = \sqrt{3}\bar{\sigma} \tag{1.35}$$

It is sometimes convenient to refer this equation to the subtangent z of the true stress–true strain curve, the condition of instability then being given by the equation

$$\frac{d\bar{\sigma}}{d\bar{\epsilon}} = \frac{\bar{\sigma}}{z} \tag{1.36}$$

In the particular case of the cylindrical closed-end vessel the value of z is $1/\sqrt{3}$. Swift (1952) and Mellor (1960) considered several other cases of biaxial stressing, but the principles are the same as for the problem discussed here. For equal biaxial tension the subtangent has the value 2, whilst for a thin spherical vessel under internal pressure it has the value $\frac{2}{3}$, which can also be shown to be the value for a thin-walled cylinder with open ends under internal pressure.

1.6.2 Compressive Plastic Instability

Compressive plastic instability manifests itself in the compressive testing of cylindrical rods, which will buckle if they are too long. Similarly, in the production of axially symmetrical components by the processes of forging or extrusion, buckling of the initial billet will also occur if it is too long in relation to its diameter.

Engineers are generally familiar with the Euler *elastic* buckling load for a long prismatic column, which is of the form

$$P_{cr} = C\pi^2 EI/l^2$$

where P_{cr} is the critical load for elastic buckling, E is Young's modulus, $I = \pi r^4/4$ for a cylindrical rod of radius r, l is column length and $C = 4$ for a fixed-ended column (for example).

A comprehensive discussion of plastic buckling is given in Johnson and Mellor (1973). They point out that a realistic estimate of the plastic buckling of short columns of the type under discussion here can be obtained by simply replacing the Young's modulus E with the plastic tangent modulus $E_p = d\bar{\sigma}/d\bar{\epsilon}_p$. The value of E_p chosen from the stress–strain curve must correspond with the value of $\bar{\sigma} = \sigma_{cr}$, of course, which may involve some iteration, unless a simple constitutive relation can be used for the stress–strain curve of the material concerned.

For the plastic buckling of short fixed-ended cylindrical columns, radius r, of the type used as billets in forging or extrusion, $I = \pi r^4/4$, E is replaced by E_p, P_{cr} by $\sigma_{cr} \times \pi r^2$, $C = 4$, and hence Euler's equation takes the form

$$l/r = \pi \sqrt{E_p/\sigma_{cr}} \tag{1.37}$$

where E_p is the value of $d\bar{\sigma}/d\bar{\epsilon}_p$ at the stress σ_{cr}. Since the current values of l, r, σ_{cr} and E_p are interrelated through the stress–strain curve and the constancy of the volume $V = \pi r^2 l$, it is possible to determine σ_{cr}.

1.6.3 Shear Instability

Buckling is specifically avoided in large plastic deformation by suitable choice of diameter/height ratio based on analyses of the type described in the previous discussion. It is more common to find failure in shear, due to intense local heating by the localized shear, which causes softening and thus facilitates further shear in the same band. This unstable situation is enhanced by high-speed working and with low-conductivity materials such as titanium. The extreme condition is known as *adiabatic shear instability*. In hot forging the local heating is sometimes visible as 45° cross patterns of brighter colour red on the sides of the billets.

References

Alexander JM (1955) A slip-line field for the hot rolling process. Proc I Mech E 169:1021

Alexander JM (1981) Strength of materials, vol I: Fundamentals. Ellis Horwood, Chichester

Alexander JM, Brewer RC, Rowe GW (1987) Manufacturing Technology, vol 1: Engineering materials; vol 2: Engineering processes. Ellis Horwood, Chichester

Baker JF, Horne MR, Heyman J (1956) The steel skeleton. Cambridge University Press

Bridgman PW (1952) Studies in large plastic flow and fracture. McGraw-Hill, New York

Christopherson DG, Oxley PLB and Palmer WB (1958) Orthogonal cutting of work-hardening material. Engineering 186:113

Crossland B (1954) The effect of fluid pressure on the shear properties of metals. Proc I Mech E 168:935

Drucker DC (1954) Limit analysis and design. Appl Mech Rev 7:421

Ford H, Alexander JM (1977) Advanced mechanics of materials, 2nd edn. Ellis Horwood, Chichester (first published by Longmans Green, London, 1963)

Green AP (1951) A theoretical investigation of the compression of a ductile material between smooth flat dies. Phil Mag 42:900

Hill R (1950) The mathematical theory of plasticity. Clarendon Press, Oxford

Johnson W (1959) Estimation of upper bound loads for extrusion and coining operations. Proc I Mech E 173:61

Johnson W, Mellor PB (1973) *Engineering Plasticity* (2nd edn of Plasticity for mechanical engineers, 1962). Van Nostrand, London

Mellor PB (1960) Plastic instability in tension. Engineer 209:517

Nadai A (1931) Plasticity. McGraw-Hill, New York

Prager W (1953) A geometrical discussion of the slip line field in plane plastic flow. Trans R Inst Technol. Stockholm 65.

Prager W (1959) Introduction to plasticity. Addison-Wesley, Reading, MA

Prager W, Hodge PG (1951) The theory of perfectly plastic solids. John Wiley, New York
Prandtl L von (1924) Spannungsverteilung in plastischen Körpern. In: Proceedings 1st International Congress on Applied Mechanics, Delft, pp. 43–54
Reuss E (1930) Berücksichtigung der elastischen Formänderung in der Plastizitätstheorie. Z Angew Math Mech 10:266–274
Rowe GW (1979) Elements of metalworking theory. Edward Arnold, London
Sopwith DG, Morrison JLM (1956) (Letter) Criteria of yield. Engineer 202:810
Swift HW (1952) Plastic instability under plane stress. J Mech Phys Solids 1:1

2 Finite Deformation of Solids

S. Cescotto

2.1 Introduction

The behaviour of a solid undergoing large strains is highly nonlinear. Many sources of nonlinearity are encountered in practice. Some of them are listed below.

2.1.1 Large Changes in Geometry

The material particles are submitted to translations, rotations and deformations of arbitrary (usually important) magnitude. A typical example is shown in Fig. 2.1 for the upsetting of a cylinder.

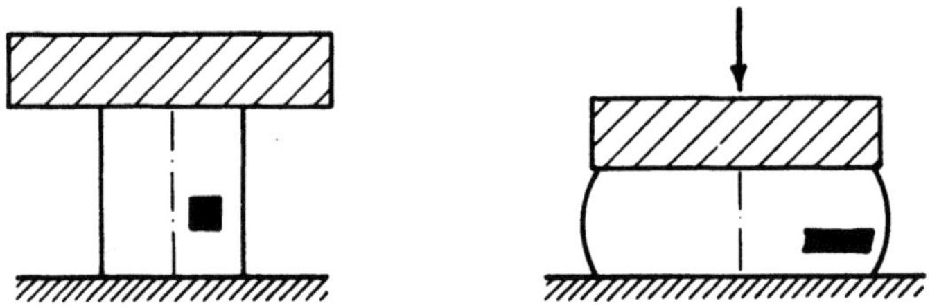

Fig. 2.1. Change in geometry during the upsetting of a cylinder.

2.1.2 Inelastic Material Behaviour

In the domain of large strains, the relationship between stress rate and strain rate is much more complex than in linear elastic solids. Usually, three typical types of behaviour of materials undergoing large strains are distinguished:

- Hyperelastic models (Fig. 2.2a) for rubber-like materials: there is a one-to-one, nonlinear relationship between stress and strain; this implies a reversible, history-independent behaviour of the solid.

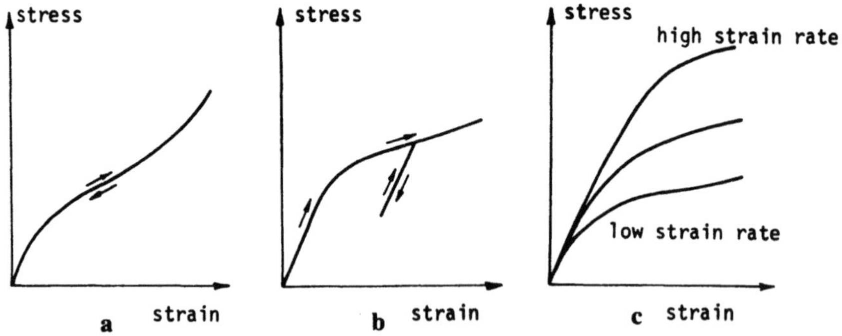

Fig. 2.2. Types of inelastic behaviour: **a** hyperelastic; **b** elastic–plastic; **c** elastic–viscoplastic.

- Elastic-plastic models (Fig. 2.2b) for materials whose behaviour can be irreversible but is independent of strain rate: this model applies for many metals and some soil materials at room temperature with reasonable accuracy.
- Elastic-viscoplastic models (Fig. 2.2c) for strain rate sensitive materials, such as metals at high temperature, soils, polymers, etc.

The limit between these models is not well defined. The same solid can be represented by any of these models according to the type of application, the forces acting on it, the desired degree of accuracy, etc. The selection of a model is a matter of judgement and experience.

2.1.3 Boundary Conditions

During the deformation process, the boundary conditions of the solid can be modified. This is typically the case when a part of the solid boundary comes into contact with another solid, such as a die in metal forming processes: if this occurs, the new boundary conditions must take account of the contact stresses (pressure and friction) between both solids (Fig. 2.3a).

Another typical situation is the appearance and development of a crack inside the deforming body which creates new internal boundaries that must be taken into account (Fig. 2.3b).

In this chapter, this last kind of nonlinearity will not be considered. We will concentrate only on the geometrical and material nonlinearities.

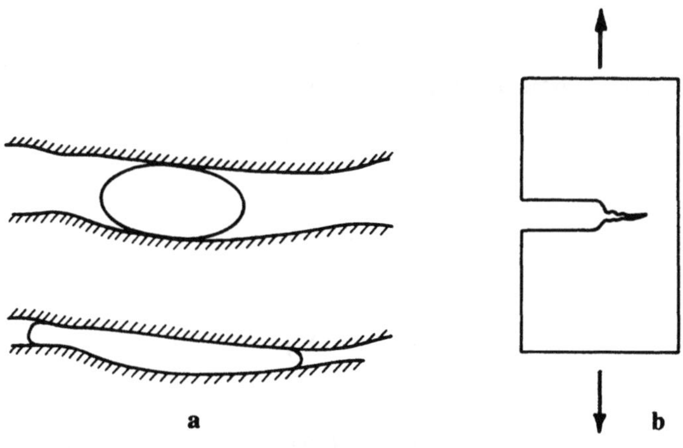

Fig. 2.3. Examples of changing boundary conditions: **a** due to changing die contact; **b** due to the formation of a crack.

2.1.4 Scope of This Chapter

The aim of this chapter is to provide an introduction to the analysis of large strains in solids by the finite-element method. First, some basic topics of solid mechanics will be recalled for the sake of completeness: metric, stresses, strains, equilibrium and the virtual work principle. Then, the different steps of the finite-element analysis of solids subjected to large strains will be introduced in the simplest way; most of the problems mentioned will be treated in some detail in the following chapters. Therefore, our approach of large strain analysis will be rather general and not specialized to a particular kind of constitutive model for the material. In other words, the developments of this chapter will be valid for a large class of constitutive equations that will be clearly specified later.

2.2 Configurations

During the deformation process, the successive configurations of the solid are labelled as follows:

- Initial configuration at time $t = 0$: Γ
- Current configuration at time t: γ

The description of the geometry of the solid during its motion can use three different kinds of coordinates (Fig. 2.4):

x^1, x^2, x^3	Cartesian coordinates of material points in the deformed configuration γ
X^1, X^2, X^3	Cartesian coordinates of material points in the undeformed configuration Γ
ξ^1, ξ^2, ξ^3	Intrinsic coordinates of material points

In the deformed configuration γ, one has

$$x^i = x^i(\xi^\alpha) \tag{2.1}$$

or inversely

$$\xi^\alpha = \xi^\alpha(x^j) \tag{2.2}$$

In the undeformed configuration Γ, one has

$$X^i = X^i(\xi^\alpha) \tag{2.3}$$

or

$$X^i = X^i(x^j) \tag{2.4}$$

Inversely:

$$\xi^\alpha = \xi^\alpha(X^j) \tag{2.5}$$

or

$$x^i = x^i(X^j) \tag{2.6}$$

2.3 Metric Structure in the Current Configuration γ

2.3.1 Covariant Base

In γ, a point P is located by the vector (Fig. 2.4)

$$\mathbf{r} = x^i \mathbf{e}_i \tag{2.7}$$

One defines:

$$\mathbf{g}_\alpha = \frac{\partial \mathbf{r}}{\partial \xi^\alpha} = \mathbf{r}_{,\alpha} \tag{2.8}$$

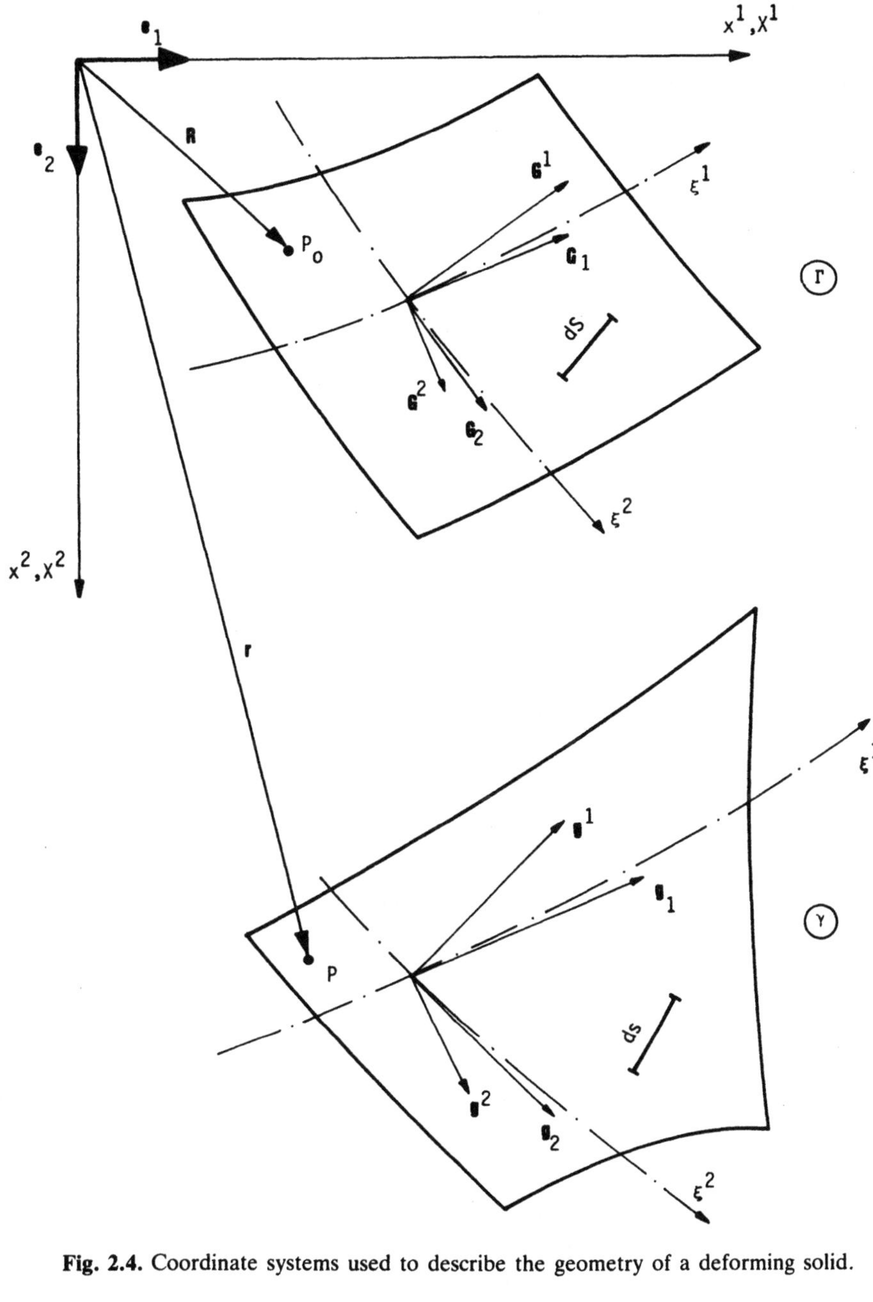

Fig. 2.4. Coordinate systems used to describe the geometry of a deforming solid.

whence

$$\mathbf{g}_\alpha = \frac{\partial x^j}{\partial \xi^\alpha}\,\mathbf{e}_j \tag{2.9}$$

The vectors $\mathbf{g}_\alpha$ form a local reference base at each point of the deformed solid. This base is named the local covariant base.

2.3.2 Contravariant Base

One defines the contravariant base vectors by

$$\mathbf{g}^{\alpha} = \frac{\partial \xi^{\alpha}}{\partial x^{j}} \mathbf{e}_j \tag{2.10}$$

2.3.3 Jacobian Matrix

One has

$$dx^s = \frac{\partial x^s}{\partial \xi^{\alpha}} d\xi^{\alpha} \tag{2.11}$$

Let

$$j_{s\alpha} = \frac{\partial x^s}{\partial \xi^{\alpha}} \tag{2.12}$$

with

$$\mathbf{j} = \begin{pmatrix} \dfrac{\partial x^1}{\partial \xi^1} & \dfrac{\partial x^1}{\partial \xi^2} & \dfrac{\partial x^1}{\partial \xi^3} \\ \dfrac{\partial x^2}{\partial \xi^1} & \dfrac{\partial x^2}{\partial \xi^2} & \dfrac{\partial x^2}{\partial \xi^3} \\ \dfrac{\partial x^3}{\partial \xi^1} & \dfrac{\partial x^3}{\partial \xi^2} & \dfrac{\partial x^3}{\partial \xi^3} \end{pmatrix} \tag{2.13}$$

whence

$$\begin{Bmatrix} dx^1 \\ dx^2 \\ dx^3 \end{Bmatrix} = \mathbf{j} \begin{Bmatrix} d\xi^1 \\ d\xi^2 \\ d\xi^3 \end{Bmatrix} \tag{2.14}$$

Inversely

$$d\xi^{\beta} = \frac{\partial \xi^{\beta}}{\partial x^s} dx^s \tag{2.15}$$

It is seen that

$$\begin{Bmatrix} d\xi^1 \\ d\xi^2 \\ d\xi^3 \end{Bmatrix} = \mathbf{j}^{-1} \begin{Bmatrix} dx^1 \\ dx^2 \\ dx^3 \end{Bmatrix} \tag{2.16}$$

with

$$\mathbf{j}^{-1} \begin{pmatrix} \dfrac{\partial \xi^1}{\partial x^1} & \dfrac{\partial \xi^1}{\partial x^2} & \dfrac{\partial \xi^1}{\partial x^3} \\ \dfrac{\partial \xi^2}{\partial x^1} & \dfrac{\partial \xi^2}{\partial x^2} & \dfrac{\partial \xi^2}{\partial x^3} \\ \dfrac{\partial \xi^3}{\partial x^1} & \dfrac{\partial \xi^3}{\partial x^2} & \dfrac{\partial \xi^3}{\partial x^3} \end{pmatrix} \tag{2.17}$$

The matrix $\mathbf{j}$ is the Jacobian matrix of the deformed configuration γ. Using the Jacobian matrix, it is seen that Eqs (2.9) and (2.10) can be written

$$\begin{Bmatrix} \mathbf{g}_1 \\ \mathbf{g}_2 \\ \mathbf{g}_3 \end{Bmatrix} = \mathbf{j}^{\mathrm{T}} \begin{Bmatrix} \mathbf{e}_1 \\ \mathbf{e}_2 \\ \mathbf{e}_3 \end{Bmatrix} \tag{2.18}$$

$$\begin{Bmatrix} \mathbf{g}^1 \\ \mathbf{g}^2 \\ \mathbf{g}^3 \end{Bmatrix} = \mathbf{j}^{-1} \begin{Bmatrix} \mathbf{e}_1 \\ \mathbf{e}_2 \\ \mathbf{e}_3 \end{Bmatrix} \tag{2.19}$$

The last two formulas yield

$$\mathbf{g}^{\alpha}\mathbf{g}_{\beta} = \delta^{\alpha}_{\beta} \tag{2.20}$$

where δ^{α}_{β} is the Kronecker symbol.

2.3.4 Metric Tensors

By definition,

$$g_{\alpha\beta} = \mathbf{g}_{\alpha} \cdot \mathbf{g}_{\beta} = \frac{\partial x^k}{\partial \xi^{\alpha}} \cdot \frac{\partial x^k}{\partial \xi^{\beta}} \tag{2.21}$$

$$g^{\alpha\beta} = \mathbf{g}^{\alpha} \cdot \mathbf{g}^{\beta} = \frac{\partial \xi^{\alpha}}{\partial x^k} \cdot \frac{\partial \xi^{\beta}}{\partial x^k} \tag{2.22}$$

$$g^{\alpha}_{\beta} = \mathbf{g}^{\alpha} \cdot \mathbf{g}_{\beta} = \frac{\partial x^k}{\partial \xi^{\beta}} \cdot \frac{\partial \xi^{\alpha}}{\partial x^k} = \delta^{\alpha}_{\beta} \tag{2.23}$$

Let

$$g_{\alpha\beta} = \begin{Bmatrix} \mathbf{g}_1 \\ \mathbf{g}_2 \\ \mathbf{g}_3 \end{Bmatrix} \cdot \langle \mathbf{g}_1\mathbf{g}_2\mathbf{g}_3 \rangle = \mathbf{j}^{\mathrm{T}}\mathbf{j} \tag{2.24}$$

$$g^{\alpha\beta} = \begin{Bmatrix} \mathbf{g}^1 \\ \mathbf{g}^2 \\ \mathbf{g}^3 \end{Bmatrix} \cdot \langle \mathbf{g}^1\mathbf{g}^2\mathbf{g}^3 \rangle = \mathbf{j}^{-1}\mathbf{j}^{-1\mathrm{T}} \tag{2.25}$$

It is seen that

$$g^{\alpha\beta} = g^{-1}_{\alpha\beta} \rightarrow g^{\alpha\gamma}g_{\gamma\beta} = \delta^{\alpha}_{\beta} \tag{2.26}$$

2.3.5 Interpretation of the Metric Tensor $\mathbf{g}_{\alpha\beta}$

Consider a line element ds in γ. One has

$$ds^2 = dx^i dx^i$$

whence

$$ds^2 = g_{\alpha\beta} d\xi^{\alpha} d\xi^{\beta} \tag{2.27}$$

This shows that the metric tensor allows one to calculate the length of a line element in curvilinear coordinates.

2.3.6 Unit Base Vectors

The vectors $\mathbf{g}_\alpha$ and $\mathbf{g}^\alpha$ are not unit vectors. Sometimes, it is necessary to define unit curvilinear base vectors by

$$\mathbf{a}_\alpha = \frac{\mathbf{g}_\alpha}{\sqrt{g_{\alpha\alpha}}} \text{ (no sum)} \tag{2.28}$$

$$\mathbf{a}^\beta = \frac{\mathbf{g}^\beta}{\sqrt{g^{\beta\beta}}} \text{ (no sum)} \tag{2.29}$$

At a given material point of the deformed solid, they constitute a reference base with the following relation:

$$\mathbf{a}_\alpha \mathbf{a}^\beta = \delta_\alpha^\beta \tag{2.30}$$

2.3.7 Line and Surface Elements

At a given point of the deformed body, consider the segments $d\mathbf{s}_1$, $d\mathbf{s}_2$, $d\mathbf{s}_3$ oriented according to the local axes ξ^1, ξ^2, ξ^3 (Fig. 2.5)

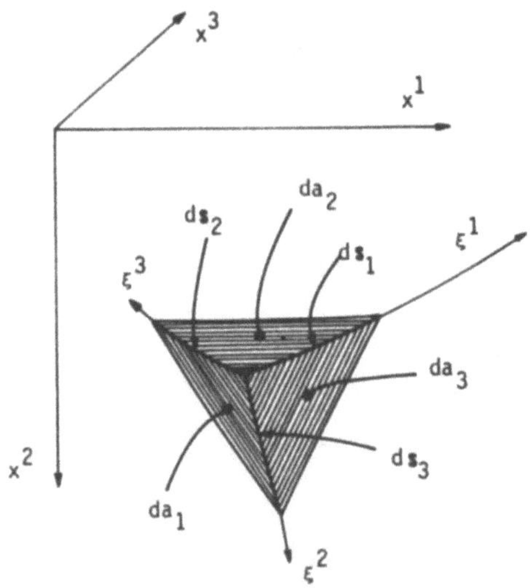

Fig. 2.5. Line and surface elements in a deformed body.

One has

$$d\mathbf{s}_\alpha = \mathbf{g}_\alpha d\xi^\alpha \text{ (no sum)} \tag{2.31}$$

$$d\mathbf{s}_\alpha = \mathbf{a}_\alpha ds_\alpha \text{ (no sum)} \tag{2.32}$$

whence

$$ds_\alpha = \sqrt{g_{\alpha\alpha}}\, d\xi^\alpha \text{ (no sum)} \qquad (2.33)$$

These segments define three curvilinear triangles whose respective areas are da_α.
One has

$$\left.\begin{aligned} da_1 &= \tfrac{1}{2}\,|d\mathbf{s}_2 \Lambda d\mathbf{s}_3| \\ da_2 &= \tfrac{1}{2}\,|d\mathbf{s}_1 \Lambda d\mathbf{s}_3| \\ da_3 &= \tfrac{1}{2}\,|d\mathbf{s}_1 \Lambda d\mathbf{s}_2| \end{aligned}\right\} \qquad (2.34)$$

Hence

$$\left.\begin{aligned} da_1 &= \tfrac{1}{2}\sqrt{g\cdot g^{11}}\, d\xi^2 d\xi^3 \\ da_2 &= \tfrac{1}{2}\sqrt{g\cdot g^{22}}\, d\xi^1 d\xi^3 \\ da_3 &= \tfrac{1}{2}\sqrt{g\cdot g^{33}}\, d\xi^1 d\xi^2 \end{aligned}\right\} \qquad (2.35)$$

where g is the determinant of $\mathbf{g}_{\alpha\beta}$.

2.3.8 Relations Between Covariant and Contravariant Bases

From Eqs (2.18) and (2.19), one deduces

$$\begin{Bmatrix} \mathbf{g}_1 \\ \mathbf{g}_2 \\ \mathbf{g}_3 \end{Bmatrix} = \mathbf{j}^T\mathbf{j} \begin{Bmatrix} \mathbf{g}^1 \\ \mathbf{g}^2 \\ \mathbf{g}^3 \end{Bmatrix} \qquad (2.36)$$

whence:

$$\begin{Bmatrix} \mathbf{g}_1 \\ \mathbf{g}_2 \\ \mathbf{g}_3 \end{Bmatrix} = g_{\alpha\beta} \begin{Bmatrix} \mathbf{g}^1 \\ \mathbf{g}^2 \\ \mathbf{g}^3 \end{Bmatrix}, \quad \mathbf{g}_\alpha = g_{\alpha\beta}\mathbf{g}^\beta \qquad (2.37)$$

Inversely:

$$\begin{Bmatrix} \mathbf{g}^1 \\ \mathbf{g}^2 \\ \mathbf{g}^3 \end{Bmatrix} = g^{\alpha\beta} \begin{Bmatrix} \mathbf{g}_1 \\ \mathbf{g}_2 \\ \mathbf{g}_3 \end{Bmatrix}, \quad \mathbf{g}^\alpha = g^{\alpha\beta}\mathbf{g}_\beta \qquad (2.38)$$

Any vector $\mathbf{v}$ related to configuration γ can be decomposed with respect to the covariant or the contravariant local base:

$$\mathbf{v} = v^\alpha \quad \mathbf{g}_\alpha = v_\alpha \mathbf{g}^\alpha \qquad (2.39)$$

From Eqs (2.37) and (2.38), one deduces:

$$v^\alpha = g^{\alpha\beta} v_\beta \qquad (2.40)$$

$$v_\alpha = g_{\alpha\beta} v^\beta \qquad (2.41)$$

2.4 Metric Structure in the Initial Configuration Γ

All the results obtained in the preceding section for configuration γ can be easily transposed to configuration Γ. It suffices to replace x by X.

2.4.1 Covariant Base

Referring once more to Fig. 2.4:

$$\mathbf{R} = X^i \mathbf{e}_i \tag{2.42}$$

$$\mathbf{G}_\alpha = \frac{\partial \mathbf{R}}{\partial \xi^\alpha} = \mathbf{R}_{,\alpha} \tag{2.43}$$

$$\mathbf{G}_\alpha = \frac{\partial X^j}{\partial \xi^\alpha} \mathbf{e}_j \tag{2.44}$$

2.4.2 Contravariant Base

$$\mathbf{G}^\alpha = \frac{\partial \xi^\alpha}{\partial X^j} \mathbf{e}_j \tag{2.45}$$

2.4.3 Jacobian Matrix

$$dX^s = \frac{\partial X^s}{\partial \xi^\alpha} d\xi^\alpha \tag{2.46}$$

$$J_{s\alpha} = \frac{\partial X^s}{\partial \xi^\alpha} \tag{2.47}$$

$$\mathbf{J} = \begin{pmatrix} \frac{\partial X^1}{\partial \xi^1} & \frac{\partial X^1}{\partial \xi^2} & \frac{\partial X^1}{\partial \xi^3} \\ \frac{\partial X^2}{\partial \xi^1} & \frac{\partial X^2}{\partial \xi^2} & \frac{\partial X^2}{\partial \xi^3} \\ \frac{\partial X^3}{\partial \xi^1} & \frac{\partial X^3}{\partial \xi^2} & \frac{\partial X^3}{\partial \xi^3} \end{pmatrix} \tag{2.48}$$

$$\begin{Bmatrix} dX^1 \\ dX^2 \\ dX^3 \end{Bmatrix} = \mathbf{J} \begin{Bmatrix} d\xi^1 \\ d\xi^2 \\ d\xi^3 \end{Bmatrix} \tag{2.49}$$

$$d\xi^\alpha = \frac{\partial \xi^\alpha}{\partial X^s} dX^s \tag{2.50}$$

$$\begin{Bmatrix} d\xi^1 \\ d\xi^2 \\ d\xi^3 \end{Bmatrix} = \mathbf{J}^{-1} \begin{Bmatrix} dX^1 \\ dX^2 \\ dX^3 \end{Bmatrix} \tag{2.51}$$

$$\mathbf{J}^{-1} = \begin{pmatrix} \dfrac{\partial \xi^1}{\partial X^1} & \dfrac{\partial \xi^1}{\partial X^2} & \dfrac{\partial \xi^1}{\partial X^3} \\ \dfrac{\partial \xi^2}{\partial X^1} & \dfrac{\partial \xi^2}{\partial X^2} & \dfrac{\partial \xi^2}{\partial X^3} \\ \dfrac{\partial \xi^3}{\partial X^1} & \dfrac{\partial \xi^3}{\partial X^2} & \dfrac{\partial \xi^3}{\partial X^3} \end{pmatrix} \tag{2.52}$$

$$\begin{Bmatrix} \mathbf{G}_1 \\ \mathbf{G}_2 \\ \mathbf{G}_3 \end{Bmatrix} = \mathbf{J}^{\mathrm{T}} \begin{Bmatrix} \mathbf{e}_1 \\ \mathbf{e}_2 \\ \mathbf{e}_3 \end{Bmatrix} \tag{2.53}$$

$$\begin{Bmatrix} \mathbf{G}^1 \\ \mathbf{G}^2 \\ \mathbf{G}^3 \end{Bmatrix} = \mathbf{J}^{-1} \begin{Bmatrix} \mathbf{e}_1 \\ \mathbf{e}_2 \\ \mathbf{e}_3 \end{Bmatrix} \tag{2.54}$$

$$\mathbf{G}^\alpha \mathbf{G}_\beta = \delta^\alpha_\beta \tag{2.55}$$

2.4.4 Metric Tensors

$$G_{\alpha\beta} = \mathbf{G}_\alpha \cdot \mathbf{G}_\beta = \frac{\partial X^k}{\partial \xi^\alpha} \cdot \frac{\partial X^k}{\partial \xi^\beta} \tag{2.56}$$

$$G^{\alpha\beta} = \mathbf{G}^\alpha \cdot \mathbf{G}^\beta = \frac{\partial \xi^\alpha}{\partial X^k} \cdot \frac{\partial \xi^\beta}{\partial X^k} \tag{2.57}$$

$$G^\alpha_\beta = \mathbf{G}^\alpha \cdot \mathbf{G}_\beta = \frac{\partial \xi^\alpha}{\partial X^k} \cdot \frac{\partial X^k}{\partial \xi^\beta} = \delta^\alpha_\beta \tag{2.58}$$

$$G_{\alpha\beta} = \begin{Bmatrix} \mathbf{G}_1 \\ \mathbf{G}_2 \\ \mathbf{G}_3 \end{Bmatrix} \cdot \langle\, \mathbf{G}_1\, \mathbf{G}_2\, \mathbf{G}_3 \,\rangle = \mathbf{J}^{\mathrm{T}}\mathbf{J} \tag{2.59}$$

$$G^{\alpha\beta} = \begin{Bmatrix} \mathbf{G}^1 \\ \mathbf{G}^2 \\ \mathbf{G}^3 \end{Bmatrix} \cdot \langle \mathbf{G}^1\, \mathbf{G}^2\, \mathbf{G}^3 \rangle = \mathbf{J}^{-1}\mathbf{J}^{-1\mathrm{T}} \tag{2.60}$$

$$G^{\alpha\beta} = G^{-1}_{\alpha\beta} \rightarrow G^{\alpha\gamma} G_{\gamma\beta} = \delta^\alpha_\beta \tag{2.61}$$

2.4.5 Interpretation of the Metric Tensor $\mathbf{G}_{\alpha\beta}$

$$dS^2 = G_{\alpha\beta}\, d\xi^\alpha d\xi^\beta \tag{2.62}$$

2.4.6 Unit Base Vectors

$$\mathbf{A}_\alpha = \frac{\mathbf{G}_\alpha}{\sqrt{G_{\alpha\alpha}}} \text{ (no sum)} \tag{2.63}$$

$$\mathbf{A}^\alpha = \frac{\mathbf{G}^\alpha}{\sqrt{G^{\alpha\alpha}}} \text{ (no sum)} \tag{2.64}$$

$$\mathbf{A}_\alpha \mathbf{A}^\beta = \delta_\alpha^\beta \tag{2.65}$$

2.4.7 Line and Surface Elements

Considering the segments $d\mathbf{S}_1$, $d\mathbf{S}_2$, $d\mathbf{S}_3$ oriented according to the local axes ξ^1, ξ^2, ξ^3 (Fig. 2.6):

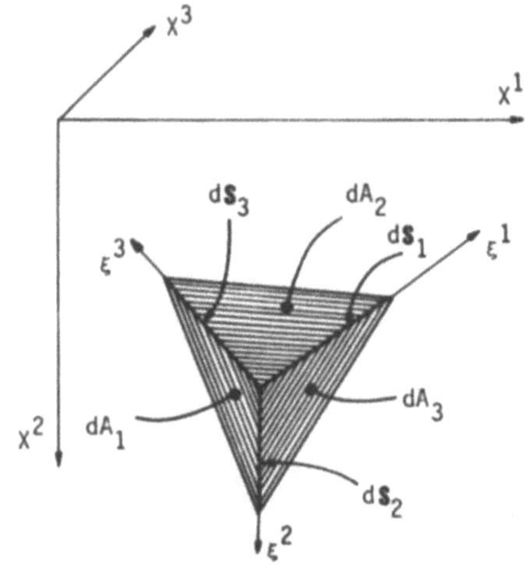

Fig. 2.6. Line and surface elements in an undeformed body.

$$d\mathbf{S}_\alpha = \mathbf{G}_\alpha d\xi^\alpha \text{ (no sum)} \tag{2.66}$$

$$d\mathbf{S}_\alpha = \mathbf{A}_\alpha ds_\alpha \text{ (no sum)} \tag{2.67}$$

$$d\mathbf{S}_\alpha = \sqrt{G_{\alpha\alpha}}\, d\xi^\alpha \text{ (no sum)} \tag{2.68}$$

$$\left.\begin{aligned} dA_1 &= \sqrt{G\cdot G^{11}}\, d\xi^2 d\xi^3 \\ dA_2 &= \sqrt{G\cdot G^{22}}\, d\xi^1 d\xi^3 \\ dA_3 &= \sqrt{G\cdot G^{33}}\, d\xi^1 d\xi^2 \end{aligned}\right\} \tag{2.69}$$

where G is the determinant of $\mathbf{G}_{\alpha\beta}$.

2.4.8 Relations Between Covariant and Contravariant Bases

$$\begin{Bmatrix} \mathbf{G}_1 \\ \mathbf{G}_2 \\ \mathbf{G}_3 \end{Bmatrix} = [G_{\alpha\beta}] \begin{Bmatrix} \mathbf{G}^1 \\ \mathbf{G}^2 \\ \mathbf{G}^3 \end{Bmatrix}, \quad \mathbf{G}_\alpha = G_{\alpha\beta}\mathbf{G}^\beta \tag{2.70}$$

$$\begin{Bmatrix} \mathbf{G}^1 \\ \mathbf{G}^2 \\ \mathbf{G}^3 \end{Bmatrix} = [G^{\alpha\beta}] \begin{Bmatrix} \mathbf{G}_1 \\ \mathbf{G}_2 \\ \mathbf{G}_3 \end{Bmatrix}, \mathbf{G}^\alpha = G^{\alpha\beta}\mathbf{G}_\beta \tag{2.71}$$

$$\mathbf{v} = V^\alpha \mathbf{G}_\alpha = V_\alpha \mathbf{G}^\alpha \tag{2.72}$$

$$\mathbf{V}^\alpha = G^{\alpha\beta}\mathbf{V}_\beta \tag{2.73}$$

$$\mathbf{V}_\alpha = G_{\alpha\beta}\mathbf{V}^\beta \tag{2.74}$$

2.5 Stresses

2.5.1 "True" Stress Vector

Let P be a material point of the deformed body (configuration γ) and let Δa be an elementary area whose orientation is determined by its unit normal (Fig. 2.7).

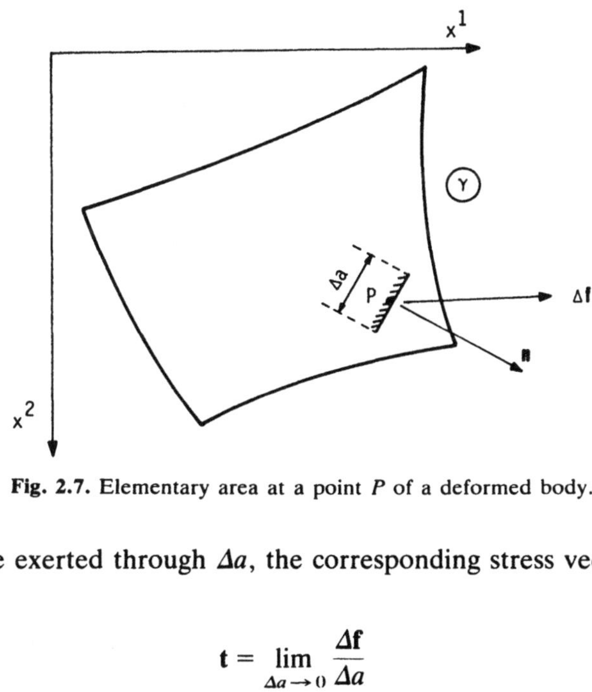

Fig. 2.7. Elementary area at a point P of a deformed body.

If $\Delta\mathbf{f}$ is the force exerted through Δa, the corresponding stress vector $\mathbf{t}$ is defined by

$$\mathbf{t} = \lim_{\Delta a \to 0} \frac{\Delta\mathbf{f}}{\Delta a} \tag{2.75}$$

It is clear that the stress vector $\mathbf{t}$ is defined per unit deformed area in the current configuration γ.

2.5.2 Cauchy Stresses

Consider the motion of an elementary material area (Fig. 2.8):

$$M_0N_0 \rightarrow MN \rightarrow M^*N^*$$

in the three successive configurations: $\Gamma \rightarrow \gamma \rightarrow \gamma^*$.

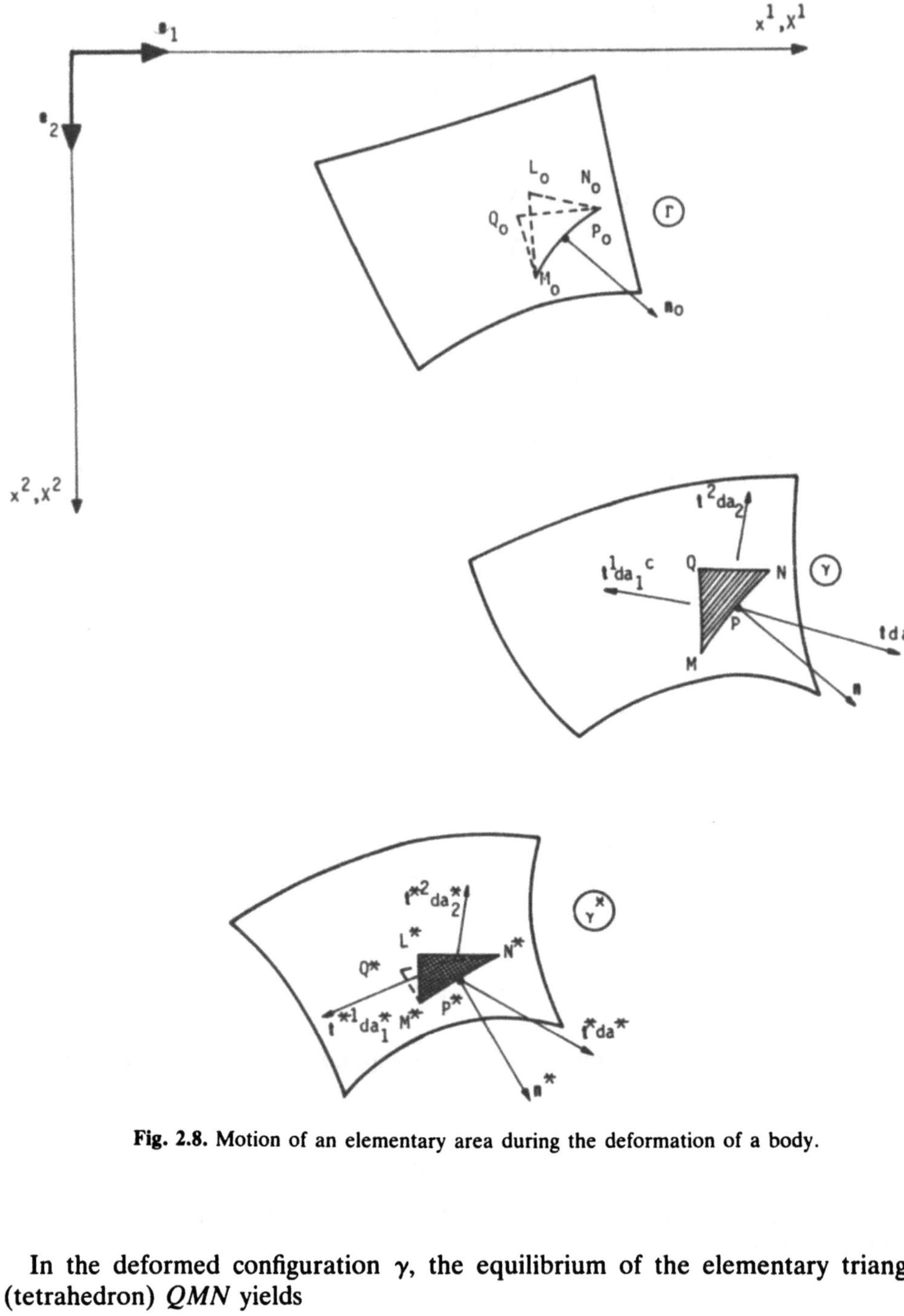

Fig. 2.8. Motion of an elementary area during the deformation of a body.

In the deformed configuration γ, the equilibrium of the elementary triangle (tetrahedron) QMN yields

$$\mathbf{t}da = \mathbf{t}^i da_i \tag{2.76}$$

Using the Cartesian components of

$$\mathbf{n} = n_i \mathbf{e}^i \tag{2.77}$$

$$da_i = n_i da \tag{2.78}$$

we obtain the equilibrium equation

$$\mathbf{t} = n_i \mathbf{t}^i \tag{2.79}$$

The nine Cartesian components of the stress vectors $\mathbf{t}^i$ are the true stresses or Cauchy stresses σ^{ij} in the deformed configuration γ,

$$\mathbf{t}^i = \sigma^{ij} \mathbf{e}_j \tag{2.80}$$

and the equilibrium of the elementary triangle (tetrahedron) can be written

$$\mathbf{t} = n_i \sigma^{ij} \mathbf{e}_j \tag{2.81}$$

The Cauchy stresses are related to the elementary areas (QN, QM) perpendicular to the global reference axes, in the deformed configuration γ.

In another deformed configuration γ^*, the area MN becomes M^*N^* and the Cauchy stresses σ^{ij} are defined with the help of the triangle (tetrahedron) $L^*M^*N^*$. It is important to notice that the areas L^*M^* and L^*N^* do not result from the deformation of QM and QN. In other words, QM and L^*M^*, for example, are not composed of the same material particles.

2.5.3 "Nominal" Stress Vector

The elementary area da (MN in γ) results from the deformation of an elementary area dA (M_0N_0 in Γ).

The equation

$$\mathbf{t}da = \mathbf{T}dA \tag{2.82}$$

defines the nominal stress vector $\mathbf{T}$ per unit undeformed area.

2.5.4 Lagrange Stresses

The Lagrange stresses P^{ij} are defined by the following equation, which is similar to Eq. (2.81):

$$\mathbf{T} = n_{0i} P^{ij} \mathbf{e}_j \tag{2.83}$$

where $\mathbf{n}_0$ is the unit normal to M_0N_0 (Fig. 2.9)

$$\mathbf{n}_0 = n_{0i} \mathbf{e}^i \tag{2.84}$$

2.5.5 "Back-Transformed" Stress Vector

Let

$$\mathbf{F} = \begin{Bmatrix} \dfrac{\partial x^1}{\partial X^1} & \dfrac{\partial x^1}{\partial X^2} & \dfrac{\partial x^1}{\partial X^3} \\ \dfrac{\partial x^2}{\partial X^1} & \dfrac{\partial x^2}{\partial X^2} & \dfrac{\partial x^2}{\partial X^3} \\ \dfrac{\partial x^3}{\partial X^1} & \dfrac{\partial x^3}{\partial X^2} & \dfrac{\partial x^3}{\partial X^3} \end{Bmatrix} \quad \text{or} \quad \mathbf{F} = \frac{\partial \mathbf{x}}{\partial \mathbf{X}} \tag{2.85}$$

be the Jacobian matrix of the mapping (2.6)

$$x^i = x^i(X^1, X^2, X^3) \tag{2.86}$$

F is often called the deformation gradient tensor and plays a central role in the definition of strain measures (see section 2.9). With the help of **F**, it is possible to define another stress vector **T**′ per unit undeformed area:

$$\mathbf{T}' = \mathbf{F}^{-1}\mathbf{T} \tag{2.87}$$

2.5.6 PK2 Stresses

The second-order Piola–Kirchhoff (PK2) stresses S^{ij} are defined by:

$$\mathbf{T}' = \mathbf{F}^{-1}\mathbf{T} = n_{0i}S^{ij}\mathbf{e}_j \tag{2.88}$$

2.5.7 Physical Interpretation of Lagrange and PK2 Stresses

A physical interpretation of Eqs (2.83) and (2.88) can provide a better understanding of the meaning of Lagrange and PK2 stresses. The material triangle $Q_0M_0N_0$ in Γ becomes QMN in γ (Fig. 2.9).

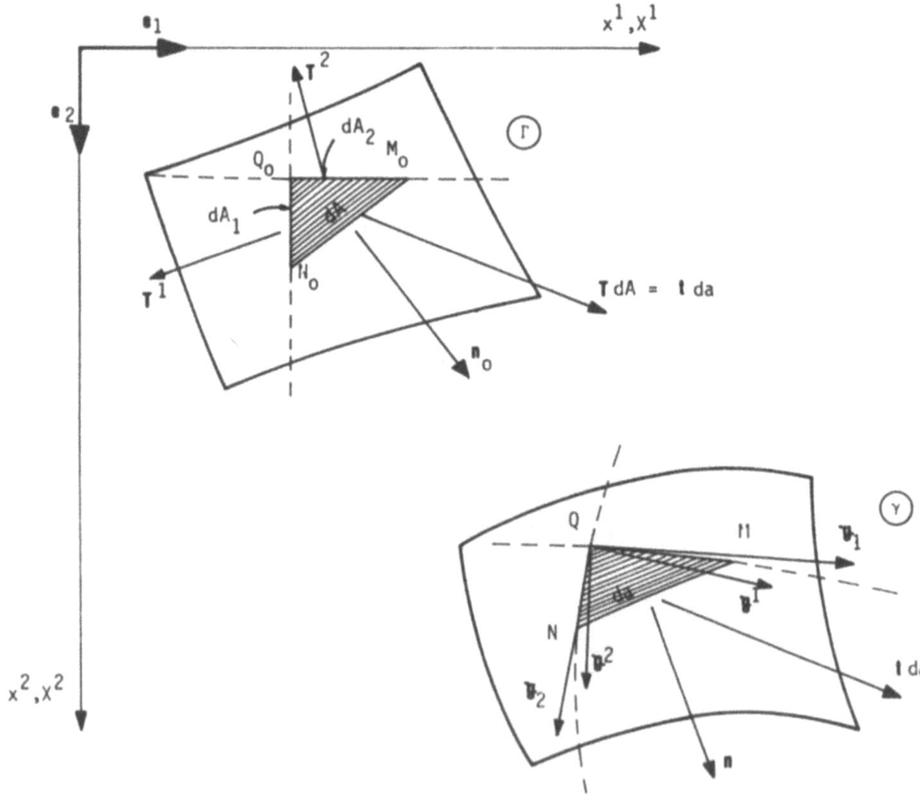

Fig. 2.9. Undeformed and deformed elementary areas and corresponding base vectors.

The elementary force $\mathbf{t}da$ is applied to MN. This vector is translated on M_0N_0 and is decomposed (by equilibrium) with the help of triangle $Q_0M_0N_0$:

$$\mathbf{t}da = \mathbf{T}dA = \mathbf{T}^i dA_i \tag{2.89}$$

In this relation, $\mathbf{T}^1$, $\mathbf{T}^2$, $\mathbf{T}^3$ are fictitious stress vectors applied to the elementary undeformed areas dA_1, dA_2, dA_3 respectively perpendicular to $\mathbf{e}_1$, $\mathbf{e}_2$, $\mathbf{e}_3$. Since

$$dA_i = n_{0i}dA \tag{2.90}$$

we get:

$$\mathbf{T} = n_{0i}\mathbf{T}^i \tag{2.91}$$

which is quite similar to Eq. (2.79).

From Eqs (2.91) and (2.83), it is seen that Lagrange stresses P^{ij} are the components of $\mathbf{T}^i$ with respect to the Cartesian base vectors ($\mathbf{e}_1$, $\mathbf{e}_2$, $\mathbf{e}_3$):

$$\mathbf{T}^i = P^{ij}\mathbf{e}_j \tag{2.92}$$

To obtain a similar interpretation of PK2 stresses, it is necessary to introduce a new metric base, $\hat{\mathbf{g}}_1$ $\hat{\mathbf{g}}_2$ $\hat{\mathbf{g}}_3$ or $\hat{\mathbf{g}}^1$ $\hat{\mathbf{g}}^2$ $\hat{\mathbf{g}}^3$:

$$\begin{Bmatrix} \hat{\mathbf{g}}_1 \\ \hat{\mathbf{g}}_2 \\ \hat{\mathbf{g}}_3 \end{Bmatrix} = \mathbf{F}^{\mathrm{T}} \begin{Bmatrix} \mathbf{e}_1 \\ \mathbf{e}_2 \\ \mathbf{e}_3 \end{Bmatrix}, \begin{Bmatrix} \hat{\mathbf{g}}^1 \\ \hat{\mathbf{g}}^2 \\ \hat{\mathbf{g}}^3 \end{Bmatrix} = \mathbf{F}^{-1} \begin{Bmatrix} \mathbf{e}_1 \\ \mathbf{e}_2 \\ \mathbf{e}_3 \end{Bmatrix} \tag{2.93}$$

From Eqs (2.93) and (2.88), it is seen that PK2 stresses S^{ij} are the components of $\mathbf{T}^i$ with respect to the metric base vectors ($\hat{\mathbf{g}}_1$, $\hat{\mathbf{g}}_2$, $\hat{\mathbf{g}}_3$):

$$\mathbf{T}^i = S^{ij}\hat{\mathbf{g}}_j \tag{2.94}$$

Hence

$$\mathbf{T} = n_{0i}S^{ij}\hat{\mathbf{g}}_j \tag{2.95}$$

These interpretations, Eqs (2.92) and (2.94), clearly show that Lagrange and PK2 stresses are referred to unit undeformed areas. In a solid undergoing large strains, their magnitudes can differ significantly from the true stresses σ^{ij} defined per unit deformed area.

Furthermore, from their definitions Eqs (2.83), (2.88), P^{ij} and S^{ij} stresses imply the use of two configurations: the deformed one γ and the initial undeformed one Γ. Sometimes, this can be a disadvantage because there is some arbitrariness in the choice of an initial undeformed configuration. For example, in soil mechanics, what is the initial state of a rock layer which has been submitted to geological forces for thousands of years? Another typical case can be found in metal forming. What is the initial state of a billet? Should we start from the initial casting?

On the contrary, Cauchy stresses are true stresses and their definition uses only the current configuration of the solid. However, it must be recalled that, during the deformation of the solid, Cauchy stresses at a given material point are not referred to the same material elementary areas and this has important consequences on the formulation of constitutive equations.

2.5.8 Material Stresses

For completeness, we introduce here several stress tensors deduced from the equilibrium of a curvilinear triangle (tetrahedron) defined with the help of the intrinsic coordinates ξ^α in the deformed configuration γ (Fig. 2.10).

The equilibrium of triangle (tetrahedron) QMN can be written

$$\mathbf{t}da = \mathbf{t}^\alpha da_\alpha \tag{2.96}$$

But, one has

$$da_\alpha = n_\alpha \sqrt{g^{\alpha\alpha}}\, da \text{ (no sum)} \tag{2.97}$$

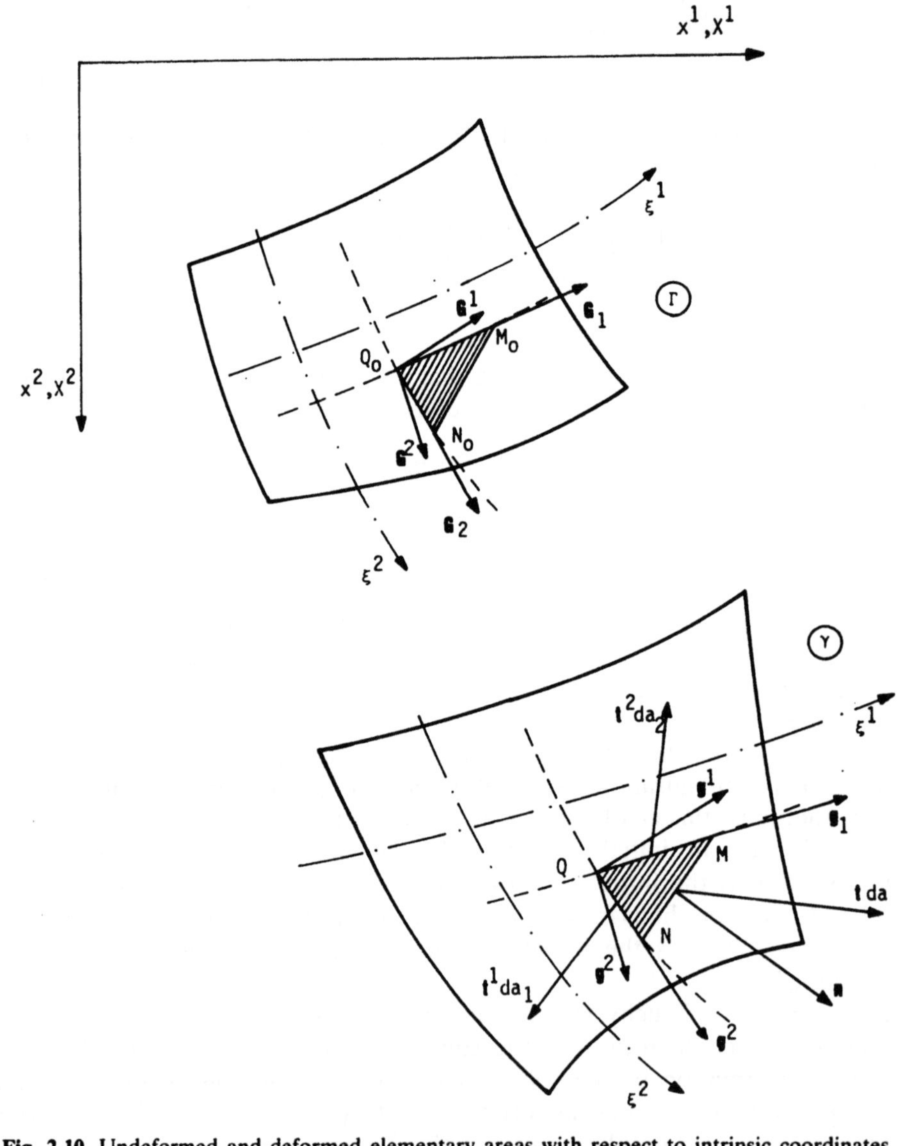

Fig. 2.10. Undeformed and deformed elementary areas with respect to intrinsic coordinates.

with

$$\mathbf{n} = n_\alpha \mathbf{g}^\alpha \tag{2.98}$$

Whence

$$\mathbf{t} = n_1 \sqrt{g^{11}}\, \mathbf{t}^1 + n_2 \sqrt{g^{22}}\, \mathbf{t}^2 + n_3 \sqrt{g^{33}}\, \mathbf{t}^3 \tag{2.99}$$

If one performs the decomposition of $\sqrt{g^{\alpha\alpha}}\, \mathbf{t}^\alpha$ with respect to $\mathbf{g}_\alpha$ or $\mathbf{g}^\alpha$, one gets respectively

$$\sqrt{g^{\alpha\alpha}}\, \mathbf{t}^\alpha = \tau^{\alpha\beta} \mathbf{g}_\beta \tag{2.100}$$

$$\sqrt{g^{\alpha\alpha}}\, \mathbf{t}^\alpha = \tau^\alpha_\beta \mathbf{g}^\beta \tag{2.101}$$

Hence

$$\mathbf{t} = \tau^{\alpha\beta} n_\alpha \mathbf{g}_\beta \tag{2.102}$$

$$\mathbf{t} = \tau^\alpha_\beta n_\alpha \mathbf{g}^\beta \tag{2.103}$$

$\tau^{\alpha\beta}$ and τ^α_β are, respectively, contravariant and mixed material stress tensors per unit deformed area. These stresses are, therefore, true stresses. The basic difference with Cauchy stresses is that $\tau^{\alpha\beta}$ and τ^α_β refer to the same material elementary areas during the deformation process. They are represented in Fig. 2.11.

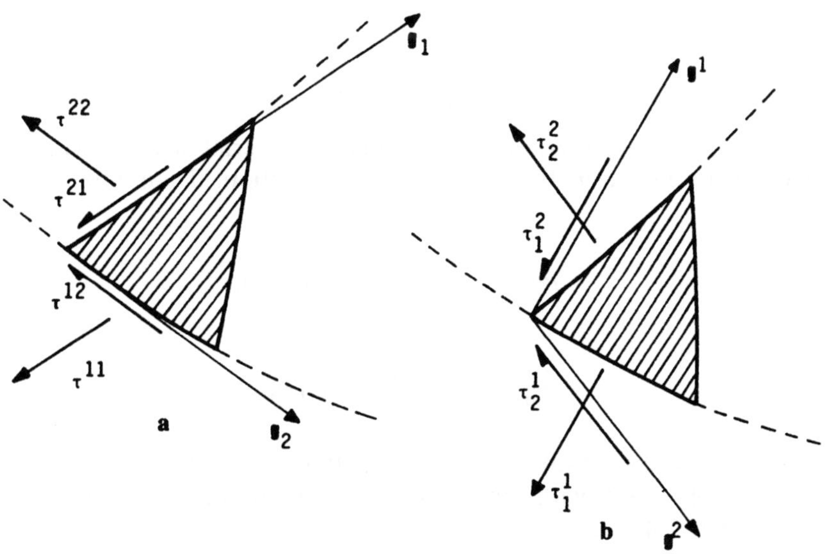

Fig. 2.11. Material stresses acting at a point of a deformed body: **a** contravariant tensor $\tau^{\alpha\beta}$; **b** mixed tensor τ^α_β.

It is also possible to define material stresses per unit undeformed area:

$$\mathbf{T} = s^{\alpha\beta} N_{0\alpha} \mathbf{g}_\beta \tag{2.104}$$

$$\mathbf{T} = s^\alpha_\beta N_{0\alpha} \mathbf{g}^\beta \tag{2.105}$$

$$\mathbf{n}_0 = N_{0\alpha}\mathbf{G}_\alpha \tag{2.106}$$

Finally, it is also possible to decompose $\sqrt{g^{\alpha\alpha}}\,\mathbf{t}^\alpha$ with respect to $\mathbf{G}_\alpha$ or $\mathbf{G}^\alpha$. This yields

$$\mathbf{t} = \pi^{\alpha\beta} n_\alpha \mathbf{G}_\beta \tag{2.107}$$

$$\mathbf{t} = \pi^{\alpha}_{\beta} n_\alpha \mathbf{G}^\beta \tag{2.108}$$

$$\mathbf{T} = t^{\alpha\beta} N_{0\alpha} \mathbf{G}_\beta \tag{2.109}$$

$$\mathbf{T} = t^{\alpha}_{\beta} N_{0\alpha} \mathbf{G}^\beta \tag{2.110}$$

Remember that n_α are the components of $\mathbf{n}$ in axes $\mathbf{g}^\alpha$, according to (2.98) and $N_{0\alpha}$ the components of $\mathbf{n}_0$ in axes $\mathbf{G}^\alpha$ according to (2.106).

2.5.9 Physical Components

The base vectors $\mathbf{G}_\alpha$, $\mathbf{G}^\alpha$, $\mathbf{g}_\alpha$, $\mathbf{g}^\alpha$ are not of unit length and may be of different physical dimension. Hence, this is also the case for the material stresses defined above.

Stresses with the same physical dimension can be obtained from the following decompositions with respect to the unit curvilinear base vectors in the deformed configuration defined in Sect. 2.3.6:

$$\mathbf{t}^\alpha = \tau^{(\alpha\beta)} \mathbf{a}_\beta \tag{2.111}$$

or

$$\mathbf{t}^\alpha = \tau^{(\alpha)}_{(\beta)} \mathbf{a}^\beta \tag{2.112}$$

$\tau^{(\alpha\beta)}$ and $\tau^{(\alpha)}_{(\beta)}$ are the physical components of material stresses with respect to curvilinear coordinates in the deformed configuration. Their relations with $\tau^{\alpha\beta}$ and τ^{α}_{β} are

$$\tau^{\alpha\beta} = \tau^{(\alpha\beta)} \sqrt{\frac{g^{\alpha\alpha}}{g_{\beta\beta}}} \quad \text{(no sum)} \tag{2.113}$$

$$\tau^{\alpha}_{\beta} = \tau^{(\alpha)}_{(\beta)} \sqrt{\frac{g^{\alpha\alpha}}{g^{\beta\beta}}} \quad \text{(no sum)} \tag{2.114}$$

as can be verified by substituting in Eqs (2.100) and (2.101).

If one decomposes $\mathbf{t}_i$ with respect to the undeformed curvilinear unit vectors defined in Sect. 2.4.6, one gets

$$\mathbf{t}^\alpha = \pi^{(\alpha\beta)} \mathbf{A}_\beta \tag{2.115}$$

or

$$\mathbf{t}^\alpha = \pi^{(\alpha)}_{(\beta)} \mathbf{A}^\beta \tag{2.116}$$

with the following relations:

$$\pi^{\alpha\beta} = \pi^{(\alpha\beta)} \sqrt{\frac{g^{\alpha\alpha}}{G_{\beta\beta}}} \quad \text{(no sum)} \tag{2.117}$$

$$\pi^{\alpha}_{\beta} = \pi^{(\alpha)}_{(\beta)} \sqrt{\frac{g^{\alpha\alpha}}{G^{\beta\beta}}} \quad \text{(no sum)} \tag{2.118}$$

2.5.10 Symmetry of Stress Tensors

Table 2.1 summarizes the definitions and the symmetry relations of the stress measures introduced in the preceding paragraphs. The symmetry relations are given without demonstration.

Table 2.1. Symmetry relations of stress tensors

Name	Symbol	Definition	Symmetry
Cauchy	σ^{ij}	$\mathbf{t} = n_i\sigma^{ij}\mathbf{e}_j$	$\sigma^{ij} = \sigma^{ji}$
Lagrange	P^{ij}	$\mathbf{T} = n_{(0)i}P^{ij}\mathbf{e}_j$	$P^{ir}\mathbf{e}_r\hat{\mathbf{g}}^j = P^{jr}\mathbf{e}_r\hat{\mathbf{g}}^i$
PK2	S^{ij}	$\mathbf{T} = n_{(0)i}S^{ij}\hat{\mathbf{g}}_j$	$S^{ij} = S^{ji}$
Material stresses	$\tau^{\alpha\beta}$	$\mathbf{t} = n_\alpha\tau^{\alpha\beta}\mathbf{g}_\beta$	$\tau^{\alpha\beta} = \tau^{\beta\alpha}$
in deformed	τ^{α}_{β}	$\mathbf{t} = n_\alpha\tau^{\alpha}_{\beta}\mathbf{g}^\beta$	$\tau^{\alpha}_{\beta} = \tau^{\beta}_{\alpha}$
curvilinear axes	$s^{\alpha\beta}$	$\mathbf{T} = N_{(0)\alpha}s^{\alpha\beta}\mathbf{g}_\beta$	$s^{\alpha\beta} = s^{\beta\alpha}$
	s^{α}_{β}	$\mathbf{T} = N_{(0)\alpha}s^{\alpha}_{\beta}\mathbf{g}^\beta$	$s^{\alpha}_{\beta} = s^{\beta}_{\alpha}$
Material stresses	$\pi^{\alpha\beta}$	$\mathbf{t} = n_\alpha\pi^{\alpha\beta}\mathbf{G}_\beta$	$\pi^{\alpha\gamma}\mathbf{G}_\gamma \cdot \mathbf{g}^\beta = \pi^{\beta\gamma}\mathbf{G}_\gamma \cdot \mathbf{g}^\alpha$
in undeformed	π^{α}_{β}	$\mathbf{t} = n_\alpha\pi^{\alpha}_{\beta}\mathbf{G}^\beta$	$\pi^{\alpha}_{\gamma}\mathbf{G}^\gamma \cdot \mathbf{g}_\beta = \pi^{\beta}_{\gamma}\mathbf{G}^\gamma \cdot \mathbf{g}_\alpha$
curvilinear axes	$t^{\alpha\beta}$	$\mathbf{T} = N_{(0)\alpha}t^{\alpha\beta}\mathbf{G}_\beta$	$t^{\alpha\gamma}\mathbf{G}_\gamma \cdot \mathbf{g}^\beta = t^{\beta\gamma}\mathbf{G}_\gamma \cdot \mathbf{g}^\alpha$
	t^{α}_{β}	$\mathbf{T} = N_{(0)\alpha}t^{\alpha}_{\beta}\mathbf{G}^\beta$	$t^{\alpha}_{\gamma}\mathbf{G}^\gamma \cdot \mathbf{g}_\beta = t^{\beta}_{\gamma}\mathbf{G}^\gamma \cdot \mathbf{g}_\alpha$
Physical components	$\tau^{(\alpha\beta)}$	$\mathbf{t} = n_\alpha\sqrt{g^{\alpha\alpha}}\,\tau^{(\alpha\beta)}\mathbf{a}_\beta$	$\tau^{(\alpha\beta)}\sqrt{\frac{g^{\alpha\alpha}}{g_{\beta\beta}}} = \tau^{(\beta\alpha)}\sqrt{\frac{g^{\beta\beta}}{g_{\alpha\alpha}}}$
	$\tau^{(\alpha)}_{(\beta)}$	$\mathbf{t} = n_\alpha\sqrt{g^{\alpha\alpha}}\,\tau^{(\alpha)}_{(\beta)}\mathbf{a}^\beta$	$\tau^{(\alpha)}_{(\beta)}\sqrt{\frac{g^{\alpha\alpha}}{g_{\beta\beta}}} = \tau^{(\beta)}_{(\alpha)}\sqrt{\frac{g^{\beta\beta}}{g_{\alpha\alpha}}}$
	$\pi^{(\alpha\beta)}$	$\mathbf{t} = n_\alpha\sqrt{g^{\alpha\alpha}}\,\pi^{(\alpha\beta)}\mathbf{A}_\beta$	$\pi^{(\alpha\beta)}\sqrt{\frac{g^{\alpha\alpha}}{G_{\beta\beta}}} = \pi^{(\beta\alpha)}\sqrt{\frac{g^{\beta\beta}}{G_{\alpha\alpha}}}$
	$\pi^{(\alpha)}_{(\beta)}$	$\mathbf{t} = n_\alpha\sqrt{g^{\alpha\alpha}}\,\pi^{(\alpha)}_{(\beta)}\mathbf{A}^\beta$	$\pi^{(\alpha)}_{(\beta)}\sqrt{\frac{g^{\alpha\alpha}}{G^{\beta\beta}}} = \pi^{(\beta)}_{(\alpha)}\sqrt{\frac{g^{\beta\beta}}{G^{\alpha\alpha}}}$

2.6 Covariant Derivatives and Gauss Theorem

2.6.1 Derivatives of Curvilinear Base Vectors

In the preceding sections, the following base vectors were developed:

$\mathbf{g}_1\mathbf{g}_2\mathbf{g}_3$	Covariant base in γ	Functions of
$\mathbf{g}^1\mathbf{g}^2\mathbf{g}^3$	Contravariant base in γ	$\xi^1\xi^2\xi^3$
$\mathbf{G}_1\mathbf{G}_2\mathbf{G}_3$	Covariant base in Γ	
$\mathbf{G}^1\mathbf{G}^2\mathbf{G}^3$	Contravariant base in Γ	

$\hat{\mathbf{g}}_1\hat{\mathbf{g}}_2\hat{\mathbf{g}}_3$ Covariant base in γ } Functions of
$\hat{\mathbf{g}}^1\hat{\mathbf{g}}^2\hat{\mathbf{g}}^3$ Contravariant base in γ } $X^1X^2X^3$

To simplify the writing, we note, for any quantitity Q,

$$\left.\begin{aligned} \frac{\partial Q}{\partial \xi^\alpha} &= Q_{,\alpha} \\ \frac{\partial Q}{\partial x^s} &= Q_{:s} \\ \frac{\partial Q}{\partial X^s} &= Q_{;s} \end{aligned}\right\} \tag{2.119}$$

Let us calculate $\mathbf{G}_{\alpha,\beta} = \partial \mathbf{G}_\alpha / \partial \xi^\beta$:

$$\mathbf{G}_{\alpha,\beta} = \frac{\partial}{\partial \xi^\beta}\frac{\partial X^k}{\partial \xi^\alpha}\mathbf{e}_k = \frac{\partial^2 X^k}{\partial \xi^\beta \partial \xi^\alpha}\mathbf{e}_k \tag{2.120}$$

Inverting Eq. (2.53) yields

$$\begin{Bmatrix} \mathbf{e}_1 \\ \mathbf{e}_2 \\ \mathbf{e}_3 \end{Bmatrix} = \mathbf{J}^{-\mathrm{T}} \begin{Bmatrix} \mathbf{G}_1 \\ \mathbf{G}_2 \\ \mathbf{G}_3 \end{Bmatrix} \tag{2.121}$$

whence

$$\mathbf{e}_k = J^{-1}_{\alpha k}\mathbf{G}_\alpha = \frac{\partial \xi^\alpha}{\partial X^k}\mathbf{G}_\alpha \tag{2.122}$$

$$\mathbf{G}_{\alpha,\beta} = \frac{\partial^2 X^k}{\partial \xi^\alpha \partial \xi^\beta}\frac{\partial \xi^\gamma}{\partial X^k}\mathbf{G}_\gamma \tag{2.123}$$

which can also be written

$$\mathbf{G}_{\alpha,\beta} = \Gamma^\gamma_{\alpha\beta}\mathbf{G}_\gamma \tag{2.124}$$

with

$$\Gamma^\gamma_{\alpha\beta} = \frac{\partial^2 X_k}{\partial \xi^\alpha \partial \xi^\beta}\frac{\partial \xi^\gamma}{\partial X_k} = X_{k,\alpha,\beta}\xi^\gamma_{:k} \tag{2.125}$$

Eq. (2.124) shows that the derivatives of the curvilinear base vectors are a linear combination of these base vectors. The coefficients of this linear combination are the Cristoffel symbols $\Gamma^\gamma_{\alpha\beta}$.

In the same way, the following relations can be established:

$$\mathbf{g}_{\alpha,\beta} = \gamma^\gamma_{\alpha\beta}\mathbf{g}_\gamma,\ \mathbf{g}^\alpha_{,\beta} = -\gamma^\alpha_{\beta\gamma}\mathbf{g}^\gamma \tag{2.126}$$

$$\mathbf{G}_{\alpha,\beta} = \Gamma^\gamma_{\alpha,\beta}\mathbf{G}_\gamma,\ \mathbf{G}^\alpha_{,\beta} = -\Gamma^\alpha_{\beta\gamma}G^\gamma \tag{2.127}$$

$$\hat{\mathbf{g}}_{i;j} = \mathscr{C}^s_{ij}\hat{\mathbf{g}}_s,\ \hat{\mathbf{g}}^i_{;j} = -\mathscr{C}^i_{js}\hat{\mathbf{g}}^s \tag{2.128}$$

$$\mathbf{g}_{\alpha,\beta} = \gamma^\gamma_{\alpha\beta\gamma}\mathbf{g}^\gamma \tag{2.129}$$

$$\mathbf{G}_{\alpha,\beta} = \Gamma_{\alpha\beta\gamma}\mathbf{G}^\gamma \tag{2.130}$$

$$\hat{\mathbf{g}}_{i;j} = \mathscr{C}_{ijs}\hat{\mathbf{g}}^s \tag{2.131}$$

$\gamma_{\alpha\beta\gamma}$, $\Gamma_{\alpha\beta\gamma}$ and $\mathscr{C}_{ijs}$ are Cristoffel symbols of first kind; $\gamma^{\gamma}_{\alpha\beta}$, $\Gamma^{\gamma}_{\alpha\beta}$ and $\mathscr{C}^{s}_{ij}$ are Cristoffel symbols of second kind.

$$\gamma_{\alpha\beta\gamma} = \frac{\partial^2 x^r}{\partial\xi^\alpha\partial\xi^\beta}\frac{\partial x^r}{\partial\xi^\gamma} \tag{2.132}$$

$$\Gamma_{\alpha\beta\gamma} = \frac{\partial^2 X^r}{\partial\xi^\alpha\partial\xi^\beta}\frac{\partial X^r}{\partial\xi^\gamma} \tag{2.133}$$

$$\mathscr{C}_{ijs} = \frac{\partial^2 x^r}{\partial X^i\partial X^j}\frac{\partial x^r}{\partial X^s} \tag{2.134}$$

$$\gamma^{\gamma}_{\alpha\beta} = \frac{\partial^2 X^r}{\partial\xi^\alpha\partial\xi^\beta}\frac{\partial\xi^\gamma}{\partial x^r} \tag{2.135}$$

$$\Gamma^{\gamma}_{\alpha\beta} = \frac{\partial^2 X^r}{\partial\xi^\alpha\partial\xi^\beta}\frac{\partial\xi^\gamma}{\partial X^r} \tag{2.136}$$

$$\mathscr{C}^{s}_{ij} = \frac{\partial^2 x^r}{\partial X^i\partial X^j}\frac{\partial X^s}{\partial x^r} \tag{2.137}$$

2.6.2 Covariant Derivatives

The components of any vector $\mathbf{v}$ can be expressed by

$$\left.\begin{aligned}
\mathbf{v} &= v^\alpha\mathbf{g}_\alpha \\
\mathbf{v} &= v_\alpha\mathbf{g}^\alpha \\
\mathbf{v} &= V^\alpha\mathbf{G}_\alpha \\
\mathbf{v} &= V_\alpha\mathbf{G}^\alpha \\
\mathbf{v} &= \mathscr{V}^i\hat{\mathbf{g}}_i \\
\mathbf{v} &= \mathscr{V}_i\hat{\mathbf{g}}^i \\
\mathbf{v} &= v_i\mathbf{e}^i = v^i\mathbf{e}_i
\end{aligned}\right\} \tag{2.138}$$

Let us calculate, for example,

$$\left.\begin{aligned}
\frac{\partial\mathbf{v}}{\partial\xi^\alpha} = \mathbf{v}_{,\alpha} &= V^{\gamma}_{,\alpha}\mathbf{G}_\gamma + V^\gamma\mathbf{G}_{\gamma,\alpha} \\
\mathbf{v}_{,\alpha} &= V^{\gamma}_{,\alpha}\mathbf{G}_\gamma + V^\gamma\Gamma^{\beta}_{\gamma\alpha}\mathbf{G}_\beta \\
\mathbf{v}_{,\alpha} &= (V^{\beta}_{,\alpha} + V^\gamma\Gamma^{\beta}_{\gamma\alpha})\mathbf{G}_\beta
\end{aligned}\right\} \tag{2.139}$$

This result is written

$$\mathbf{v}_{,\alpha} = V^\beta\|_\alpha\mathbf{G}_\beta \tag{2.140}$$

with:

$$V^\beta\|_\alpha = V^\beta_{,\alpha} + V^\gamma \Gamma^\beta_{\gamma\alpha} \tag{2.141}$$

The quantity $V^\beta\|_\alpha$ is the covariant derivative of V^β in the curvilinear base $\mathbf{G}_1\mathbf{G}_2\mathbf{G}_3$. Similarly, one can write

$$\left.\begin{array}{lll} \mathbf{v}_{,\alpha} = v^\beta|_\alpha \mathbf{g}_\beta & \text{with} & v^\beta|_\alpha = v^\beta_{,\alpha} + v^\gamma \gamma^\beta_{\gamma\alpha} \\ \mathbf{v}_{,\alpha} = V^\beta\|_\alpha \mathbf{G}_\beta & \text{with} & V^\beta\|_\alpha = V^\beta_{,\alpha} + V^\gamma \Gamma^\beta_{\gamma\alpha} \\ \mathbf{v}_{;\sigma} = \mathcal{V}^p|_s \hat{\mathbf{g}}_p & \text{with} & \mathcal{V}^p|_s = \mathcal{V}^p_{;s} + \mathcal{V}^i \mathcal{C}^p_{is} \end{array}\right\} \tag{2.142}$$

$$\left.\begin{array}{lll} \mathbf{v}_{,\alpha} = v_\beta|_\alpha \mathbf{g}^\beta & \text{with} & v_\beta|_\alpha = v_{\beta,\alpha} - v_\gamma \gamma^\gamma_{\beta,\alpha} \\ \mathbf{v}_{,\alpha} = V_\beta\|_\alpha \mathbf{G}^\beta & \text{with} & V_\beta\|_\alpha = V_{\beta,\alpha} - V_\gamma \Gamma^\gamma_{\beta\alpha} \\ \mathbf{v}_{;\sigma} = \mathcal{V}_p|_s \hat{\mathbf{g}}^p & \text{with} & \mathcal{V}_p|_s = \mathcal{V}_{p;s} - \mathcal{V}_i \mathcal{C}^i_{ps} \end{array}\right\} \tag{2.143}$$

It can be shown that covariant derivatives follow the same rules as ordinary derivatives. For example,

$$\begin{array}{rl} (a^\alpha b^\beta)\|_\gamma = & a^\alpha b^\beta\|_\gamma + b^\beta a^\alpha\|_\gamma \\ (\lambda a_\alpha)\|_\gamma = & \lambda_{,\gamma} a_\alpha + \lambda a_\alpha\|_\gamma \end{array} \tag{2.144}$$

2.6.3 Gauss Theorem

Consider a solid of volume v with exterior surface a. Let $\mathbf{n}$ be the outward normal to this surface. One has

$$\left.\begin{array}{l} \int_v \operatorname{grad} \phi dv = \int_a \phi \mathbf{n} da \\ \int_v \operatorname{div} \mathbf{v} dv = \int_a \mathbf{v} \cdot \mathbf{n} da \end{array}\right\} \tag{2.145}$$

These equalities can be expressed using components. For example,

$$\left.\begin{array}{rl} \operatorname{grad} \phi = & \phi_{,\alpha} \mathbf{g}^\alpha \\ \operatorname{div} \mathbf{v} = & v^\alpha|_\alpha = \frac{1}{\sqrt{g}} (\sqrt{g}\, v^\alpha)_{,\alpha} \\ \mathbf{n} = & n_\alpha \mathbf{g}^\alpha \end{array}\right\} \tag{2.146}$$

whence

$$\int_v \phi_{,\alpha} dv = \int_a \phi n_\alpha da \tag{2.147}$$

$$\int_v v^\alpha|_\alpha dv = \int_v \frac{1}{\sqrt{g}} (\sqrt{g} v^\alpha)_{,\alpha} dv = \int_a v^\alpha n_\alpha da \tag{2.148}$$

Also, with respect to the initial geometry,

$$\left.\begin{aligned} \mathbf{v} &= V^\alpha \mathbf{G}_\alpha \\ \mathbf{n}_0 &= N_{0\alpha} \mathbf{G}^\alpha \\ \operatorname{div} \mathbf{v} &= V^\alpha \|_\alpha = \frac{1}{\sqrt{G}} (\sqrt{G} V^\alpha)_{,\alpha} \end{aligned}\right\} \tag{2.149}$$

whence

$$\int_V V^\alpha \|_\alpha dV = \int_V \frac{1}{\sqrt{G}} (\sqrt{G} V^\alpha)_{,\alpha} \, dV = \int_A V^\alpha N_{0\alpha} dA \tag{2.150}$$

Using Cartesian coordinates x^i of the deformed geometry gives

$$\left.\begin{aligned} \mathbf{v} &= v^r \mathbf{e}_r \\ \mathbf{n} &= n_r \mathbf{e}^r \\ \operatorname{div} \mathbf{v} &= v^r_{;r} \\ \int_v v_{;r} dv &= \int_a n_r v^r da \end{aligned}\right\} \tag{2.151}$$

Using Cartesian coordinates X^i of the undeformed geometry gives

$$\left.\begin{aligned} \mathbf{v} &= v^r \mathbf{e}_r \\ \mathbf{n}_0 &= n_{0r} \mathbf{e}^r \\ \operatorname{div} \mathbf{v} &= v^r_{;r} \\ \int_V v^r_{;r} dV &= \int_A n_{0r} v^r dA \end{aligned}\right\} \tag{2.152}$$

2.7 Infinitesimal Volume Equilibrium Equations

2.7.1 Applied Forces

The deformed solid is at equilibrium under the action of

- Body forces per unit mass $\mathbf{f}$
- Surface tractions per unit deformed area $\mathbf{p}$

If ρ is the mass per unit deformed volume, the body force per unit deformed volume is $\rho\mathbf{f}$.

2.7.2 Surface Equilibrium Equations

The surface equilibrium equations are simply

$$\mathbf{p} = \mathbf{t} \quad \text{or} \quad \mathbf{P} = \mathbf{T} \tag{2.153}$$

where $\mathbf{t}$ and $\mathbf{T}$ are the stress vectors per unit deformed and undeformed areas and where $\mathbf{P}$ is a fictitious surface traction per unit deformed area defined by

$$\mathbf{p}da = \mathbf{P}dA \tag{2.154}$$

Hence, the components of (2.153) with respect to any coordinate system are obtained easily. It suffices to replace **t** by **p** or **T** by **P** in the definitions of the different stresses (summarized in Sect. 2.5.10).

2.7.3 Volume Equilibrium Equations

Consider an arbitrary volume element v of the deformed solid. Let a be its bounding surface with unit outward normal **n**. This volume is at equilibrium under the action of body forces $\rho\mathbf{f}$ and stress vectors **t** applied to it, through surface a, by the rest of the body. Hence, the expression of equilibrium is

$$\int_v \rho\mathbf{f}dv + \int_a \mathbf{t}da = 0 \tag{2.155}$$

The surface integral can be transformed with the help of Gauss's theorem. For example,

$$\mathbf{f} = f^\alpha \mathbf{g}_\alpha \tag{2.156}$$

$$\mathbf{t} = n_\alpha \tau^{\alpha\beta}\mathbf{g}_\beta \tag{2.157}$$

whence

$$\int_v \rho f^\beta dv + \int_a n_\alpha \tau^{\alpha\beta} da = 0 \tag{2.158}$$

$$\int_v \rho f^\beta dv + \int_v \tau^{\alpha\beta}|_\alpha dv = 0 \tag{2.159}$$

Since this equation is valid for any volume **v**, one has

$$\tau^{\alpha\beta}|_\alpha + \rho f^\beta = 0 \tag{2.160}$$

A similar calculation can be done with any stress definition. With the notations

$$\left.\begin{aligned} \mathbf{f} &= f^\alpha \mathbf{g}_\alpha = f_\alpha \mathbf{g}^\alpha \\ \mathbf{f} &= F^\alpha \mathbf{G}_\alpha = F_\alpha \mathbf{G}^\alpha \\ \mathbf{f} &= \mathcal{F}^i \hat{\mathbf{g}}_i = \mathcal{F}_i \hat{\mathbf{g}}^i \\ \mathbf{f} &= f^i \mathbf{e}_i \end{aligned}\right\} \tag{2.161}$$

one can obtain the results summarized in Table 2.2.

2.8 The Virtual Work Principle – Conjugated Virtual Strains

2.8.1 External Virtual Work

Let $\delta\mathbf{u}$ be a virtual displacement field. By definition, it is infinitesimal and we assume that it is kinematically admissible (i.e. it respects the continuity and the

Table 2.2. Equilibrium equations of stress tensors

Name	Symbol	Volume equilibrium equation
Cauchy	σ^{ij}	$\sigma^{ij}_{;i} + \rho f^j = 0$
Lagrange	P^{ij}	$P^{ij}_{;i} + \rho_0 f^j = 0$
PK2	S^{ij}	$S^{ij}_{;i} + \rho_0 \mathcal{F}^j = 0$
Material stresses in deformed curvilinear axes	$\tau^{\alpha\beta}$	$\tau^{\alpha\beta}_{\alpha} + \rho f^\beta = 0$
	τ^{α}_{β}	$\tau^{\alpha}_{\beta}\vert_\alpha + \rho f_\beta = 0$
	$s^{\alpha\beta}$	$s^{\alpha\beta}\Vert_\alpha + \rho_0 f^\beta = 0$
	s^{α}_{β}	$s^{\alpha}_{\beta}\Vert_\alpha + \rho_0 f_\beta = 0$
Material stresses in undeformed curvilinear axes	$\pi^{\alpha\beta}$	$\pi^{\alpha\beta}\vert_\alpha + \rho F^\beta = 0$
	π^{α}_{β}	$\pi^{\alpha}_{\beta}\vert_\alpha + \rho F_\beta = 0$
	$t^{\alpha\beta}$	$t^{\alpha\beta}\Vert_\alpha + \rho_0 F_\beta = 0$
	t^{α}_{β}	$t^{\alpha}_{\beta}\Vert_\alpha + \rho_0 F_\beta = 0$
Material components	$\tau^{(\alpha\beta)}$	$\left(\tau^{(\alpha\beta)}\sqrt{\frac{g^{\alpha\alpha}}{g_{\beta\beta}}}\right)\Big\vert_\alpha + \rho f^\beta = 0$
	$\tau^{(\alpha)}_{(\beta)}$	$\left(\tau^{(\alpha)}_{(\beta)}\sqrt{\frac{g^{\alpha\alpha}}{g_{\beta\beta}}}\right)\Big\vert_\alpha + \rho f_\beta = 0$
	$\pi^{(\alpha\beta)}$	$\left(\pi^{(\alpha\beta)}\sqrt{\frac{g^{\alpha\alpha}}{g_{\beta\beta}}}\right)\Big\vert_\alpha + \rho F^\beta = 0$
	$\pi^{(\alpha)}_{(\beta)}$	$\left(\pi^{(\alpha)}_{(\beta)}\sqrt{\frac{g^{\alpha\alpha}}{g^{\beta\beta}}}\right)\Big\vert_\alpha + \rho F_\beta = 0$

support conditions of the solid). The corresponding virtual work done by the body forces $\rho\mathbf{f}$ and the applied surface tractions per unit deformed area $\mathbf{p}$ is

$$\delta W_E = \int_v \rho\mathbf{f}\cdot\delta\mathbf{u}dv + \int_a \mathbf{p}\cdot\delta\mathbf{u}da \qquad (2.162)$$

Using

$$\rho dv = \rho_0 dV \qquad (2.163)$$

$$\mathbf{p}da = \mathbf{P}dA \qquad (2.164)$$

this equation becomes

$$\delta W_E = \int_V \rho_0\mathbf{f}\cdot\delta\mathbf{u}dV + \int_A \mathbf{P}\cdot\delta\mathbf{u}dA \qquad (2.165)$$

In Eq. (2.162), the current configuration γ is taken as reference to compute the external virtual work, while in Eq. (2.165), the initial, undeformed configuration Γ is the reference configuration. An obvious extension is obtained by taking any known configuration Γ_R as reference (Fig. 2.12). In this case, we have

$$\rho dv = \rho_R dV_R \qquad (2.166)$$

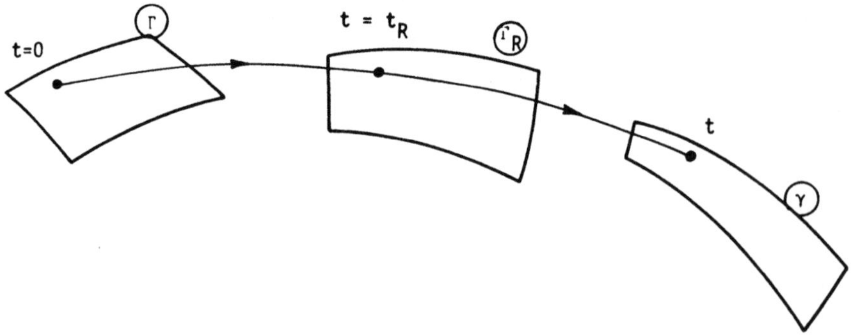

Fig. 2.12. Various configurations of a deforming body.

$$\mathbf{p}da = \mathbf{P}_R dA_R \tag{2.167}$$

$$\delta W_E = \int_{V_R} \rho_R \mathbf{f} \cdot \delta \mathbf{u} dV_R + \int_{A_R} \mathbf{P}_R \cdot \delta \mathbf{u} dA_R \tag{2.168}$$

2.8.2 Internal Virtual Work

The virtual displacements $\delta\mathbf{u}$ create virtual strains in the solid. The corresponding virtual work done by the stresses existing in the body is called the internal virtual work δw_I. Its general expression is:

$$\delta W_I = \int_{\text{volume}} (\text{stresses}) \cdot (\text{conjugated virtual strains})\, d(\text{volume}) \tag{2.169}$$

According to the chosen stress tensor and reference configuration, the expression of the conjugated virtual strains will be different.

2.8.3 Particular Forms of the Virtual Work Principle

2.8.3.1 Deformed Curvilinear Axes; Material Stresses $\tau^{\alpha\beta}$

One obtains

$$\delta W_E = \int_v \rho \mathbf{f} \cdot \delta \mathbf{u} dv + \int_a \mathbf{p} \cdot \delta \mathbf{u} da \tag{2.170}$$

From

$$\mathbf{p} = \mathbf{t} = n_\alpha \tau^{\alpha\beta} \mathbf{g}_\beta \tag{2.171}$$

one gets

$$\delta W_E = \int_v \rho f^\beta \mathbf{g}_\beta \cdot \delta \mathbf{u} dv + \int_a n_\alpha \tau^{\alpha\beta} \mathbf{g}_\beta \cdot \delta \mathbf{u} da \tag{2.172}$$

Applying the Gauss theorem:

$$\delta W_{\mathrm{E}} = \int_v \rho f^\beta \mathbf{g}_\beta \cdot \delta \mathbf{u} dv + \int_v (\tau^{\alpha\beta} \mathbf{g}_\beta \cdot \delta \mathbf{u}) |_\alpha dv \tag{2.173}$$

$$\delta W_{\mathrm{E}} = \int_v (\rho f^\beta + \tau^{\alpha\beta}|_\alpha) \mathbf{g}_\beta \cdot \delta \mathbf{u} dv + \int_v \tau^{\alpha\beta} (\mathbf{g}_\beta \cdot \delta \mathbf{u}) |_\alpha dv \tag{2.174}$$

The first term vanishes because of the volume equilibrium equation. Hence, there remains

$$\delta W_{\mathrm{E}} = \int_v \tau^{\alpha\beta} (\mathbf{g}_\beta \cdot \delta \mathbf{u}) |_\alpha dv \tag{2.175}$$

Term $(\mathbf{g}_\beta \cdot \delta \mathbf{u}) |_\alpha$ represents the virtual strain conjugated to $\tau^{\alpha\beta}$. If one defines

$$\eta_{\alpha\beta} = \tfrac{1}{2} (g_{\alpha\beta} - G_{\alpha\beta}) \tag{2.176}$$

it is easily verified that

$$\tau^{\alpha\beta} (\mathbf{g}_\beta \cdot \delta \mathbf{u}) |_\alpha = \delta \eta_\alpha \cdot \tau^{\alpha\beta} \tag{2.177}$$

Indeed, one has, successively

$$\delta \eta_{\alpha\beta} = \tfrac{1}{2} \delta g_{\alpha\beta} = \tfrac{1}{2} (\delta \mathbf{g}_\alpha \cdot \mathbf{g}_\beta + \delta \mathbf{g}_\beta \cdot \mathbf{g}_\alpha) \tag{2.178}$$

$$\delta \mathbf{g}_\alpha = \delta \mathbf{r}_{,\alpha} = \delta(\mathbf{R} + \mathbf{u})_{,\alpha} = \delta \mathbf{u}_{,\alpha} \tag{2.179}$$

$$\delta \mathbf{u}_{,\alpha} = \delta u_\gamma |_\alpha \mathbf{g}^\gamma \tag{2.180}$$

Whence

$$\begin{aligned} \delta \eta_{\alpha\beta} &= \tfrac{1}{2} (\delta u_\gamma |_\alpha \mathbf{g}^\gamma \cdot \mathbf{g}_\beta + \delta u_\gamma |_\beta \mathbf{g}^\gamma \cdot \mathbf{g}_\alpha) \\ &= \tfrac{1}{2} (\delta u_\beta |_\alpha + \delta u_\alpha |_\beta) \end{aligned} \tag{2.181}$$

By symmetry of $\tau^{\alpha\beta}$

$$\tau^{\alpha\beta} \delta \eta_{\alpha\beta} = \tau^{\alpha\beta} \delta u_\beta |_\alpha \tag{2.182}$$

On the other hand,

$$(\mathbf{g}_\beta \cdot \delta \mathbf{u}) |_\alpha = (\mathbf{g}_\beta \cdot \delta u_\gamma \mathbf{g}^\gamma) |_\alpha = \delta u_\beta |_\alpha \tag{2.183}$$

This proves equality (2.177).

Replacing in Eq. (2.175) yields

$$\delta W_{\mathrm{I}} = \delta W_{\mathrm{E}} = \int_v \tau^{\alpha\beta} \delta \eta_{\alpha\beta} dv \tag{2.184}$$

Hence, $\eta_{\alpha\beta}$ are the strains conjugated to the material stresses $\tau^{\alpha\beta}$.

2.8.3.2 Deformed Curvilinear Axes; Material Stresses $s^{\alpha\beta}$

From

$$\delta W_{\mathrm{E}} = \int_{V} \rho_0 \mathbf{f} \cdot \delta \mathbf{u} dV + \int_{A} \mathbf{P} \cdot \delta \mathbf{u}\, dA \tag{2.185}$$

with

$$\mathbf{P} = \mathbf{T} = N_{0\alpha} s^{\alpha\beta} \mathbf{g}_{\beta} \tag{2.186}$$

one obtains, by an analogous development

$$\delta W_{\mathrm{I}} = \delta W_{\mathrm{E}} = \int_{V} s^{\alpha\beta} \delta \eta_{\alpha\beta} dV \tag{2.187}$$

2.8.3.3 Undeformed Curvilinear Axes; Material Stresses $\pi^{\alpha\beta}$

One obtains

$$\delta W_{\mathrm{I}} = \delta W_{\mathrm{E}} = \int_{V} \pi^{\alpha\beta} \delta (U_{\beta}\|_{\alpha}) dV \tag{2.188}$$

2.8.3.4 Undeformed Curvilinear Axes; Material Stresses $t^{\alpha\beta}$

One obtains

$$\delta W_{\mathrm{I}} = \delta W_{\mathrm{E}} = \int_{V} t^{\alpha\beta} \delta (U_{\beta}\|_{\alpha}) dV \tag{2.189}$$

2.8.3.5 Global Cartesian Axes; Cauchy Stresses σ^{ij}

One obtains

$$\delta W_{\mathrm{E}} = \int_{v} \rho \mathbf{f} \cdot \delta \mathbf{u} dv + \int_{a} \mathbf{p} \cdot \delta \mathbf{u} da \tag{2.190}$$

$$\mathbf{p} = \mathbf{t} = n_i \sigma^{ij} \mathbf{e}_j ; \mathbf{f} = f^j \mathbf{e}_j ; \delta \mathbf{u} = \delta u_j \mathbf{e}^j \tag{2.191}$$

$$\delta W_{\mathrm{E}} = \int_{v} \rho f^j \delta u_j dv + \int_{a} n_i \sigma^{ij} \delta u_j da \tag{2.192}$$

$$\delta W_{\mathrm{E}} = \int_{v} (\rho f^j + \sigma^{ij}_{:i}) \delta u_j dv + \int_{v} \sigma^{ij} (\delta u_j)_{:i} dv \tag{2.193}$$

where the first integral vanishes because of the volume equilibrium equation. Let

$$\delta \epsilon_{ij} = \frac{1}{2} [(\delta u_i)_{:j} + (\delta u_j)_{:i}] = \frac{1}{2} \left[\frac{\partial(\delta u_i)}{\partial x^j} + \frac{\partial(\delta u_j)}{\partial x^i} \right] \tag{2.194}$$

where the symbols δ and $\partial/\partial x^i$ cannot be permuted. By symmetry of σ^{ij}, one gets

$$\delta W_{\mathrm{I}} = \delta W_{\mathrm{E}} = \int_v \sigma^{ij} \delta \epsilon_{ij} dv \tag{2.195}$$

2.8.3.6 Lagrange Stresses P^{ij}

One obtains

$$\delta W_{\mathrm{I}} = \delta W_{\mathrm{E}} = \int_V P^{ij} \delta(u_{j;i}) dV \tag{2.196}$$

with

$$\delta(u_{j;i}) = \delta\left(\frac{\partial u_j}{\partial X^i}\right) = \frac{\partial}{\partial X^i}(\delta u_j) \tag{2.197}$$

2.8.3.7 PK2 Stresses S^{ij}

By analogy with Sect. 2.8.3.1, one gets

$$\delta W_{\mathrm{I}} = \delta W_{\mathrm{E}} = \int_V S^{ij} \delta E_{ij} dV \tag{2.198}$$

where E_{ij} is the Green strain tensor defined by

$$E_{ij} = \tfrac{1}{2}(\hat{\mathbf{g}}_{ij} - \delta_{ij}) \tag{2.199}$$

where

$$\hat{\mathbf{g}}_{ij} = \hat{\mathbf{g}}_i \cdot \hat{\mathbf{g}}_j \tag{2.200}$$

is the metric tensor corresponding to the base vectors $\hat{\mathbf{g}}_i$ and δ_{ij} is the Kronecker symbol.

2.8.4 Summary of Formulations and Conjugated Strains

Table 2.3 summarizes the preceding results.

It is clear that the virtual work principle

$$\delta W_{\mathrm{E}} = \delta W_{\mathrm{I}} \tag{2.201}$$

is merely an expression of the equilibrium of the body. In this equation, any one of the formulae (2.162), (2.165), or (2.168) can be used for δW_{E} and any one of the seven expressions recalled in Table 2.3 can be used for δW_{I}. Furthermore, all the expressions of δW_{I} referring to the undeformed configuration Γ (2.187, 2.189, 2.196, 2.198) could obviously be extended to an arbitrary reference configuration Γ_{R}.

In any case, the expression of equilibrium of a solid by the virtual work principle only requires the definition of stress tensors and their conjugated virtual strain tensors. Virtual strain tensors can often be obtained by taking the variation of strain tensors but this is not a necessary condition. Indeed, it appears that the virtual strain tensor $\delta\epsilon_{ij}$ conjugated to σ^{ij} is not the variation of any existing strain tensor.

Table 2.3. Conjugated strain, conjugated virtual strain and internal virtual work for stress tensors

Stress name	Symbol	Conjugated strain	Conjugated virtual strain	Internal virtual work
Cauchy	σ^{ij}	—	$\delta\epsilon_{ij} = \frac{1}{2}[(\delta u_i)_{:j} + (\delta u_j)_{:i}]$	$\delta W_I = \int_v \sigma^{ij}\delta\epsilon_{ij}dv$
Lagrange	P^{ij}	$u_{j;i} = \frac{\partial}{\partial X^i}(u_j)$	$\delta(u_{j;i})$	$\delta W_I = \int_V P^{ij}\delta(u_{j;i})dV$
PK2	S^{ij}	$E_{ij} = \frac{1}{2}(\hat{g}_{ij} - \delta_{ij})$	δE_{ij}	$\delta W_I = \int_V S^{ij}\delta E_{ij}dV$
Material stresses in deformed curvilinear axes	$\tau^{\alpha\beta}$	$\eta_{\alpha\beta} = \frac{1}{2}(g_{\alpha\beta} - G_{\alpha\beta})$	$\delta\eta_{\alpha\beta}$	$\delta W_I = \int_v \tau^{\alpha\beta}\delta\eta_{\alpha\beta}dv$
	$s^{\alpha\beta}$	$\eta_{\alpha\beta} = \frac{1}{2}(g_{\alpha\beta} - G_{\alpha\beta})$	$\delta\eta_{\alpha\beta}$	$\delta W_I = \int_V s^{\alpha\beta}\delta\eta_{\alpha\beta}dV$
Material stresses in undeformed curvilinear axes	$\pi^{\alpha\beta}$	$U_{\beta\Vert\alpha} = U_{\beta.\alpha} - U_\gamma\Gamma^\gamma_{\alpha\beta}$	$\delta(U_{\beta\Vert\alpha})$	$\delta W_I = \int_v \pi^{\alpha\beta}\delta(U_{\beta\Vert\alpha})dv$
	$t^{\alpha\beta}$	$U_{\beta\Vert\alpha} = U_{\beta.\alpha} - U_\gamma\Gamma^\gamma_{\alpha\beta}$	$\delta(U_{\beta\Vert\alpha})$	$\delta W_I = \int_V t^{\alpha\beta}(U_{\beta\Vert\alpha})dV$

2.9 Strains

2.9.1 Introduction

The geometrical transformation

$$x^i = x^i(X^1, X^2, X^3) \tag{2.202}$$

of the solid from its initial configuration to the current one can be considered as a mapping of Γ into γ. The Jacobian matrix of this mapping

$$\mathbf{F} = \frac{\partial \mathbf{x}}{\partial \mathbf{X}} \tag{2.203}$$

is called the deformation gradient as already mentioned in Sect. 2.5.5, and is of fundamental importance in the definition of strain measures.

In particular, if the motion between Γ and γ reduces to a rigid body translation **S**,

$$\mathbf{x} = \mathbf{X} + \mathbf{S} \tag{2.204}$$

the deformation gradient reduces to the unit tensor **I**:

$$\mathbf{F} = \mathbf{I} \tag{2.205}$$

On the other hand, if this motion consists of a rigid body rotation **R**,

$$\mathbf{x} = \mathbf{R}\mathbf{X} \tag{2.206}$$

the deformation gradient coincides with the rotation tensor **R**:

$$\mathbf{F} = \mathbf{R} \tag{2.207}$$

In order to be an acceptable measure of strain, a strain tensor should at least fulfil the following requirements.

- *Objectivity*. A strain tensor should vanish if the deformation gradient **F** reduces to the unit tensor **I** or a rotation tensor **R**. Physically, this condition means that if a material particle has only been submitted to a translation plus a rotation from γ to Γ, the strains of this particle are zero. In particular, if the whole solid has only been translated and rotated from Γ to γ, the strains vanish everywhere in the solid. Obviously, it is also the case if γ coincides with Γ (no motion), which means that the strain tensor vanishes in the reference configuration.

  Finally, the objectivity condition also implies that the strain tensor is indifferent from a translation and a rotation of the reference frame (which is equivalent to a translation and a rotation of the solid in a fixed frame).
- *Regularity*. The inverse of a strain tensor should exist. Physically, this means that two (or more) different deformations of a material particle should not produce the same strain tensor. This condition also implies that the values of the strain tensor cannot become infinite.

It is clear that additional restrictions could be added, such as conditions of isotropy of symmetry. However, they are not compulsory, and in any case, they leave enough freedom to define many different strain measures. Some of them are reviewed hereafter.

2.9.2 Strain Tensors Conjugated to Lagrange, PK2 and Material Stresses

In the development of the virtual work principle, we have introduced several strain measures:

$$u_{j;i} = \frac{\partial u_j}{\partial X^i} \tag{2.208}$$

$$E_{ij} = \tfrac{1}{2}(\hat{g}_{ij} - \delta_{ij}) \tag{2.209}$$

$$\eta_{\alpha\beta} = \tfrac{1}{2}(g_{\alpha\beta} - G_{\alpha\beta}) \tag{2.210}$$

$$U_{\beta\|\alpha} = U_{\beta,\alpha} - U_\gamma \Gamma^\gamma_{\alpha\beta} \tag{2.211}$$

It is easily seen that E_{ij} and $\eta_{\alpha\beta}$ satisfy the conditions of objectivity and regularity.

For example, if a material particle is submitted to a translation and a rotation without deformation, the local covariant bases in Γ and γ only differ by a rotation, which implies that the metric tensors $g_{\alpha\beta}$ and $G_{\alpha\beta}$ are equal and $\eta_{\alpha\beta}$ vanishes.

However, it can be shown that $u_{j;i}$ and $U_{\beta\|\alpha}$ are not objective strain measures. For example, if the motion from Γ to γ is a rigid body rotation, Eq. (2.206), the displacement field is:

$$\mathbf{u} = \mathbf{x} - \mathbf{X} = (\mathbf{R} - \mathbf{I})\,\mathbf{X} \tag{2.212}$$

Hence:

$$\frac{\partial \mathbf{u}}{\partial \mathbf{X}} = \mathbf{R} - \mathbf{I} \neq \mathbf{0} \tag{2.213}$$

Therefore, $u_{j;i}$ and $U_{\beta\|\alpha}$ cannot be accepted as valid strain measures.

2.9.3 Strain Tensors Based on the Polar Decomposition

If the deformation from Γ to γ is arbitrary, it is possible to perform the following decomposition of the deformation tensor

$$\mathbf{F} = \mathbf{R}\mathbf{U} \tag{2.214}$$

where $\mathbf{R}$ is the rotation tensor and $\mathbf{U}$ is the right stretch tensor. They have the following properties:

$$\mathbf{R}^{\mathrm{T}} = \mathbf{R}^{-1} \tag{2.215}$$

$$\mathbf{U} = \mathbf{U}^{\mathrm{T}} \tag{2.216}$$

This decomposition Eq. (2.214) is called the polar decomposition. Physically, it can be interpreted as if the motion of the material particle from Γ to γ were composed of a stretch $\mathbf{U}$ followed by a rotation $\mathbf{R}$. If this motion reduces to a translation or a rotation, $\mathbf{U}$ becomes the unit tensor $\mathbf{I}$. Therefore, many objective strain tensors can be constructed as functions of $\mathbf{U}$. It suffices to choose regular functions that vanish for $\mathbf{U} = \mathbf{I}$.

The best known tensors of this kind are

- Green strain tensor $\mathbf{E}^{\mathrm{G}} = \tfrac{1}{2}(\mathbf{U}^2 - \mathbf{I})$ (2.217)

- Biot strain tensor $\mathbf{E}^{B} = \mathbf{U} - \mathbf{I}$ (2.218)
- Natural strain tensor $\mathbf{E}^{N} = \ln \mathbf{U}$ (2.219)
- Hill strain tensor $\mathbf{E}^{H} = \mathbf{I} - \mathbf{U}^{-1}$ (2.220)
- Karni strain tensor $\mathbf{E}^{K} = \frac{1}{2}(\mathbf{I} - \mathbf{U}^{-2})$ (2.221)

Among these, the Green strain tensor is the most popular. It can also be written as

$$\mathbf{E}^{G} = \frac{1}{2}(\mathbf{U}^{2} - \mathbf{I}) = \frac{1}{2}(\mathbf{F}^{T}\mathbf{F} - \mathbf{I}) = \frac{1}{2}(\mathbf{C} - \mathbf{I}) \tag{2.222}$$

where $\mathbf{C}$ is called the material deformation tensor.

From Eq. (2.222), it is clear that the strain tensor $\mathbf{E}^{G}$ defined here coincides with the one defined by Eqs (2.199) and (2.200). In the case of one-dimensional uniform stretching, these strain measures are compared in Fig. 2.13. It is seen that, if the strains are infinitesimal, they all coincide.

2.10 Discretized Equilibrium Equations

2.10.1 Discretization of the Solid

The equilibrium equations of the solid are highly nonlinear, as explained in the introduction. Therefore, their analytical solution is impossible, except for very simple academic problems.

The finite-element method is well suited to solve these equations. The solid is divided into finite elements (Fig. 2.14).

For the sake of simplicity, we will only consider isoparametric kinematically admissible finite elements.

In each element, the displacement field is interpolated by

$$u_i = \sum_{I=1}^{N} \Phi^I(\xi,\eta,\zeta)u_i^I \tag{2.223}$$

$$\delta u_i = \sum_{I=1}^{N} \Phi^I(\xi,\eta,\zeta)\delta u_i^I \tag{2.224}$$

where u_i^I are the components of the displacements of node I with respect to a global Cartesian reference base (Fig. 2.15).

2.10.2 Discretized Equilibrium Equations

As usual in nonlinear analysis, the deformation process is followed step by step. At each step, the equilibrium of the solid must be satisfied. The equilibrium of any deformed configuration γ is expressed by the virtual work principle:

$$\delta W_{\text{I}} = \delta W_{\text{E}} \tag{2.225}$$

In Sects 2.8.1 and 2.8.3, we have established many different forms of the internal and external virtual work. Mathematically, they are strictly equivalent. In other words, any one of the expressions of δW_{I} can be combined with any expression of δW_{E} in the virtual work principle. Hence, the description of the equilibrium of the solid can take many different forms. Among these, the most popular are:

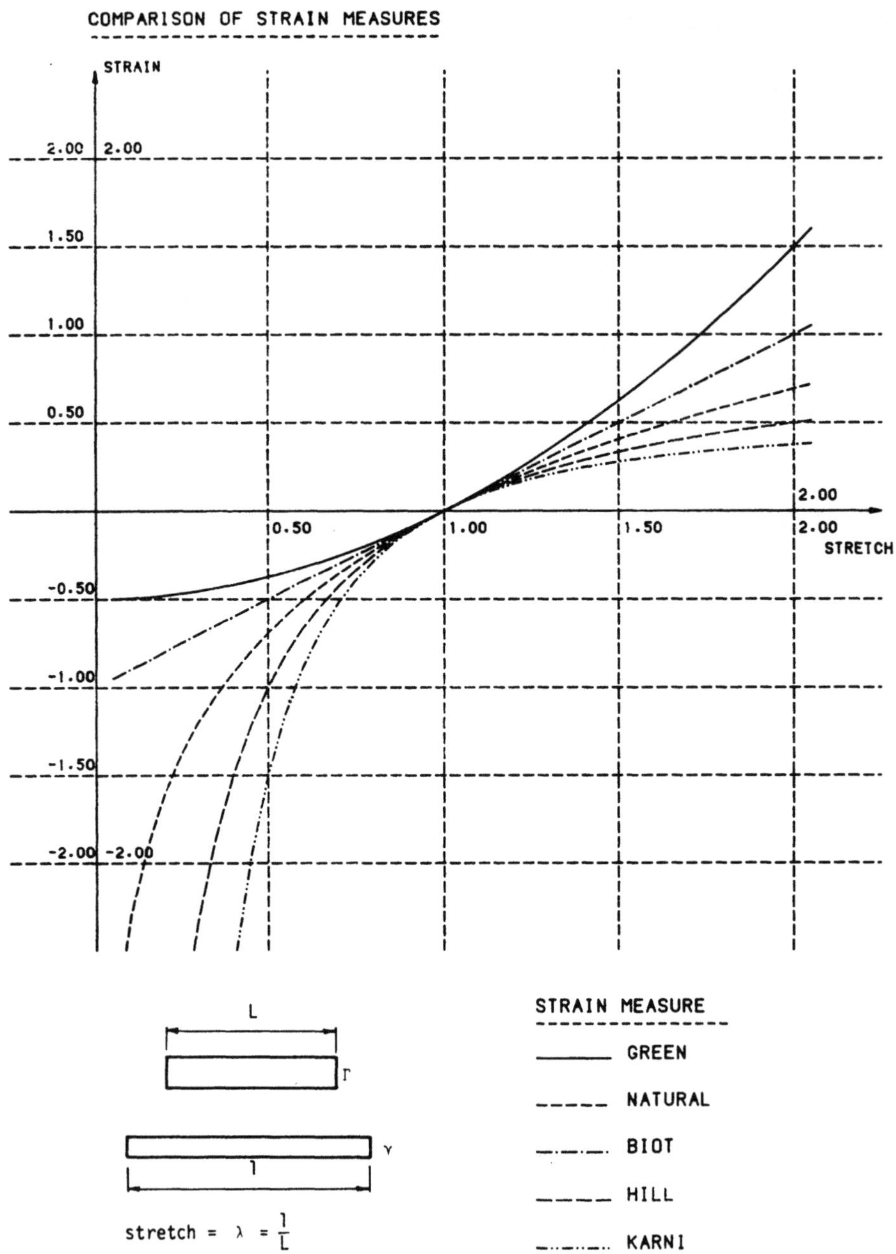

Fig. 2.13. Comparison of strain measures.

- The total Lagrangian description in which the initial configuration is taken as reference configuration: the integrals appearing in δW_{I} and δW_{E} are computed over the initial volume V and the initial surface A of the solid.
- The updated Lagrangian description in which the reference configuration is the last known equilibrium configuration of the solid.
- The current configuration description in which the reference configuration is the current state of the solid.

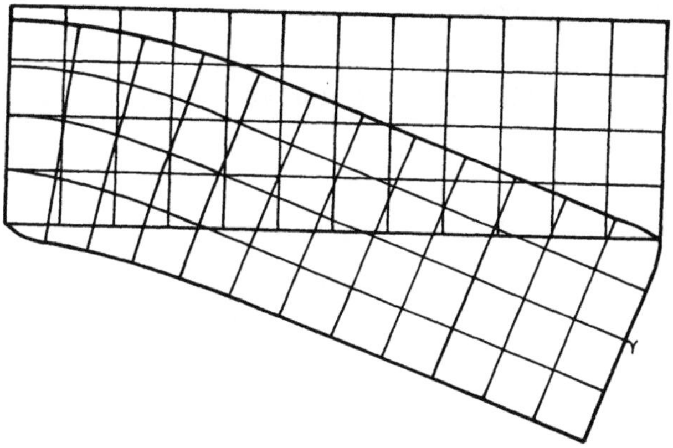

Fig. 2.14. Finite-element model.

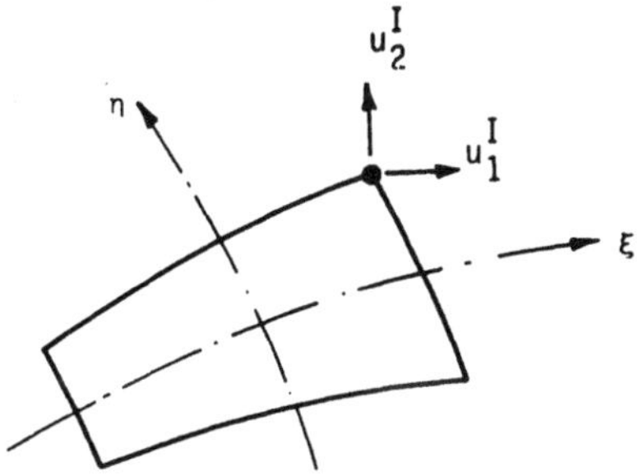

Fig. 2.15. Displacement of a node of a finite element.

It is worth saying that there is no universal agreement on this terminology but this is not important. What matters is that all different formulations of the virtual work principle must give the same numerical results since they are different mathematical expressions of the same physical phenomenon: the equilibrium of the deformed solid. The choice of a particular formulation may be influenced, for example, by the kind of constitutive equation for the material or may be a matter of personal convenience.

Finally, we insist again on the fact that the definition of strain tensors is not necessary to express equilibrium. Only virtual strains are required in the virtual work principle.

To emphasize these points, we will develop the discretized equilibrium equation starting from the following expression of the virtual work principle

$$\int_v \sigma^{ij}\delta\epsilon_{ij}dv = \int_v \rho f^i \delta u_i dv + \int_a p^i \delta u_i da \tag{2.226}$$

which uses the current configuration γ as reference, but this choice does not constitute a limitation. Similar developments can be done with any particular formulation of the virtual work principle.

We will also consider only the two-dimensional (plane strain) state in order to shorten the notations.

First, we rewrite Eq. (2.226) under matrix form:

$$\int_v = \langle\sigma\rangle \{\delta\epsilon\} dv = \int_v \rho\langle f\rangle\langle\delta u\rangle dv + \int_a \langle p\rangle \{\delta u\} da \tag{2.227}$$

with

$$\langle\sigma\rangle = \langle\sigma_{11}; \sigma_{12}; \sigma_{21}; \sigma_{22}\rangle \tag{2.228}$$

$$\langle\delta\epsilon\rangle = \left\langle \frac{\partial \delta u_1}{\partial x^1}; \frac{\partial \delta u_1}{\partial x^2}; \frac{\partial \delta u_2}{\partial x^1}; \frac{\partial \delta u_2}{\partial x^2} \right\rangle \tag{2.229}$$

$$\langle f\rangle = \langle f^1; f^2\rangle \tag{2.230}$$

$$\langle p\rangle = \langle p^1; p^2\rangle \tag{2.231}$$

$$\langle\delta u\rangle = \langle\delta u_1; \delta u_2\rangle \tag{2.232}$$

Using Eq. (2.224), we get

$$\{\delta u\} = [H]\{\delta U\} \tag{2.233}$$

$$\{\delta\epsilon\} = [B]\{\delta U\} \tag{2.234}$$

with

$$[H] = \begin{bmatrix} \Phi^1 & 0 & \dots & \Phi^N & 0 \\ 0 & \Phi^1 & \dots & 0 & \Phi^N \end{bmatrix} \tag{2.235}$$

$$[B] = \begin{bmatrix} \frac{\partial \Phi^1}{\partial x^1} & 0 & \dots & \frac{\partial \Phi^N}{\partial x^1} & 0 \\ 0 & \frac{\partial \Phi^1}{\partial x^2} & \dots & 0 & \frac{\partial \Phi^N}{\partial x^2} \\ \frac{\partial \Phi^1}{\partial x^1} & 0 & \dots & \frac{\partial \Phi^N}{\partial x^1} & 0 \\ 0 & \frac{\partial \Phi^1}{\partial x^2} & \dots & 0 & \frac{\partial \Phi^N}{\partial x^2} \end{bmatrix} \tag{2.236}$$

$$\langle\delta U\rangle = \langle\delta u_1^1; \delta u_2^1 \dots \delta u_1^N \delta u_2^N\rangle \tag{2.237}$$

Replacing in Eq. (2.227) yields

$$\int_v \langle\sigma\rangle [B]\{\delta U\} dv = \int_v \rho\langle f\rangle [H]\{\delta U\} dv + \int_a \langle p\rangle [H]\{\delta U\} da \tag{2.238}$$

Or, since the $\{\delta U\}$ are arbitrary,

$$\int_v \langle\sigma\rangle [B] dv = \int_v \rho\langle f\rangle [H] dv + \int_a \langle p\rangle [H] da \tag{2.239}$$

This is the discretized expression of the equilibrium of the solid.[1] It is quite

[1] In fact, the equilibrium equation for the discretized solid is obtained by assembly of equations (2.239) for all the elements. But since this assembly operation is absolutely classical, we will skip it in this text.

similar to the case of infinitesimal elastic analysis, except for the fact that all the computations are performed in the current deformed configuration γ.

Usually, one writes

$$\{F\} = \int_v \rho [H]^T \{f\} dv + \int_a [H]^T \{p\} da \tag{2.240}$$

for the column matrix containing the nodal (equivalent) forces. Then, for Eq. (2.232) one also writes

$$\{R\} = \int_v [B]^T \{\sigma\} dv - \{F\} = 0 \tag{2.241}$$

If the current configuration is out of equilibrium, $\{R\}$ does not vanish. It is called the vector of residuals or vector of out-of-balance forces.

2.11 Solution of the Discretized Equilibrium Equations

2.11.1 Constitutive Equation of the Solid

In this section, we assume that the material behaviour can be represented by a set of constitutive equations of the form

$$\sigma^{\nabla ij} = f_\sigma(\sigma^{ij}, q^i, L_{ij}) \tag{2.242}$$

$$\dot{q}^i = f_q(\sigma^{ij}, q^i) \tag{2.243}$$

where the q^i are state variables (scalar, vectors, tensors, . . .) and

$$L_{ij} = \frac{\partial v^i}{\partial x^j} = \frac{\partial \dot{x}^i}{\partial x^j} \tag{2.244}$$

are the velocity gradients with respect to the current coordinates in the deformed configuration.

The symmetric part, **D**, is called the rate of deformation tensor and the skew-symmetric part **W** is the spin tensor.

This theory assumes that the entire stress–strain history of a material particle can be summarized by the values of a finite number of scalar, vectorial or tensorial variables q^i called state variables.

The three types of material behaviour (hyperelastic, elastic–plastic, and elastic–viscoplastic) mentioned in Sect. 2.1.2 can be modelled by this kind of constitutive equation.

2.11.2 Step-by-Step Solution

Since the stresses of the solid usually depend on the entire stress–strain history of the particle, the solution of Eq. (2.241) must follow the equilibrium path of the solid in a classical step-by-step procedure.

Let $\gamma_{(M)}$ be a known equilibrium configuration of the solid at time t_M. This means that the coordinates $x^i_{(M)}$, stresses of $\sigma^{ij}_{(M)}$ and state variables $q^i_{(M)}$ have already been computed and satisfy Eq. (2.241).

The next step of the analysis is to obtain another equilibrium configuration $\gamma_{(N)}$ at time $t_N = t_M + \Delta t$. This configuration can be obtained by the following iterative procedure (Fig. 2.16).

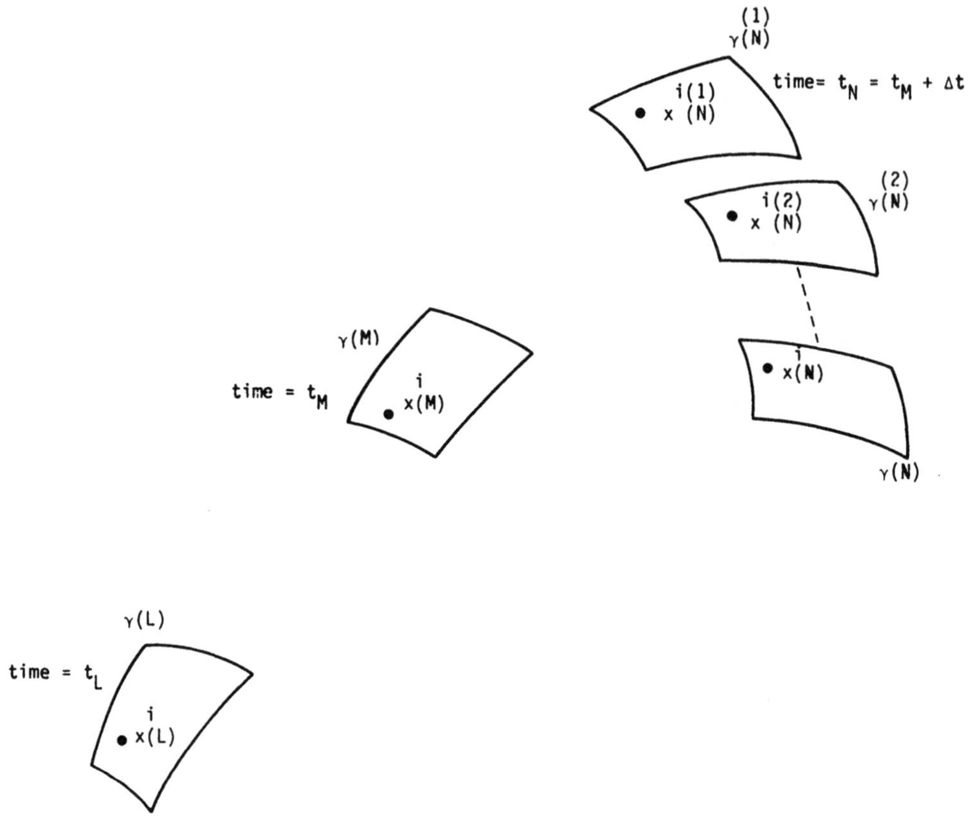

Fig. 2.16. Configurations at successive times.

1. Find a first approximation for the coordinates $x^{i(1)}_{(N)}$ of material points in $\gamma^{(1)}_{(N)}$.
2. Choose a strain path between $\gamma_{(M)}$ and $\gamma^{(1)}_{(N)}$.
3. Integrate the constitutive equation along this strain path in order to compute the stresses $\sigma^{ij(1)}_{(N)}$ and state variables $q^{i(1)}_{(N)}$ in $\gamma^{(1)}_{(N)}$.
4. Compute the out-of-balance forces $\{R\}^{(1)}_{(N)}$ by (2.241); if they are small enough, the configuration obtained is considered as an equilibrium configuration and the iterative procedure is finished; otherwise, perform point 5.
5. Compute the tangent stiffness matrix and compute a correction $d\mathbf{x}$ such that the new coordinates

$$x^{(2)}_{(N)} = \mathbf{x}^{(1)}_{(N)} + d\mathbf{x} \tag{2.245}$$

produce a configuration $\gamma^{(2)}_{(N)}$ closer to equilibrium and go back to point 2.

Details of these topics are given in the following paragraphs.

2.11.3 First Approximation for $\gamma_{(N)}$

There are many possible choices to compute the first approximation for the coordinates $\mathbf{x}_{(N)}^{(1)}$ of material points. Some are listed below:

1. $\gamma_{(N)}^{(1)}$ is identical to $\gamma_{(M)}$. This is probably the simplest choice. It may be useful if the loading changes sharply (is reversed, for example) at time t_N.
2. $\gamma_{(N)}^{(1)}$ is extrapolated from the preceding equilibrium configurations. For example, a linear extrapolation gives

$$\mathbf{x}_{(N)}^{(1)} = \mathbf{x}_{(M)} + \frac{\mathbf{x}_{(M)} - \mathbf{x}_{(L)}}{t_M - t_L} \Delta t \tag{2.246}$$

   where $\gamma_{(L)}$ is the equilibrium configuration obtained before $\gamma_{(M)}$. Obviously parabolic or cubic extrapolations are also possible if the coordinates of two or three equilibrium configurations preceding $\gamma_{(M)}$ are recorded. This choice is interesting if the loading is fairly continuous.
3. $\gamma_{(N)}^{(1)}$ is obtained with the help of the tangent stiffness matrix in $\gamma_{(M)}$. If $[K_{\mathrm{T}(M)}]$ is the tangent stiffness matrix of $\gamma_{(M)}$ (it will be developed in Section 2.11.7), the nodal coordinates $\{x_{(N)}^{(1)}\}$ of $\gamma_{(N)}^{(1)}$ are given by

$$\{x_{(N)}^{(1)}\} = \{x_{(M)}\} + \{\Delta x\} \tag{2.247}$$

   where

$$\{\Delta F\} = \{F(N)\} - \{F(M)\} \tag{2.248}$$

   and $\{\Delta x\}$ is the solution of the system

$$\{\Delta F\} = [K_{\mathrm{T}(M)}]\{\Delta x\} \tag{2.249}$$

This choice is usually efficient for any kind of loading but requires the solution of a system of equations (2.249). The coordinates $\mathbf{x}_{(N)}^{(1)}$ are then interpolated in each finite element by an equation similar to Eq. (2.223).

2.11.4 Choice of a Strain Path Between $\gamma_{(M)}$ and $\gamma_{(N)}$

Knowing the coordinates $x_{(M)}^i$, the stress $\sigma_{(M)}^{ij}$ and state variables $q_{(M)}^i$ in $\gamma_{(M)}$, and knowing the coordinates $x_{(N)}^i$ in a configuration $\gamma_{(N)}$ at time t_N (here $\gamma_{(N)}$ can be any one of the approximate configurations $\gamma_{(N)}^{(1)}, \gamma_{(N)}^{(2)}, \ldots$ or the equilibrium configuration itself at time t_N), there remains to compute the stresses and state variables in $\gamma_{(N)}$. However, the path between $\gamma_{(M)}$ and $\gamma_{(N)}$ is not yet defined.

Indeed, many different assumptions are possible, and since the constitutive equation is path dependent, the computed stresses and state variables are influenced by this choice. Some usual hypotheses are listed hereafter.

2.11.4.1 Constant Velocity Path

It is assumed that each material particle follows a rectilinear trajectory at constant velocity between $\gamma_{(M)}$ and $\gamma_{(N)}$. Hence, the coordinates at time t between t_M and t_N are given by

$$\mathbf{v}_{MN} = \frac{1}{\Delta t}(\mathbf{x}_{(N)} - \mathbf{x}_{(M)}) \tag{2.250}$$

$$\mathbf{x} = \mathbf{x}_{(M)} + \mathbf{v}_{MN}(t - t_M) t_M \leq t \leq t_N \tag{2.251}$$

In this hypothesis, the velocity gradients with respect to current coordinates, the rate of deformation tensor and the spin tensor are not constant along this path. In particular, if $\gamma_{(M)}$ and $\gamma_{(N)}$ differ only by a rigid-body rotation (Fig. 2.17) this path induces strains in the intermediate configurations.

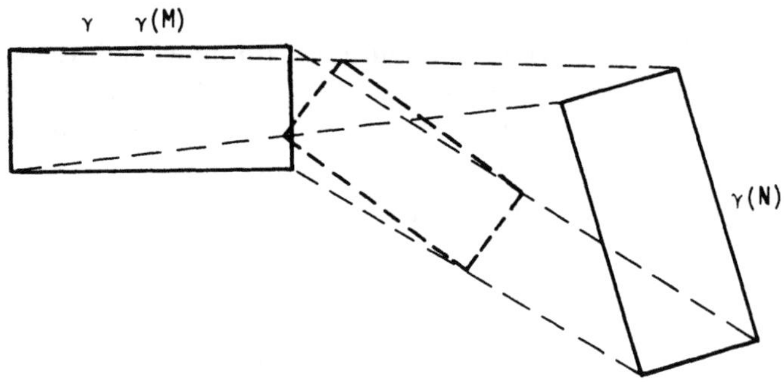

Fig. 2.17. Rigid-body rotation.

For inelastic constitutive equations, this produces unwanted stresses in $\gamma_{(N)}$. For example, with an elastoplastic material behaviour, the first half of the path (Fig. 2.17) corresponds to an elastoplastic loading while the second half is an elastic unloading. This means that the constant velocity strain path is not incrementally objective. However, this path is widely used. It can be shown (Charlier 1987) that, if the rotations between $\gamma_{(M)}$ and $\gamma_{(N)}$ are moderate, the spurious stresses remain small. Further, it is also possible to use special integration schemes for stresses that restore the incremental objectivity (Hughes and Winget 1980).

2.11.4.2 Constant Velocity Gradient Path

The transformation between $\gamma_{(M)}$ and $\gamma_{(N)}$ can be performed according to a nonrectilinear trajectory of material particles in such a way that the velocity gradients with respect to current coordinates remain constant (Godinas and Cescotto 1984).

The basic ideas are the following. Let $\mathbf{F}_{(MN)}$ be the Jacobian matrix of the mapping $\gamma_{(M)} \rightarrow \gamma_{(N)}$ and let $\mathbf{F}(t)$ be the Jacobian matrix of the mapping $\gamma_{(M)} \rightarrow \gamma(t)$ where $\gamma(t)$ is the configuration of the solid at any time t between t_M and t_N.

$$\mathbf{F}_{(MN)} = \frac{\partial \mathbf{x}_{(N)}}{\partial \mathbf{x}_{(M)}} \tag{2.252}$$

$$\mathbf{F}(t) = \frac{\partial \mathbf{x}(t)}{\partial \mathbf{x}_{(M)}} \tag{2.253}$$

The velocity gradients in $\gamma(t)$ are

$$\mathbf{L}(t) = \frac{\partial \mathbf{v}(t)}{\partial \mathbf{x}(t)} = \frac{\partial \mathbf{v}(t)}{\partial \mathbf{x}_{(M)}} \frac{\partial \mathbf{x}_{(M)}}{\partial \mathbf{x}(t)} = \dot{\mathbf{F}}(t)\mathbf{F}(t)^{-1} \tag{2.254}$$

The goal is to maintain **L** constant between $\gamma_{(M)}$ and $\gamma_{(N)}$. The eigenvalues of **L** are given by

$$\mathbf{L} = \mathbf{\Phi}^{-1}\boldsymbol{\rho}\mathbf{\Phi} \text{ with } \boldsymbol{\rho} \text{ diagonal} \tag{2.255}$$

$$\boldsymbol{\rho} = \mathbf{\Phi}\mathbf{L}\mathbf{\Phi}^{-1} = \mathbf{\Phi}\dot{\mathbf{F}}(t)\mathbf{\Phi}^{-1}\mathbf{\Phi}\mathbf{F}(t)^{-1}\mathbf{\Phi}^{-1} \tag{2.256}$$

Let

$$\mathbf{F}'(t) = \mathbf{\Phi}\mathbf{F}(t)\mathbf{\Phi}^{-1} \text{ with } \mathbf{F}'(t) \text{ diagonal} \tag{2.257}$$

$$\dot{\mathbf{F}}'(t) = \mathbf{\Phi}\dot{\mathbf{F}}(t)\mathbf{\Phi}^{-1} \tag{2.258}$$

$$\mathbf{F}'^{-1}(t) = \mathbf{\Phi}\mathbf{F}(t)^{-1}\mathbf{\Phi}^{-1} \tag{2.259}$$

whence

$$\boldsymbol{\rho} = \dot{\mathbf{F}}'(t)\mathbf{F}'(t)^{-1} \tag{2.260}$$

This is a system of decoupled differential equations

$$\left.\begin{aligned} \rho_1 &= \dot{F}'_1(t)/F_1(t) \\ \rho_2 &= \dot{F}'_2(t)/F_2(t) \\ \rho_3 &= \dot{F}'_3(t)/F_3(t) \end{aligned}\right\} \tag{2.261}$$

whence

$$\left.\begin{aligned} F'_1(t) &= e^{\rho_1(t-t_M)} \\ F'_2(t) &= e^{\rho_2(t-t_M)} \\ F'_3(t) &= e^{\rho_3(t-t_M)} \end{aligned}\right\} \tag{2.262}$$

and

$$\mathbf{F}(t) = \mathbf{\Phi}^{-1}\mathbf{F}'(t)\mathbf{\Phi} \tag{2.263}$$

At time $t = t_M$, $\mathbf{F}(t_M) = \mathbf{I}$. This condition is verified by Eq. (2.263).

At time $t = t_N$, $\mathbf{F}(t_N) = \mathbf{F}_{(MN)}$. Hence

$$\mathbf{F}_{(MN)} = \mathbf{\Phi}^{-1}\mathbf{F}'(t_N)\mathbf{\Phi} \tag{2.264}$$

which means that $\mathbf{F}'(t_N)$ is the diagonal matrix of eigenvalues of $\mathbf{F}_{(MN)}$ and $\mathbf{\Phi}$ the corresponding eigenvectors.

Applying the development to the two-dimensional (plane strain) case gives the following results:

$$\mathbf{L} = C_1 \begin{bmatrix} \dfrac{F_{11} - F_{22}}{2} & F_{12} \\ F_{21} & \dfrac{F_{22} - F_{11}}{2} \end{bmatrix} + C_2 \begin{bmatrix} 1 & 0 \\ 0 & 1 \end{bmatrix} \tag{2.265}$$

$$A = \sqrt{\left(\frac{F_{11}-F_{22}}{2}\right)^2 + F_{12}F_{21}}, \quad C_1 = \frac{\rho_1-\rho_2}{2A}, \quad C_2 = \frac{\rho_1+\rho_2}{2} \tag{2.266}$$

$$\rho_1 = \frac{F_{11}+F_{22}}{2} + A, \quad \rho_2 = \frac{F_{11}+F_{22}}{2} - A \tag{2.267}$$

where F_{11}, F_{12}, F_{21}, F_{22} are the components of $\mathbf{F}_{(MN)}$. The eigenvalues ρ_1 and ρ_2 may be imaginary numbers in some cases, but the coefficients C_1 and C_2 are always real numbers (Godinas and Cescotto 1984).

If $\gamma_{(N)}$ is obtained from $\gamma_{(M)}$ by a rigid body rotation, $\mathbf{F}_{(MN)}$ is a rotation matrix and $\mathbf{L}$ becomes antisymmetric. This implies that the rate of deformation tensor $\mathbf{D}$ vanishes in this case. This path is, therefore, incrementally objective.

2.11.4.3 Final Rotation Path

If the polar decomposition of (MN) is performed,

$$\mathbf{F}_{(MN)} = \mathbf{R}_{(MN)}\mathbf{U}_{(MN)} \tag{2.268}$$

the following path can be chosen (Braudel, Abouaf and Chenot 1986). For $t_M \leq t < t_N$, the deformation path consists of the stretching $\mathbf{U}_{(MN)}$ alone, while at time t_N, the rotation $\mathbf{R}_{(MN)}$ is applied instantaneously. If $\gamma_{(N)}$ is obtained from $\gamma_{(M)}$ by a rigid body rotation, $\mathbf{U}_{(MN)}$ vanishes. Therefore, no deformation is produced between t_M and t_N and the rotation takes place at time t_N. This path is also incrementally objective.

2.11.5 Integration of the Constitutive Equation Along a Path

Once a path between $\gamma_{(M)}$ and $\gamma_{(N)}$ is chosen, the stresses $\sigma^{ij}_{(N)}$ are computed by integration of Eqs (2.292), (2.293) along this path. Usually, the integration must be performed numerically.

Many integration techniques are available. Here, we only give very rough indications about this. Eqs (2.292) and (2.293) constitute a system of first order differential equations of the form

$$\{\dot{y}\} = \{f(\{y\},t)\} \tag{2.269}$$

with the initial condition

$$\{y\} = \{y_{(M)}\} \text{ at } t = t_M \tag{2.270}$$

2.11.5.1 Explicit Scheme

$$\{y_{(N)}\} = \{y_{(M)}\} + \{\dot{y}_{(M)}\}\Delta t \tag{2.271}$$

This is the simplest scheme, but its numerical stability requires the use of small time intervals Δt.

2.11.5.2 Implicit Scheme

$$\{y_{(N)}\} = \{y_{(M)}\} + [(1-\theta)\{\dot{y}_{(M)}\} + \theta\{\dot{y}_{(N)}\}]\Delta t \tag{2.272}$$

This scheme is stable for any Δt provided $\theta \geq 0.5$, but is iterative since

$$\{\dot{y}_{(N)}\} = \{f(\{y_{(N)}\}, t_N)\} \tag{2.273}$$

appears on the right-hand side of Eq. (2.272).

2.11.5.3 Substep Schemes

The time interval Δt is divided into subintervals dt. For example, equal subintervals may be chosen, but it is also possible to choose the size of each subinterval in such a way as to reach a desired accuracy for the numerical integration (Desbordes, El Mouatassim and Touzot 1987). Passing from subinterval J to subinterval $J+1$ can be done by an explicit scheme or an implicit scheme. It is also possible to use the following classical modification of the implicit scheme:

$$\{y(J+1)\} = \{y(J)\} + [(1-\theta)\{\dot{y}(J)\} + \theta\{\dot{y}(J+1)\}]dt \tag{2.274}$$

where $\{\dot{y}(J+1)\}$ is computed approximately by

$$\{\dot{y}(J+1)\} = \{\dot{y}(J)\} + \left[\frac{\partial f}{\partial y}(J)\right][\{y(J+1)\} - \{y(J)\}] \tag{2.275}$$

to avoid iterations. This yields

$$\left[[I] - \theta dt\left[\frac{\partial f}{\partial y}(J)\right]\right]\{\{y(J+1)\} - \{y(J)\}\} = \{\dot{y}(J)\}dt \tag{2.276}$$

2.11.6 Out-of-Balance Forces and Incidence Constitutive Tensor

The computed stresses $\boldsymbol{\sigma}_{(N)}$ depend on the path chosen between $\gamma_{(M)}$ and $\gamma_{(N)}$, as well as on the integration scheme along this path. Usually, they are not in equilibrium with the applied forces: the out-of-balance forces

$$\{R\} = \int_v [B]^{\mathrm{T}}\{\sigma\}dv - \{F\} \tag{2.277}$$

are not equal to zero.

It is then necessary to search for another configuration closer to equilibrium. More precisely, assume that, in the iterative procedure, we have reached a configuration $\gamma_{(N)}^{(p)}$ after p iterations. The corresponding computed stresses are $\sigma_{(N)}^{(p)}$. Let $d\mathbf{x}$ be a correction of coordinates producing a new configuration $\gamma_{(N)}^{(p+1)}$ (Fig. 2.18).

$$\mathbf{x}_{(N)}^{(p+1)} = \mathbf{x}_{(N)}^{(p)} + d\mathbf{x} \tag{2.278}$$

closer to equilibrium.

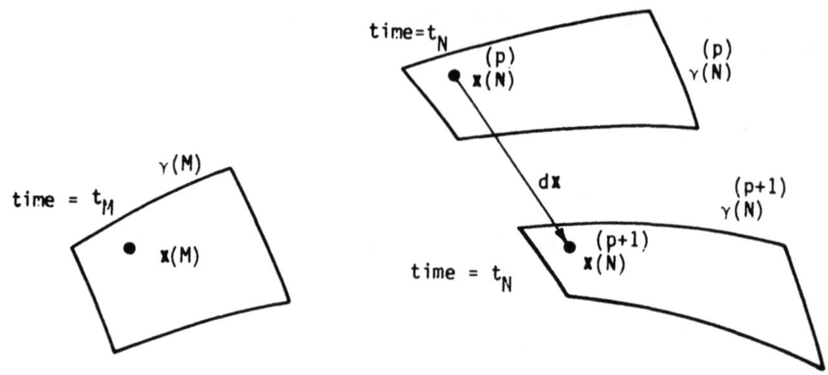

Fig. 2.18. Configurations during iteration.

The corresponding stresses $\sigma_{(N)}^{(p+1)}$ must be computed along the path $\gamma_{(M)} \rightarrow \gamma_{(N)}^{(p+1)}$ and not along the path $\gamma_{(M)} \rightarrow \gamma_{(N)}^{(p)} \rightarrow \gamma_{(N)}^{(p+1)}$, which is not only physically meaningless, but furthermore, would require an instantaneous transformation between $\gamma_{(N)}^{(p)}$ and $\gamma_{(N)}^{(p+1)}$, which is impossible for elastic-viscoplastic materials.

The stress correction

$$d\boldsymbol{\sigma} = \boldsymbol{\sigma}_{(N)}^{(p+1)} - \boldsymbol{\sigma}_{(N)}^{(p)} \tag{2.279}$$

induced by the coordinate correction $d\mathbf{x}$, also depends on the choice of strain path and on the numerical scheme used to compute stresses.

Let $d\mathbf{L}$ be the modification of velocity gradients induced by $d\mathbf{x}$.

We define the incidence constitutive tensor $\mathbf{C}_\mathrm{T}$ by

$$d\boldsymbol{\sigma} = \mathbf{C}_\mathrm{T} d\mathbf{L} \Delta t \tag{2.280}$$

This tensor must be distinguished from the elastoplastic constitutive tensor $\mathbf{C}^{\mathrm{ep}}$ often used in finite element small strain elastoplastic computations. The latter is defined by

$$\dot{\boldsymbol{\sigma}} = \mathbf{C}^{\mathrm{ep}} \dot{\boldsymbol{\epsilon}} \tag{2.281}$$

or, multiplying by dt,

$$d\boldsymbol{\sigma} = \mathbf{C}^{\mathrm{ep}} d\boldsymbol{\epsilon} \tag{2.282}$$

In this equation, $d\boldsymbol{\sigma}$ is the stress increment produced by a strain increment along a physical strain path. On the contrary, in Eq. (2.280) $d\mathbf{L}$ is a perturbation of the velocity gradients caused by the perturbation of coordinate $d\mathbf{x}$ between two configurations $\gamma_{(N)}^{(p)}$ and $\gamma_{(N)}^{(p+1)}$ through which the solid never goes because they are out of equilibrium. Furthermore, Eq. (2.280) is also valid for elastic-viscoplastic materials, while Eq. (2.282) is not.

Hence, the incidence constitutive tensor $\mathbf{C}_\mathrm{T}$ indicates how the computed stresses $\boldsymbol{\sigma}_{(N)}^{(p)}$ are modified by a perturbation of the configuration $\gamma_{(N)}^{(p)}$. This tensor depends on

- The constitutive equation of the material
- The strain path between $\gamma_{(M)}$ and $\gamma_{(N)}^{(p)}$
- The numerical integration scheme along the path

Except for very simple integration schemes, its analytical form is usually unreachable, but it can be computed by numerical perturbation of the components of **L**. In general, it is nonsymmetric (Cescotto and Charlier 1985).

2.11.7 Tangent Stiffness Matrix

Let $\{dx\}$ be the vector of nodal coordinate increments corresponding to $d\mathbf{x}$. The corresponding increment of the out-of-balance forces, $\{dR\}$ is related to $\{dx\}$ by

$$\{dR\} = [K_T]\,\{dx\} \tag{2.283}$$

where $[K_T]$ is the tangent stiffness matrix. This equation means that the tangent stiffness matrix is the derivative of $\{R\}$ with respect to $\{x\}$. Its expression can be obtained by differentiating (2.277):

$$[K_T]\,\{dx\} = d\left\{\int_v [B]^T\{\sigma\}dv\right\} - d\{F\} \tag{2.284}$$

Care must be taken when evaluating the differential, because the integral is taken over $\gamma_{(N)}$, a changing configuration.

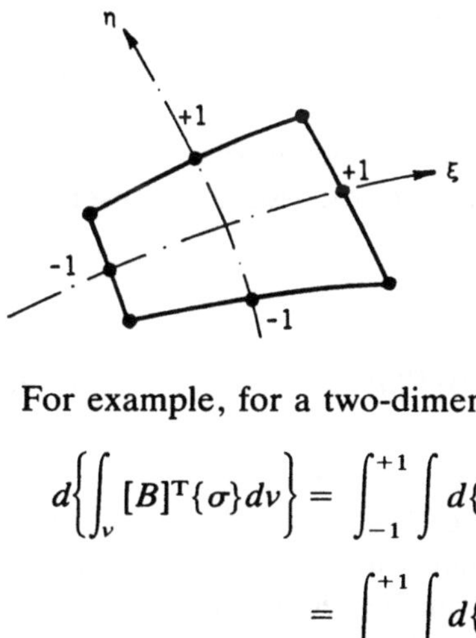

Fig. 2.19. Two-dimensional isoparametric element.

For example, for a two-dimensional isoparametric element (Fig. 2.19), one has

$$d\left\{\int_v [B]^T\{\sigma\}dv\right\} = \int_{-1}^{+1}\int d\{[B]^T\{\sigma\}|J|\}d\xi d\eta$$

$$= \int_{-1}^{+1}\int d\{|J|[B]^T\}\{\sigma\}d\xi d\eta + \int_{-1}^{+1}\int |J|[B]^T d\{\sigma\}d\xi d\eta \tag{2.285}$$

where $|J|$ is the Jacobian of the isoparametric transformation. A lengthy but straightforward calculation gives

$$\int_{-1}^{+1}\int d\{|J|[B]^T\}\{\sigma\}d\xi d\eta = \left[\int_v [B]^T[\sigma_0][B]dv\right]\{dx\} \tag{2.286}$$

where

$$[\sigma_0] = \begin{bmatrix} 0 & 0 & -\sigma_{12} & \sigma_{11} \\ 0 & 0 & -\sigma_{22} & \sigma_{12} \\ \sigma_{12} & -\sigma_{11} & 0 & 0 \\ \sigma_{22} & -\sigma_{12} & 0 & 0 \end{bmatrix} \tag{2.287}$$

The other term is

$$\int_{-1}^{+1}\int |J|[B]^{\mathrm{T}}d\{\sigma\}d\xi d\eta = \Delta t\int_{v}[B]^{\mathrm{T}}[C_{\mathrm{T}}]\{dL\}dv$$

$$= \left[\int_{v}[B]^{\mathrm{T}}[C_{\mathrm{T}}][B]dv\right]\{dx\} \tag{2.288}$$

where

$$\langle dL\rangle = \langle dL_{11}\, dL_{12}\, dL_{21}\, dL_{22}\rangle \tag{2.289}$$

Finally, one has

$$d\left\{\int_{v}[B]^{\mathrm{T}}\{\sigma\}dv\right\} = \left[\int_{v}[B]^{\mathrm{T}}\{[\sigma_0] + [C_{\mathrm{T}}]\}[B]dv\right]\{dx\} \tag{2.290}$$

Remember that all these terms are computed in configuration $\gamma_{(N)}^{(p)}$ obtained after p iterations. There remains to evaluate $d\{F\}$ in Eq. (2.284). This term represents the perturbation of applied loads induced by $d\{x\}$. If the loading is conservative, it is not affected by a perturbation of the coordinates of $\gamma_{(N)}^{(p)}$. However, if the loading is geometry dependent (follower forces, pressure, or friction loads), the perturbation $\{dx\}$ creates an increment $\{dF\}$ given by

$$\{dF\} = [K_{\mathrm{NC}}]\{dx\} \tag{2.291}$$

where $[K_{\mathrm{NC}}]$ is the well-known stiffness matrix for nonconservative loads. Its form depends on the type of loading. For simplicity, we will assume a conservative loading for which this matrix disappears.

Hence, from Eqs (2.284) and (2.290), one gets

$$[K_{\mathrm{T}}] = \int_{v}[B]^{\mathrm{T}}\{[\sigma_0] + [C_{\mathrm{T}}]\}[B]dv \tag{2.292}$$

In general, $[K_{\mathrm{T}}]$ is not symmetric since $[\sigma_0]$ and $[C_{\mathrm{T}}]$ are not.

However, it can be shown (Cescotto and Charlier 1985), with the help of some approximations valid for very small strain increments during the step, that a more classical and symmetric tangent stiffness matrix can be recovered. The stiffness matrix allows us to compute an increment $d\{x\}$ of coordinates and such that $\gamma_{(N)}^{(p+1)}$ is closer to equilibrium than $\gamma_{(N)}^{(p)}$. Indeed, the out-of-balance forces in $\gamma_{(N)}^{(p+1)}$ can be approximated by a first-order development

$$\{R_{(N)}^{(p+1)}\} \simeq \{R_{(N)}^{(p)}\} + \left[\frac{\partial R^{(p)}}{\partial x_{(N)}}\right]\{dx\}$$

$$= \{R_{(N)}^{(p)}\} + [K_{\mathrm{T}}]\{dx\} \tag{2.293}$$

and must be kept to zero to ensure equilibrium. This gives

$$\{dx\} = -[K_{\mathrm{T}}]^{-1}\{R_{(N)}^{(p)}\} \tag{2.294}$$

Because of the approximation in Eq. (2.293), the new configuration $\gamma_{(N)}^{(p+1)}$ will be slightly out of equilibrium and additional iterations may be required.

2.12 Conclusions

This chapter has presented, in general terms, the main aspects of the analysis of large strains in solids. The differences with the nonlinear analysis in the domain of small strains are not of a fundamental nature since the basic tool is the same: the virtual work principle which expresses the equilibrium of the solid.

However, due to the presence of large strains, some topics require special care:

- Definition of stress tensors because true stresses differ significantly from nominal stresses
- Definition of conjugated virtual strains
- Formulation of constitutive equations
- Choice of objective strain paths between two consecutive configurations
- Numerical procedures such as the integration of the constitutive equations between two neighbouring configurations and the correct derivation of the tangent stiffness matrix

Some of these points will be developed further in the following chapters.

References

Few references are given because the main part of this text simply recalls basic notions of solid mechanics and numerical techniques applied to finite-element analysis. Concerning stresses, strains and the virtual work principle, classical textbooks like those of Malvern, Truesdell, etc. may be referred to. But it is worth mentioning three reports that constitute an interesting summary of these notions.

Curnier A and Rakotomanana L (1984) Generalized strain and generalized stress measures. Internal report, Laboratoire de Mécanique Appliquée, Département de Mécanique, EPFL

Sidoroff F (1982) Cours sur les grandes déformations. Rapport GRECO No 51, CNRS, Sophia-Antipolis

Cescotto S (1981) Contraintes et déformations conjuguées. Internal report no 117, MSM, Université de Liège

The other references mentioned in the text of this chapter are concerned with more specific points.

Braudel HJ, Abouaf M, Chenot JL (1986) An implicit and incremental formulation for the solution of elastoplastic problems by the finite element method. Comput Struct 22(5):801–814

Cescotto S and Charlier R (1985) Numerical simulation of elasto-visco-plastic large strains of metals at high temperature. In: Proceedings 8th Conference on SMIRT, Brussels, 19–23 August 1985

Charlier R (1987) Approche unifiée de quelques problèmes non linéaires de mécanique des milieux continus par la méthode des éléments finis (grandes déformations des métaux et des sols, contact unilatéral de solides, conduction thermique et écoulement en milieux poreux). PhD thesis, Université de Liège

Desbordes O, El Mouatassim M, Touzot G (1987) Local numerical integration of large strain elasto-plastic constitutive laws. In: Second international conference and short course on constitutive laws for engineering material theory and applications, Tucson, Arizona, 5–10 January 1987

Godinas A and Cescotto S (1984) Calcul des gradients de vitesse constants au cours d'un pas. Internal report No 159, MSM, Université de Liège

Hughes TJR and Winget J (1980) Finite rotation effects in numerical integration of rate constitutive equations arising in large-deformation analysis. Int J Num Meth Engng 15:1862–1867

3 Isotropic Plasticity

P. Bate

3.1 Introduction

The mechanical behaviour of solids is very complex, and in order to analyse the behaviour, approximations have to be made. Sometimes these approximations stem from a degree of ignorance about real behaviour, but most often they are necessary if process and design modelling is to be achieved within the limits of present computer technology. The commonest explicit approximation made is that of isotropy. In an isotropic model, the properties of a material do not depend upon the direction in which they are measured. More specifically, the properties are invariant to transformations involving rigid body rotations or spatial inversions (reflections). It is then impossible to determine any special directions in an isotropic ideal material by measuring its properties.

3.2 Constitutive Equations and Isotropy

To see why isotropy is an attractive approximation to make in the field of general plastic, or inelastic, behaviour and also to lay down some basic concepts used in describing such behaviour, the general form used for modelling inelastic material needs to be considered. This formalism appears fairly complex because of its generality, but simpler expressions will eventually be derived.

The requirement is a form of relationship between the deformation occurring and the forces associated with that deformation. It is better to think of the deformation and forces being associated, rather than the forces causing the deformation, because this viewpoint maintains complementarity in terms of the work involved in the deformation. In continuous media, the more simple geometric aspects of the problem can be removed by considering stresses rather than forces, and by using strain measures to describe the deformation. The most convenient strain measure to use in the case of inelastic behaviour is strain rate, sometimes called deformation rate. This is the time rate of change of any strain measure, in the current configuration, and its use reflects the absence of any special configuration in the inelastic case. It is defined as the symmetric part of the velocity gradient:

$$\dot{\epsilon}_{ij} = \frac{1}{2}\left(\frac{\partial v_i}{\partial x_j} + \frac{\partial v_j}{\partial x_i}\right) \tag{3.1}$$

where v are the velocities and x the current spatial coordinates. Because the current configuration is used for reference, the true, Cauchy, stress, σ_{ij}, is convenient. These two measures are complementary in terms of the current volume specific work rate of the inelastic deformation, given by

$$\dot{W}^{\mathrm{p}} = \sigma_{ij}\,\dot{\epsilon}_{ij}^{\mathrm{p}} \tag{3.2}$$

where the superscript indicates that only the inelastic contribution is being considered. The use of time rates does not imply any necessary effect of time scale, though this often exists; it is really used for convenience.

On the basis outlined, the behaviour can be formalized as a pair of complementary constitutive equations which have the following functional form:

$$\sigma_{ij} = \sigma(\dot{\epsilon}_{kl}, T, \Lambda^1, \Lambda^2, \ldots, \Lambda^n) \tag{3.3a}$$

$$\dot{\epsilon}^{\mathrm{P}}_{ij} = \dot{\epsilon}^{\mathrm{P}}(\sigma_{kl}, T, \Lambda^1, \Lambda^2, \ldots, \Lambda^n) \tag{3.3b}$$

The variable T is temperature, and Λ^1, ..., Λ^n are internal, or state, variables. These state variables are quantities which define the current microstructural state of the material, as far as that influences the mechanical behaviour. They could, in principle, be measured by examining the microscopic configuration of the material. In practice this would be somewhat difficult, of course. To complete the material model, the evolution of the state variables has to be described, formally by giving their time rates of change as a function of the whole current variable set:

$$\dot{\Lambda}^i = \dot{\Lambda}^i(\sigma_{ij}, \dot{\epsilon}_{kl}, T, \Lambda^1, \Lambda^2, \ldots, \Lambda^n) \tag{3.4}$$

Sometimes these evolutions can be included directly into constitutive equations. One important case involves elastic behaviour. In any real material, the stress which exists is associated with elastic, recoverable, strains. The elastic state of a material is associated with definite features of its microscopic structure – namely the relative displacements of atomic scale components. These can even be measured by using, say, x-ray diffraction. There is a good case for including stress as a state variable for this reason. It would then have an evolution equation, which for the case of a linear isotropic elastic (Hookean) material, would serve to describe the whole of the mechanical behaviour. Unless elastic strains are ignored, the occurrence of stress as a constitutive equation result, and its rate as an evolution equation result due to its elastic association, can create a complication.

In the general case, the state variables are allowed to be tensor quantities. The constitutive and evolution equations are tensor functions of tensors, and can then be complicated expressions. The effect of rigid body rotations would also need to be considered in the evolution equations. However, by making the isotropic assumption much of the complexity is eliminated. Firstly, apart from the special case of stress or associated elastic strain, all the state variables can be reduced to scalars. This is because there must be no inheritance of directionality from previous deformation – at least, from the inelastic part. Secondly, the remaining tensor quantities involved, stress and inelastic strain rate, can be simplified considerably.

The simplification of the stress and inelastic strain rate variables, and of the functional relationship between them, results from the representation theorem for an isotropic tensor function of a tensor. This states that the two related tensors must have the same principal axes and furthermore, that the function is restricted to the following form, given in this example as the inelastic strain rate principal components:

$$\dot{\epsilon}^{\mathrm{P}}_i = A + B\,\sigma_i + C\,\sigma_i^2 \tag{3.5a}$$

where A, B and C are scalar functions of invariants of stress, and state variables:

$$
\left.\begin{aligned}
A &= A\,(\sigma_1 + \sigma_2 + \sigma_3, T, \Lambda^1, \Lambda^2, \ldots, \Lambda^n) \\
B &= B\,(\sigma_1^2 + \sigma_2^2 + \sigma_3^2, T, \Lambda^1, \blacktriangle^2, \ldots, \Lambda^n) \\
C &= C\,(\sigma_1^3 + \sigma_2^3 + \sigma_3^3, T, \Lambda^1, \Lambda^2, \ldots, \Lambda^n)
\end{aligned}\right\} \tag{3.5b}
$$

A proof of this representation theorem is given in Spencer (1980). The evolution equations for the state variables can be simplified, as they must now be scalar functions of invariants of stress and inelastic strain rate, current state variables and temperature.

The inelastic constitutive and evolution equations are then simplified considerably. They now relate three stress values to three strain-rate values. Either principal values or invariants can be used. Principal values are useful for graphic representation, and invariants are often used for concise representation in equations.

3.3 Incompressibility and Planar Representation

An approximation which simplifies matters even more is that of inelastic incompressibility. In a great many cases, both observation and consideration of the microscopic mechanisms of inelastic deformation show that hydrostatic pressure has very little effect, and that volume changes occurring in a deformation are purely elastic. So, in these cases, both the inelastic strain rates and the stresses associated with them are deviatoric, or shear-like, in nature. Cases where this approximation is poor are usually those where the material is porous. There are important materials with significant porosity, for example many soils and metals fabricated by powder sintering. For the sake of the advantage gained by assuming incompressibility, these will be ignored for the time being.

Incompressibility reduces the number of independent quantities in the inelastic stress and strain-rate tensors for the isotropic case from three to two. They can then be represented very conveniently on plane diagrams. The plane in principal stress or strain-rate space used for incompressible deformation has a normal vector inclined equally to the three principal axes. This vector direction represents a pure hydrostatic pressure or volume change. The plane is often referred to as the π plane. It is shown in Fig. 3.1 for stress space.

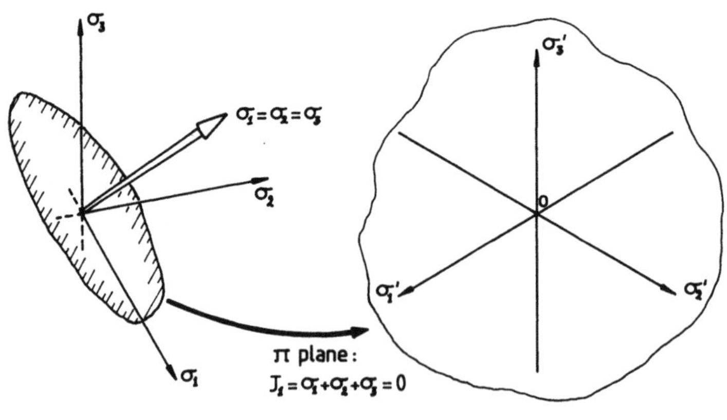

Fig. 3.1. π plane in stress space.

Any plane with a normal corresponding to hydrostatic pressure or volume strain rate is equivalent, of course. There is no special set of orthogonal axes in the π plane, and it is usual to show the three principal deviatoric quantities, each at 120° to the others, as shown. Because the stress and strain rate principal axes are coincident in isotropy, vectors representing both quantities can be presented on the same diagram.

3.4 The Yield Surface and Plastic Potential

The concepts which are needed now for developing models of isotropic inelastic deformation are those of the yield surface and the plastic potential. To understand these, it is useful to consider, initially, what might be called a unit mechanism. This is an inelastic deformation mechanism which is defined by a certain strain rate and a certain specific work rate. The details of what the unit mechanism might mean in terms of microscopic processes are not important for present purposes, although understanding these details is ultimately useful.

A unit mechanism, as defined, will be represented by an infinitely long straight line in the π plane, as shown in Fig. 3.2. This is because the work rate associated with the unit mechanism is given by the scalar product of the deviatoric stress and the inelastic strain rate vectors, which have parallel base vectors.

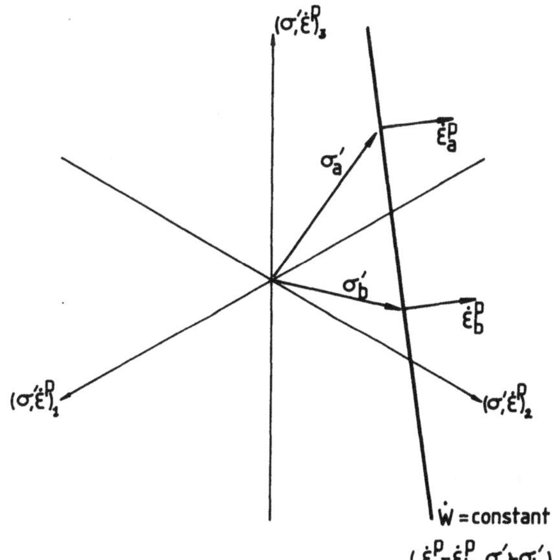

Fig. 3.2. A unit mechanism represented on the π plane.

Two significant points about this representation are that only stresses which lie on the line can operate the specified unit mechanism, and the inelastic strain rate vector is normal to the line.

There will be many, in fact a theoretically unlimited number, of such unit mechanisms, for different strain-rate vectors. If those mechanisms which produce the same specific work rate are considered, then the envelope of the lines representing them will form a polygon in stress space, as shown in Fig. 3.3.

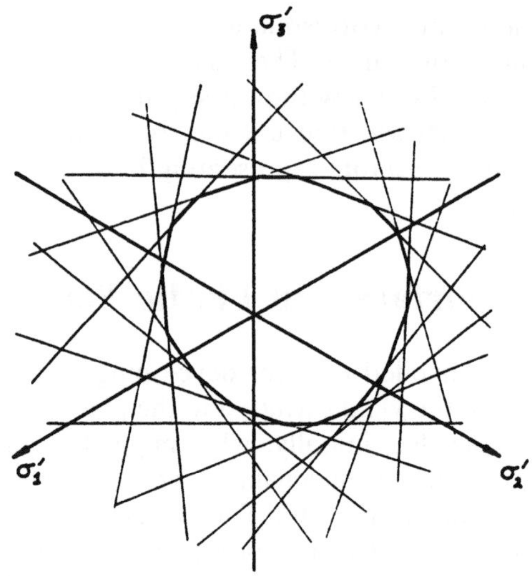

Fig. 3.3. Unit mechanisms for different strain rate vectors.

This shape is a generalized form of a yield surface. Because the surface is defined in two dimensions only here, it is sometimes referred to as a yield locus. Stresses which lie inside the surface cannot operate any of the unit mechanisms and inelastic deformation, at the particular work rate involved, does not occur. Stresses outside the yield surface can only be attained if there are unit mechanisms which become operative at the higher stress with a compensating increase in the work rate. In classical plasticity, the material is assumed to be rate insensitive. This means that there is only one yield surface for all work rates in a constant state condition. In effect the real time parts of the strain-rate and work-rate terms can cancel, there being no effect of actual rate on stress. Even with the rate insensitive case, the occurrence of inelastic deformation can change the size and, possibly, the shape and position of the yield surface. This is because of the evolution of the material state, as formalized in Eq. (3.4). The shape of the yield surface is severely restricted, however, and this will be elaborated below.

The yield surface described above, resulting from the envelope of unit mechanism lines, also serves as a plastic potential. This simply means that the inelastic strain rate vector is normal to the surface at any point (or stress) lying on the surface. This normality principle arises naturally from the unit mechanism development. Formally, if the yield surface is given as a function of the stresses for the current material state in the following way:

$$f(\sigma_{ij}) = 0 \tag{3.6}$$

then the inelastic strain rate components are given by

$$\dot{\epsilon}_{ij}^{\mathrm{p}} = \dot{\lambda} \frac{\partial f}{\partial \sigma_{ij}} \tag{3.7}$$

where $\dot{\lambda}$ is a scalar, with dimensions of inverse time, which is determined by the current material state and the working rate associated with the yield surface.

The normality rule, Eq. (3.7), together with adequate formulation for the yield surface, Eq. (3.6), and the scalar function $\dot{\lambda}$, provides the constitutive equation for inelastic strain rate.

There is no fundamental reason why the yield surface and the plastic potential should be equivalent. This equivalence is known as associated plasticity, where Eq. (3.7) is the flow rule associated with the yield surface, f. Associated plasticity arises here from the unit mechanism argument, but there was an implicit assumption which led to the lines being straight in Figs 3.2 and 3.3. This was that the state of stress did not affect the relationship between work rate and strain rate, or equivalently that stresses orthogonal to the strain rate had no effect. Nearly every microscopic mechanism which is known to be involved in inelastic deformation will, in fact, lead to associated plasticity, and because of this and the resulting complication of the models, there would have to be strong evidence of nonassociated plasticity before the concept was abandoned in any particular situation.

One special case which is often called nonassociated is worth further consideration here. This is where a hydrostatic pressure, while producing a purely elastic volume strain, influences the deviatoric stress magnitudes involved in inelastic deformation. If this occurs, the yield surface size depends upon elastic terms. As discussed above, elastic quantities can be included as state variables. The use of hydrostatic stress as a state variable in this case would effectively convert the situation to one of associated plasticity. This type of approach might be possible in other cases as well.

The development of a yield surface and plastic potential as the envelope of straight lines restricts the possible shape it can be. Simply, it can never be concave. Further restrictions on shape result from the isotropic assumption. The shape, in the π plane, must have at least three-fold symmetry because the surface must be invariant to the choice of the ordering of the principal deviatoric stress axes. This can be extended to a six-fold symmetry by considering the spatial inversion aspect of isotropic invariance. Graphically, this simply means that the mirror image shearing deformations shown in Fig. 3.4 have exactly equal but opposite shear stress associated with them for material in equivalent microstructural states.

The symmetry and lack of concavity mean that the isotropic yield surface must contain, and be centred on, the origin of the π plane (zero deviatoric stress). In

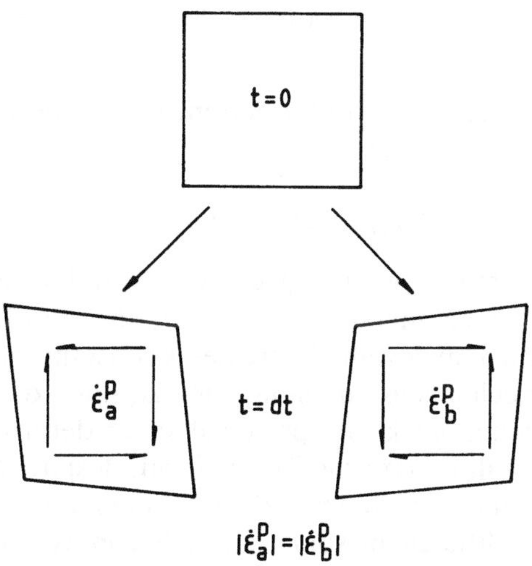

Fig. 3.4. Mirror image shearing deformations.

fact, deformation often causes an effective shift in the yield surface, a phenomenon known as kinematic hardening which is experimentally observed as the Bauschinger effect. However, only changes in size and, within the limits set out above, shape of the yield surface during deformation are allowed with the isotropic assumption. This restricts attention to the phenomenon known as isotropic hardening: this term includes the possible occurrence of softening during deformation as well.

3.5 The Tresca and von Mises Yield Criteria

The bounds on possible yield surface shape in the π plane are shown in Fig. 3.5. They are regular hexagons. The simplest shape between these bounds, the circle, is also shown. The inner bound hexagon and the circle shown in Fig. 3.5 correspond to historically significant yield criteria that are associated, usually, with the names of Tresca and von Mises, respectively. Both have been used extensively.

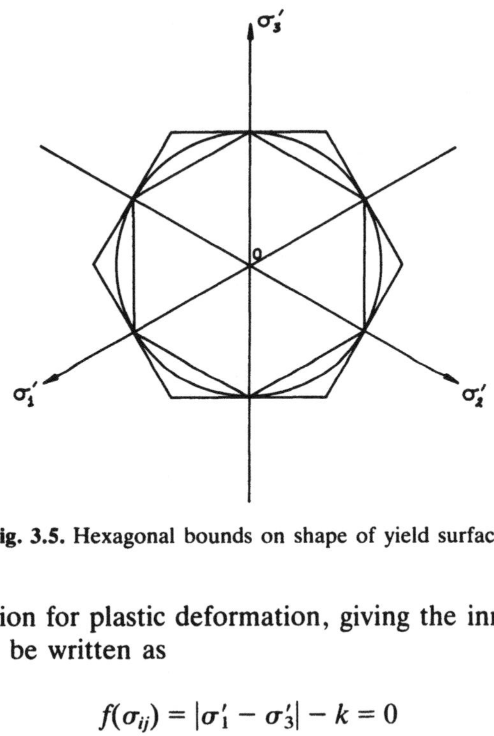

Fig. 3.5. Hexagonal bounds on shape of yield surface.

The Tresca criterion for plastic deformation, giving the inner hexagonal shape in the π plane, can be written as

$$f(\sigma_{ij}) = |\sigma'_1 - \sigma'_3| - k = 0 \tag{3.8}$$

This means that, for material with a given state variable set and, in the general rate-sensitive case, a given inelastic work rate, the absolute difference of the largest and smallest principal deviatoric stresses is constant for inelastic deformation to occur. The unit mechanisms, defining the six sides of the hexagon, can be identified as the optimum work-rate principal shear deformations.

There are some difficulties with the Tresca form, despite its apparent simplicity. First, the minimum and maximum principal deviatoric stresses need to be identified; this can be difficult in some cases. Secondly, and more significantly, the use of this form as a plastic potential leads to significant problems. This is

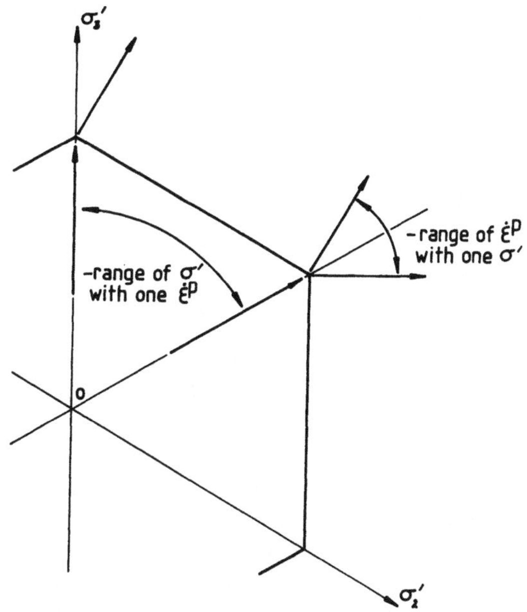

Fig. 3.6. Indeterminacy of stress and strain rate with Tresca criterion.

because the shape contains vertices which lead to an indeterminacy, illustrated in Fig. 3.6. Any strain rate vector not exactly normal to a side of the hexagon corresponds to a stress at a vertex, so one state of stress can give a whole range of strain rates. Similarly, a strain rate exactly normal to the yield surface can be associated equally well with any stress state on that particular side of the hexagon.

The indeterminacy encountered when using the Tresca criterion with associated plasticity has meant that the von Mises criterion, giving the circle in Fig. 3.5, has been used extensively in isotropic plasticity models. Because it is a smooth curve, an infinite number of unit mechanisms needs to be involved, but this is not a serious problem because a large number would give a very close approximation. The von Mises criterion can be expressed very concisely by using the second invariant of the deviatoric stress:

$$f(\sigma_{ij}) = J_2' - k^2 = 0 \tag{3.9}$$

where the invariant is given, in terms of the principal deviatoric stresses, by

$$J_2' = \sigma_1'^2 + \sigma_2'^2 + \sigma_3'^2 \tag{3.10a}$$

or in general components of the stress deviator by

$$J_2' = \sigma_{ij}' \sigma_{ij}' \tag{3.10b}$$

The scalar, k, has a similar meaning to that of the same symbol in Eq. (3.8). It is squared in Eq. (3.9) to maintain its dimensionality as that of stress. The yield criterion expressed in the dimensions of stress is simply

$$f(\sigma_{ij}) = \sqrt{J_2'} - k = 0 \tag{3.11}$$

The normality principle gives a very simple result with the von Mises criterion, which in general components is

$$\dot{\epsilon}_{ij}^{P} = \dot{\lambda} \frac{\sigma_{ij}}{\sqrt{J_2'}} \tag{3.12a}$$

or in principal components:

$$\dot{\epsilon}_{i}^{P} = \dot{\lambda} \frac{\sigma_{i}}{\sqrt{J_2'}} \tag{3.12b}$$

This means that the inelastic strain rate vector and its associated deviatoric stress vector are parallel, and both are radial. Comparing this result with the general isotropic form given above, Eq. (3.5a), we can see that the terms in hydrostatic stress and powers of stress above squared have been eliminated. The first elimination is because incompressibility of inelastic deformation was assumed, and the stresses replaced by their deviatoric parts. The second elimination is characteristic of the von Mises criterion.

Inelastic compressibility, mentioned above, is sometimes important. The von Mises criterion can be extended fairly readily to this case, by including the hydrostatic stress term to give a yield criterion of the form

$$f(\sigma_{ij}) = \alpha J_1 + \sqrt{J_2'} - k = 0 \tag{3.13}$$

The normal to the yield surface is now inclined to the π plane, so that inelastic volume strain rate will occur.

3.6 Isotropic Plastic and Viscoplastic Equations

Using the von Mises criterion has then, given a very simple form for isotropic inelastic deformation. The material model is not complete because the evolution and influence of the material state has not been considered. Often very simple assumptions are made which depend upon a constant state, i.e. no evolution, or a state which depends in a simple way upon the total inelastic work done over the relevant part of the materials deformation history.

If the material state is constant, then two extreme behaviours can be identified in Eq. (3.12). First, setting the scalar term $\dot{\lambda}$ equal to the square root of J_2' divided by twice a material constant, μ gives

$$\dot{\epsilon}_{ij}^{P} = \frac{\sigma_{ij}'}{2\mu} \tag{3.14}$$

This is the special case of a Newtonian fluid, with μ being the viscosity. Inversion is straightforward, to give

$$\sigma_{ij}' = 2\mu\, \dot{\epsilon}_{ij}^{P} \tag{3.15}$$

This model is for a material where deviatoric stress depends linearly upon inelastic strain rate. The other extreme case of a constant state material is where there is no effect of rate. To obtain this form, replace $\dot{\lambda}$ in Eq. (3.12) by the square root of the second invariant of the inelastic strain rate, $\dot{I}_2^p$, given by

$$\dot{I}_2^p = d\dot{\epsilon}_1^{p2} + d\dot{\epsilon}_2^{p2} + d\dot{\epsilon}_3^{p2}$$
$$= d\dot{\epsilon}_{ij}^p \, d\dot{\epsilon}_{ij}^p \tag{3.16}$$

which leads to

$$\dot{\epsilon}_{ij}^p = \sqrt{\frac{\dot{I}_2^p}{J_2'}}\, \sigma_{ij}' \tag{3.17}$$

The time cancels between the strain rates and the invariant to give a rate insensitive form. As it stands, Eq. (3.17) is very unsatisfactory. The only information it conveys is the normality principle for the von Mises case. If used in conjunction with the yield criterion Eq. (3.11), then the stresses at which inelastic deformation occur would also be known. There is no restriction, however, upon the rate at which the inelastic deformation takes place. This indeterminacy can only be resolved by considering other aspects of the whole deformation system; essentially velocity boundary conditions would have to be specified.

This second extreme case given is an ideal plastic material. If velocity boundary conditions cannot be adequately defined in order to make the deformation determinate, then this approximation is useless. One way of modifying this material model is to admit rate sensitivity, but not to the extent of reducing the idealisation to the Newtonian fluid. Such material models are called viscoplastic. The form of Eq. (3.12) which may be used is

$$\dot{\epsilon}_{ij}^p = \dot{\lambda}\left(\sqrt{J_2'}\right) \frac{\sigma_{ij}'}{\sqrt{J_2'}} \tag{3.18}$$

with $\dot{\lambda}$ now some nonlinear function of the second deviatoric stress invariant. One such function is the power law:

$$\dot{\lambda} = K\,\sqrt{J_2'}^{\,M} \tag{3.19}$$

This gives the result

$$\dot{\epsilon}_{ij}^p = K\,J_2'^{(M-1)/2}\,\sigma_{ij}' \tag{3.20}$$

where K is a constant with the dimensionality required by the equation. If M is zero then the indeterminate, rate-insensitive, form is returned. Alternatively, $M = 1$ gives linear viscous flow. M is known as a rate sensitivity exponent. This simple type of constant state viscoplastic material model has proven quite useful in approximate analysis of some metalworking processes, and in simulating the creep of components loaded at relatively high temperatures for extended times. Because (3.20) is determinate in general, it can be inverted to give an expression for the deviatoric stress in terms of the inelastic strain rates:

$$\sigma_{ij}' = K^{1/M}\,\dot{I}_2^{p\,(1-M)/2M}\,\dot{\epsilon}_{ij}^p \tag{3.21}$$

This type of development can be carried out for functions other than the simple power law, and state variables and temperature can be included.

A rate-independent development of Eq. (3.12) can be undertaken which is determinate as long as inelastic deformation causes the yield surface to increase in size. A simple assumption, mentioned above, is that this depends on the total inelastic work done during the deformation. This inelastic work can be described as a state variable, with the following evolution equation:

$$\dot{W}^{\mathrm{p}} = \sigma_i' \, \dot{\epsilon}_i^{\mathrm{p}} \tag{3.22}$$

which is exactly equivalent to Eq. (3.2) for the isotropic case where the principal axes of the two tensors coincide. Also, with the von Mises criterion and associated plasticity, the work rate is given by the following function of invariants:

$$\dot{W}^{\mathrm{p}} = \sqrt{J_2' \dot{I}_2^{\mathrm{p}}} \tag{3.23}$$

The vectors representing the deviatoric principal stress and the inelastic strain rate are parallel in the π plane, and the second invariants give measures of the magnitudes of those vectors.

Because a rate insensitive material model is intended, time derivatives can be discarded and differentials used. Then Eq. (3.23) becomes

$$dW^{\mathrm{p}} = \sqrt{J_2'} \, d\sqrt{I_2^{\mathrm{p}}} \tag{3.24}$$

where

$$d\sqrt{I_2^{\mathrm{p}}} = \sqrt{\dot{I}_2^{\mathrm{p}}} \, dt \tag{3.25}$$

The size of the yield surface is represented by $\sqrt{J_2'}$, and this depends on the inelastic work so that

$$\frac{d\sqrt{J_2'}}{dW^{\mathrm{p}}} = \frac{1}{\sqrt{J_2'}} \frac{d\sqrt{J_2'}}{d\sqrt{I_2^{\mathrm{p}}}} \tag{3.26}$$

The differential, rate-insensitive, form of the normality rule with the von Mises yield surface is, from Eq. (3.17):

$$d\epsilon_{ij}^{\mathrm{p}} = d\sqrt{I_2^{\mathrm{p}}} \, \frac{\sigma_{ij}'}{\sqrt{J_2'}} \tag{3.27}$$

The differential form of the relationship between the yield surface size and the inelastic work, Eq. (3.26), can now be used to eliminate the strain invariant term, giving

$$d\epsilon_{ij}^{\mathrm{p}} = \frac{1}{H_{\mathrm{W}}} \frac{d\sqrt{J_2'}}{\sqrt{J_2'}} \frac{\sigma_{ij}'}{\sqrt{J_2'}} \tag{3.28}$$

where H_{W} is the rate of change of yield surface size with inelastic work. From Eq. (3.26), it is clear that this is simply related to the increase in yield surface size with the invariant of inelastic strain, so that

$$d\epsilon_{ij}^{\mathrm{p}} = \frac{d\sqrt{J_2}}{H_{\mathrm{W}}} \frac{\sigma'_{ij}}{\sqrt{J'_2}} \tag{3.29}$$

with

$$H_w = \frac{d\sqrt{J'_2}}{d\sqrt{I_2^{\mathrm{p}}}} \tag{3.30}$$

The form given in Eq. (3.29) is most frequently used, because H_ϵ can be measured in simple tests. For example, the gradient of the uniaxial flow stress against logarithmic (true) strain in a tensile test is 1.5 H_ϵ. The use of equivalent uniaxial test stress is very common. The relationship between the uniaxial stress and the invariant measure of yield surface size is simply

$$\sigma_u = \sqrt{\frac{3}{2} J'_2} \tag{3.31}$$

With all the assumptions, this means that for material deforming inelastically, knowledge of the current stress and the data from a simple test gives the infinitesimal inelastic strain increment in terms of an infinitesimal increment in yield surface size, and so of stress. For finite increments some integration would be required, and usually this would have to be done numerically.

3.7 Isotropic Elastoplastic Equations

So far, only the inelastic part of the strain rate has been considered. For a full description, the elastic part also needs to be included. Sometimes this recoverable strain is ignored because of its limited possible magnitude. Material models which make this assumption are called rigid-plastic or rigid-viscoplastic. When such models are used, the problems arising from the incompressibility assumed for inelastic deformation, and from the possibility of unloading so that the stress becomes incapable of producing inelastic deformation, have to be overcome by techniques outside the scope of this chapter.

Inclusion of elastic strains in the model has several advantages, apart from overcoming the incompressibility problem. It allows the modelling of unloading, which can be useful when residual stresses are important after deformation. It also provides a very useful basis for inversion of the rate-insensitive equation, (3.29), to give the stress rates in terms of the total strain rates. This, in turn, can be extended to rate-sensitive material models.

The usual assumption made in elastoplastic material models is that the total strain rate is the simple sum of elastic and inelastic rates:

$$\dot{\epsilon}_{ij} = \dot{\epsilon}_{ij}^{\mathrm{e}} + \dot{\epsilon}_{ij}^{\mathrm{p}} \tag{3.32}$$

In the isotropic case, the elastic strain rates are given by

$$\dot{\epsilon}_{ij}^{\mathrm{e}} = \frac{\dot{\sigma}'_{ij}}{2G} + \frac{\delta_{ij}\,\dot{\sigma}_{kk}}{3K} \tag{3.33a}$$

or

$$\dot{\epsilon}_i^e = \frac{\dot{\sigma}_i'}{2G} + \frac{\dot{\sigma}_k}{3K} \tag{3.33b}$$

in principal components. For many materials, the elasticity is linear and in these cases, the shear and bulk elastic moduli, G and K, are material constants. To get the total strain for rate-insensitive associated von Mises plasticity, Eq. (3.29) is returned to rate form for convenience and added to Eq. (3.33a) to give

$$\dot{\epsilon}_{ij} = \frac{\dot{\sigma}_{ij}'}{2G} + \frac{\delta_{ij}\,\dot{\sigma}_{kk}}{3K} + \frac{\sqrt{\dot{J}_2'}}{H_\epsilon}\frac{\sigma_{ij}'}{\sqrt{J_2'}} \tag{3.34}$$

This equation can be inverted to give the stress rate in terms of the current stress, hardening rate and total strain rate:

$$\dot{\sigma}_{ij} = 2G\dot{\epsilon}_{ij} + \frac{(K-2G)}{3}\,\delta_{ij}\,\dot{\epsilon}_{mm} - 2G\,\frac{\sigma_{ij}'\,\sigma_{kl}'}{S}\,\dot{\epsilon}_{kl} \tag{3.35}$$

where

$$S = J_2'\left(1 + \frac{H_\epsilon}{2G}\right) \tag{3.36}$$

Details of this inversion are given in Yamada et al. (1968), although the process was effectively carried out in Hill (1950, Chap. 3). The inversion given does not become indeterminate when H_ϵ is zero. Eq. (3.35) must be integrated to give results for any finite deformation. This integration is usually done numerically, and very small deformation step sizes are usually required for accurate results. The definition of the stress rate in (3.35) is not trivial, various types of stress rate exist because of geometric effects. Usually, the Jaumann (corotational) stress rate is taken to be correct in this case.

In the strict terminology, Eq. (3.35) is an evolution equation of a state variable: elastic stress. Geometrically, it can be described in the following way. An infinitesimal strain is applied, and the resulting increase in stress calculated corresponding to the first two parts of the equation. The new stress will exceed the bounds of the yield surface, assuming inelastic deformation is occurring, otherwise this is the end of the process. The new stress values are returned to the yield surface in a direction normal to the initial stress point, but with the yield surface expanded to account for any hardening. This hardening, which can be negative, or softening, is assumed to be linear for the infinitesimal deformation. The procedure is shown in Fig. 3.7.

3.8 Finite Deformation Increments

All of the relationships given apply only to current rates and configurations. In practice, equations are required which can be used for deformations which, even though they may be quite small in extent, are finite. Sometimes the total problem can be broken down into small enough increments for equations such as (3.35)

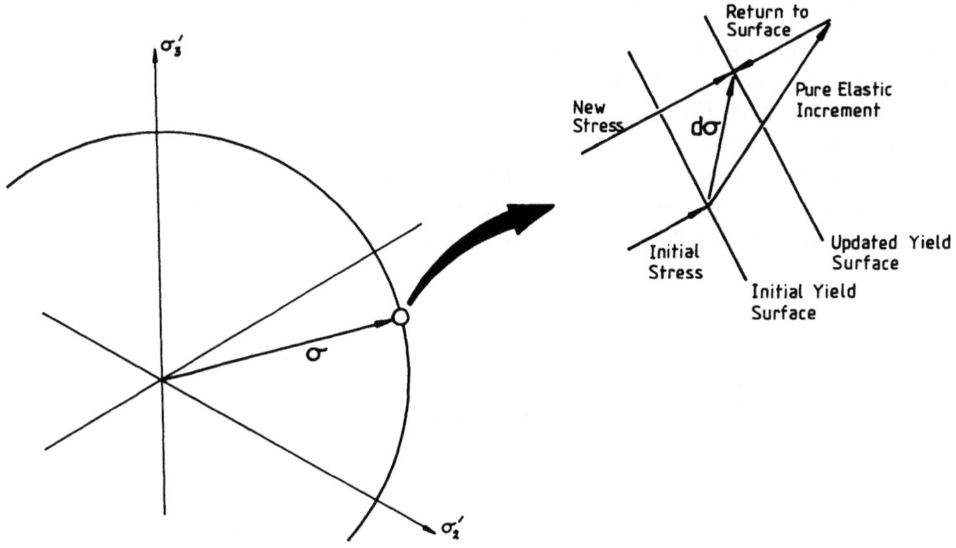

Fig. 3.7. Change in stress during an elastic-plastic increment.

to be used directly. This can be inefficient. A technique has been developed which is of quite general applicability for isotropic plasticity including elastic effects (Nagtegaal and Veldpaus 1982), and can take into account rate sensitivity.

As with the viscoplastic models, such as Eq. (3.21), the constitutive equation for stress is admitted to be nonlinear. Some finite deformation increment of the stress in the current configuration is given by

$$\Delta\sigma_{ij} = \sigma(\Delta\epsilon_{kl}, \Delta t, T, \Lambda^1, \Lambda^2, \dots, \Lambda^n) \tag{3.37}$$

where $\Delta\epsilon$ is a suitable measure of the finite strain increment, which takes place in the interval Δt. $\Delta\epsilon$ will usually be a nonlinear function of the displacement gradients. A useful measure for reasonable increment sizes is the quadratic approximation to the true, logarithmic, strain:

$$\Delta\epsilon_{ij} \cong \frac{1}{2}\left\{\frac{\partial U_i}{\partial x_j} + \frac{\partial U_j}{\partial x_i} + \frac{1}{2}\left[\frac{\partial U_i}{\partial x_k}\frac{\partial U_k}{\partial x_j} + \frac{\partial U_k}{\partial x_i}\frac{\partial U_j}{\partial x_k}\right] + \left[\frac{\partial U_k}{\partial x_i}\frac{\partial U_k}{\partial x_j} - \frac{\partial U_i}{\partial x_k}\frac{\partial U_j}{\partial x_k}\right]\right\} \tag{3.38}$$

where U is the finite displacement. The average strain rate is approximated as

$$\dot{\epsilon}_{ij} \cong \frac{\Delta\epsilon_{ij}}{\Delta t} \tag{3.39}$$

The evolution of the state variables has to be described for finite deformations as well:

$$\Delta\Lambda^i = \Lambda\,(\sigma_{ij}, \Delta\epsilon_{kl}, \Delta t, T, \Lambda^1, \Lambda^2, \dots, \Lambda^n) \tag{3.40}$$

The stresses existing at the start of the deformation will have to be updated for any rigid body rotation. Other state variables are reduced to scalars for the isotropic case so these are not affected in that way. Linear approximation of the rotation produces the same order of error as the quadratic approximation for strain.

Assuming the state variables can be described after the deformation increment, the problem remains of calculating the stress, Eq. (3.37). For the simplest case, that of a von Mises yield surface and associated plasticity, an extension of the procedure shown in Fig. 3.7, known as radial return, can be used. The strain is initially assumed to be totally elastic. The stress state associated with this elastic situation is then returned to the updated yield surface in a radial direction in deviatoric stress space. With the von Mises criterion, such radial directions are normal to the yield surface. The geometry in the π plane is shown in Fig. 3.8.

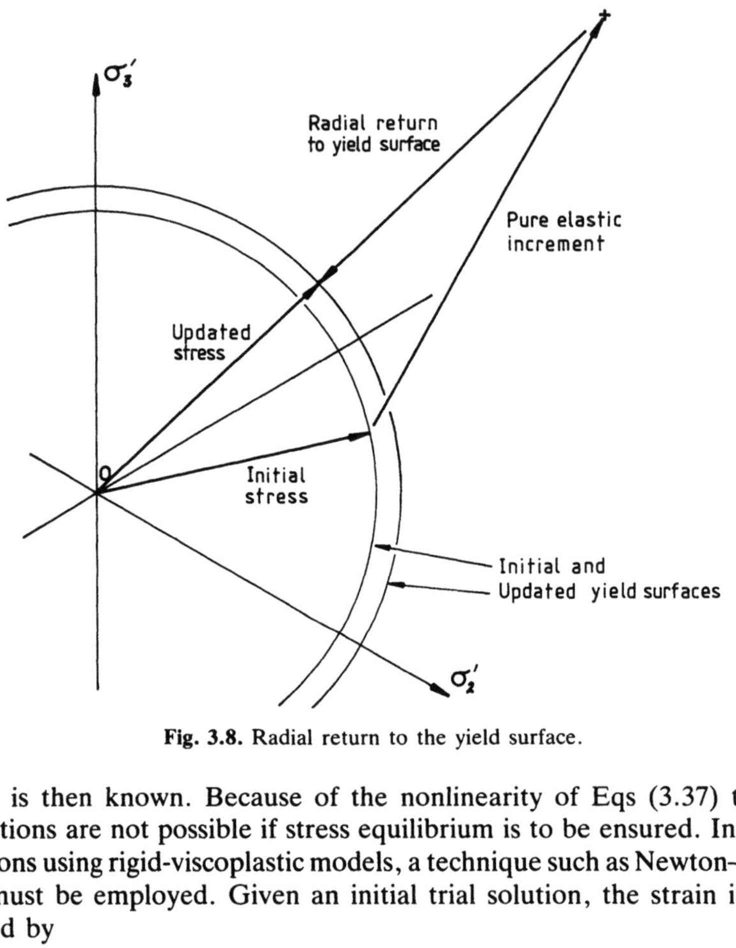

Fig. 3.8. Radial return to the yield surface.

The stress is then known. Because of the nonlinearity of Eqs (3.37) to (3.40), direct solutions are not possible if stress equilibrium is to be ensured. In common with solutions using rigid-viscoplastic models, a technique such as Newton–Raphson iteration must be employed. Given an initial trial solution, the strain increment is corrected by

$$\Delta\epsilon_{ij} = \Delta\epsilon_{ij} - \left(\frac{\partial\sigma_{kl}}{\partial\Delta\epsilon_{ij}}\right)^{-1} \delta\sigma_{kl} \tag{3.41}$$

where $\delta\sigma$ is the imbalance, or error, in stress resulting from the trial strain increment. The derivative terms, the moduli are given by

$$\frac{\partial\sigma_{ij}}{\partial\Delta\epsilon_{kl}} = 2G^*\,\delta_{ik}\,\delta_{jl} + \frac{(K-2G^*)}{3}\,\delta_{ij}\,\delta_{kl}$$

$$-2G^*\,\frac{\sigma'_{ij}\,\sigma'_{kl}}{S^*} \tag{3.42}$$

which is similar to the form given in Eq. (3.35), but with an effective elastic shear modulus,

$$G^* = R\,G \tag{3.43}$$

and an effective "hardening" term,

$$S^* = J_2'\left(1 + \left[R\left(\frac{H_{\Delta W^*}}{2G + H_{\Delta W^*}}\right) - H_{\Delta W^*}\right]\right) \tag{3.44}$$

where R is the ratio of the second invariants of the deviatoric stresses after and before the radial return to the yield surface from the pure elastic state. The other modified term is

$$H_{\Delta W^*} = \frac{d\sqrt{J_2'}}{d\Delta\sqrt{I_2}} \tag{3.45}$$

which is the current effect of changing the strain increment on the size of the yield surface. Rate effects can be included in this term.

Usually the Newton–Raphson process (Eq. 3.41) would be applied to the internal forces and displacements of a whole system, and the effects of geometry would then be included. Iterative methods are of quite general applicability when nonlinear constitutive equations are involved, and also prove useful for dealing with nonlinearities which result from geometric aspects of finite deformation.

The isotropic plasticity models described briefly above have been used almost exclusively in numerical simulation of metal working processes. There is still considerable scope for the development of isotropic models, especially the inclusion of state variables which not only reflect more accurately the real microstructural state, but also give better modelling in complex deformations. Sometimes however, the isotropic assumption has to be abandoned in order to get the desired accuracy. The extension of the yield surface, into five- or six-dimensional stress space, and the introduction of more complex expressions for its shape and evolution, has then to be considered. The area of anisotropic plasticity is the subject of the next chapter.

References

Hill R (1950) The mathematical theory of plasticity. Oxford University Press

Nagtegaal JG, Veldpaus FE (1982) Analysis of metal forming problems with an improved finite element plasticity formulation. In: Numerical methods in industrial forming processes. Pineridge Press, Swansea

Spencer AJM (1980) Continuum mechanics. Longman, London

Yamada Y, Yoshmura N, Sakurai T (1968) Plastic stress–strain matrix for the solution of elastic-plastic problems. Int J Mech Sci 10:343–352

4 Anisotropic Plasticity

P. Van Houtte

4.1 Introduction

In this chapter, the summation convention is used unless stated otherwise; subscripts denote indices, superscripts are not indices but further specifications; they sometimes refer to the coordinate system in which a vector or tensor component is expressed.

Many physical properties of materials can only be defined or measured when reference is made to a direction in space. Consider for example the apparent Young's modulus:

$$E = \sigma/\epsilon \tag{4.1}$$

where σ and ϵ are respectively the normal stress and the elastic strain in a tensile test (in the elastic range) on a bar cut from a piece of material in a certain direction characterized by a unit vector $\mathbf{u}$ (Fig. 4.1). A priori there is no reason why the value of E should be independent of $\mathbf{u}$. But the material could be such that the measured values of E would appear not to vary with $\mathbf{u}$. In such a case, the material would be called "*isotropic* with respect to the apparent Young's modulus E". It is necessary to specify for which physical property the material is isotropic; the metal tungsten, for example, is as nearly isotropic as possible with respect to the apparent Young's modulus, but not for other properties such as the yield stress (Nye 1969). On the other hand, when some physical property of the material would appear to be a function of $\mathbf{u}$, the material would be called anisotropic with respect to this property.

The extent to which a material property can be anisotropic depends very much on its microstructure. Crystalline materials are fundamentally anisotropic. Yet material properties that can be described by a second-rank tensor, such as the electrical conductivity, are isotropic for all cubic crystals (not for other crystals) (Nye 1969). Elastic properties are described by a stiffness or compliance tensor, which is a fourth-rank tensor. It need not be isotropic, not even for cubic crystals; nevertheless, it is nearly isotropic in cases such as aluminium or tungsten (Nye 1969). So when the elastic properties of a single crystal of these metals are already almost isotropic, then of course those of a polycrystalline material that consists of millions of such crystals, all with different crystal orientations, would be even more isotropic. It is clear that any anisotropy of a property of a crystal is smeared out and attenuated in a polycrystalline material, unless all the crystallites of which it consists have the same crystal orientation with respect to the external coordinate system $\langle x_i \rangle$ (Fig. 4.1).

So in many cases, it is reasonable to assume that an engineering material (which is often polycrystalline) is isotropic with respect to some physical property. The assumption of isotropy is quite a comfortable one; it makes many calculation procedures much easier. It is one of the very first simplifying assumptions made by most users of the mathematical theory of plasticity. The anisotropy of the plastic properties of engineering materials is, however, much more pronounced than that of their elastic properties. Some examples will be given below.

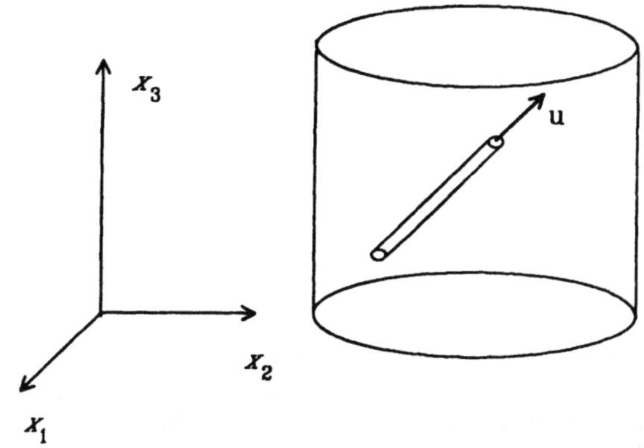

Fig. 4.1. A bar cut in a direction **u** from a cast ingot of a magnesium alloy. The apparent Young's modulus E is calculated from the result of a tensile test on the bar. E will be a function of **u**. The coordinate system $\langle x_1\ x_2\ x_3\rangle$ is called the external system; it is fixed with respect to the ingot.

4.2 Examples of the Plastic Anisotropy of Engineering Materials

4.2.1 *r*-Values of Steel Sheets

Imagine a steel sheet of the type used for the production of car body parts (Fig. 4.2). x_1 is the rolling direction (RD), x_2 the transverse direction and x_3 the normal to the steel sheet. One may cut a tensile test sample from such a sheet. Let α be the angle between the rolling direction and x'_1, the longitudinal axis of the tensile test sample. x'_2 is the transverse direction of the test sample; x'_3 coincides

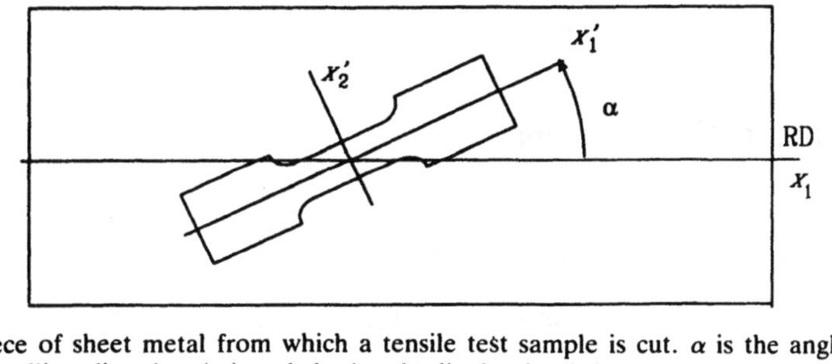

Fig. 4.2. A piece of sheet metal from which a tensile test sample is cut. α is the angle between the rolling direction (x_1) and the longitudinal axis of the test sample (x'_1).

with x_3. Suppose now that the tensile test is stopped at a reference strain, e.g. 10% *plastic* strain in longitudinal direction. The r value is the ratio of the transverse to the thickness strain:

$$r = \epsilon'_{22}/\epsilon'_{33} \tag{4.2}$$

in which ϵ'_{22} and ϵ'_{33} are expressed in the coordinate system $\langle x'_1\,x'_2\,x'_3\rangle$.

In the case of an isotropic material, there should be no difference between the plastic response in transverse direction and in thickness direction. So a value of 1 would be expected for r. In practice, however, r varies between 1.4 and 2.5

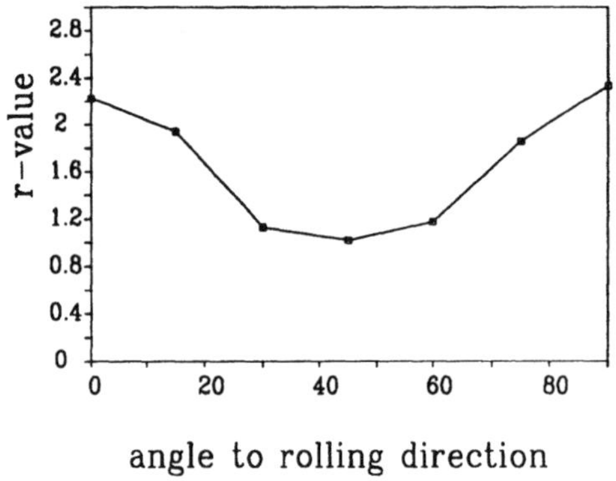

Fig. 4.3. Variation of r with α (Fig. 4.2) for a commercial deep drawing steel sheet (Material X3 from Welch and Bunge 1986).

for the steel qualities that are used for the deep drawing of car body parts; moreover, it is a function of the angle α, as shown by Fig. 4.3.

It is useful to define an *instantaneous* strain ratio as well:

$$\rho = \dot{\epsilon}'_{22}/\dot{\epsilon}'_{33} \tag{4.3}$$

in which $\dot{\epsilon}'_{22}$ and $\dot{\epsilon}'_{33}$ are plastic strain rates. ρ applies to a given state of the material. It is difficult to measure, whereas it is much easier to measure r. ρ and r are not necessarily equal, because the former usually varies with strain (Welch et al. 1983, Lake et al. 1988).

The average value of r as a function of α is strongly correlated with the deep drawability of the steel sheet. This is the reason why car manufacturers require a high average r from steel makers. It is usually estimated on the basis of measurements for $\alpha = 0°$, $\alpha = 45°$ and $\alpha = 90°$:

$$r_m = (r_0 + 2r_{45} + r_{90})/4 \tag{4.4}$$

4.2.2 Earing in Cup Drawing

The variation of r (or ρ) as a function of α is also an important anisotropy parameter. It is usually characterized by Δr:

$$\Delta r = (r_0 - 2r_{45} + r_{90})/2 \tag{4.5}$$

Δr is correlated with the tendency of a sheet material to form ears in cup drawing (Van Houtte et al. 1987). Cup drawing is a standard test for measuring deep drawability. A circular piece of sheet material is clamped between two rings while a flat punch is pressed through the opening in the rings (Fig. 4.4). The clamp pressure between the ring-shaped dies is supposed to prevent wrinkling of the circular blank; the pressure should not be too high so that the material can still move in a radial direction despite the friction. The pressure is by no means high enough to prevent thickening of the sheet.

It appears that the cups that are formed have ondulated rims (Fig. 4.5). The ondulations are reproducible, i.e. they do not depend on random misadjustments of the machine. They are caused by the *plastic anisotropy* of the sheet material. It has been shown that they are correlated to Δr or $\Delta \rho$, although there are other

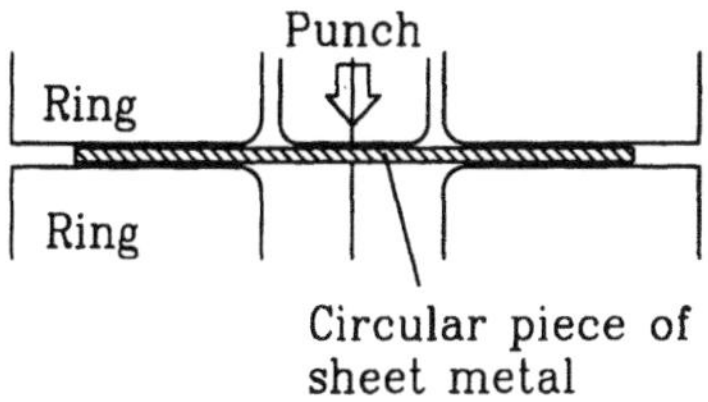

Fig. 4.4. Transverse section of the cup-drawing process. The punch is moving downwards. This causes an inwards radial flow of the material flow between the rings. Finally a cup is formed.

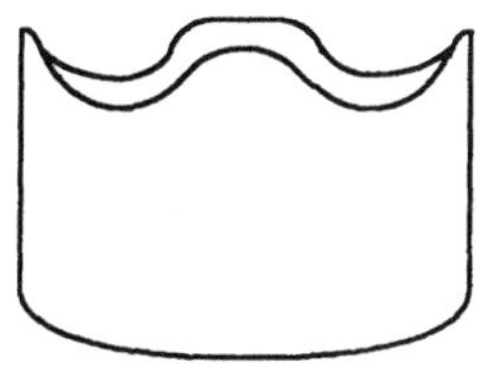

Fig. 4.5. A drawn cup. The ondulations of the rim are caused by plastic anisotropy. A case with four "ears" is quite common for aluminium alloys.

anisotropy parameters with which the correlation is better (Van Houtte et al. 1987).

One of the steps of the industrial fabrication processes of food cans and beverage cans from steel sheet or aluminium alloy sheet is in fact a cup drawing operation. "Earing" then is an undesired effect because the ears have to be removed in order to produce a usable can. So material suppliers are required by can manufacturers to minimize the earing tendency of their material.

4.2.3 Forming Processes with Partially Unconstrained Material Flow

The two examples given in the previous section are two cases of a deformation test in which the deformation is *not fully constrained*: in a tensile test, the transverse and thickness strains are not directly imposed during the test; only the longitudinal strain is. In cup drawing (Fig. 4.4), the dies and the punch do not prevent deviations from a purely radial material flow: the rings prevent wrinkling, but not thickening of the sheet; lateral deviations from radial flow are possible.

The experience of these two cases has shown that plastic anisotropy (whatever its cause) can influence the flow pattern significantly as long as the boundary conditions imposed by the dies, the clamps etc. are satisfied. This is a general conclusion: the flow pattern of any forming process that leaves some freedom to the material flow will be influenced by anisotropy. Other examples are: upsetting of rods; flat rolling of bars; rolling of thick slabs. In these forming operations, the external shapes of the products may be influenced by plastic anisotropy. However, the flow field and the resulting stress distribution will be influenced in other forming operations as well, regardless of whether the external shapes of the products reveal it or not. This may be particularly important in forging operations.

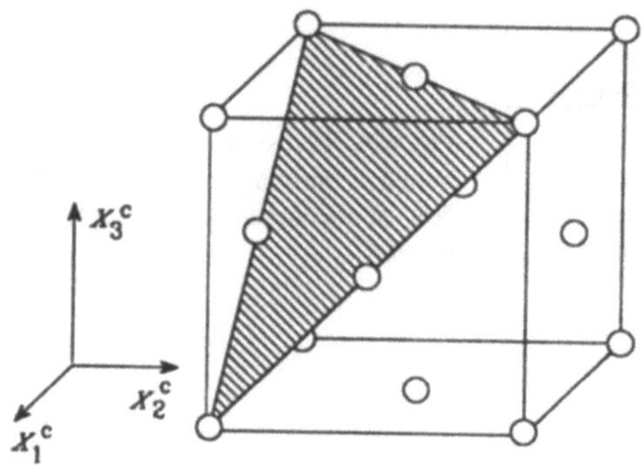

Fig. 4.6. Unit cell of a fcc crystal lattice. The shaded triangle is part of one of the slip planes of the (1 $\bar{1}$ 1) family. The sides of the triangle are parallel to the [0 1 1], [1 0 $\bar{1}$] and [1 1 0] slip directions.

4.3 The Fundamental Plastic Anisotropy of Crystalline Materials

4.3.1 Slip Systems

Fig. 4.6 shows a unit cell of a *face centred cubic* (fcc) crystal lattice. This is the crystal structure of each grain of metals such as aluminium or copper. On an atomic scale, the plastic deformation of such crystal is made possible by the motion of dislocations on certain crystallographic planes and in certain directions, called *slip planes* and *slip directions*. In cubic metals, a slip plane is identified by its Miller indices ($h\ k\ l$), from which the unit vector **n** of the normal on the plane can be derived as follows:

$$n_1 = h/s$$
$$n_2 = k/s$$
$$n_3 = l/s$$
$$\text{with } s = (h^2 + k^2 + l^2)^{1/2} \tag{4.6}$$

The slip direction is also defined by Miller indices [$u\ v\ w$]. They represent a crystallographic direction *in the slip plane*. For cubic metals, the unit vector **b** in the slip direction is given by

$$b_1 = u/s'$$
$$b_2 = v/s'$$
$$b_3 = w/s'$$
$$\text{with } s' = (u^2 + v^2 + w^2)^{1/2} \tag{4.7}$$

The vectors **n** and **b** must be perpendicular to each other. A particular combination of a slip plane and slip direction is called a *slip system*. In fcc metals, there are 12 slip systems:

- (1 1 1) with [0 1 $\bar{1}$], [1 0 $\bar{1}$] or [1 $\bar{1}$ 0]
- ($\bar{1}$ 1 1) with [0 1 $\bar{1}$], [1 0 1] or [1 1 0]
- (1 $\bar{1}$ 1) with [0 1 1], [1 0 $\bar{1}$] or [1 1 0]
- (1 1 $\bar{1}$) with [0 1 1], [1 0 1] or [1 $\bar{1}$ 0]

These 12 systems are crystallographically equivalent. Together they are called the {1 1 1} ⟨1 1 0⟩ slip systems. Fig. 4.6 shows a unit cell of a fcc crystal with one of the slip planes shaded. Note that the slip planes are close-packed planes.

The lattice of the ferrite phase in low carbon steel is of the *body centred cubic* (bcc) type. The slip directions are the four ⟨1 1 1⟩ directions. This type of crystal structure does not have close-packed planes. The slip planes are of the {1 1 0} and {1 1 2} type. There are 24 slip systems:

- [1 1 1] with $(0\ 1\ \bar{1})$, $(1\ 0\ \bar{1})$, $(1\ \bar{1}\ 0)$
  and $(\bar{2}\ 1\ 1)$, $(1\ \bar{2}\ 1)$, $(1\ 1\ \bar{2})$
- $[\bar{1}\ 1\ 1]$ with $(0\ 1\ \bar{1})$, $(1\ 0\ 1)$, $(1\ 1\ 0)$
  and $(2\ 1\ 1)$, $(\bar{1}\ \bar{2}\ 1)$, $(\bar{1}\ 1\ \bar{2})$
- $[1\ \bar{1}\ 1]$ with $(0\ 1\ 1)$, $(1\ 0\ \bar{1})$, $(1\ 1\ 0)$
  and $(\bar{2}\ \bar{1}\ 1)$, $(1\ 2\ 1)$, $(1\ \bar{1}\ \bar{2})$
- $[1\ 1\ \bar{1}]$ with $(0\ 1\ 1)$, $(1\ 0\ 1)$, $(1\ \bar{1}\ 0)$
  and $(\bar{2}\ 1\ \bar{1})$, $(1\ \bar{2}\ \bar{1})$, $(1\ 1\ 2)$

Some authors have also reported {1 2 3} planes as slip planes (for a reference list, see Gil Sevillano et al. 1980). In hexagonal closed packed metals such as α-titanium, the close-packed plane is the basal plane (0 0 0 1), which is the most important slip plane. The ⟨1 0 −1 0⟩ directions are the three slip directions. In addition to these, various other slip systems have been reported in the literature.

A sufficiently high shear stress acting on a slip system can cause a massive motion of dislocations on the slip plane and in the slip direction. This causes a permanent deformation of the crystal which has the nature of a simple shear, at least to an observer in a reference system $\langle x_i^c \rangle$ (Fig. 4.6) fixed to the crystal lattice. If the crystal lattice rotates, the "observer" has to rotate with it. The slip plane is the shear plane and the slip direction is the shear direction. The process changes the shape of the crystallites (and may cause a rigid-body rotation, *and* a rotation of the crystal lattice, which are not identical) without altering the nature of the crystal lattice. As long as the dislocations are moving, the material is in the plastic state. The microscopic velocity gradient caused by dislocations moving on a single slip system s is a tensor of which the components are

$$\dot{f}_{ij} = K_{(s)ij}\dot{\gamma}_{(s)} + \dot{\omega}_{ij}^L \tag{4.8}$$

In this equation and the next, the index s of the slip system is put between parentheses to indicate that Einstein's summation convention is not used. $\mathbf{K}_s$ is the tensorial product $\mathbf{b}_{(s)}\mathbf{n}_{(s)}$:

$$K_{sij} = b_{(s)i}\, n_{(s)j} \tag{4.9}$$

in which $b_{(s)i}$ is the component i of the vector $\mathbf{b}_s$ (of slip system s) etc. $\dot{\omega}_{ij}^L$ is an asymmetric velocity gradient tensor that represents the rate of rotation of the crystal lattice. This microscopic feature also determines $\dot{f}_{ij}$. $\dot{\gamma}_s$ is a quantity that has the nature of a shear strain rate and that is called the "slip rate" associated to slip system s.

Single slip is an exceptional case. In case of multiple slip, the microscopic velocity gradient is given by

$$\dot{f}_{ij} = K_{sij}\dot{\gamma} + \dot{\omega}_{ij}^L \tag{4.10}$$

in which this time a summation is made over all slip systems. The symmetrical part of the tensorial equation (4.10) can be written separately:

$$\dot{\epsilon}_{ij} = M_{sij}\dot{\gamma}_s \tag{4.11}$$

with

$$M_{sij} = [(K_s)_{ij} + (K_s)_{ji}]/2 \tag{4.12}$$

It can easily be shown (Van Houtte 1988) that the resolved shear stress "felt" by a particular slip system is given by

$$\tau_s^r = M_{sij}\,\sigma_{ij} \tag{4.13}$$

when the microscopic stress σ_{ij} is acting on a particular crystallite.

The basic constitutive law of crystal plasticity is the *generalized Schmid* law (see e.g. Backofen 1972). The dislocations will move on a particular slip system (and hence the crystal will be in the plastic state) when the resolved shear stress reaches a critical level τ_s^c. A mathematical expression for this is

$$-\tau_{(-s)}^c \leq \tau_s^r \leq \tau_s^c \tag{4.14}$$

The equality sign holds for the plastic condition; the "smaller than" sign for the elastic condition; other stress states are in principle impossible. $\tau_{(-s)}^c$ stands for the absolute value of the critical shear stress for slip in the negative direction on slip sysem s. It is often (but not always) equal to τ_s^c. The critical shear stresses τ_s^c and $\tau_{(-s)}^c$ depend weakly on the *strain rate* and moderately on the *strain history* of the material. Generally their values increase when the material is deformed (see e.g. Aernoudt et al. 1987). This phenomenon is called *work hardening*. Thermal treatments may undo this phenomenon by causing physical processes such as *recovery* or *recrystallization*. When the deformation takes place at sufficiently high temperatures, these processes may occur during the deformation itself.

Crystallographic slip is the main mechanism for plastic deformation in crystalline materials. Other mechanisms are:

- Mechanical twinning, observed in hcp metals and in cubic fcc metals with low stacking fault energy such as brass and austenitic stainless steel
- Stress-induced martensitic transformation, for example observed in austenitic stainless steel
- Grain boundary sliding, which sometimes occurs at high temperatures and/or for small grain sizes, and which is believed to be the cause of *superplastic* behaviour

The occurrence of such special mechanisms makes the behaviour of the material less predictable and more difficult to model. They will not be considered further in this text.

4.3.2 The Yield Locus of Single Crystals

Let us provisorily neglect strain rate sensitivity and consider the critical resolved shear stresses as constants at a given time t. Combining Eqs (4.13) and (4.14) leads to the following equation:

$$-\tau^c_{(-s)} \leq M_{sij}\,\sigma_{ij} \leq \tau^c_s \tag{4.15}$$

which defines a set of two hyperplanes in a six-dimensional stress space. All admissible stress states must either be on these hyperplanes or between them. There is a condition such as Eq. (4.15) for each slip system. Combining all of them gives the set of all admissible stress states in stress space. What does the resulting surface look like? First, let us remark that

$$M_{s11} + M_{s22} + M_{s33} = 0 \tag{4.16}$$

which only expresses the fact that $\mathbf{n}_s$ and $\mathbf{b}_s$ are perpendicular (see Eqs (4.13) and (4.10)). In geometrical terms, Eq. (4.16) means that the normals to the hyperplanes are perpendicular to the vector π, i.e. the vector from the origin of stress space to a point that represents a unit hydrostatic stress and which has the coordinates

$$\sigma_{11} = \sigma_{22} = \sigma_{33} = 1,\ \sigma_{23} = \sigma_{31} = \sigma_{12} = 0 \tag{4.17}$$

This means that the hyperplanes themselves are perpendicular to the π-plane (the plane normal to the vector π). The latter is a five-dimensional subspace that contains all the deviatoric stresses. These are the only ones that matter when it comes to finding out whether a certain stress causes plastic deformation or not. Indeed, any modification of the hydrostatic component of the stress merely causes a translation parallel to the hyperplanes. Note that the well-known von Mises or Tresca yield loci also consist of generatrices that are perpendicular to the π-plane. So in order to study the yield locus, it is sufficient to study its intersection with the π-plane. This intersection has the aspect of a hyperpolyhedron, or a "hyper"-facet-eye in five dimensions (Fig. 4.7). The vertices are points in which

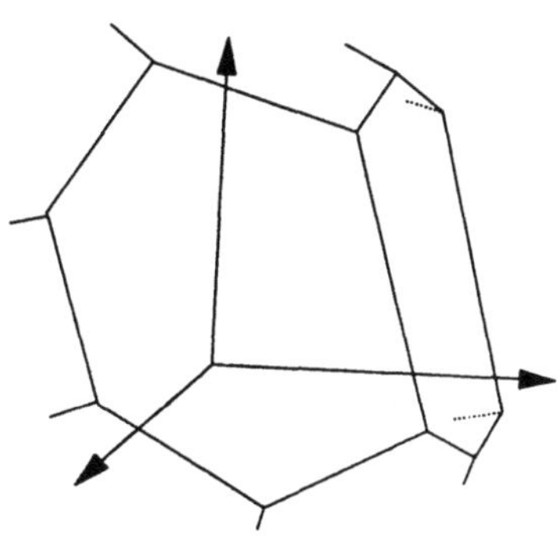

Fig. 4.7. An example of a single crystal yield locus with a facet-eye aspect. For reasons of simplicity, it is assumed that the subspace of the deviatoric stresses is three-dimensional.

in principle five hyperplanes intersect; the corresponding stress states are such that the resolved shear stresses simultaneously reach their critical levels on five slip systems, hence activating these. Bishop and Hill (1951) have listed the 56 vertices that they have found for fcc metals, assuming the same value for all critical resolved shear stresses. In this particular case, some of the vertices happened to be the intersections not of five, but of six or even eight hyperplanes. The points on the facets lie on one hyperplane only; these stress states activate only one slip system. Finally, there are also "hyper-edges" representing sets of stress states for which 2, 3 or 4 slip systems are active.

Fig. 4.8 shows a plane section of the yield locus of a single crystal of fcc material, assuming the same τ^c on all slip systems. The material would be isotropic

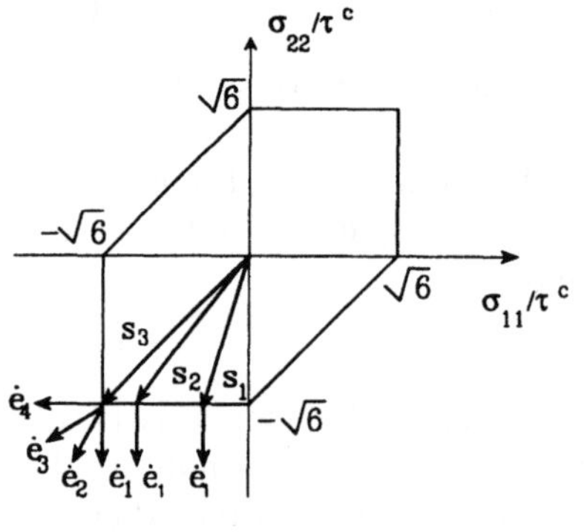

Fig. 4.8. $\sigma_{33} = \sigma_{23} = \sigma_{31} = \sigma_{12} = 0$ section of the yield locus of a single crystal with a fcc lattice. The ⟨1 0 0⟩ crystal axes coincide with the $\langle x_i \rangle$ axes of the coordinate system.

with respect to plasticity when it would have the same yield locus after expressing the stresses in different (rotated) coordinate systems in physical space. Figs 4.9 and 4.10 show such "rotated" yield loci. It is seen that they are quite different. This shows clearly that such crystals are anisotropic with respect to plastic deformation.

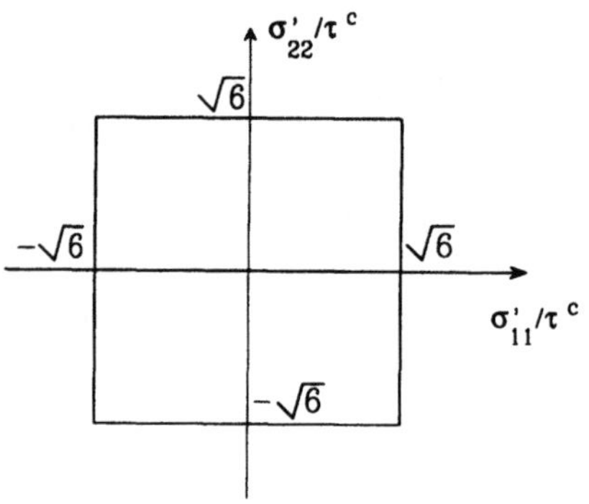

Fig. 4.9. The yield locus of the same crystal as in Fig. 4.8, but expressed in the reference system $\langle x_i' \rangle$, obtained from $\langle x_i \rangle$ by a 45° rotation around x_3.

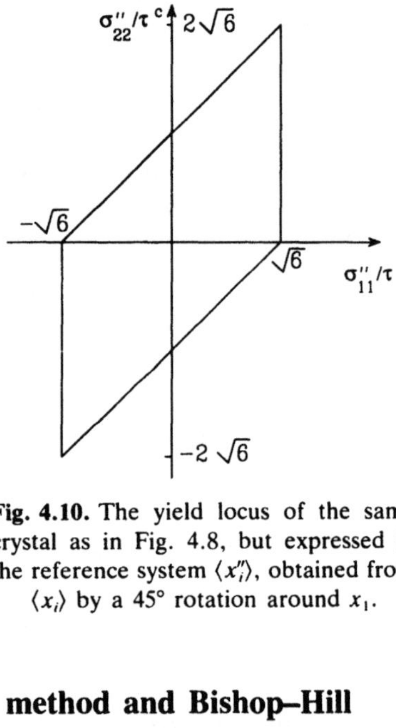

Fig. 4.10. The yield locus of the same crystal as in Fig. 4.8, but expressed in the reference system $\langle x_i'' \rangle$, obtained from $\langle x_i \rangle$ by a 45° rotation around x_1.

4.3.3 Stress–Strain Relationship: Taylor method and Bishop–Hill method

The stress–strain relationship is based on the *maximum work principle* (Hill 1950). This leads to the normality rule in smooth portions of the yield locus. In order to define "normality" in six-dimensional stress–strain space, it is necessary to give the formulas for stress and for strain rate vectors. Let $\dot{\mathbf{e}}$ and $\mathbf{s}$ be vectors that respectively represent a strain rate tensor $\dot{\varepsilon}_{ij}$ and a stress tensor σ_{ij}:

$$\left.\begin{aligned}
\dot{e}_1 &= \dot{\epsilon}_{11} & s_1 &= \sigma_{11} \\
\dot{e}_2 &= \dot{\epsilon}_{22} & s_2 &= \sigma_{22} \\
\dot{e}_3 &= \dot{\epsilon}_{33} & s_3 &= \sigma_{33} \\
\dot{e}_4 &= \sqrt{2}\dot{\epsilon}_{23} & s_4 &= \sqrt{2}\sigma_{23} \\
\dot{e}_5 &= \sqrt{2}\dot{\epsilon}_{31} & s_5 &= \sqrt{2}\sigma_{31} \\
\dot{e}_6 &= \sqrt{2}\dot{\epsilon}_{12} & s_6 &= \sqrt{2}\sigma_{12}
\end{aligned}\right\} \tag{4.18}$$

These conversion formulas present the advantage that they are identical for strain rates and for stresses and that the plastic power dissipation per unit volume is given by the scalar product of the stress and the strain rate vectors:

$$p = \sigma_{ij}\,\dot{\epsilon}_{ij} = s_p\,\dot{e}_p \tag{4.19}$$

p is an index that runs from 1 to 6. Two vectors are perpendicular when their scalar product is zero. Note that all vectors $\dot{\mathbf{e}}$ that represent strain rates are contained by the π-plane, since $\dot{e}_1 + \dot{e}_2 + \dot{e}_3 = 0$ (the volume remains constant during plastic deformation).

Application of the normality rule on the intersection of the yield locus of a single crystal with the π-plane leads to the following cases:

- If the stress is on one of the facets (s_1 or s_2 in Fig. 4.8), the strain rate vector is normal to this facet ($\dot{e}_1$ in Fig. 4.8). One slip system is active.
- If the stress corresponds to a vertex (s_3 in Fig. 4.8), the direction of the strain rate vector must be on, or contained by, the pyramid formed by the normals on the facets that intersect in the vertex ($\dot{e}_2$ and $\dot{e}_3$ in Fig. 4.8). The number of active slip systems is equal to the number of facets that intersect in the vertex.
- If the stress belongs to a "hyper-edge" (a subspace that is the intersection of two to four facets), then the direction of the strain rate vector must also be on or contained by the pyramid formed by the normals on the facets involved.

One can also reverse the question: suppose that a strain rate vector $\dot{\mathbf{e}}$ is chosen in six-dimensional space. What would then be the plastic stress vector? It is seen that it is rather improbable that the strain rate vector would be normal to one of the facets, or even to one of the subspaces that correspond to intersections of less than five facets. (The probability that this would happen is zero if the strain rate vector is arbitrarily chosen.) Only the vertices are left. The plastic stress can then be found by trying out all vertices. Let $\mathbf{s}^{\mathrm{V}}$ be the stress that corresponds to a vertex. For each vertex, one calculates the virtual plastic power dissipation per unit volume:

$$p^{\mathrm{V}} = \dot{\mathbf{e}} \cdot \mathbf{s}^{\mathrm{V}} = \dot{e}_p\, s_p^{\mathrm{V}} \tag{4.20}$$

The vertex for which p^{V} is maximum will correspond to the correct plastic stress according to the maximum work principle. This procedure is called the *Bishop–Hill method* (Bishop and Hill 1951). It can easily be carried out if a precalculated list of vertex stresses is available. There are 56 vertices in the case of fcc metals (Bishop and Hill 1951). The method also identifies the active slip systems. Note that the stress that is found does not depend on the *length* of the prescribed strain rate vector, but only on its *direction*.

The Taylor method is an alternative way of solving the same problem. Let $\dot{\epsilon}_{ij}$ be the strain rate tensor that corresponds to $\dot{\mathbf{e}}$. So $\dot{\mathbf{e}}$ and $\dot{\epsilon}_{ij}$ are prescribed. Eq. (4.11) can then be seen as a set of equations in which the slip rates $\dot{\gamma}_s$ are unknown. It is convenient to rewrite these equations as follows:

$$\dot{\epsilon}_{ij}/\dot{\epsilon}_0 = M_{sij}\,(\dot{\gamma}_s/\dot{\epsilon}_0) \tag{4.21}$$

in which $\dot{\epsilon}_0$ is some scalar measure of the magnitude of the local strain rate tensor $\dot{\epsilon}_{ij}$. It could be defined as the von Mises equivalent strain rate (but this is not necessary):

$$\dot{\epsilon}_{\mathrm{vM}} = \left(\frac{2}{3}\,\dot{\epsilon}_{ij}\,\dot{\epsilon}_{ij}\right)^{1/2} = \left(\frac{2}{3}\,\dot{\mathrm{e}}_p\,\dot{\mathrm{e}}_p\right)^{1/2} \tag{4.22}$$

Eq. (4.21) is a set of linear equations in the unknowns $(\dot{\gamma}_s/\dot{\epsilon}_0)$. The number of unknowns is the number of slip systems. The values that will be found for them will only depend on the ratios $\dot{\epsilon}_{ij}/\dot{\epsilon}_0$, since only these appear in the equations. In other words: the solutions will be functions of the strain mode (= direction of the strain rate vector), not of the magnitude of the strain rate (= length of the strain rate vector). Eq. (4.22) represents only five independent equations, because the tensors involved are symmetrical, and their traces are zero. In general there will be more unknowns than equations. So Taylor (1938) introduced an additional criterion to choose among the multiple solutions:

$$P^* = \tau_s^{\mathrm{c}}\,|\dot{\gamma}_s| = \mathrm{Min} \tag{4.23a}$$

which is equivalent to

$$P^*/\dot{\epsilon}_0 = \tau_s^{\mathrm{c}}\,|\dot{\gamma}_s/\dot{\epsilon}_0| = \mathrm{Min} \tag{4.23b}$$

This means that the frictional power dissipation in the slip systems must be minimal. The combined equations (4.21) and (4.23b) can be solved efficiently by means of linear programming (Van Houtte 1988). The *simplex multipliers* that are generated during this procedure give the plastic stress. Compared to the Bishop–Hill method, this method has the advantage that a precalculated list of the vertices of the single crystal is not required. It has been demonstrated by several authors (see e.g. Bishop and Hill 1951 and Van Houtte 1988) that the Bishop–Hill method and the Taylor method are strictly equivalent to each other; they can be regarded as a primal-dual set of linear programming problems.

4.3.3.1 Taylor Factor

The Taylor method and the Bishop–Hill method both allow the calculation of P, the plastic power dissipation per unit volume (P^{V} or P^*, see Eqs (4.20) and (4.23b)), and of the plastic stress. The Taylor factor M can be defined where the same value τ^{c} is adopted for the critical resolved shear stresses on all slip systems:

$$M = P/(\dot{\epsilon}_0\,\tau^{\mathrm{c}}) = (\sigma'_{kl}/\tau^{\mathrm{c}})\,(\dot{\epsilon}_{kl}/\dot{\epsilon}_0) \tag{4.24}$$

This factor is a function of the strain mode ($\dot{\epsilon}_{ij}/\dot{\epsilon}_0$), not of the magnitude of the strain rate $\dot{\epsilon}_0$.

4.3.4 Strain Rate Sensitivity

In the present context, a crystal would be said to have a nonzero strain rate sensitivity when the plastic stress would not only depend on the strain *mode*, but also on the *magnitude* of the local strain rate. It is sometimes observed that the stress (measured in a tensile test for example) depends on the plastic strain rate. An approximation of such behaviour would be (neglecting the elastic range):

$$\sigma = K\,\dot{\epsilon}^m \tag{4.25}$$

σ and $\dot{\epsilon}$ are uniaxial data in this equation. K is a factor that comprises all other influences. m is the strain rate sensitivity exponent. At room temperature, it has a very low value for most metals: of the order of 0.01. Eq. (4.25) then represents a behaviour that is very close to a rigid–plastic behaviour. In the hot deformation range, the value of m may reach values such as 0.1 to 0.3.

Some authors (see e.g. Toth et al. 1988 or Molinari et al. 1987) have proposed a model that assumes a similar strain rate sensitivity for each individual slip system. It is assumed that each slip system is active, regardless of the level of the resolved shear stress τ_s^r acting on it. There is a relation between the latter and the slip rate $\dot{\gamma}_s$:

$$\tau_s^r = \tau_0 \operatorname{sgn}(\dot{\gamma}_s) \left|\frac{\dot{\gamma}_s}{\dot{\gamma}_0}\right|^m \tag{4.26}$$

in which τ_0 and $\dot{\gamma}_0$ are constants. No summation over s is done in this equation. Note that this model assumes that the resolved shear stress acting on a particular slip system is only related to the slip rate on the system itself, and not to those on other slip systems. The advocates of the method do not give a physical justification of this assumption.

Substituting τ_s^r by its value (Eq. (4.13)) and inverting Eq. (4.26) leads to

$$\dot{\gamma}_s/\dot{\epsilon}_0 = \operatorname{sgn}(M_{skl}\,\sigma'_{kl})\,|M_{sij}\,(\sigma'_{ij}/\tau_0)\,(\dot{\gamma}_0/\dot{\epsilon}_0)^m|^{1/m} \tag{4.27}$$

(no summation over s). In Eq. (4.21), $\dot{\gamma}_s/\dot{\epsilon}_0$ can be substituted by this expression. A set of five nonlinear equations in the components of the deviatoric stress tensor σ'_{ij} is then obtained. It can be solved by numerical techniques (Leffers et al. 1988). The structure of Eqs (4.21) and (4.27) shows, that the solution may be written as follows:

$$\sigma'_{ij} = \tau_0\,(\dot{\epsilon}_0/\dot{\gamma}_0)^m\,W_{ij} \tag{4.28}$$

in which W_{ij} is some function of the strain mode $\dot{\epsilon}_{kl}/\dot{\epsilon}_0$. This shows clearly that the plastic stress resulting from this model depends on the magnitude of the strain rate $\dot{\epsilon}_0$. Note also that the concept of a yield locus does not exist in the case where strain rate sensitivity is adopted. But it is still possible to construct a "pseudo-yield locus" valid for all strain rates with the same magnitude $\dot{\epsilon}_0$. The size of this surface is proportional to $\dot{\epsilon}_0^m$; its shape is unaffected by $\dot{\epsilon}_0$. For small values of m such as 0.01, these surfaces will, however, all be close together and they will be quite close to the conventional Bishop–Hill yield locus; they will not have sharp vertices, but rounded ones with a small radius of curvature.

It is possible to define a "pseudo-Taylor factor" as follows:

$$M = P/(\dot{\epsilon}_0 \tau^c) = (\sigma'_{kl}/\tau^c)\,(\dot{\epsilon}_{kl}/\dot{\epsilon}_0)$$
$$= W_{kl}\,(\dot{\epsilon}_{kl}/\dot{\epsilon}_0) \tag{4.29}$$

in which

$$\tau^c = \tau_0\,(\dot{\epsilon}_0/\dot{\gamma}_0)^m \tag{4.30}$$

is a "pseudo-critical resolved shear stress". M is a function of the strain mode $\dot{\epsilon}_{ij}/\dot{\epsilon}_0$, but not of the magnitude of the strain rate $\dot{\epsilon}_0$.

4.4 Anisotropic Polycrystalline Materials

Many engineering materials are polycrystalline: they consist of a very large number of microscopic crystallites. This is the case for metals. The plastic anisotropy of such materials will be discussed in this section. Unless stated otherwise, it will be assumed that only one phase is present.

4.4.1 Crystallographic Texture

The crystallites of a polycrystal do not all have the same crystallographic orientation with respect to the external axes of the body that is considered. If all crystal orientations were equally present, then the crystalline anisotropy would be "smeared out" and the material as a whole would be isotropic. But such cases are rare. The usual case is that some crystal orientations occur more than others; some may be not present at all. It is then said that the material has a crystallographic texture; some or all of its physical properties will be anisotropic.

It would be beyond the scope of this chapter to give an extensive treatise on crystallographic textures. We refer to the specialized literature for this (Hatherly and Hutchinson 1979, Bunge 1982). Some concepts need, however, to be introduced in order to make the analysis below understandable.

4.4.1.1 Measurement of Textures

There are various experimental techniques for studying crystallographic textures (Hatherly and Hutchinson 1979). The most widely spread is the use of a *texture goniometer*. This is a variant of an X-ray diffractometer that is able to "see" the presence of crystal lattice planes with a certain lattice spacing d that are perpendicular to a particular direction **y**. **y** is defined with respect to the external axes of a sample (Fig. 4.11). The use of monochromatic radiation makes it possible to select the type of diffracting planes by adjusting the Bragg angle. Suppose that the $\{1\ 1\ 1\}$ planes are selected (for fcc metals). The instrument will then produce a signal that after some simple processing is proportional to the volume fraction of the crystallites that happen to have a $\{1\ 1\ 1\}$ plane perpendicular to the direction **y**. The value of that signal is then marked on the point that

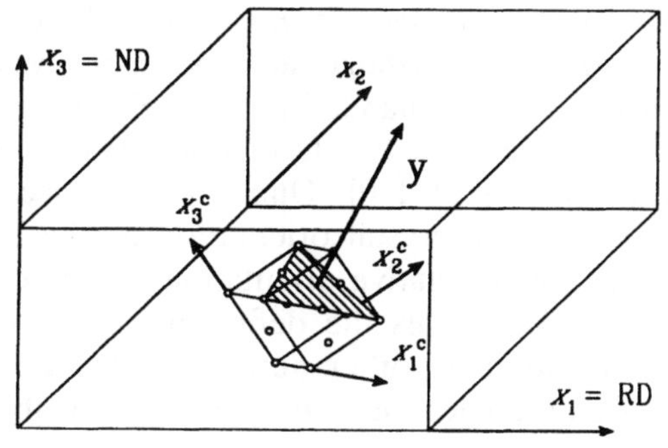

Fig. 4.11. The external reference system of a rolled sheet: x_1 = rolling direction; x_2 = transverse direction; x_3 = normal direction. x_1^c, x_2^c and x_3^c are the ⟨1 0 0⟩ axes of one of its crystallites. **y** is a vector normal to a diffracting plane.

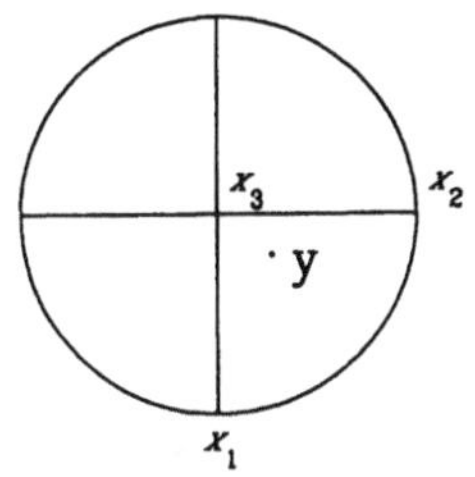

Fig. 4.12. The direction **y** of Fig. 4.11 can be represented by a point in a pole figure, which is a stereographic projection in the Ox_1x_2 plane.

represents **y** in a stereographic projection, called a *pole figure* (Fig. 4.12). This is done for all possible directions **y**; a complete function $P_{111}(\mathbf{y})$ is obtained. It is usually represented by means of contour lines in a pole figure. Fig. 4.13 shows examples of the {1 1 0} and {2 0 0} pole figures of a rolled steel sample. Functions like this ($P_{hkl}(\mathbf{y})$ in general) must be constant for materials without texture, for which the volume fraction of all crystallite orientations is equal.

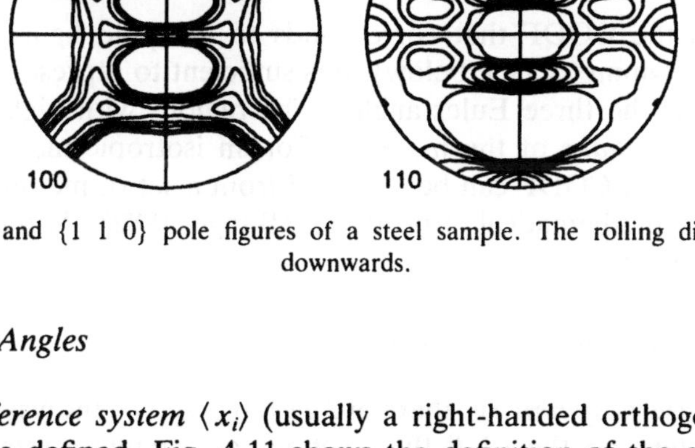

Fig. 4.13. {2 0 0} and {1 1 0} pole figures of a steel sample. The rolling direction is pointing downwards.

4.4.1.2 Euler Angles

An *external reference system* ⟨x_i⟩ (usually a right-handed orthogonal coordinate system) must be defined. Fig. 4.11 shows the definition of the axes in the case of a rolled sheet. A *crystal reference system* $\langle x_i^c \rangle$ must be chosen as well for each crystallite. It must also be an orthogonal coordinate system. If $\langle x_i \rangle$ is right-handed, then $\langle x_i^c \rangle$ has to be right-handed as well. For cubic crystals, the $\langle 1\ 0\ 0 \rangle$ directions are used to define $\langle x_i^c \rangle$. For hexagonal crystals, one could choose x_3^c to be $[0\ 0\ 0\ 1]$ and x_1^c to be $[2\ \bar{1}\ \bar{1}\ 0]$. Once these choices have been made, it becomes possible to define the crystal orientation as the relative orientation of $\langle x_i^c \rangle$ with respect to $\langle x_i \rangle$. Such relative orientation can be described by means of three Euler angles. Fig. 4.14 gives the definition of the Euler angles that are used by Bunge (1982) and most other authors in the field of crystallographic texture. The set of Euler angles (ϕ_1, Φ, ϕ_2) is often abbreviated to a single symbol g.

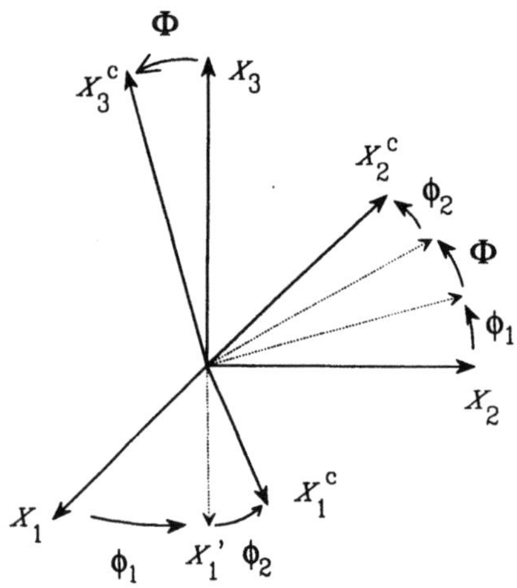

Fig. 4.14. Definition of the Euler angles that rotate $\langle x_i \rangle$ into $\langle x_i^c \rangle$. The first rotation is over ϕ_1 from x_1. The second is over Φ from x_1'. The third is over ϕ_2 from x_3^c.

4.4.1.3 *Crystallite Orientation Distribution Function (CODF)*

The CODF $f(g)$ of a single phase polycrystalline material is the statistical distribution function that gives the volume fraction $dV(g)/V$ of the crystallites that have a particular crystal orientation within dg around g:

$$dV(g)/V = f(g)\, dg \tag{4.31}$$

in which (Bunge 1982)

$$dg = (\sin\Phi\, d\phi_1\, d\Phi\, d\phi_2)/(8\pi^2) \tag{4.32}$$

Fig. 4.15 shows the CODF that corresponds to the pole figures given by Fig. 4.13. Because of symmetry (see below), it is sufficient to represent only the range from 0° to 90° for the three Euler angles. The regions with high values indicate the *preferred orientations* of the material. For an isotropic material, $f(g)$ would be 1 for all g. Such a CODF can be obtained from a set of measured pole figures by using certain mathematical procedures (Bunge 1982). Software for this is commercially available.

4.4.1.4 *Crystal and Sample Symmetry*

Because of the *crystal symmetry*, there are 24 fully equivalent ways to put a right-handed $\langle x_i^c \rangle$ system on the $\langle 1\ 0\ 0 \rangle$ directions of a cubic crystal. Accordingly,

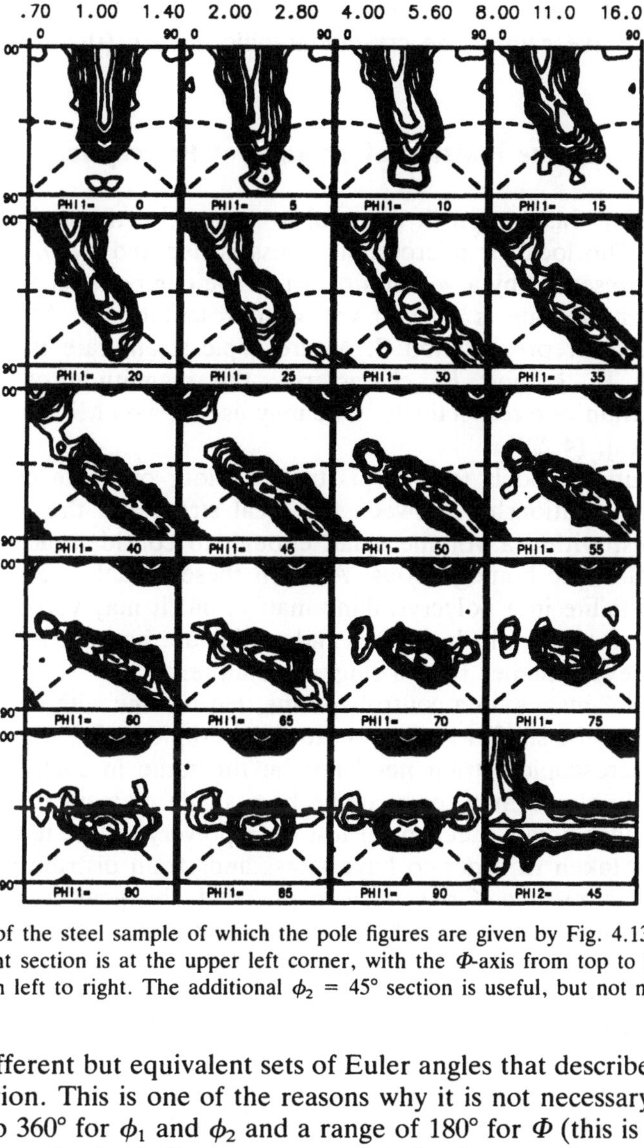

Fig. 4.15. CODF of the steel sample of which the pole figures are given by Fig. 4.13. The origin of each ϕ_1 = constant section is at the upper left corner, with the Φ-axis from top to bottom and the ϕ_2-axis from left to right. The additional $\phi_2 = 45°$ section is useful, but not necessary.

there are 24 different but equivalent sets of Euler angles that describe a particular crystal orientation. This is one of the reasons why it is not necessary to consider a range of 0° to 360° for ϕ_1 and ϕ_2 and a range of 180° for Φ (this is theoretically the maximal range). A similar property exists for hexagonal metals. Another reduction of Euler space is due to the *sample symmetry*. It is often observed that the physical properties of a polycrystalline sample have some symmetry although being anisotropic.

The properties of a rolled sheet, for example, usually do not change after a 180° rotation around the rolling direction, the transverse direction or the normal direction. Such symmetry is called *orthorhombic sample symmetry*. The pole figure is one of the physical properties; so it must reflect these symmetries. The fact that the normal direction is a twofold axis of rotation means that the centre of the pole figure (Fig. 4.13) is a centre of symmetry. The fact that the rolling direction is a twofold axis of rotation means that the trace of the plane that is perpendicular to it (the *horizontal* direction in Fig. 4.13) acts as a symmetry axis of the drawing. Similarly for the transverse direction. This type of symmetry is *not* due to crystal symmetry, but is a property of the statistical distribution of the crystal orientations (CODF). Its cause is to be sought in the symmetry of the processes that have caused the crystallographic texture (the rolling process for example).

4.4.2 Models for the Plastic Deformation of Polycrystals

Before beginning this section, it is useful to summarize the various symbols that will be used. The local or microscopic stress is denoted by σ_{ij}; in stress–strain space, it is represented by a vector **s** of which the components are given by Eqs. (4.18). Local strain rate is $\dot{\epsilon}_{ij}$ with vector representation $\dot{\mathbf{e}}$. Macroscopic stress is Σ_{ij} with vector representation **S**. Macroscopic strain rate is $\dot{E}_{ij}$ with vector representation $\dot{\mathbf{E}}$. $\dot{\epsilon}_0$ and $\dot{E}_0$ characterize the magnitude of the local and macroscopic strain rate respectively. One may use the von Mises equivalent strain rate for this (Eq. (4.22)).

The constitutive models for the plastic behaviour of single crystals discussed above give the relationship between the local stress and the local strain rate. "Local" means: within a volume small enough to consider the distributions of stress and strain rate homogeneous. As such these models are still valid for an individual crystallite in a polycrystalline matrix; but it may very well be possible that the volume within which stresses and strains are homogeneously distributed is very small, even smaller than a single crystallite.

"Macroscopic" means: in a control volume that is large with respect to a single crystallite. It may be small with respect to the workpiece as a whole. In the latter case, the "macroscopic" strain need not be the same in each control volume. Such a control volume may for example be a single element in a finite-element analysis. Consider now a mechanical test on a polycrystalline test piece in which special care is taken in order to have stress and strain distributions that are as homogeneous as possible. On a macroscopic level, this goal may very well seem achieved; on a microscopic level however, the distribution of stresses and strains will be inhomogeneous. Sachs (1928) proposed to use an approximation in which a homogeneous distribution of the *local stresses* is assumed. This model proved to be quite unsatisfactory because it strongly violates the requirement of geometrical compatibility at the grain boundaries (see e.g. Gil Sevillano et al. 1980). Most authors prefer the theory of Taylor (1938) who assumed that the *local strains* are homogeneously distributed. This is believed to be a much better approximation of the behaviour of a real material. According to this theory, the local velocity gradient $\dot{f}_{ij}$ is equal to the macroscopic velocity gradient $\dot{F}_{ij}$:

$$\dot{f}_{ij} = \dot{F}_{ij} \tag{4.33}$$

The symmetric part of this equation leads to a similar relationship between the local strain rate $\dot{\epsilon}_{ij}$ and the macroscopic strain rate $\dot{E}_{ij}$:

$$\dot{\epsilon}_{ij} = \dot{E}_{ij} \tag{4.34}$$

So the Taylor theory leads to the knowledge of the local strain rate. The latter can be used to calculate the local stress σ_{ij} using one of the methods described above (Bishop-Hill, Taylor or strain rate sensitivity method; Eqs (4.11)–(4.30)). Before doing so, it is best to convert $\dot{\epsilon}_{ij}$ into $\dot{\epsilon}^c_{ij}$, i.e. to express the local strain rate tensor with respect to the crystal reference system. (The formula for the transformation matrix has for example been published by Gil Sevillano et al.

1980 and Bunge 1982.) In that case, the factors M_{sij} used in Eqs (4.11)–(4.30) can be calculated once and for all since they are constants in the crystal reference system. This means that although the local strain rate tensor is constant, its representation $\dot{\epsilon}_{ij}^{c}$ will depend on the crystal orientation g and hence be different for each crystallite. The local stress is also obtained in the crystal reference system (σ_{ij}^{c}); it can be converted back to the external reference system (σ_{ij}). There is no reason why σ_{ij} would be the same for all crystallites; in other words, it will be a function of g. This means that the Taylor theory does not satisfy the stress equilibrium conditions at the grain boundaries.

Several new theories have been proposed that try to overcome this problem. The most important among these are the *relaxed constraints theories* and the *self-consistent models* (see e.g. Eqs (4.17)–(4.18)). They abandon Taylor's strict assumption of homogeneous strain, an assumption that achieves perfect geometrical compatibility at the cost of the stress equilibrium. However, for many practical applications the Taylor theory is still a sufficiently good model (Leffers et al. 1988). To avoid unnecessary complexity, only the Taylor theory will be used in the rest of this chapter.

The Taylor theory makes it possible to calculate a Taylor factor for an individual crystallite with an orientation g when the macroscopic strain rate tensor is known. Eq. (4.34) gives the microscopic strain rate $\dot{\epsilon}_{ij}$; it becomes $\dot{\epsilon}_{ij}^{c}$ after re-expression with respect to the reference system $\langle x_{ij}^{c} \rangle$. $\dot{\epsilon}_{ij}^{c}$ depends on g. It allows the calculation of the Taylor factor M using either the Taylor, Bishop–Hill or the strain-rate sensitivity method (Eqs (4.24) or (4.29)). In such context, M will be a function of the strain mode, essentially a direction in strain space. We choose to describe this direction by the ratio $\dot{\mathbf{E}}/\dot{E}_0$ which is a vector of which the length does not depend on the magnitude of the strain rate. M is also a function of the crystallite orientation g. The equations show that the plastic power dissipation per unit volume in a crystallite with orientation g is given by

$$P(\dot{\mathbf{E}},g) = \dot{E}_0 \, \tau^{c} \, M(\dot{\mathbf{E}}/\dot{E}_0,g) \tag{4.35}$$

The equation shows that for a particular strain mode $\dot{\mathbf{E}}/\dot{E}_0$ and strain rate $\dot{E}_0$, the power needed for the plastic deformation of crystallites with high Taylor factors will be larger than for crystallites with a low Taylor factor. The crystal orientations of the former are called the *hard orientations*, of the latter the *soft orientations*. In the case of rolling of fcc metals, the Taylor factor of the hardest grains is about 2.5 times higher than those of the softest grains. This ratio can be much higher for hexagonal metals. The Taylor factor in the above equation is a purely geometrical factor, representing the influence of the crystallographic anisotropy. The factor τ^{c} (the critical resolved shear stress in the case of Taylor's method or the Bishop–Hill method) is the physical factor, which must account for effects such as the Hall–Petch law, work hardening, temperature dependence, strain rate sensitivity (see e.g. Eq. (4.25)) etc.

The average plastic power dissipation per unit volume of the polycrystal as a whole is given by (Van Houtte 1987)

$$P_{\mathrm{m}}(\dot{\mathbf{E}}) = \int f(g) \, P_{\mathrm{m}} (\dot{\mathbf{E}},g) \, dg \tag{4.36a}$$

or

$$P_{\mathrm{m}}(\dot{\mathbf{E}}) = \dot{E}_0 \, \tau^{c} \, M_{\mathrm{m}} (\dot{\mathbf{E}}/\dot{E}_0) \tag{4.36b}$$

in which

$$M_m(\dot{\mathbf{E}}/\dot{E}_0) = \int f(g)\, M(\dot{\mathbf{E}}/\dot{E}_0, g)\, dg \tag{4.37}$$

is called the average Taylor factor. It depends on the crystallographic texture ($f(g)$ is the CODF). The series expansion method, also called the harmonic method (Bunge 1982), has allowed the development of extremely efficient computer codes that make it possible to calculate the average Taylor factors for thousands of different strain modes.

The expression for $P_m(\dot{\mathbf{E}})$ derived above (Eq. (4.36b)) assumed a value of τ^c which is independent of g. This assumption is in principle no longer valid after a certain plastic strain, since work hardening itself will be different for each crystallite. How this can be dealt with can be found in specialized papers (Gil Sevillano et al. 1980, Aernoudt et al. 1987).

4.4.3 Yield Loci of Polycrystalline Materials

Consider a polar representation $S(\mathbf{u})$ of a yield locus of a polycrystalline material (Fig. 4.16a). $\mathbf{u}$ is a unit vector in stress space that defines a direction. S is the length of the vector OA. The complete yield locus can be constructed if one finds a means for calculating the function $S(\mathbf{u})$. The stress vector (see Eq. (4.18)) that corresponds to a particular direction $\mathbf{u}$ would be given by

$$\mathbf{S} = S(\mathbf{u})\, \mathbf{u} \tag{4.38}$$

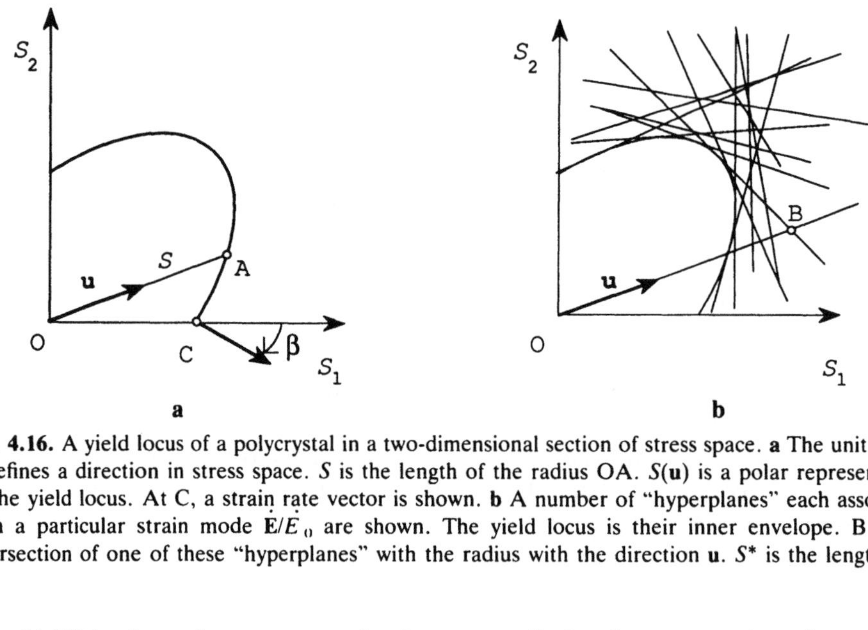

Fig. 4.16. A yield locus of a polycrystal in a two-dimensional section of stress space. **a** The unit vector **u** defines a direction in stress space. S is the length of the radius OA. $S(\mathbf{u})$ is a polar representation of the yield locus. At C, a strain rate vector is shown. **b** A number of "hyperplanes" each associated with a particular strain mode $\dot{\mathbf{E}}/\dot{E}_0$ are shown. The yield locus is their inner envelope. B is the intersection of one of these "hyperplanes" with the radius with the direction **u**. S^* is the length OB.

Eq. (4.36b) gives the average plastic power dissipation per unit volume as a function of the average Taylor factor. This leads to the following equation for the plastic stress **S**:

$$\mathbf{S} \cdot \dot{\mathbf{E}} = \dot{E}_0\, \tau^c\, M_m(\dot{\mathbf{E}}/\dot{E}_0) \tag{4.39}$$

In which $\dot{\mathbf{E}}$ with components $\dot{E}_i$ represents the macroscopic strain rate of the polycrystal. Assume for a moment that $\dot{\mathbf{E}}$ were known and **S** not. Then M_m can be calculated. All stresses **S*** that satisfy Eq. (4.39) constitute a hyperplane in stress space. The strain rate vector $\dot{\mathbf{E}}$ is perpendicular to it. The plastic stress **S** that corresponds to $\dot{\mathbf{E}}$ is a point of the hyperplane. Hence according to the normality rule, the yield locus must be tangent to the hyperplane at the location of **S**. Fig. (4.16b) shows several such hyperplanes. Since it must be convex, the yield locus can only be the inner envelope of all these hyperplanes. Let us now come back to the problem of finding S, the distance from the origin O to the yield locus in a given direction **u**. Fig. (4.16b) shows that S is the smallest of all nonnegative S^*, with S^* the length of OB, B being the intersection of a radius with direction **u** and one of the hyperplanes. It is a priori not known which hyperplane minimizes S^*. From Eqs (4.38) and (4.39),

$$S^*(\mathbf{u}) = \frac{\tau^c M_m(\dot{\mathbf{E}}/\dot{E}_0)}{\mathbf{u} \cdot (\dot{\mathbf{E}}/\dot{E}_0)} \qquad (4.40)$$

This expression can be evaluated for each hypothetical strain mode $\dot{\mathbf{E}}/\dot{E}_0$. The hyperplane that minimizes S^* is the one that corresponds to the strain mode that minimizes the right hand side of Eq. (4.40). Van Houtte (1987) has described a method that calculates M_m for all strain modes $\dot{\mathbf{E}}/\dot{E}_0$ in the π plane with a certain angular resolution. The method uses Eq. (4.37) and requires the knowledge of the CODF of the material. The set of values of M_m that is obtained is in itself a representation of the yield locus. Plane sections are constructed by minimization of S^* for a number of radii in the section. The minimization consists in calculating S^* for each of the individual M_m values that have been obtained and choosing the smallest among them. The points that are found for each radius are connected by straight line segments. Fig. 4.17 shows an example of such a yield locus for an 3004-aluminium alloy, using the {1 1 1} + ⟨1 1 0⟩ slip systems. This

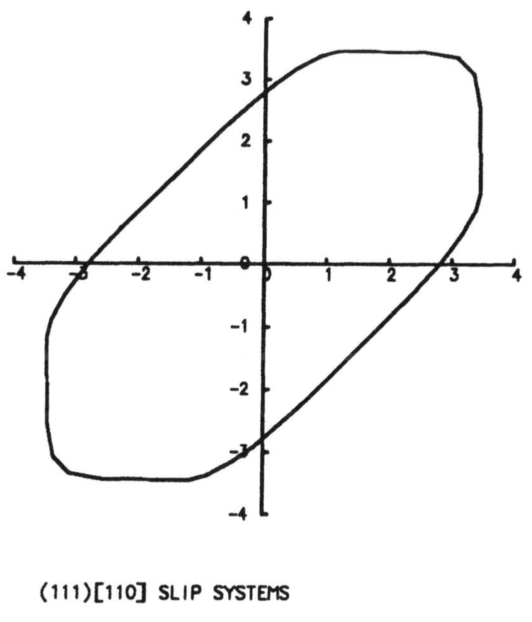

Fig. 4.17. Geometrical representation of a plane section of the yield locus of a rolled Al-3004 alloy sample. The horizontal axis shows σ_{11}/τ^c values, the vertical axis σ_{22}/τ^c values, in a reference system 45° rotated around x_3 (α = 45°, see Fig. 4.2).

"geometrical" method produces satisfactory graphical representations of two-dimensional sections of yield loci, but does not lead to an analytical expression of the yield locus which could be used in finite-element software (Van Houtte et al. 1989).

In order to check its validity, the geometrical method was applied on polycrystals with the following hypothetical texture:

$$f(g) = f_0 \exp[-(\psi/\psi_0)^2] \tag{4.41}$$

in which ψ represents the angular distance between g and the crystal orientation (0 0 1) [1 0 0], for which the ⟨1 0 0⟩ axes of the crystal coincide with the external system. f_0 and ψ_0 are constants. In the limit for $\psi_0 = 0$, this would be the "texture" of a (0 0 1) [1 0 0]-oriented single crystal. The yield locus has been calculated for $\psi_0 = 1°$ and $\psi_0 = 16.5°$. Figs 4.18 and 4.19 show the yield loci obtained for $\alpha = 45°$ (Fig. 4.2), to be compared to Fig. 4.9. The yield locus shown by the latter is a perfect square. Fig. 4.18 comes quite close to that. The broader dispersion of the texture rounds the corners off in Fig. 4.19.

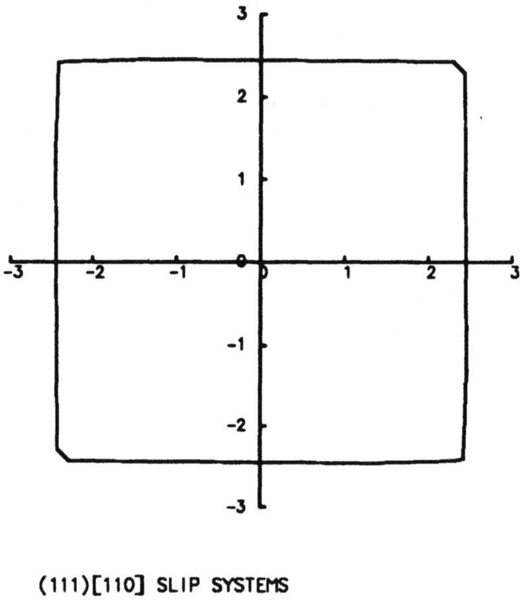

Fig. 4.18. Geometrical representation of a plane section of the yield locus of an fcc crystal with a hypothetical texture (Eq. (4.41) with $\psi_0 = 1°$). The horizontal axis shows σ'_{11}/τ^c values, the vertical axis σ'_{22}/τ^c values, in a reference system 45° rotated around x_3 ($\alpha = 45°$, see Fig. 4.2).

4.4.4 The Calculation of *r* Values from the Crystallographic Texture

The r value of a textured polycrystal can be estimated by calculating its instantaneous counterpart ρ. Eq. (4.40) is used to this purpose. The r value is a result of a uniaxial tensile test so the components of **u** are {1, 0, 0, 0, 0, 0}. Such a loading path will cause a strain rate which is a priori unknown. For simplicity, it is assumed that no shear strains will be generated. The nonzero components of the strain rate tensor then would be

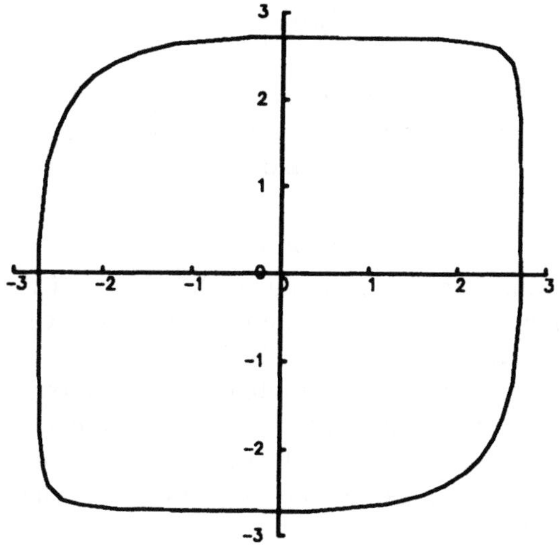

Fig. 4.19. As Fig. 4.18, ψ_0 = 16.5°.

$$\left.\begin{aligned} \dot{E}_{11} &= \dot{E}_1 = \dot{E}_0 \\ \dot{E}_{22} &= \dot{E}_2 = -q\dot{E}_0 \\ \dot{E}_{33} &= \dot{E}_3 = -(1-q)\dot{E}_0 \end{aligned}\right\} \tag{4.42}$$

The contraction coefficient q is a priori unknown. Eq. (4.3) shows, that the instantaneous plastic anisotropy ratio ρ is equal to

$$\rho = q/(1-q) \tag{4.43}$$

For this particular choice of $\dot{\mathbf{E}}$ and $\mathbf{u}$, the denominator of Eq. (4.40) becomes 1. The equation reduces to

$$S^* = \tau^c M_m(q) \tag{4.44}$$

Thus the minimum of S^* is found when the minimum of $M_m(q)$ is found. One could, for example, calculate $M_m(q)$ for 11 values of q between 0 and 1. The value of q that corresponds to the minimum then gives ρ by means of Eq. (4.43).

The value of q can also be obtained from a yield locus. C in Fig. 4.16a represents a uniaxial stress state. The vector that describes the strain mode is normal to the yield locus in C. q is equal to minus the ratio $\dot{E}_2/\dot{E}_1$, or minus $\tan\beta$. ρ can be obtained from q (Eq. (4.43)).

4.5 Analytical Expressions for Anisotropic Yield Loci

4.5.1 Hill's Yield Locus

The well-known von Mises isotropic yield locus is described by the following equation:

$$\tfrac{3}{2} S'_p S'_p = \sigma_0^2 \tag{4.45}$$

in which the vector S'_p represents the deviatoric part of the macroscopic stress tensor Σ_{ij}. σ_0 is the yield stress in a uniaxial tensile test. This is a particular quadratic equation. A more general quadratic equation would be

$$A_{pq} S'_p S'_q = 1 \tag{4.46}$$

Such a yield locus would not be invariant to the choice of the reference system, as Eq. (4.45) is. Thus it represents an anisotropic yield locus. Now suppose that the crystallographic texture is the only effect that causes anisotropy, and that it is characterized by an orthorhombic sample symmetry (see Sect. 4.4.1). Then Eq. (4.46) should at least be invariant to 180° rotations around the coordinate axes. This can be analysed and leads to certain conditions for the coefficients A_{pq}. As a result of this, Eq. (4.46) can be reduced to what is known as Hill's yield locus (Hill 1950):

$$\begin{aligned} F(S_2-S_3)^2 + G(S_3-S_1)^2 + H(S_1-S_2)^2 \\ + 2L\,S_4^2 + 2M\,S_5^2 + 2N\,S_6^2 = 1 \end{aligned} \tag{4.47}$$

For sheet metal forming applications, it is often assumed that there is *planar isotropy*, i.e. physical properties that are measured along directions in the plane of the sheet are independent of these directions. According to this assumption, r and ρ will be independent of α (Fig. 4.2). Another conclusion is that $F = G$, since the axes x_1 and x_2 are interchangeable (x_3 is the normal to the sheet). The associated flow rule leads to

$$\left.\begin{aligned} \dot{E}_2 &= 2\dot{\lambda}\,[F(S_2-S_3) - H(S_1-S_2)] \\ \dot{E}_3 &= 2\dot{\lambda}\,[F(2S_3-S_1-S_2)] \end{aligned}\right\} \tag{4.48}$$

The instantaneous anisotropy ratio ρ is equal to $\dot{E}_2/\dot{E}_3$ for the case of a uniaxial tensile test in the direction x_1. S_1 is the only stress component that is different from zero in such a test; Eqs (4.48) then lead to

$$\rho = \dot{E}_2/\dot{E}_3 = H/F \tag{4.49}$$

Suppose now that one considers only stress states for which $S_3 = S_4 = S_5 = 0$, which is a quite reasonable assumption in the case of a thin sheet. x_3 then is a principal axis of stress. The two other principal axes are in the plane of the sheet. Since planar isotropy is assumed, they can be chosen as x_1 and x_2 axes. This means that $S_6 = \Sigma_{12} = 0$. Hill's equation now reduces to (using Eq. (4.49))

$$S_1^2 + S_2^2 - [2\rho/(\rho+1)]\,S_1 S_2 = \sigma_0^2 \tag{4.50}$$

Fig. 4.20 shows the influence of the ρ value on the shape of a yield locus of this kind. It is seen that the long axis of the ellipse is strongly affected by ρ, not the short axis. This is used to explain the beneficial influence of a high ρ value on the drawability of cups (Backofen 1972).

The yield loci given by Eqs (4.48) and (4.50) did not prove very successful in describing the anisotropic behaviour of engineering materials. The yield loci of such materials cannot be described accurately by means of a quadratic expression. Even when they could, it would be quite difficult to determine the six coefficients

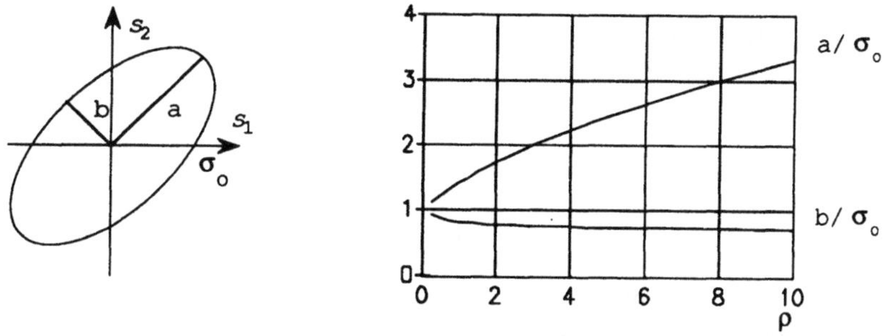

Fig. 4.20. The influence of the instantaneous anisotropy ratio ρ (of which r is an estimate) on the principal axes of a Hill-type yield locus assuming planar isotropy.

from the results of mechanical tests. Lequeu et al. (1987) proposed a method to derive the coefficients from the crystallographic texture of a material. These authors performed a least squares fit of a Hill-type yield locus to a single crystal yield locus expressed in the crystal reference system. Figs 4.8–4.10 show that such approximation can only be very crude. For each crystallite of the polycrystal, this approximate yield locus is then re-expressed in the external reference system. It is then attempted to combine the various yield loci of the individual crystallites to obtain a Hill-type yield locus for the polycrystal. For this operation, the authors propose either the Sachs method, which is quite questionable, or the Taylor method, which they do not explain in detail but describe as "complicated". In any case it is clear that this so-called "CMTP method" can only produce a quadratic approximation of an anisotropic yield locus.

4.5.2 Yield Loci Based on Analytical Fits of the Average Taylor Factors

Van Houtte et al (1989) proposed to use a "type II" plastic potential function to derive an analytical representation of an anisotropic yield locus. Such a plastic potential is written in terms of the strain rate vector $\dot{\mathbf{E}}$ as an independent variable:

$$Q(\dot{\mathbf{E}}) = \dot{E}_{\text{vM}}\, G(\dot{\mathbf{E}}/\dot{E}_{\text{vM}}) \tag{4.51}$$

$\dot{E}_{\text{vM}}$ is the von Mises equivalent strain rate calculated from the macroscopic strain rate tensor (Eq. (4.22)); it also depends on $\dot{\mathbf{E}}$. G is a function of the strain mode only. In order to simplify the formulas, the following variables are introduced:

$$V_p = \dot{E}_p/\dot{E}_{\text{vM}} \tag{4.52}$$

The components of the stress vector are derived as follows:

$$S_p = \tau^{\text{c}} \frac{\partial Q(\dot{\mathbf{E}})}{\partial \dot{E}_p} \tag{4.53}$$

which can be elaborated further:

$$S_p/\tau^{\text{c}} = \tfrac{2}{3} V_p G + \frac{\partial G}{\partial V_q}[\delta_{pq} - \tfrac{2}{3} V_p V_q] \tag{4.54}$$

The right-hand side of this expression only depends on the strain mode. Strain-rate sensitivity can be brought in by making τ^{c} a function of $\dot{E}_{\text{vM}}$ (see for example Eq. (4.30) with $\dot{E}_{vM}$ as a particular choice for $\dot{\epsilon}_0$). Eqs (4.51)–(4.54) can be used to demonstrate the following property:

$$\dot{E}_p S_p = \tau^c Q \tag{4.55a}$$

or

$$\dot{E}_p S_p = \tau^c \dot{E}_{vM} G(\dot{\mathbf{E}}/\dot{E}_{vM}) \tag{4.55b}$$

The left-hand side is equal to the plastic power dissipation per unit volume. In the case of a polycrystalline material, comparison of Eq. (4.55b) and Eq. (4.36b) shows that G can be identified with the average Taylor factor M_m.

So far, Eq. (4.53) seems a parametric description of a surface in stress space, in which the components $\dot{E}_p$ are the free parameters. Such a surface can however only be a yield locus when the normality rule is satisfied. This will be demonstrated on the basis of Eq. (4.40), which expresses the normality rule.[1] Suppose that one has used Eq. (4.53) to find the stress vector S_p^0 that is associated to a particular strain-rate tensor $\dot{E}_p^0$. **u** can be calculated from this stress:

$$u_p = S_p^0/S^0 \text{ with } S^0 = (S_q^0 S_q^0)^{1/2} \tag{4.56}$$

Let us now reverse the problem. Consider the u_p as fixed, and the $\dot{E}_p$ as variables. It must be shown that $\dot{E}_p^0$ is the value for these variables for which S^* of Eq. (4.40) is minimal. That means that the partial derivatives of S^* to $\dot{E}_p$ must take the value zero for $\dot{E}_p = \dot{E}_p^0$. These partial derivatives are

$$\frac{\partial S^*}{\partial \dot{E}_p} = \left[\frac{\partial(\tau^c \dot{E}_{vM} M_m)}{\partial \dot{E}_p}(u_q \dot{E}_q) - (\tau^c \dot{E}_{vM} M_m) u_p\right] \Big/ (u_q \dot{E}_q)^2 \tag{4.57}$$

For $\dot{E}_p = \dot{E}_p^0$, the condition that they must be zero reduces to

$$\frac{\partial(\tau^c \dot{E}_{vM} M_m)}{\partial \dot{E}_p}(u_q \dot{E}_q^0) - (\tau^c \dot{E}_{vM} M_m) u_p = 0 \tag{4.58}$$

or, using Eqs (4.51), (4.53) and (4.56), knowing that $G = M_m$ and assuming that τ^c does not depend on $\dot{E}_p$:[2]

$$S_p^0 [(S_q^0 \dot{E}_q^0) - \tau^c \dot{E}_{vM} G]/S^0 = 0 \tag{4.59}$$

It is seen from Eq. (4.55b) that the left-hand side is indeed zero. So the formalism of Eqs (4.51)–(4.53) leads to a surface in stress space that can be regarded as a yield locus.

[1]Eq. (4.40) was derived for polycrystalline materials; the demonstration can, however, be generalized to any material for which the average plastic power dissipation per unit volume can be written by an expression which has the form of Eq. (4.36b), i.e. a product of some function of $\dot{E}_{vM}$ with some function of the strain mode.

[2]In the case where τ^c depends on $\dot{E}_{vM}$ (see for example Eq. (4.30), with $\dot{E}_0 = \dot{E}_{vM}$), a true yield locus does not exist, although a pseudo yield locus can be defined for constant values of $\dot{E}_{vM}$ (see Sect. 4.3.4). In such a case one has to choose a particular value for $\dot{E}_{vM}$. This fixes τ^c and along with it the position of the hyperplanes given by Eq. (4.39). The condition that the partial derivatives of Eq. (4.57) must be zero merely serves the purpose of choosing between these hyperplanes. So it can be calculated assuming that τ^c is a constant.

The formalism will now be used to obtain an approximate analytical expression for the yield locus. Let G for example be given by a sixth-order series expansion of the variables V_p (Eq. (4.52)), which describe the strain mode:

$$G = a_{pq} V_p V_q + b_{pqrs} V_p V_q V_r V_s + c_{pqrstu} V_p V_q V_r V_s V_t V_u \tag{4.60}$$

In the example, the series expansion is even, which means that the origin of stress space will be a centre of symmetry of the yield locus.

The coefficients of an expression such as Eq. (4.60)) can be found by fitting G to a set of values of the average Taylor factor M_m, calculated from the CODF of a textured polycrystal as explained above, with the restriction that $\dot{E}_{vM}$ must be the conventional choice for $\dot{\epsilon}_0$.

It is highly recommended in this kind of work to do as much as one can to reduce the number of coefficients to be determined. A first step would be to use a different basis in stress–strain space. In the original basis as defined by the Eqs (4.18), the basis vectors $\mathbf{u}_1$, $\mathbf{u}_2$ and $\mathbf{u}_3$ lie on the coordinate axes for s_1, s_2 and s_3. They should be replaced by a set of basis vectors $\mathbf{u}_1'$, $\mathbf{u}_2'$ and $\mathbf{u}_3'$ of which the first two are in the π-plane and perpendicular to each other, while the third is normal to the π-plane. $\mathbf{u}_4$, $\mathbf{u}_5$ and $\mathbf{u}_6$ need not be changed. Suppose that plastic strain rate and strain mode vectors (V_p) are expressed in this new basis. Their third component will always be zero, since they all lie in the π-plane (as long as it is assumed that the volume remains constant). In Eq. (4.60), these third components are always multiplied with constants of which one of the indices is 3. As a result, the values of such coefficients do not matter; a zero value can be assigned to them. The third component of the stress vector (component normal to the π-plane, i.e. the hydrostatic stress) that results from Eq. (4.54) will then always be zero; so all stresses obtained will be deviatoric stresses. In other words, the change of basis proposed above directly leads to the intersection of the yield locus with the π-plane. This procedure, combined with the exploitation of the intrinsic symmetry of the coefficient matrices, allows a reduction of the number of independent coefficients in Eq. (4.60) to 296. This number can be reduced further if sample symmetry is present.

Once the series expansion is known, the expressions Eq. (4.54) can be used to find the plastic stress that corresponds to a given strain mode. The method is very fast, since based on analytical expressions, and is suitable for use in finite-element codes (Pillinger et al. 1989). The inverse problem: finding the strain mode when the stress mode $\mathbf{u}$ is known, is solved by minimizing S^* in Eq. (4.40). This is done on the basis of the analytical expression for G, taken as an approximation for M_m. The minimum is found by means of a pseudo-Newtonian algorithm. This method is also useful in finite-element codes, or can be used to construct plane sections of yield loci.

Fig. 4.21 shows the yield locus obtained with a fourth-order series expansion for G, for the same case as in Fig. 4.19. A yield locus such as shown by Fig. 4.18 cannot be reproduced properly by this method; the texture is too sharp. Fig. 4.22 shows an example of a yield locus of a polycrystalline sample (a cold rolled Al 3004-alloy). The reproduction (using a sixth-order series expansion) is excellent.

It can be concluded that analytical yield loci based on average Taylor factors can be obtained for polycrystalline materials of which the CODF is known. They make it possible to take plastic anisotropy into account in finite-element calculations without increasing the calculation time too much.

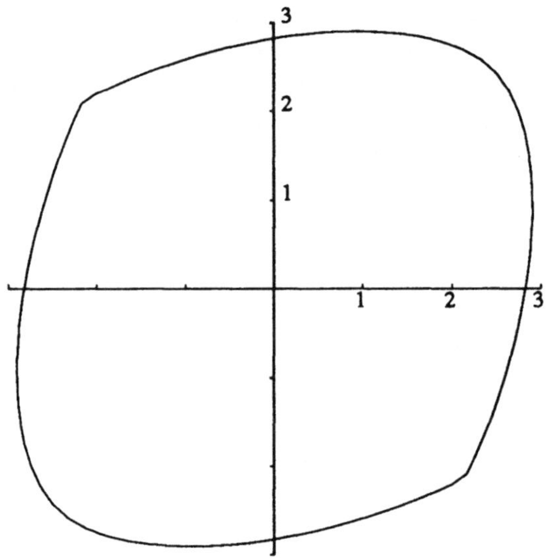

Fig. 4.21. As Fig. 4.19, but using an analytical approximation based on a fourth-order series expansion.

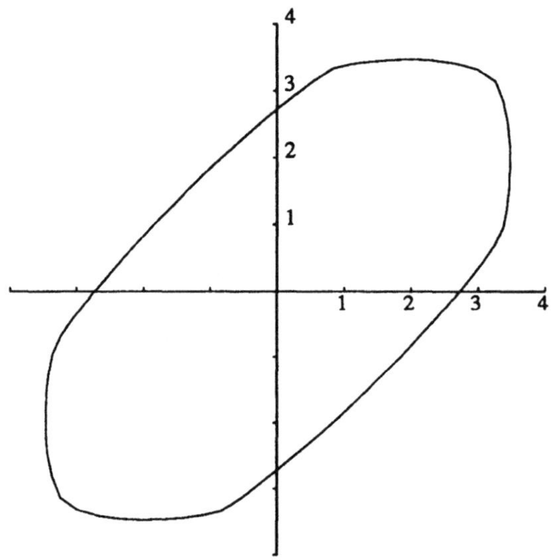

Fig. 4.22. Same as Fig. 4.17, but using an analytical approximation based on a sixth-order series expansion.

References

Aernoudt E, Gil Sevillano J, Van Houtte P (1987) Structural background of yield and flow. In: Andersen SI et al. (eds) Constitutive relations and their physical basis (Proceedings 8th Risö International Symposium on Metallurgy and Materials Science). Risö National Laboratory, Roskilde, Denmark pp. 1–38

Backofen WA (1972) Deformation processing. Addison-Wesley, Reading, Massachussetts

Bishop JFW, Hill R (1951) A theoretical derivation of the plastic properties of a polycrystalline face-centred metal. Phil Mag 42:1298–1307

Bunge HJ (1982) Texture analysis in materials science. Butterworths, London

Gil Sevillano J, Van Houtte P, Aernoudt E (1980) Large strain work hardening and textures. Prog Mater Sci 25:69–412

Hatherly M, Hutchinson WB (1979) Introduction to textures in metals. Institution of Metallurgists, London

Hill R (1950) The mathematical theory of plasticity. Clarendon Press, Oxford

Lake JSH, Willis DJ, Fleming HG (1988) The variation of plastic anisotropy during straining. Met Trans 19A:2805–2817

Leffers T, Asaro RJ, Driver JH, Kocks UF, Mecking H, Tomé G, Van Houtte P (1988) Deformation textures: simulation principles. In: Kallend JS, Gottstein G (eds) Proceedings 8th international conference on textures and materials, Santa Fe, 20–25 September 1987, The Metallurgical Society, Warrendale, Pennsylvania, pp 265–272

Lequeu Ph, Gilormini P, Montheillet F, Bacroix B, Jonas JJ (1987) Yield surfaces for textured polycrystals – II analytical approach. Acta Metall 35:1159–1174

Molinari A, Canova GR, Ahzi A (1987) A self consistent approach of the large deformation polycrystal viscoplasticity. Acta Metall 35:2983–2994

Nye JF (1969) Physical properties of crystals. Clarendon Press, Oxford

Pillinger I, Hartley P, Sturgess CEN, Van Bael A, Van Houtte P, Aernoudt E (1989) Finite-element analysis of anisotropic material deformation. In: Thompson EG, Wood RD, Zienkiewicz OC, Samuelsson A (eds) Supplement, proceedings 3rd international conference on numerical methods in industrial forming processes (Numiform 89), Fort Collins, Colorado, 26–30 June 1989. AA Balkema, Rotterdam

Sachs G (1928) Zur Ableitung einer Fliessbedingung. Z Verein Deutscher Ing 72:734–736

Taylor GI (1938) Plastic strain in metals. J Inst Met 62:307–324

Toth LS, Gilormini P, Jonas JJ (1988) Effect of rate sensitivity on the stability of torsion textures. Acta Metall 36:3077–3091

Van Houtte P (1984) Some recent developments in the theories for deformation texture prediction. In: Brakman CM, Jongenburger P, Mittemeijer EJ (eds) Proceedings 7th international conference on textures of materials (ICOTOM 7). Netherlands Society for Material Science, Zwijndrecht, pp. 7–23

Van Houtte P (1987) Calculation of the yield locus of textured polycrystals using the Taylor and the relaxed Taylor theory. Textures Microstruct 7:29–72

Van Houtte P (1988) A comprehensive mathematical formulation of an extended Taylor–Bishop–Hill model featuring relaxed constraints, the Renouard–Wintenberger theory and a strain rate sensitivity model. Textures Microstruct 8–9:313–350

Van Houtte P, Cauwenberg G, Aernoudt E (1987) Analysis of the earing behaviour of aluminium 3004 alloys by means of a simple model based on yield loci calculated from orientation distribution functions. Mater Sci Engng 95:115–124

Van Houtte P, Mols K, Van Bael A, Aernoudt E (1989) Application of yield loci calculated from texture data. Textures Microstruct 11:23–39

Welch PI, Bunge HJ (1986) Influence of grain size on plastic anisotropy in low carbon steels. Mater Sci Technol 2:354–362

Welch PI, Ratke L, Bunge HJ (1983) Consideration of anisotropy parameters in polycrystalline metals. Z Metall 73:233–237

Part II

Review of Numerical Techniques in Material Deformation Processes

5 The Upper Bound Method

A.N. Bramley and F.H. Osman

5.1 Introduction

Limit analysis has been proposed as a means of obtaining solutions to many otherwise intractable metal-forming problems. In its complete form, this approach seeks to establish two expressions for forming loads, one (a so-called lower bound) which is definitely an underestimate and the other (a so-called upper bound) which is an overestimate. Since it is generally sufficient to ensure that the calculated load for a forming operation is indeed sufficient to perform the operation, then interest naturally centres on the upper bound technique. If used correctly, *realistic overestimates* are obtained.

5.2 The Upper Bound Theorem

An upper bound solution involves a consideration of the conditions which have to be fulfilled by the strain increments and the associated strain rate in a plastically deforming medium and does not concern itself with stress equilibrium. An important concept involved is that of a *kinematically admissible velocity* field. This is a distribution of generalized particle velocities which is kinematically compatible within itself and with the externally imposed velocities at boundaries, i.e. it satisfies the condition of incompressibility. The total amount of the work done by the externally applied forces in a metal-forming operation is used to overcome the resistance to plastic deformation and the frictional resistance to relative motion on the tool–workpiece interfaces. These two components can be calculated from a velocity field. The upper bound is formulated in accordance with the theorem proposed by Prager and Hodge (1951) which states that the kinematically admissible velocity field which minimizes the work done is the actual velocity field. From such a field the actual work done can thus be derived, although in practice the actual field may never be completely determined. The theorem states that the actual power is less than or equal to P where

$$P = \int_V \bar{\sigma}\dot{\bar{\varepsilon}} dv + \int_{S_s} \tau \dot{u}_i \, ds + \int_{S_c} T_i \, \dot{u}_i \, ds - \int_{S_t} T_i \, \dot{u}_i ds \tag{5.1}$$

The first term of the right-hand side is due to the internal homogeneous deformation throughout the deforming volume V. The second term expresses the power dissipated due to the velocity discontinuities along the surface S_s. The third term considers the friction losses over the tool–material interface S_c. The last term is due to the external body traction, such as applied tensions in rolling and drawing, over the surface S_t. P is the upper bound computed power.

The general procedure when formulating a solution is to divide the deformation zone into one or more assumed zones throughout each of which the velocity is continuous. In adjacent zones a different velocity distribution may exist whilst across the interfaces and also on the tool–workpiece interfaces a tangential velocity discontinuity may occur. The best velocity distribution is the one which minimizes the value of P.

5.3 Deformation Power (P_1)

Consider an elemental cube of material of volume δv, face area δa and side length δl subjected to a principal stress system σ_1, σ_2, σ_3 which causes corresponding strain increments $\delta\epsilon_1$, $\delta\epsilon_2$ and $\delta\epsilon_3$. The strain increments are related to the deviatoric stresses σ_1', σ_2' and σ_3' via Levy–von Mises equations. Thus $\delta\epsilon_1 = (3\delta\bar{\epsilon}/2\bar{\sigma})\sigma_1'$ where $\bar{\sigma}$ and $\delta\bar{\epsilon}$ are the equivalent stress and strain increment respectively. The work done on the cube is derived as follows:

$$
\begin{aligned}
\text{Work done} &= \text{force} \times \text{displacement} \\
&= (\text{stress} \times \text{area}) \times (\text{length} \times \text{strain}) \\
&= \sum(\sigma_i \delta a)\,(\delta l \delta\epsilon_i) \text{ for } i = 1 \text{ to } 3 \\
&= \tfrac{3}{2}\,\delta\bar{\epsilon}/\bar{\sigma}\,\delta v\,\sigma_i\sigma_i' \text{ (adopting the usual summation convention)}
\end{aligned}
$$

$$\text{Where } \sigma_i' = \sigma_i - \sigma_{mm}/3 \tag{5.2}$$

After substitution,

$$
\begin{aligned}
\text{Work done} &= \tfrac{3}{2}\,\delta\bar{\epsilon}/\bar{\sigma}\,\delta v\,\tfrac{1}{3}\,2\bar{\sigma}^2 \\
&= \bar{\sigma}\delta\bar{\epsilon}\delta v
\end{aligned}
\tag{5.3}
$$

Integrating over the deforming volume,

$$\text{Total work done} = \int \bar{\sigma}\, d\bar{\epsilon}\, dv \tag{5.4}$$

$$\text{Therefore the total power } P_1 = \int \bar{\sigma}(d\bar{\epsilon}/dt)\, dv$$

or

$$P_1 = \int \bar{\sigma}\dot{\bar{\epsilon}} dv \tag{5.5}$$

$\dot{\bar{\epsilon}}$ at any point in the volume is evaluated from the velocity field, u, v, w (the velocities in the x, y, z coordinate sysem) as follows:

$$\dot{\bar{\epsilon}} = \frac{\sqrt{2}}{3}\left[(\dot{\epsilon}_x-\dot{\epsilon}_y)^2 + (\dot{\epsilon}_y-\dot{\epsilon}_z)^2 + (\dot{\epsilon}_z-\dot{\epsilon}_x)^2 + \tfrac{3}{2}(\dot{\phi}_{xy}^2 + \dot{\phi}_{yz}^2 + \dot{\phi}_{xz}^2)\right]^{1/2} \tag{5.6}$$

where $\dot{\epsilon}_x = \partial u/\partial x$ and $\dot{\phi}_{xy} = \partial u/\partial y + \partial v/\partial x$ etc.

Since the material is in a state of plastic flow the equivalent stress must correspond to the current yield stress, Y. If the further assumption of no hardening is made then $\bar{\sigma}$ is a constant and

$$\bar{\sigma} = Y \text{ or } 2k \text{ (according to Tresca criterion)} \tag{5.7}$$

or

$$\bar{\sigma} = Y \text{ or } \sqrt{3}k \text{ (according to von Mises criterion)} \tag{5.8}$$

where k is the shear yield stress.

Hence, the power integral P_1 can be evaluated.

5.3.1 Example: Frictionless Plane-Strain Forging

A simple example of application to frictionless plane-strain forging depicted in Fig. 5.1 is as follows. The platens approach each other with unit velocity and the deformation is assumed to be uniform, i.e. a parallel velocity field exists:

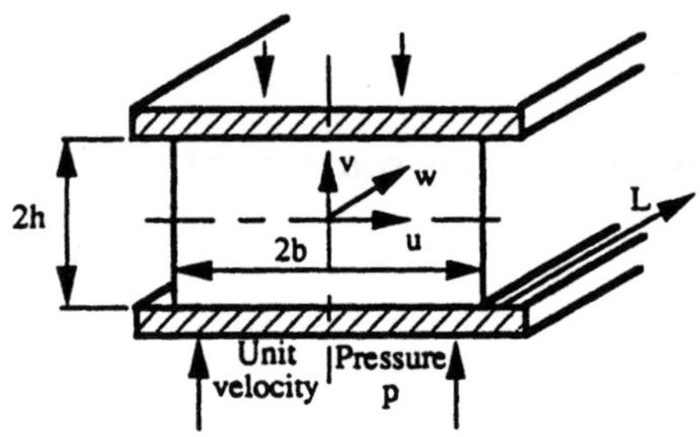

Fig. 5.1. Plane-strain forging.

$$\left.\begin{aligned} u &= \frac{x}{h} \times 1 \\ v &= -\frac{y}{h} \times 1 \\ w &= 0 \end{aligned}\right\} \tag{5.9}$$

Note having chosen v and a parallel velocity field, then u follows from the volume constancy condition $\dot{\epsilon}_{ii} = 0$, and

$$\left.\begin{aligned} \dot{\epsilon}_x &= \frac{\partial u}{\partial x} = \frac{1}{h} \\ \dot{\epsilon}_y &= \frac{\partial v}{\partial y} = -\frac{1}{h} \\ \dot{\epsilon}_z &= \dot{\phi}_{xy} = \dot{\phi}_{yz} = \dot{\phi}_{xz} = 0 \end{aligned}\right\} \tag{5.10}$$

Therefore

$$\begin{aligned} \dot{\bar{\epsilon}} &= \frac{\sqrt{2}}{3}\sqrt{\frac{6}{h^2}} \\ &= \frac{2}{\sqrt{3}}\frac{1}{h} \end{aligned} \tag{5.11}$$

Therefore

Deformation power $$P_1 = 4L\bar{\sigma}\int_0^b \int_0^h \frac{2}{\sqrt{3}}\frac{1}{h}\,dxdy$$

$$= 4L\bar{\sigma}\frac{2}{\sqrt{3}}b \qquad (5.12)$$

The operation is frictionless, and no velocity discontinuities are proposed. Hence $P_2 = P_3 = 0$ and the external rate of working P induced by the applied pressure p is given by $2p2bL$.

Applying the upper bound theorem

$$2p2bL \leq 4L\bar{\sigma}\frac{2}{\sqrt{3}}b$$

Therefore

$$p \leq 2\frac{\bar{\sigma}}{\sqrt{3}} \qquad (5.13)$$

Reference to the plane-strain compression test will confirm the validity of this solution.

5.4 Power Dissipation due to Velocity Discontinuities (P_2)

Flow of material through converging dies as in extrusion or drawing is a useful vehicle with which the principles of the technique can be demonstrated. Experimental observations indicate that stream lines of flow are very much as suggested by Fig. 5.2.

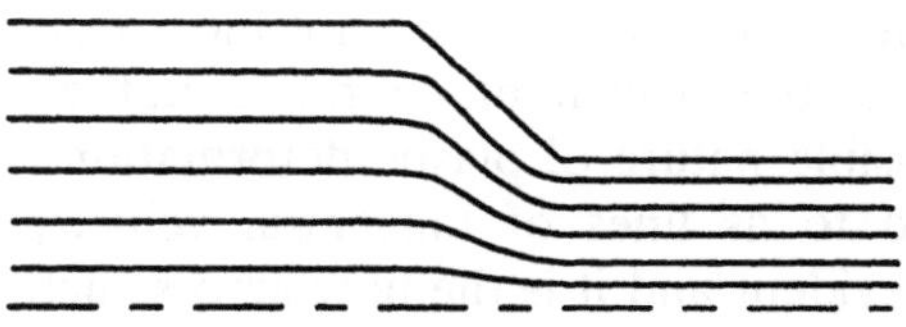

Fig. 5.2. Stream lines in converging flow.

Not surprisingly such a continuously transforming pattern is difficult to describe quantitatively and it becomes necessary to effect a simplification of the flow such as in Fig. 5.3.

In region A the material is flowing into the die in a rigid manner i.e. no plastic deformation occurs. On the boundary between A and B (also B and C) there is a sudden change in direction of flow. In the practical case this change of flow direction takes place over a finite distance although it can be surprisingly sudden. As the material crosses these boundaries, it is in fact being sheared and the energy required for this is provided by the extrusion/drawing force. Across region

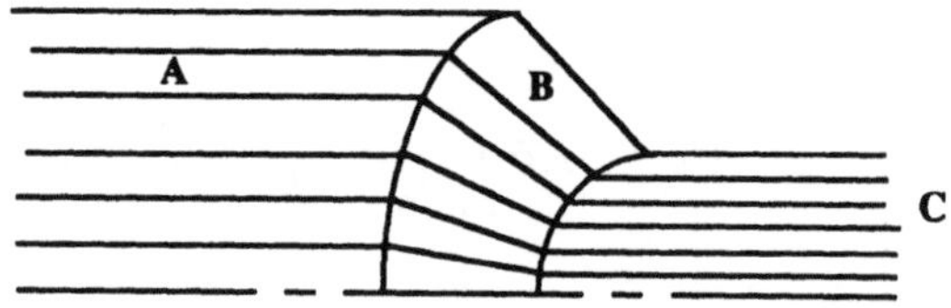

Fig. 5.3. Simplification of a converging flow field.

B the material will, in general, flow in a smooth/continuous manner just as the material within the gauge length of a tensile specimen deforms. The deformation energy required for this can be calculated using a velocity distribution which varies linearly through the die. Initially this issue can be avoided considering cases of plane-strain deformation only. Under plane-strain conditions the flow can be accounted for by having region B behave as a rigid region as shown in Fig. 5.4. Region B could also be subdivided into a number of rigid regions separated by discontinuities.

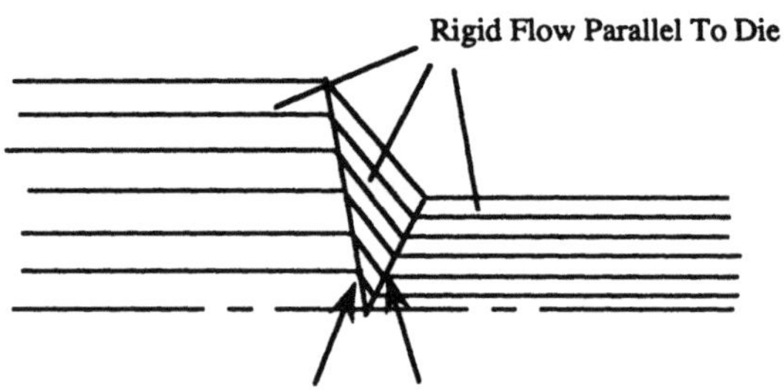

Fig. 5.4. Triangular field for converging flow.

The majority of upper bound solutions for plane-strain conditions can be effected by this technique of dividing the deformation zone into a number of regions of rigid flow separated by planes of shear. Such regions are necessarily of triangular shape since a region defined by more than three sides would be incompatible with the concept of rigid flow. For example, in the field of Fig. 5.5 it is not possible to identify a direction for rigid flow which would not incur flow away from or into both the die and centre line, which of course is not an acceptable proposition due to the incompressible nature of plastic deformation.

The planes of sudden shear are referred to as lines of tangential velocity discontinuity for reasons which will become evident and it is the flow across such a plane which is considered next.

A simple analogy to this is an idealized plane-strain lathe cutting operation as depicted in Fig. 5.6a. XX can be likened to a plane of velocity discontinuity. A

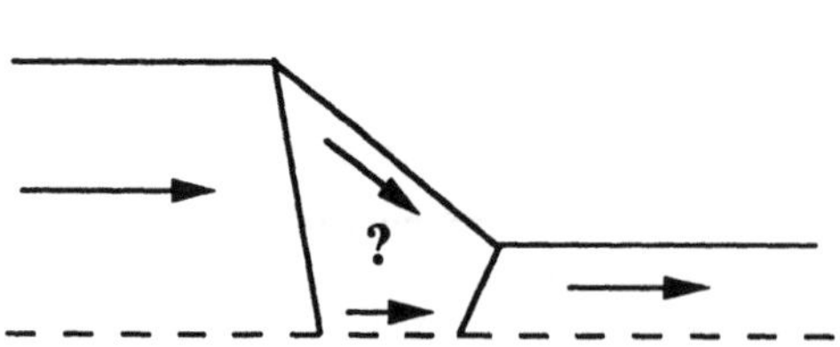

Fig. 5.5. Invalid rigid block field.

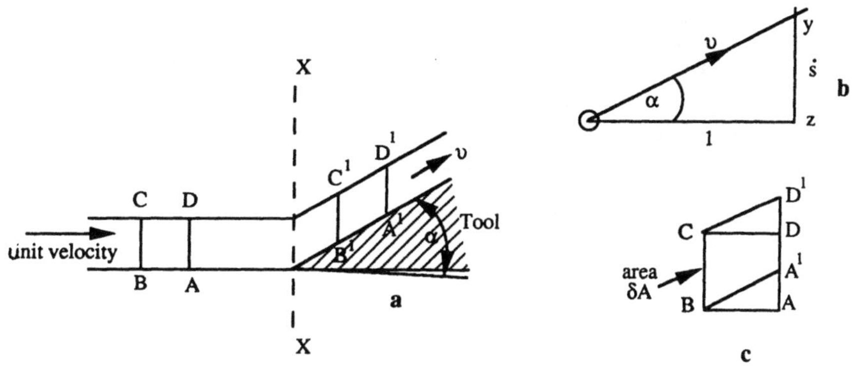

Fig. 5.6. Power dissipation over a velocity discontinuity. **a** Cutting tool analogy; **b** hodograph; **c** element deformation.

rigid element ABCD approaches XX with unit horizontal velocity, this is depicted in the hodograph Fig. 5.6b by oz. Due to the action of the cutting tool, the element distorts into the form A^1 B^1 C^1 D^1. This distortion is assumed to occur instantaneously and the element then travels at an angle α alongside the tool face, again as a rigid body. The velocity, v, must have a horizontal component equal to unity to satisfy the incompressibility condition. The vector zy in the hodograph thus represents a change in the tangential component of velocity, i.e. the element sustains a tangential velocity discontinuity $\dot{s}$ when crossing XX. The rate at which energy is dissipated in this operation can be calculated from Fig. 5.6c, where the deformed and undeformed elements are superimposed. The material is assumed to flow plastically with a shear yield stress k.

$$\text{Work done} = \text{force} \times \text{displacement}$$

$$= k\, dA \times AA^1 \tag{5.14}$$

$$\text{Rate of working} = k\, dA \frac{AA^1}{AB} \tag{5.15}$$

where AB is a unit length of element travelling at unit velocity. Comparing the similar triangles in Fig. 5.6b,c,

$$\text{rate of working or power} = k\, dA\, \dot{s} \tag{5.16}$$

Therefore the total power due to all such discontinuities is given by

$$P_2 = \int k|\dot{s}|\, dA \tag{5.17}$$

5.4.1 Example: Plane-Strain Indentation – Upper Bound Solution

This problem has been solved by the slip-line field technique and it is by correlation with that solution that the velocity field of Fig. 5.7a is suggested as being an approximately equivalent straight-line field. The physical interpretation of the indentation field is as follows.

The whole zone ACDEB flows plastically and the metal below ACDE remains rigid. As the punch moves downwards with unit velocity, any particle in the

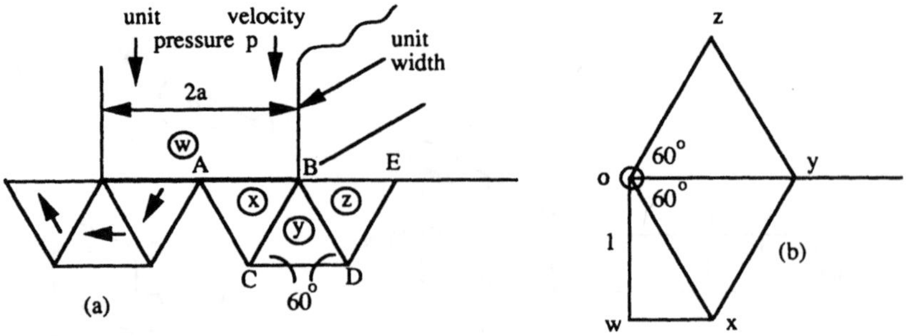

Fig. 5.7. Plane-strain indentation – upper bound solution. **a** Velocity field; **b** hodograph.

region ABC will move down with a vertical component of unity, but is constrained to slide parallel to the rigid boundary AC on which shear occurs. These velocities can be drawn on a hodograph (Fig. 5.7b) where ow represents the velocity of the punch and ox the velocity of the region ABC. As a particle crosses the boundary BC, its velocity is changed by shear parallel to BC after which it is constrained to move parallel to CD. The velocity vector for triangle BCD, i.e. oy is thus found by the intersection of a horizontal line drawn through o and a line parallel to CB drawn through x. A further change in velocity occurs on crossing BD after which the material flows parallel to DE. The vector for BDE is thus found from the intersection of a line from o parallel to DE and a line from y parallel to DB. This completes the hodograph. Note that all the particles in a particular zone bounded by tangential velocity discontinuities have the same velocity. Therefore, the energy dissipation calculations will only be concerned with the second term of Eq. (5.1), i.e.

$$P_1 = P_3 = 0 \tag{5.18}$$

and

$$P_2 = k \times 1\,[(\mathrm{AC}ox) + (\mathrm{BC}xy) + (\mathrm{CD}oy) + (\mathrm{BD}yz) + (\mathrm{DE}oz)] \times 2 \tag{5.19}$$

with

$$ox = \frac{2}{\sqrt{3}} \tag{5.20}$$

and

$$xy = oy = yz = oz = ox \tag{5.21}$$

$$\mathrm{AC} = \mathrm{BC} = \mathrm{CD} = \mathrm{BD} = \mathrm{DE} = a \tag{5.22}$$

$$\therefore \quad P_2 = 2k\frac{2}{\sqrt{3}}5a \tag{5.23}$$

The external rate of working by applied pressure p is given by ($p \times 2a \times 1$).

Applying the upper bound theorem,

$$p \times 2a \times 1 \leq 2k \frac{2}{\sqrt{3}} 5a$$

$$\therefore \quad \frac{p}{2k} \leq 2.89 \tag{5.24}$$

This is a higher figure than 2.57 obtained by using the more exact slip-line technique. However, it should be realized that the field of equilateral triangles is not necessarily the optimum. In fact, if general isosceles triangles are used and the apex angle varied to obtain the minimum power, then a value of 2.83 can be obtained.

5.4.2 Example: Plane Strain Extrusion Through Square Dies – Upper Bound Solution

This example demonstrates how a pseudo-independent variable can be incorporated in the velocity field and subsequently used to optimize the field and achieve a minimum power.

The proposed field is formed by two velocity discontinuities as shown in Fig. 5.8a. The pattern is uniquely defined by the angle θ. Particles approach the die

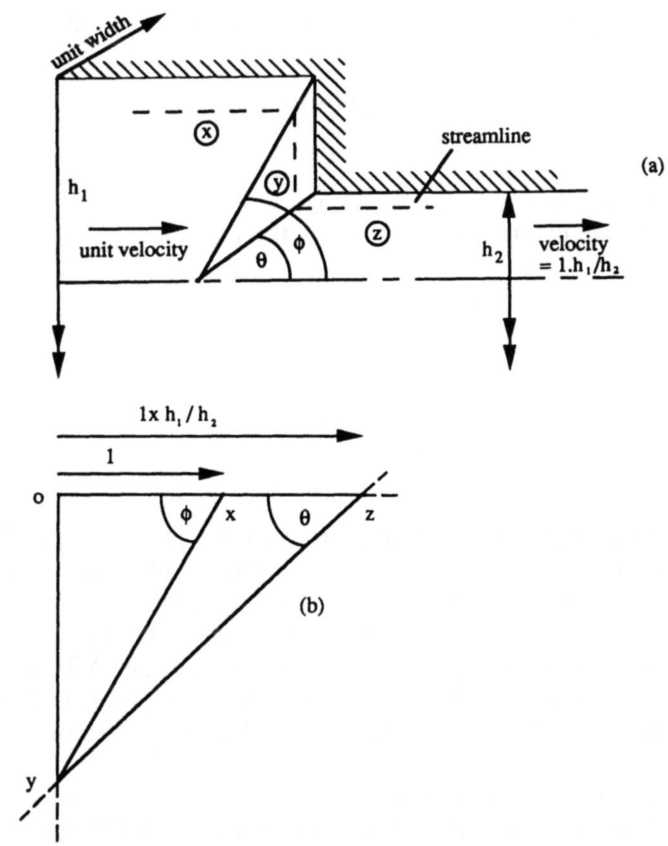

Fig. 5.8. Plane-strain extrusion through square dies. **a** Velocity field; **b** hodograph.

with unit velocity, are then sheared across a velocity discontinuity, travel parallel to the die, are sheared across a second discontinuity and then emerge with a velocity $1 \times h_1/h_2$. The hodograph drawn in Fig. 5.8b confirms that this is a kinematically admissible field. The horizontal unit velocity in region x is depicted on the hodograph as ox. In region y, the material is constrained to flow parallel to the die face. Additionally, when crossing from x to y it suffers a discontinuity at an angle ϕ. Thus y is located in the hodograph. A line drawn through y inclined at θ to the horizontal corresponds to the discontinuity between y and z. The extruded product is constrained to move horizontally. This is represented by a horizontal line drawn from o. This intersects to define z. oz is of magnitude $1 \times h_1/h_2$ thus confirming the field to be kinematically admissible. P_1 and P_3 are again considered zero.

so $$P_1 = P_3 = 0 \tag{5.25}$$

and $$P_2 = k\frac{1}{\cos\phi}\frac{h_1}{\sin\phi} + k\frac{h_1}{h_2\cos\theta}\frac{h_2}{\sin\theta} \tag{5.26}$$

This reduces to

$$P_2 = \frac{h_1 k(1+n)(1+n\tan^2\theta)}{\tan\theta} \tag{5.27}$$

where $n = h_1/h_2$.

A minimum value for this expression can be obtained by differentiating with respect to θ which yields

$$P_2 = \frac{2k(1+n)h_1}{\sqrt{n}} \tag{5.28}$$

as a minimum value when $\theta = \tan^{-1}(1/\sqrt{n})$.

Relating this to the external work done by the extrusion pressure p, we obtain

$$(ph_1 \times 1) \leq \frac{2k(1+n)h_1}{\sqrt{n}} \tag{5.29}$$

Hence

$$p/2k \leq \frac{(1+n)}{\sqrt{n}} \tag{5.30}$$

This simple velocity field overestimates the pressure given by the slip-line field technique by about 30%. A more realistic approach is to consider the possibility of dead metal zone forming in the corner of the die, thus converting the square die into a perfectly rough wedge-shaped die of unspecified angle. For a given reduction, the field shown in Fig. 5.9 is assumed to operate. ABC is a dead zone and the field is defined by the angles α and θ. The shear power is then calculated as before for the two velocity discontinuities and also the friction power for the interface AB.

The results obtained are summarized in Fig. 5.10 where the minima with respect to both α and θ can be observed. A final value for $p/2k$ of 2.55 is obtained which compares favourably with the 2.48 obtained from the slip-line field technique.

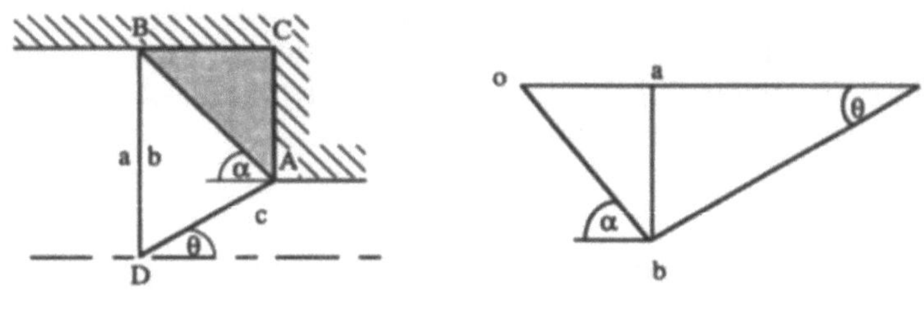

Fig. 5.9. Plane-strain extrusion through square dies with dead metal zone. **a** Velocity field; **b** hodograph.

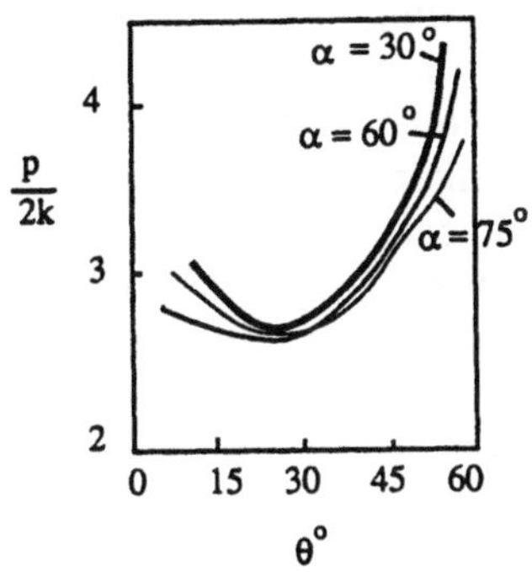

Fig. 5.10. Plane-strain extrusion with dead metal zone – effect of pseudo-independent variables on upper bound for forming pressure.

5.5 Power Dissipation due to Friction (P_3)

The upper bound analysis of plane-strain deformation divides the deforming zone into blocks which can be considered as elements sliding against each other or as regions through which material flows in a rigid manner. Material adjacent to the tools is restricted to follow the die profile and therefore frictional resistance exists at the tool–material interface. There are two extreme conditions, firstly where material is sliding along smooth surfaces and secondly where the frictional condition is too rough to allow the material to slide along the tool–material interface. Under frictionless conditions there will be no resistance to metal flow and hence P_3 (min) is zero. When friction is extremely high, the frictional effect is imagined to be confined to a very thin layer at the tool–material interface. Friction stress in this case will exceed the material yield stress in shear and whilst the material is inseparable from the tool, internal shear occurs in the vicinity of the friction layer. The power dissipated is therefore

$$P_{3\max} = \int_s k\, \Delta v\, ds \tag{5.31}$$

where Δv is the relative velocity along the interface s.

At intermediate friction conditions the stresses developed at the interface are less than the current material shear stress. Hence, the power dissipated due to friction can be expressed by

$$P_3 = mk \int_s \Delta v\, ds \tag{5.32}$$

where m is the friction factor. The friction factor varies from zero under frictionless conditions to unity under full sticking. Intermediate values of m can be obtained experimentally by conducting a ring test.

5.6 Upper Bound Solutions for Axisymmetric Problems

The preceding simple examples have been chosen in order that the reader can be easily familiarized with the technique. Further examples can be found in the texts by Thomsen et al. (1965), Avitzur (1968) and Johnson and Mellor (1973). In conditions of axial symmetry, the calculations become more complex due to the appearance of the ϵ_θ term in the deformation power integral P_1. A number of specific solutions have, however, appeared in the literature. Clearly a generalized approach would be desirable and it was probably with this in mind that Kudo (1960a–d, 1961) developed his unit deforming region concept. He produced "manual" solutions for a range of rectangular section elements. The particular problem to be addressed would be broken down into a number of these elements and the individual power dissipation rates calculated and summed for the whole process. The plastically deforming material was assumed to flow through the elements, i.e. the elements themselves did not distort. Subsequently this approach was further developed by McDermott and Bramley (1974), Cramphorn and Bramley (1977), Osman and Bramley (1979) and Thornton and Bramley (1980). Additional elements of triangular cross-section were developed and the calculations were computerized. The term Upper Bound Elemental Technique (UBET) was coined to describe the procedure.

5.7 UBET

The UBET analysis is based on the concept of subdividing the plastically deforming shape into a series of singly connected elements of triangular and rectangular cross-section. Each of these element types has a general solution which incorporates a parallel velocity field together with tangential velocity discontinuities on its surface. Thus together, the elements compose a very general type of velocity field for the whole workpiece. Currently, the procedure has been restricted to conditions of axial symmetry where the elements constitute annular rings. The velocity fields together with associated boundary velocity discontinuities, flow stress and friction data, then enable an upper bound to the power dissipation and hence forging loads to be calculated. The computation of the boundary velocity discontinuities between adjacent elements requires separate treatment depending on the particular combination of elements which are involved. Full details of these elemental power calculations are given by Osman (1981).

The subdivision into elements is carried out automatically by imposing a variable spacing orthogonal mesh but is arranged to satisfy the requirement that the elements be singly connected, that very large numbers of elements are not generated and that the boundaries of the deforming region coincide with the grid intersection points. This is achieved by orthogonally projecting the outline defining coordinate points and reflecting them as shown in Fig. 5.11. Except for simple

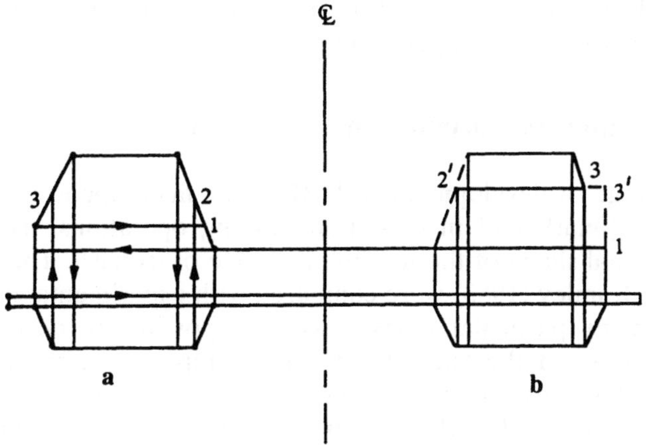

Fig. 5.11. Automatic subdivision of section into elements and modifications to profile to reduce number of elements.

shapes, this will generally lead to an excessive and possibly infinite number of elements. This difficulty is overcome by modifying the outline in order to bring closely adjacent points together or by adding additional points. Since the subdivision is producing an element structure from which it is a simple matter to calculate the volume, it is useful to check that the effect of altering the outline does not cause the volume to diminish with the consequent risk that an upper bound may not be obtained. Fig. 5.11 shows an application of the technique where the reflected points 2 and 3 are caused to move into alignment. The boundary was then chosen to run from 3 to a new point 3′ to avoid the reduction in volume which would arise if points 3 to 1 were connected directly; all of which is done automatically.

Before numerical data can be established, it is clearly necessary to determine boundary velocities on the elements which, for elements with free surfaces, effectively describe the metal flow. These velocities are determined by a computer-based optimizing routine which seeks to establish the particular set of velocities which will give a minimum value for the overall rate of energy dissipation. Because of the high degree of generality associated with the elemental subdivision and with the boundary velocities, it is held that the optimization process will yield a set of velocities very close to the actual velocity field. This is consistent with the concept that the plastic flow occurs in such a way as to minimize the power requirements.

In order to keep the computational time down to a level at which interactive use is possible, the initial die and workpiece configurations are described by a series of straight lines which will generally result in an approximation of the shapes. Likewise, the elemental subdivision as described earlier can be adjusted so that it produces a "reasonable" number of elements. There is, therefore, a "squared-off" appearance to the simulation. A variable flow stress accommodating strain and/or strain rate effects, if these are known, can readily be incorporated into the simulation. Strain hardening is accounted for by maintaining a record of the strain induced by the shear discontinuities on the element boundaries and the general straining within the elements but not that due to friction on the tool/workpiece boundaries. Strain rate sensitivity is readily accommodated since the power calculations are made in terms of local velocities and strain rates. An equivalent strain is then computed and used to update the current flow stress.

5.7.1 Application to Forming Load Prediction

At an early stage of development of UBET, the main application was to predict the maximum (final) load in closed-die forging operations. This was done by analysing the final shape of the forging, inclusive of its flash. The results obtained by Islam and Bramley (1983) were compared with experimental data on industrial forgings. In many particular cases it was not possible to be certain about the precise dimensions of the "actual" forgings, temperature, material deformation speed etc., but the overall results shown in Fig. 5.12 were nevertheless quite encouraging and certainly sufficient to stimulate further developments into flow prediction. The detailed analysis of an aluminium forging shown in Fig. 5.13 shows an application of the method to predicting the complete load trace of the process. Clearly from these data, energy requirements can be determined. Fig. 5.13 also indicates the CPU time used for this analysis in two cases where the optimizing routine was allowed to run for a different number of iterations. Clearly the solution is converging quite rapidly and also the absolute CPU time is sufficiently small to justify interactive use by forging process designers.

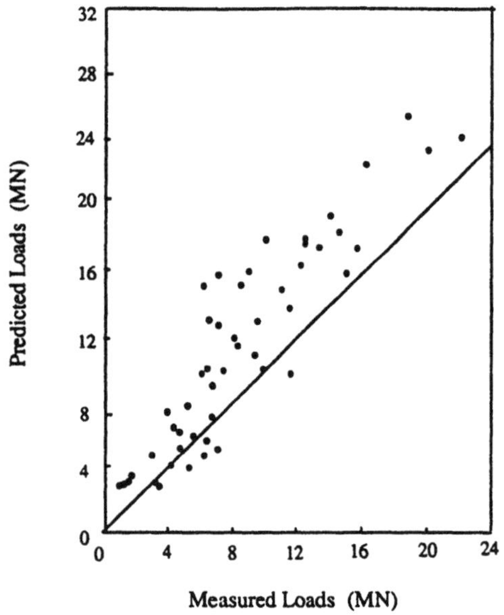

Fig. 5.12. Case studies for forging load predictions using UBET.

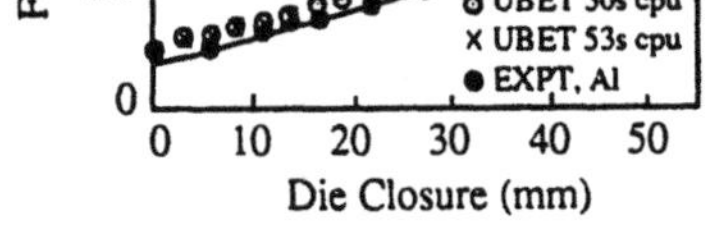

Fig. 5.13. UBET analysis of an aluminium forging.

5.7.2 Application to Metal Flow Prediction

Flow simulation can be achieved by analysing the initial configuration of the dies and workpiece whether this be a preform or a billet and then allowing the dies to move through an appropriate increment of displacement during which the free boundaries execute a displacement corresponding to the determined optimum velocities on their associated elements.

The application of UBET for the determination of optimum billet aspect ratio and flash dimensions was demonstrated by taking a somewhat artificial forging with excessively deep cavities (Osman et al. 1984, Ghobrial et al. 1985a,b). This shape, shown in Fig. 5.14, was hot forged in aluminium HE30. Fig. 5.15 shows the effect of billet aspect ratio on the degree of die cavity filling and corresponding energy requirement when the flash dimension is kept constant. Friction data were determined independently from ring tests and assumed to be constant over the die face. An interesting result obtained in this particular case is that the degree of die filling appears to pass through a maximum as the aspect ratio of the billet increases. Fig. 5.16 shows the effect of flash geometry on filling of the same forging and confirms the often quoted advice that a flash land width to thickness ratio should be between 2 and 4.

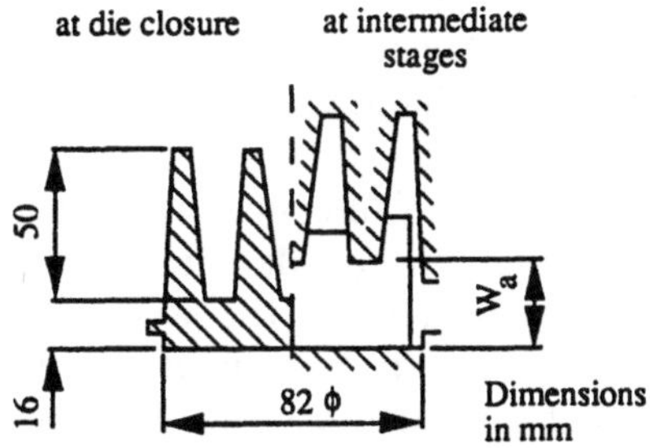

Fig. 5.14. Experimental shape for flow prediction studies. (Dimensions in mm.)

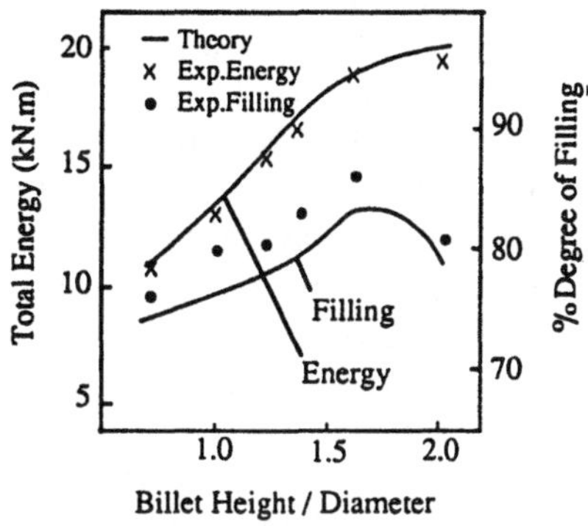

Fig. 5.15. Effect of billet aspect ratio on die filling.

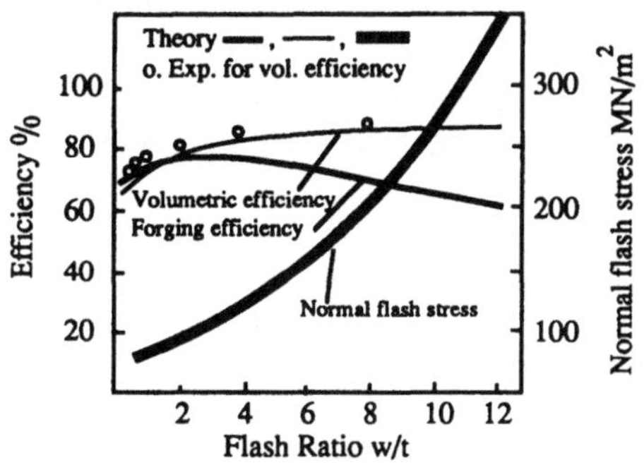

Fig. 5.16. Effect of flash geometry on die filling and flash land stress; w = flash width, t = flash thickness.

Below a value of 2 the curve is very steep and a significant filling effect can be gained. After a value of 4, the curve is horizontal and little benefit is to be gained by making the flash more restrictive. Energy dissipated in the flash can be interpreted as a flash load and stress. Such a land stress has to be sustained by the tooling and is also seen in Fig. 5.16 to rise rapidly after a flash ratio of 4.

It has not yet proved possible to accommodate the influence of friction on metal flow. Friction is in effect only taken into account as an additional independent term. As a consequence, metal flow prediction is not very realistic near the tool/workpiece surface. A proposal by Yang et al. (1985) to overcome this problem has not yet been incorporated in a comprehensive program.

5.7.3 Application to Forging Preform Design

A preforming stage is frequently necessary in closed die forging operations in order to ensure filling of the die cavities but its determination is not a particularly straightforward matter. As far as the author can determine, the design of a preform shape is very much a matter of guessing a likely intermediate shape which resembles a "rounded-off" version of the final forging and looks as though it would form the final shape without too much difficulty. Such a procedure not infrequently fails to produce satisfactory results by simply not resulting in a filled cavity or by causing an excessive billet volume to be necessary. In aerospace applications, preforming is sometimes associated with the need to achieve controlled amounts of deformation in certain parts of the forging and this imposes a further constraint on the methodology which has to be adopted.

Using the UBET a variety of methods was investigated by Ghobrial et al. (1985b), the most promising of which is the so-called "reversing technique" (Osman et al. 1984). The procedure arises from the concept of simply reversing the flow simulation by starting from the final desired shape with the die velocities reversed in such a way that during decrementation, the dies move apart while the material inside the cavity flows inward and in the first instance separating from the deepest die cavity. The cavity is assumed to be filled just before die closure in a satisfactory forging operation. The surfaces of the forging in the "last-to-fill" cavity is treated as a free boundary. The definition of the deepest cavity does, however, require care and the most satisfactory procedure seems to be one where the original final forging is interpreted as a billet and is imagined to be placed between the open dies, simultaneously satisfying any requirements of billet location etc. Cavity depths are then measured from the top and bottom surfaces of the billet. Optimization of the velocity field is then carried out in the usual way and the resultant field is assumed to apply for a small decrement of movement of the dies. The resultant shape change is then passed to the next decrement and the process repeated until the forging becomes an acceptable starting shape.

Fig. 5.17 shows the effectiveness of the "reverse method". The reversal is stopped at a web thickness of about 36 mm when the dies and workpiece separate. The experimental points show how the forward process taken from that point with a preform shape then fills the die satisfactorily. During the reversal phase it is of course not possible to accommodate the effects of strain or strain rate and this in some cases may lead to errors and the need for slight iteration when subsequently followed by the forward simulation incorporating the material data.

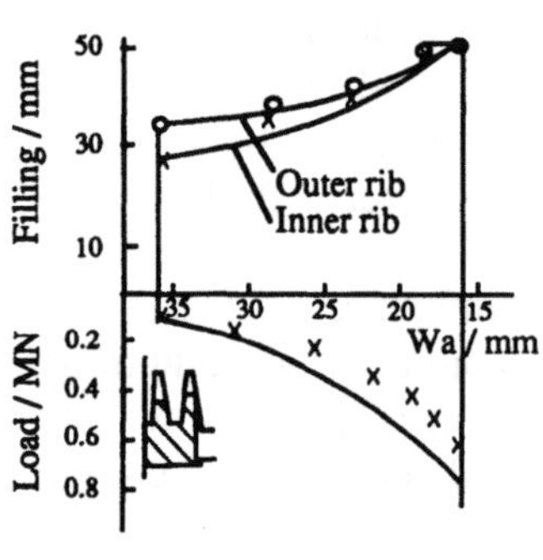

Fig. 5.17. "Reverse method" for preform design.

5.7.4 Local Die Pressure Prediction

This is a useful application of UBET which leads to a prediction of the pressure profile over the tool surfaces. This can be used subsequently for calculating elastic deflection of tool surfaces or general tooling stress levels. At first sight it would appear to be impossible to use upper bound based methods which may violate stress equilibrium conditions. However, Kiuchi and Murata (1980) proposed an "imaginary element" technique which stimulated the development of the more realistic "pressure element" technique by Christensen et al. (1986). In this procedure an extra element is introduced into the subdivision at a point where it is required to determine the local pressure. The element is then given in turn a local positive and negative change of velocity Δv from which two overall power calculations P^+ and P^- are considered. If P is the power without the introduction of the pressure element then the local pressure q required at the pressure element can be determined in the two cases as follows:

$$q^+ = \frac{(P^+ - P)}{a(+\Delta v)} \quad \text{and} \quad q^- = \frac{(P^- - P)}{a(-\Delta v)} \tag{5.33}$$

The actual surface pressure on area a is then calculated as the average of q^+ and q^-:

$$q = \frac{1}{2}\left\{\frac{P^+ - P}{a(+\Delta v)} + \frac{P^- - P}{a(-\Delta v)}\right\} \tag{5.34}$$

i.e.

$$q = \frac{P^+ - P^-}{2a\Delta v} \tag{5.35}$$

An example of the type of pressure distribution which can be predicted by this technique is shown in Fig. 5.18 for a simple forging configuration.

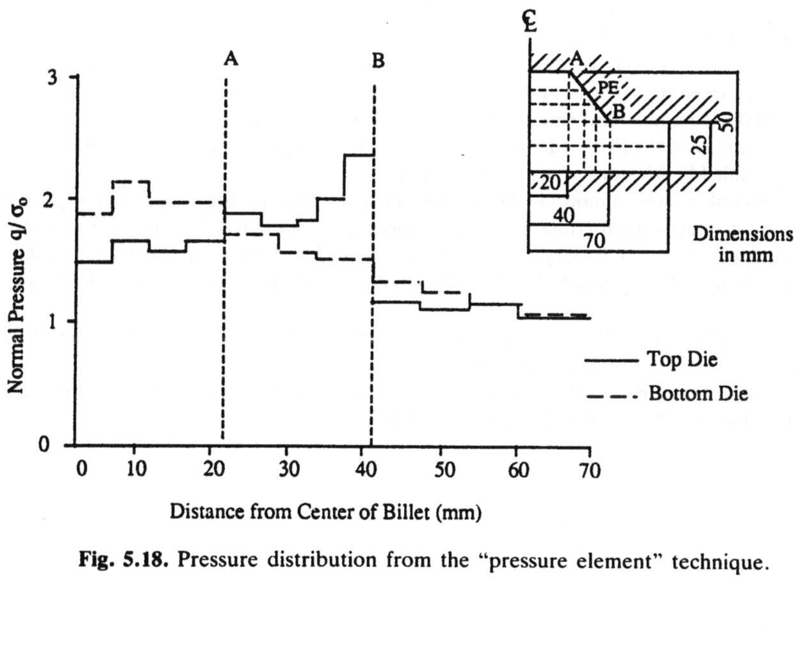

Fig. 5.18. Pressure distribution from the "pressure element" technique.

5.8 Conclusion

The upper bound technique is very powerful in providing rapid approximate answers to a range of metal-forming problems which can be constructed into plane-strain conditions. As axisymmetric or general three-dimensional problems are encountered such manual methods are inappropriate and the generalized computer-based UBET is recommended. It trades accuracy with speed and as such needs to be used with considerable care. However, it does lend itself to being used by die designers without the need for knowledge of theoretical plasticity and because of this has gained many enthusiasts.

References

Avitzur B (1968) Metal forming, processes and analysis. McGraw-Hill

Christensen P, Bay N, Bramley AN, Osman FH (1986) Calculation of local surface stresses in drop forging using UBET. In: Proceedings 14th NAMRC, Minneapolis, USA, pp 286–293

Cramphorn AS, Bramley AN (1977) Computer aided forging design with UBET. In: Proceedings 18th MTDR conference, London, pp 717–724

Ghobrial MI, Osman FH, Bramley AN (1985a) Forging flash design with UBET. In: Proceedings 13th NAMRC, Los Angeles, USA, pp 151–157

Ghobrial MI, Osman FH and Bramley AN (1985b) Forging preform design – UBET based methods. In: Proceedings 25th MTDR Conference, pp 465–472

Islam, MN, Bramley AN (1983) Parametric analysis of antisymmetric forgings. Metallurgia 50:181–184

Johnson W, Mellor PB (1973) Engineering plasticity. Van Nostrand

Kiuchi M, Murata Y (1980) Simulation of contact pressure distribution on tool surface by UBET. In: Proceedings 21st MTDR conference, Swansea

Kudo H (1960a) An upper-bound approach to plane-strain forging and extrusion – I. Int J Mech Sci 1:57–83

Kudo H (1960b) An upper-bound approach to plane-strain forging and extrusion – II. Int J Mech Sci 1:229–252

Kudo H (1960c) An upper-bound approach to plane-strain forging and extrusion – III. Int J Mech Sci 1:366–368

Kudo H (1960d) Some analytical and experimental studies of axi-symmetric cold forging and extrusion – I. Int J Mech Sci 2:102–127

Kudo H (1961) Some analytical and experimental studies of axi-symmetric cold forging and extrusion – II. Int J Mech Sci 3:91–117

McDermott RP, Bramley AN (1974) An elemental upper-bound technique for general use in forging analysis. In: Proceedings 15th MTDR conference, Birmingham, UK, pp 437–443.

Osman FH (1981) Computerized simulation of forging processes. PhD Thesis, Leeds, UK

Osman FH, Bramley AN (1979) Metal flow prediction in forging and extrusion using UBET. In: Proceedings 20th MTDR conference, Birmingham, UK, pp 51–59

Osman FH, Bramley AN, Ghobrial MI (1984) Forging and preform design using UBET. In: Proceedings 1st international conference on technology of plasticity, Tokyo, pp 563–568.

Prager W, Hodge PG (1951) Theory of perfectly plastic solids. Chapman and Hall, London

Thomsen EG, Yang CT, Kobayashi S (1965) Mechanics of plastic deformation in metal processing. Macmillan

Thornton JR, Bramley AN (1980) An approximate method for predicting metal flow in forging and extrusion operations. Proc I Mech E 194:9–15

Yang DY, Kim JH, Lim CR (1985) An arbitrarily inclined triangular UBET element and its application to combined forging. Trans ASME J Engng Ind 107:134–135

6 Boundary-Element Analysis of Forming Processes

N. Phan-Thien and R.I. Tanner

6.1 Introduction

The numerical analysis of problems in forming can be accomplished by a variety of techniques; one has finite differences, finite elements, boundary elements, and other methods. Each method has some advantage over the others in certain classes of problems, and here we review the use of boundary-element methods in some slow Newtonian and non-Newtonian (or rheological) forming problems. The term non-Newtonian fluid means a material not described by the Navier–Stokes equations. The simplest non-Newtonian fluid class includes materials where the viscosity depends on the scalar invariants of the rate-of-deformation tensor **D**. In some cases metal deformation can be so modelled (Karabin and Smelser 1989). A much greater challenge is posed by fluids with memory (or viscoelasticity) and part of this chapter is devoted to this category of materials; molten plastics are perhaps the most important examples of viscoelastic fluids.

In comparing the various numerical methods for fluid mechanics, we note that the finite-difference and finite-element methods automatically find the velocity and pressure or other fields all over the fluid domain. The boundary-element method does not necessarily do this. For elasticity problems, it is often necessary to compute the fields everywhere in order to locate, for example, the maximum stress points. In flow problems, we are often most interested in the gross features of the flow, for example, the ultimate shape of the product, or the overall pressure drop or drag, and the precise pointwise details of the fields are of less interest. Under these circumstances the boundary-element method has some advantages, which will be discussed below. The work of Rizzo (1967) on elasticity, followed by that of Cruse (1969), opened the way to the use of the method in viscous creeping flow (Youngren and Acrivos 1975). We now consider the equations to be solved.

6.2 The Behaviour of Non-Newtonian Materials

An incompressible Newtonian fluid has a constitutive equation of the form (Batchelor 1967)

$$\boldsymbol{\sigma} = -p\mathbf{I} + 2\eta(T)\mathbf{D} \tag{6.1}$$

where $\boldsymbol{\sigma}$ is the stress tensor, p is the pressure, $\mathbf{I}$ is the unit tensor, $\eta(T)$ is the viscosity (which may be a function of temperature, T, but is not dependent on the instantaneous shear rate in the fluid) and $\mathbf{D}$ is the *rate-of-strain* tensor given by

$$\mathbf{D} = \tfrac{1}{2}\{\nabla \mathbf{u} + (\nabla \mathbf{u})^{\mathrm{T}}\} \tag{6.2}$$

where ∇ is the gradient operator and $\mathbf{u}$ is the velocity vector. The viscosity is defined in a simple shearing flow, which is assumed to be in the xy plane. If the direction of motion lies on the x axis, then the viscosity is given by $\eta(T) = \sigma_{xy}/\dot{\gamma}$, where $\dot{\gamma}$ is the shear rate $(\partial u_x/\partial y)$ and σ_{xy} is the shear stress. Thus the shear stress in a Newtonian fluid depends only on the instantaneous velocity field, and the viscosity is constant under isothermal conditions; the pressure part of the total stress depends on the boundary conditions as well as on the motion; it is not determined by the velocity field alone.

The range of non-Newtonian behaviour and the associated flow phenomena is very large, and a complete description is well beyond the scope of this chapter. A broad description of the behaviour in both steady and unsteady flow situations, together with mathematical models, can be found for example in Bird et al. (1987), Walters (1975), Larson (1988) and Tanner (1988).

A non-Newtonian liquid may be *viscoelastic*, which reflects the fact that the liquid exhibits both viscous properties and elastic properties. We note, however, that not all non-Newtonian models possess elastic properties, as there are many examples of inelastic liquids having a shear-rate-dependent viscosity (Bird et al. 1987). In some flow geometries the elastic properties are of little significance and only the viscous properties need to be modelled. Thus the flow of molten plastic through processing equipment or within moulds can sometimes be adequately predicted by modelling only the viscous behaviour of the material. On the other hand there are a variety of processing situations in which the elastic properties cannot be neglected. Examples of this class of flow include the polymer extrudate swell phenomenon and the process of melt fibre drawing.

6.2.1 Viscous Behaviour

When the elastic properties of the liquid are negligible or elastic effects in the flow can be neglected, then a simple viscous model will be used. In this case the viscosity will be given as a function of both temperature and instantaneous shear rate, $\dot{\gamma} = \sqrt{2\mathrm{tr}(\mathbf{D}^2)}$, where "tr" represents the trace operation. The constitutive model can then be written in the form

$$\boldsymbol{\sigma} = -p\mathbf{I} + 2\eta(T,\dot{\gamma})\mathbf{D} \tag{6.3}$$

This model is often referred to as the *Generalized Newtonian Fluid*, and includes the Newtonian fluid as a particular case. Once again the shear stress depends only on the instantaneous flow field and no memory is present. Polymer melts and polymer solutions usually exhibit *shear-thinning* (or *pseudo-plastic*) properties, where the viscosity reduces as the shear rate increases; in some cases a yield stress is present. In the latter case no motion occurs until the stress in the material exceeds a yield value, as in plasticity.

6.2.2 Viscoelastic Behaviour

Elastic behaviour in a liquid is indicated by a tendency for the liquid to recoil after a deformation has been imposed. Such fluids therefore have a memory of the original configuration and are often termed *memory fluids*. However, if the deformation is imposed and maintained for some time the elastic stresses eventually relax, and the tendency to recoil diminishes. The fluids are therefore said to have a *fading memory*; recent configurations have a greater influence than more distant configurations. The stress in the liquid is then dependent not only on the instantaneous flow field but also on the strain history, and the constitutive models must take this into account. A *relaxation time* is often introduced to quantify the elasticity (or memory) of such liquids.

The constitutive equation in a viscoelastic liquid can be written in the form

$$\boldsymbol{\sigma} = -p\mathbf{I} + 2\eta_s\mathbf{D} + \boldsymbol{\tau} \tag{6.4}$$

where $2\eta_s\mathbf{D}$ represents a Newtonian contribution and $\boldsymbol{\tau}$ is referred to as the *extra stress* tensor. In a polymer melt or a concentrated polymer solution the Newtonian component can be neglected. The extra stress tensor is then given by additional integral or differential models, which represent both the viscous and elastic behaviour of the polymer component. A great variety of viscoelastic models has been proposed and the reader is referred to the references cited above for a more complete summary. Some examples will be given below.

6.2.3 Elongational Flows and Shear Flows

An interesting manifestation of elastic properties is observed in an elongational flow. In this flow the fluid sample undergoes uniaxial extension. The resistance to elongation can be represented in terms of an *elongational viscosity* (or *Trouton viscosity*) η_T analogous to the shear viscosity described above. If the elongation is occurring in the x_1 direction then the elongational viscosity is given by $\eta_T(\dot{\epsilon}) = (\sigma_{11}-\sigma_{22})/\dot{\epsilon}$, where $\dot{\epsilon}$ is the strain rate $\partial u_x/\partial x$ (and $\sigma_{22} = \sigma_{33}$ in this flow).

It is a simple matter to show that for a Newtonian liquid the elongational viscosity is exactly three times the shear viscosity. In the case of a polymer melt however, the steady-state elongational viscosity initially increases as the strain rate is increased, then reduces at higher strain rates. Thus the behaviour in elongation is strikingly different to the behaviour in shear. This difference is illustrated in Fig. 6.1, which shows the steady-state shear and elongational viscosity

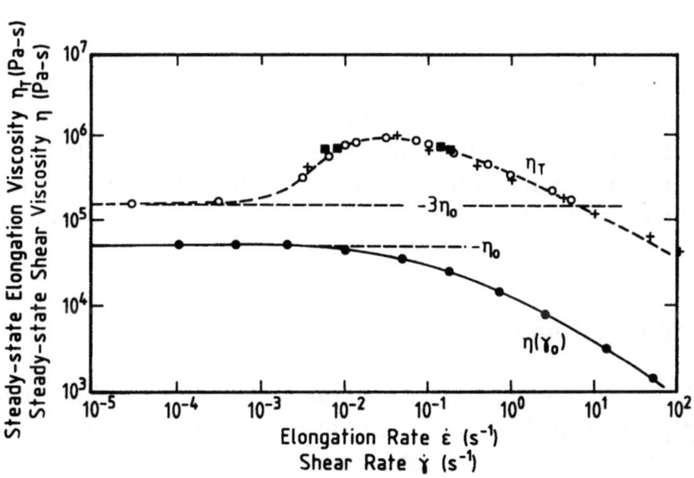

Fig. 6.1. Steady-state shear viscosity and elongational viscosity for a low density polyethylene at 150°C: ●, shear data from Meissner (1975); – – –, elongational data from Laun and Munstedt (1978).

for a sample of low-density polyethylene at 150°C (Meissner 1975, Laun and Munstedt 1978). Since many polymer processing operations involve elongational flow, or elongational-like flow, the elongational properties of the polymer are of importance. For example, the elongational behaviour has been found to have a strong influence on the swelling of an extruded jet of viscoelastic liquid (Bush et al. 1985).

In conclusion, it is important to stress that the above account is a very brief summary of the material properties and some forms of response in steady-state flows. There are a great many other flow phenomena observed in both steady flow and unsteady flow situations, and the reader is referred to the quoted references for further details.

6.3 Governing Equations

The previous section outlined the forms of non-Newtonian constitutive behaviour that have been included in boundary-element models of polymer flow. In this section the complete set of governing equations is presented.

6.3.1 Conservation Equations

The momentum and continuity equations governing the flow of an *incompressible* fluid are (Batchelor 1967)

$$\nabla \cdot \boldsymbol{\sigma} - \rho \frac{D\mathbf{u}}{Dt} + \rho\mathbf{f} = \mathbf{0}, \nabla \cdot \mathbf{u} = 0 \tag{6.5}$$

where ρ is the fluid density, $\mathbf{f}$ is a body force (per unit mass) and D/Dt is the *material derivative* defined by

$$\frac{D}{Dt} \equiv \frac{\partial}{\partial t} + \mathbf{u} \cdot \nabla \tag{6.6}$$

The importance of inertia effects relative to viscous effects in a flowing fluid are measured by the *Reynolds number*. In most polymer forming situations the Reynolds number is quite small ($\ll 1$) and the flow is dominated by viscous effects: the mass acceleration $\rho\, D\mathbf{u}/Dt$ and the body force $\rho\mathbf{f}$ can both be ignored, and these simplifications will be assumed in this chapter.

When temperature variations in the flow are significant, then the effects of temperature on the fluid properties must be included in the analysis. The temperature field can be determined by solution of the energy equation (Tanner 1988).

6.3.2 Constitutive Equations

If the fluid is to be modelled as a generalized Newtonian fluid then the total stress is simply given by Eq. (6.3). Several models have been proposed for the relationship between viscosity and shear rate (Bird et al. 1987). The simplest of these is the *power-law fluid*:

$$\eta(\dot{\gamma}) = K\,\dot{\gamma}^{n-1} \tag{6.7}$$

The nature of the viscous behaviour represented by the model depends on the ranges of the power-law index. In general terms these are:

- $n < 1$ Shear thinning (pseudoplastic)
- $n = 1$ Newtonian
- $n > 1$ shear thickening (dilatant)

Figure 6.1 clearly shows that the viscosity of the polymer melt approaches an upper limit as the shear rate approaches zero. This limit is the *zero-shear-rate viscosity* of the fluid. Note, however, that when $n < 1$ the power-law model produces an unbounded viscosity as the shear rate approaches zero. In practice this does not represent a severe problem, since the shear stress $\eta(\dot{\gamma})\dot{\gamma}$ approaches zero at small shear rates. The power-law model is therefore frequently used to model the flow of polymer materials. The simplicity of the model is a distinct advantage, as the two parameters are relatively easily determined by fitting the model to viscosity–shear-rate data. Both the parameters K and n are constant at a given temperature.

A more complicated (and more realistic) behaviour is represented by the *Carreau* model, which takes the form

$$\eta(\dot{\gamma}) = \eta_\infty + (\eta_0 - \eta_\infty)[1 + (\Gamma\dot{\gamma})^2]^{\frac{n-1}{2}} \tag{6.8}$$

In this case the model incorporates a high-shear-rate asymptote for the viscosity, η_∞, and a low-shear-rate asymptote, η_0, which overcomes the conceptual difficulties in the power-law model. In addition, the parameter Γ has the units of time and can be referred to as *time constant*. This should not be confused with a relaxation time used to quantify fluid memory, since the generalized Newtonian fluid cannot model elastic properties.

If the full viscoelastic nature of the fluid is to be modelled then more complicated models are required (Larson 1988). The total stress in the fluid is given by Eq. (6.4). The relationships for the extra stress $\boldsymbol{\tau}$ can be in *integral* form or *differential* form, and relate the stress field to the strain history of the fluid. There are a great variety of both integral and differential models, and only representative examples are given here.

A simple form of integral model is the Lodge–Maxwell model (Tanner 1988)

$$\boldsymbol{\tau} = \int_{-\infty}^{t} \left(\sum_{k=1}^{\infty} \frac{\eta_k}{\lambda_k^2} e^{-(t-t')/\lambda_k} \right) \mathbf{c}^{-1}(t')\, dt' \tag{6.9}$$

In this, η_k is the *molecular* viscosity (i.e. the contribution to the viscosity due to the polymer component) corresponding to the relaxation time λ_k, and the summation reflects the presence of a relaxation spectrum. The relaxation spectrum for a particular liquid can be measured using a variety of unsteady test procedures (Ferry 1980). For practical purposes the spectrum is usually discretized before implementation of the model. The term $\mathbf{c}^{-1}(t')$ is a relative strain tensor that links the deformation of a fluid particle at the past time t' to the deformation of the particle at the present time t. An integral of this sort can be viewed as an integral along a particle path, or stream line, in the flow.

If the differential models are to be used we must first represent the extra stress as the sum of components contributed by all the relaxation modes in the relaxation spectrum,

$$\boldsymbol{\tau} = \sum_{k=1}^{\infty} \boldsymbol{\tau}^k \tag{6.10}$$

and then supply differential models for the individual $\boldsymbol{\tau}^k$.

A commonly employed form of differential model is the *(upper) convected* Maxwell model:

$$\lambda_k \frac{\Delta \boldsymbol{\tau}^k}{\Delta t} + \mathbf{R}^k = \mathbf{0}\,, \quad \mathbf{R}^k = \boldsymbol{\tau}^k - 2\eta_k \mathbf{D} \tag{6.11}$$

in which λ_k is the relaxation time corresponding to the kth relaxation mode. The differential term is the *upper convected derivative* defined as

$$\frac{\Delta \boldsymbol{\tau}}{\Delta t} = \frac{\partial \boldsymbol{\tau}}{\partial t} + \mathbf{u} \cdot \nabla \boldsymbol{\tau} - \mathbf{L}\,\boldsymbol{\tau} - \boldsymbol{\tau}\,\mathbf{L}^{\mathrm{T}} \tag{6.12}$$

where $\mathbf{L} = (\nabla\,\mathbf{u})^{\mathrm{T}}$ is the *velocity gradient tensor* (note $L_{ij} = \partial u_i/\partial x_j$ here). Different forms of model behaviour result from different choices for the function $\mathbf{R}^k$.

The *Phan-Thien–Tanner model* (Phan-Thien and Tanner 1977) arises if we take

$$\mathbf{R}^k = \left\{1 + \epsilon \frac{\lambda_k}{\eta_k} \operatorname{tr}(\boldsymbol{\tau}^k)\right\} \boldsymbol{\tau}^k + \lambda_k \zeta (\mathbf{D}\,\boldsymbol{\tau}^k + \boldsymbol{\tau}^k\,\mathbf{D}) - 2\eta_k \mathbf{D} \tag{6.13}$$

In this, η_k is the zero-shear-rate viscosity of the kth relaxation mode and $\operatorname{tr}(\boldsymbol{\tau})$ represents the trace of $\boldsymbol{\tau}$. The parameters ϵ and ζ are of order 0.01 and 0.1 respectively. The elongational behaviour of the model is determined by ϵ, while ζ affects the shear viscosity behaviour. If $\epsilon = \zeta = 0$ then the model reverts to the differential form of the Maxwell model of Eq. (6.11). In this case, the model is referred to as the *Maxwell fluid* if η_s in Eq. (6.4) is zero (a polymer melt), or the *Oldroyd-B fluid* if η_s is nonzero (a polymer solution). The Phan-Thien–Tanner model (with both ϵ and ζ nonzero) exhibits more realistic viscous and elongational behaviour than that of the Oldroyd-B or Maxwell Forms (Bush et al. 1985). Another model that has been proposed is the *Leonov* (1976) model (for which see Larson 1988).

In Fig. 6.2 we illustrate the steady shear viscosity and the steady elongational viscosity functions of the Maxwell and Phan-Thien–Tanner models. Note that the elongational viscosity of the Maxwell model is unbounded. This appears to have a significant (adverse) effect on our ability to make calculations with this model (Bush et al. 1985). Note also that the steady shear viscosity of the Maxwell model is constant. The other models exhibit a more realistic shear-thinning behaviour and a bounded elongational viscosity, however they fail to represent the thinning observed at high strain rates in Fig. 6.1. It is a simple matter to modify the Phan-Thien–Tanner model to account for this behaviour if necessary (Tanner 1988).

All of the above equations apply to the isothermal case. If temperature variations in the flow are significant then the effects of temperature on the fluid properties must be included. This results in modification of the equations, and the introduction of additional relations (see Sugeng et al. 1987). We make no attempt in this chapter to summarize these modifications.

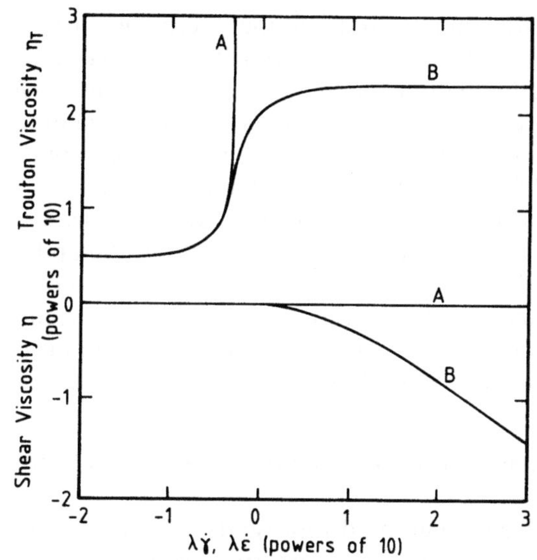

Fig. 6.2. Steady-state shear viscosity and elongational viscosity functions for two single relaxation mode models, with relaxation time λ, under isothermal conditions: A, Maxwell model; B, Phan-Thien–Tanner model with $\epsilon = 0.01$ and $\zeta = 0$ (Eq. 6.13); The high strain rate asymptotes for the elongational viscosity in case B is controlled by ϵ. In all cases, the zero shear rate viscosity is one unit.

6.3.3 Boundary Conditions

The mathematical description of the problem is not complete until boundary conditions have been properly specified. For the solution of Eq. (6.5), velocity or traction components must be prescribed in each of the coordinate directions over the whole of the boundary surface. Without such conditions, a unique solution cannot be found. The energy equation also needs boundary conditions on temperature and/or heat flux.

On the other hand, Eq. (6.11) represents a set of hyperbolic differential equations, which require the stresses to be prescribed on the sections of the boundary through which fluid enters the computational domain. If the integral form, Eq. (6.9), is to be used instead, then the stress tensor of a given fluid particle at a given time must be known, which reduces the integral in Eq. (6.9) to an integral over the strain history of the particle since that time.

If the problem involves a free surface, then two types of boundary conditions must be satisfied simultaneously on the surface. Firstly, the traction forces acting on the surface must be known; these will frequently be taken as zero if surface tension effects and surface drag are negligible. Secondly, the free surface is a stream surface, so that $\mathbf{u} \cdot \mathbf{n} = 0$ on the surface. The geometry of the free surface must be found to satisfy both sets of boundary conditions simultaneously. This necessitates the use of an iterative procedure.

6.4 Boundary Integral Methods

Having defined the governing equations, we can now examine the boundary integral formulations used to solve non-Newtonian flow problems. In this section the formulations are summarized without reference to specific applications. Discussion of applications will be reserved for the next section.

Since we are dealing with a nonlinear problem, an iterative procedure will be required. It is sometimes convenient to solve the momentum and constitutive equations by different means, which will cause us to compute the stress field and velocity field in different steps. We need to include the possibility that the geometry of the problem is not known in advance, as would be the case in problems involving a free surface, which necessitates the updating of the geometry during each cycle of iteration. Separate solution of the momentum, constitutive, energy and boundary update equations allows different techniques to be used in each step, which permits some optimization of the solution process. Conversely, it limits the range of iterative schemes that can be used.

6.4.1 Integral Representations of Governing Equations

The flow of a non-Newtonian fluid represents a nonlinear material problem, and as such must usually be solved as a perturbation to a base Newtonian flow (Bush et al. 1984, 1985). To this end, we write the stress tensor in the form

$$\boldsymbol{\sigma} = -p\mathbf{I} + 2\eta_{\mathrm{N}}\,\mathbf{D} + \boldsymbol{\epsilon} \qquad (6.14)$$

where η_{N} is a constant Newtonian viscosity and $\boldsymbol{\epsilon}$, which represents the non-Newtonian component of the extra stress tensor, describes the nonlinear part of the problem. In the case of the generalized Newtonian liquid, Eq. (6.3),

$$\boldsymbol{\epsilon} = 2(\eta(\dot{\gamma}) - \eta_{\mathrm{N}})\mathbf{D} \qquad (6.15)$$

where η_{N} is arbitrary. In the case of the viscoelastic model, Eq. (6.4),

$$\eta_{\mathrm{N}} = \eta_{\mathrm{s}} + \eta_0, \quad \boldsymbol{\epsilon} = \boldsymbol{\tau} - 2\eta_{\mathrm{N}}\,\mathbf{D} \qquad (6.16)$$

where η_0 is the zero-shear-rate viscosity of the viscoelastic model. This is given by

$$\eta_0 = \sum_{k=1}^{\infty} \eta_k \qquad \text{(Phan-Thien–Tanner model)} \qquad (6.17)$$

When Eq. (6.14) is substituted into the momentum Eq. (6.5) the result can be written in the form

$$\boldsymbol{\nabla}\cdot\boldsymbol{\sigma}^{\mathrm{N}} + \boldsymbol{\nabla}\cdot\boldsymbol{\epsilon} = \mathbf{0} \qquad (6.18)$$

where $\boldsymbol{\sigma}^{\mathrm{N}}$ represents the Newtonian component of the total stress tensor: $\boldsymbol{\sigma}^{\mathrm{N}} = -p\mathbf{I} + 2\eta_{\mathrm{N}}\mathbf{D}$. Thus the stress component $\boldsymbol{\epsilon}$ gives rise to a *pseudo-body force* term. The development of an integral equation then follows well-known methods previously described in elasticity or fluid mechanics (Brebbia 1980, Banerjee and Butterfield 1981, Bush and Tanner 1983). However, in this case we are manipulating only the Newtonian component of the total stress tensor and the boundary traction terms that are generated in the process represent the Newtonian component of the total traction force. That is, we generate $\mathbf{t}^{\mathrm{N}} = \boldsymbol{\sigma}^{\mathrm{N}}\cdot\mathbf{n}$ where the total traction force is

$$\mathbf{t} = \boldsymbol{\sigma}\cdot\mathbf{n} = \mathbf{t}^{\mathrm{N}} + \boldsymbol{\epsilon}\cdot\mathbf{n}\,, \qquad (6.19)$$

and **n** is the outward-directed unit normal vector at the surface. The resulting equation can be written using Cartesian tensor notation as

$$c_{ij}(P)u_j(P) = \int_\Gamma t_j^N(Q)\, u_{ij}^*(P,Q)d\Gamma - \int_\Gamma u_j(Q)\, t_{ij}^*(P,Q)d\Gamma$$

$$+ \int_\Omega \left\{\rho b_j(Q)u_{ij}^*(P,Q) + \frac{\partial \epsilon_{jk}}{\partial x_k}(Q)u_{ij}^*(P,Q)\right\}d\Omega \qquad (6.20)$$

This equation applies to a region Ω with boundary Γ, and relates the velocity $u_j(P)$ at a point P (on the boundary or in the domain) to the boundary values of velocity and traction and the extra stress field in the domain. The function $c_{ij}(P)$ depends on the location of the point P. If P lies in the domain then $c_{ij}(P) = \delta_{ij}$, where δ_{ij} is the Kronecker delta function ($\delta_{ij} = 1$ if $i = j$, $\delta_{ij} = 0$ if $i \neq j$). On the other hand, if P lies on the boundary, $c_{ij}(P) = \delta_{ij}/2$ if the boundary is smooth. The function $u_{ij}^*(P,Q)$ is the fundamental solution to the Stokes equation; it is called the *Stokeslet*. This is just the Kelvin fundamental solution (see Brebbia 1980) specialized for the case of an incompressible medium, and represents the velocity component in the x_j direction at a point Q due to a unit point force applied in the x_i direction at the point P. The Stokeslet is assumed to act in an infinite medium of Newtonian liquid with viscosity η_N. The function $t_{ij}^*(P,Q)$ is the corresponding traction field.

We can recover the total traction by integrating the domain integral containing the pseudo-body force, $\nabla \cdot \boldsymbol{\epsilon}$, by parts to give

$$c_{ij}(P)u_j(P) = \int_\Gamma t_j(Q)\, u_{ij}^*(P,Q)d\Gamma - \int_\Gamma u_j(Q)\, t_{ij}^*(P,Q)d\Gamma$$

$$+ \int_\Omega \left\{\rho b_j(Q)\, u_{ij}^*(P,Q) - \epsilon_{jk}(Q)\frac{\partial u_{ij}^*}{\partial x_k}(P,Q)\right\} d\Omega \qquad (6.21)$$

If the fluid is Newtonian then $\boldsymbol{\epsilon} = \mathbf{0}$ and Eq. (6.21) reverts to the classical boundary integral equation used by Youngren and Acrivos (1975), Ladyzhenskaya (1963) and Bush and Tanner (1983) to treat creeping viscous flow. Alternatively, the creeping viscous flow problem can be formulated in terms of derived variables (Wu and Rizk 1979, Bézine and Bonneau 1981, Coleman 1981, Kelmanson 1983). If the fluid inertia is not to be neglected, then the domain integral remains and an iterative procedure is required. Again both primitive variable and derived variable formulations have been used to effect (Skerget and Alujevic 1985; Onishi et al. 1984, 1985; Tasaka and Onishi 1985). If external body force loading corresponds to centrifugal loading or conservative forces, such as gravitation, then the part of the domain integral corresponding to these forces can be converted into equivalent boundary integrals (Cruse et al. 1974, Danson 1981). In this chapter we are interested in non-Newtonian flow problems; and the domain integral then only depends on the non-Newtonian stress components $\boldsymbol{\epsilon}$. For a given stress field, Eq. (6.21) can be discretized, modified according to the relevant boundary conditions, and solved for the velocity field.

Eq. (6.21) represents a primitive variable formulation of the momentum and continuity equations. A novel derived variable formulation has also been developed and utilized for non-Newtonian flow problems by Coleman (1981, 1984). The formulation described applies to planar (x,y) creeping rows, in which the stream function, ψ, and the *Airy stress function*, ϕ, are utilized. In addition, the complex variable $z = x + iy$ and the complex potential $\chi = \phi + 2\eta_N \psi i$ are introduced.

If the primitive variable formulation of Eq. (6.21) is to be used, then the stream function, if required, must be found after the velocity field has been calculated.

Finally, it should be noted that the equations listed above can be adapted to axisymmetric problems by transforming to cylindrical polar coordinates and making use of singular ring load solutions rather than point load solutions (Cruse et al. 1974, Kermanidis 1975).

6.4.2 Solution of the Constitutive Equations

For the solution of the constitutive equations we utilize different methods altogether. In the past the differential constitutive equations have been solved by finite-element methods (Crochet and Keunings 1982). However, an alternative approach that has been used successfully consists of integration of the differential equations along particle paths. To see how this can be done we define a variable, s, which is the distance measured along the particle path from some reference position. Then, the derivative of the stress tensor with respect to s is

$$\frac{d\boldsymbol{\tau}^k}{ds} = \frac{\partial \boldsymbol{\tau}^k}{\partial x_j}\frac{\partial x_j}{\partial s} = \frac{\partial \boldsymbol{\tau}^k}{\partial x_j}\frac{u_j}{|\mathbf{u}|} = \frac{1}{|\mathbf{u}|}\mathbf{u}\cdot\nabla\boldsymbol{\tau}^k \tag{6.22}$$

where $|\mathbf{u}|$ represents the magnitude of the velocity tensor. Under conditions of steady flow, the particle paths therefore become the characteristic curves of the partial differential equations in (6.11), and the equations can be integrated as a set of ordinary differential equations along the particle paths. A variety of well known techniques, such as Runge–Kutta or Gear algorithms, already exist for this purpose. Proper specification of the problem requires the stresses to be prescribed as boundary conditions on the inflow boundary.

If the integral constitutive models are to be used instead, then we must integrate over the time history of the fluid particle of interest, as indicated by Eq. (6.9). This time integral can be viewed again as an integral along the path of the particle. Integral models have been incorporated in both boundary-element (Coleman 1984) and finite-element (Luo and Tanner 1986) solutions of the momentum equations.

These methods of solution are in some ways superior to the more conventional finite-element approach to the solution of Eq. (6.11), in that more accurate calculation of the stresses near a stress singularity can be obtained without the need to include a fine mesh discretization. On the other hand, the finite-element approach can handle problems involving recirculating regions (having closed stream lines) without difficulty. Methods relying on tracking of particle paths are less easily adapted to this class of problem, since there is no beginning or end to the closed stream lines. However, effective algorithms for this purpose are now being developed.

Another means of circumventing this limitation, without going to finite-element approaches, has been proposed and applied by Sugeng et al. (1988). This approach could best be called a *pseudo-time-stepping method*. The formulation is achieved by retaining the time derivative in Eq. (6.11) and writing the equation in the form

$$\frac{\partial \boldsymbol{\tau}^k}{\partial t} = (\mathbf{L}\,\boldsymbol{\tau}^k + \boldsymbol{\tau}^k\,\mathbf{L}^{\mathrm{T}} - \mathbf{u}\cdot\nabla\,\boldsymbol{\tau}^k) - \frac{1}{\lambda_k}\mathbf{R}^k \tag{6.23}$$

A first-order Euler finite difference is then used to replace the time derivative, which allows the stress at time $t + \Delta t$, $\boldsymbol{\tau}^k(t+\Delta t)$, to be written as

$$\boldsymbol{\tau}^k(t+\Delta t) = \boldsymbol{\tau}^k(t) + \Delta t\left\{\mathbf{L}\,\boldsymbol{\tau}^k(t) + \boldsymbol{\tau}^k(t)\,\mathbf{L}^{\mathrm{T}} - \mathbf{u}\cdot\nabla\,\boldsymbol{\tau}^k(t) - \frac{1}{\lambda_k}\mathbf{R}^k(t)\right\} \tag{6.24}$$

After computing the updated velocity field, the corresponding stress field can be obtained simply by performing one of these *pseudo-time steps*, and the process repeated until a steady-state solution has been achieved. The partial derivatives of the stress tensor on the right-hand side of Eq. (6.24) are estimated by differentiation of interpolating functions in the domain. The potential source of error represented by this last step necessitates the use of fine meshes. Furthermore, the time-stepping approach is slow to converge as relatively small time steps are required. However, since the emphasis on particle tracking has been removed, the approach can be applied to problems involving recirculating flows without difficulty. Also, in the creeping flow case, one can clearly adapt this method for transient problems in viscoelastic fluids.

Having defined the relevant equations in the foregoing discussion, it remains to discretize the integral equations and generate solutions. Discretization requires breaking the boundary into a number of boundary elements and dividing the domain into a number of domain cells. After selecting suitable approximating functions on these elements and cells, and incorporating the necessary boundary conditions, systems of simultaneous equations can be generated which will yield the unknown boundary data. Once the boundary solution has been found, the internal solutions can be obtained by reapplication of the integral equations. In this manner the velocity field can be found by application of Eq. (6.21) and the stream function can be found by integrating a Poisson equation.

6.5 Applications to Forming Problems

The foregoing discussion gives a summary of the boundary-element formulations that have been used to solve non-Newtonian flow problems. In this section a variety of solutions are presented which illustrate the application of the method to this class of problems.

6.5.1 Extrudate Swell and Die Design

This problem has been the subject of several numerical studies, using both finite-element methods and boundary-element methods (see Tanner 1988). The extrusion of a fluid from a long circular die is shown in Fig. 6.3, together with the necessary boundary conditions. One of the difficult features of this problem is the presence of a stress singularity at the exit lip of the die. This necessitates the use of a fine mesh in this region to achieve accurate results. The finite-element method is

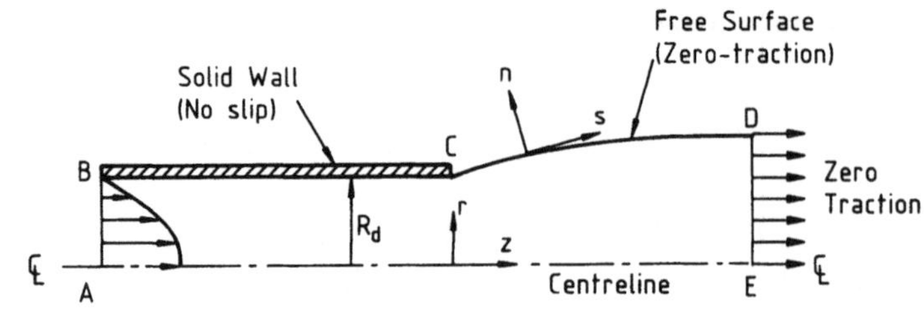

Fig. 6.3. The axisymmetric extrudate swell problem.

subject to pressure field fluctuations in the region of the singularity and these may lead to instability. Since the pressure field is not computed in the boundary-element approach, this problem does not arise with this method.

To illustrate the method a solution of Bush (1987) is reproduced here. In this case the integral equation employed for solution of the velocity field is Eq. (6.21). The constitutive equation used is the Phan-Thien–Tanner model (with $\eta_s = 0$, $\epsilon = 0.01$ and $\zeta = 0$), with a single relaxation mode. This was solved by integration of the corresponding set of ordinary differential equations along the streamlines using a fourth order Runge–Kutta algorithm, as outlined by Eq. (6.22). We use the Weissenberg number, $\lambda\dot{\gamma}_w$, where $\dot{\gamma}_w$ is the shear rate at the die-wall far upstream, as a measure of the importance of elasticity in the flow. The flow is taken to be isothermal.

The discretization used 44 constant boundary elements and 259 linear domain cells. The upstream and downstream boundaries for the computational domain were taken to be at $z_{min} = -3.2R_d$ and $z_{max} = 4R_d$. The swell ratio χ, jet radius/R_d, is plotted as a function of Weissenberg number in Fig. 6.4. Some results by Bush et al. (1985) which used both finite-element and boundary-element codes to approach the Oldroyd-B case ($\epsilon = 0$) by using small values of ϵ (0.01, 0.005 and 0.00) may also be noted.

Phuoc and Tanner (1980) discovered that if there is a significant difference in temperature between an extruded jet of *Newtonian* liquid and the surroundings,

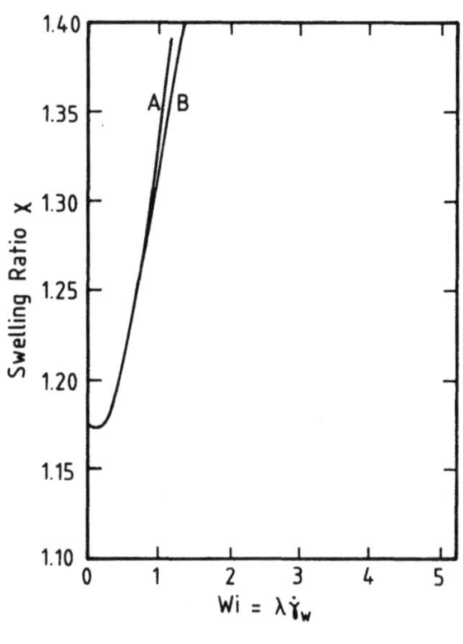

Fig. 6.4. Swell ratio as a function of Weissenberg number, Wi, for the axisymmetric jet computed by a boundary element method (Bush 1987); the Weissenberg number Wi is defined as $\lambda\dot{\gamma}_w$, where $\dot{\gamma}_w$ is the wall shear rate far upstream.

then the viscosity variations in the fluid due to temperature variations can lead to significant swelling (~70%). The numerical method used in this work was the finite-element method. Sugeng et al. (1987) have extended this study to investigate the effects of temperature variations in viscoelastic liquids. In this latter study, the boundary-element formulation of Eq. (6.21) was used to solve the momentum equations, while the finite-element formulation of Phuoc and Tanner (1980) was used to solve the energy equation. The reason for choosing the finite-element method for the latter purpose is its ability to handle the strong convective terms present in the energy equation. The constitutive equation was solved by integration along stream lines as above. The simulation therefore represents an interesting mixture of individual techniques, chosen for their particular advantages in particular applications.

Some results for the simulation of the isothermal extrusion of molten polyethylene from a circular die were given by Luo and Tanner (1986, 1988) using a finite-element method coupled with an integral constitutive model. Some experimental measurements for this problem are available (Meissner 1975) and the simulations agree well with the measurements especially at lower extrusion rates. The success of this work has prompted an investigation into the use of differential constitutive models for the solution of the problem (Bush 1987). The simulations all agree very well and closely represent the experimental data, except at an apparent shear rate of 10 s^{-1}. The reason for this discrepancy between the predicted and experimental results is not totally clear; however, a likely cause is the onset of wall slip in the experiments. This is known to reduce the swelling relative to the case of no slip (Luo and Tanner 1988). Wall slip was not modelled in the numerical work; see Phan-Thien (1988) for some recent work in this area.

The fact that several different constitutive model types and different solution techniques yielded similar results, when applied to the same problem, is a very encouraging outcome indeed.

The work reported above is confined to plane or axisymmetric flows. Of greater practical interest is the computation of flows through noncircular dies and the design of such dies. The pioneering work in this field is that of Bush and Phan-Thien (1985) and Tran-Cong and Phan-Thien (1988a), who studied the Newtonian case for square and triangular dies. Recently Tran-Cong and Phan-Thien (1988b) have extended these results to the viscoelastic case and to die design (Tran-Cong and Phan-Thien 1988b). These computations are very time consuming and further study of computer techniques are needed to cut down computational times. We believe that parallel processing techniques have much to offer here. Fig. 6.5

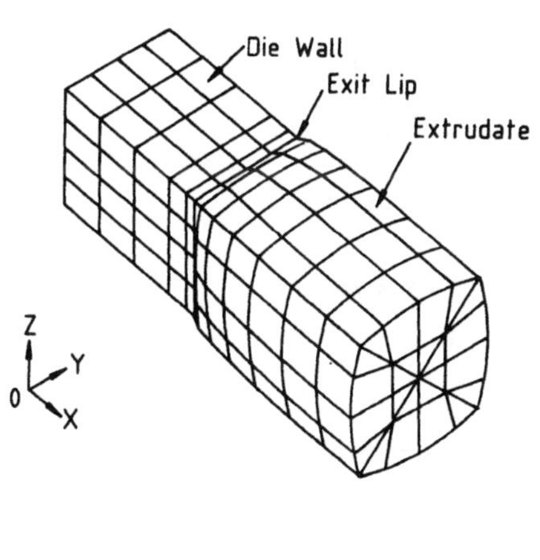

Fig. 6.5. Isometric view of final extrudate shape (square die, Wi = 0.9) (Maxwell fluid.)

shows some results obtained for a square die at a Weissenberg number (Wi) of 0.9; here Wi = $\lambda\dot{\gamma}_w$, where $\dot{\gamma}_w$ is the shear rate at the centre of a side far upstream; the swelling and shape change is clear. When we turn to the question of die design to produce a given shape of extrudate, we note that the exact solution is often not available even for the Newtonian fluid case. If the die cross-section is uniform along its length, it is very difficult to specify an exact inlet profile. Because of this inexactly prescribed velocity profile the die length must be long so that the flow could become fully developed before it reaches the exit lip. Even if this condition is met, convergence of a simple numerical iterative process is not guaranteed. This is due to the nature of our iterative scheme: we fix the prescribed velocity profile and iterate on the die cross-section; the flow rate therefore fluctuates between iterations. If these fluctuations are large, then the iterative procedure diverges. It is possible to rescale the velocity profile to keep the flow rate constant between iterations. However, a much better method, which is recommended, is the "taper" design approach. In this approach, a simple inlet cross-section is chosen so that a fixed inlet velocity profile can be prescribed. The cross-section of the inlet part of the die (about 2/5 of the die length) is not updated and thus the flow rate remains constant; this method is also the usual practice in die-making and it is numerically much simpler to implement. At the downstream section of the discretized flow domain, the plug flow condition is specified, i.e. zero traction in the flow direction and zero radial velocity components.

The above analysis was used to design dies for square and equilateral triangular (Newtonian) extrudates.

In practice, extrusion dies usually have a gradual transition from a simple cross-section (e.g. square, rectangular or circular) to the final profile forming section. Therefore, a tapered design was investigated. The objective of the design is to produce a non-uniform die with a known inlet velocity profile to produce the square extrudate. To achieve this, the die is divided into three segments. The inlet segment is of square cross-section and is fixed between iterations. The length of this segment should be reasonably long so that the prescribed velocity profile is exact. We found that a length of two units for the inlet segment is adequate. The next segment is 1.6 units long, which provides a tapered transition between the inlet segment and the final exit segment. The exit segment is of uniform cross-section and its length is 1.4 units (one unit is the length of a side). The swelling of the die profile relative to the required extrudate profile is +2.6% across the corners and −18.4% at the midside points (Fig. 6.6).

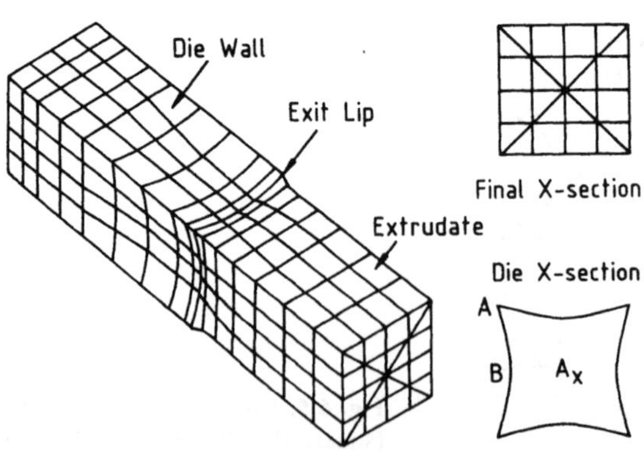

Fig. 6.6. "Taper" die design: Newtonian fluid (square cross-section extrudate).

In another example, the design of extrusion dies for a triangular extrudate is attempted. The "taper" design produces a die profile with a swelling of +7.9% at the corners (point A) and −24.9% at the midside points (point B); see Fig. 6.7. There is a slight bulging at point C which is not noticeable in the scale used in this figure.

Our die profile for a square extrudate compares well with some existing dies in use (Michaeli 1984, Fig. 5.87, p 249), but we have not been able to find any reference to a die profile that produces a triangular extrudate.

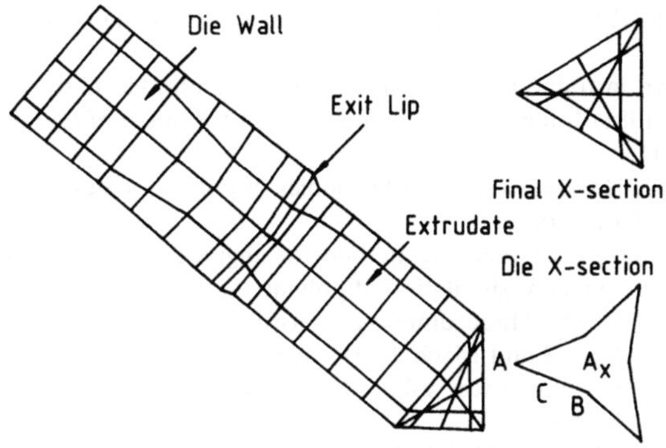

Fig. 6.7. "Taper" die design: Newtonian fluid (triangular cross-section extrudate).

6.5.2 Calendering of a Sheet

Another problem that has recently been solved (Zheng and Tanner 1988) by the pseudo-transient boundary element method is the calendering problem, where a thick mass of polymer is rolled into a sheet. The novelty of this problem resides in the fact that the point of separation is not known a priori. The final sheet thickness in the viscoelastic case is subject to a small amount of swelling after leaving the nip. A very fine discretization of the boundary is needed here; the elements on the boundary must not be greater in length than a fraction of the gap width. When this precaution is observed good agreement (in the Newtonian case) with lubrication theory is found. There is considerable recirculation at the inlet side of the rollers and hence the pseudo-time mode of computing given above (Eq. (6.24)) was used with the Phan-Thien–Tanner model. Due to the need to search for the separation point as part of the location of the free surface, the computations were long.

6.6 Conclusion

The applications summarized in this chapter indicate that the boundary-element method is a powerful tool in the solution of strongly nonlinear flow problems. A wide variety of formulations and implementations is possible. It would not be appropriate to try to identify either of the two methods, boundary elements or finite elements, as being the more powerful. In the linear case (Newtonian creeping flow) it appears that the boundary-element method is often faster for a given accuracy (Tanner 1988), but this is not always agreed (Karabin and Smelser 1989). In the nonlinear case it seems more prudent to state that the two methods are complementary, and neither should be discarded. In some circumstances, especially in computing tractions, the boundary-element method will be preferable, while in others the finite-element method will be appropriate. Perhaps the best formulation will be achieved by a mixing of the methods. At the very least, the two methods provide alternative solutions to the same problem, which is often an advantage.

References

Banerjee PK, Butterfield R (1981) Boundary element methods in engineering science. McGraw-Hill, London

Batchelor GK (1967) An introduction to fluid dynamics. Cambridge University Press, Cambridge

Bézine G, Bonneau D (1981) Integral equation method for the study of two dimensional stokes flows. Acta Mech 41:197–209

Bird RB. Armstrong RC, Hassager O (1987) Dynamics of polymeric liquids, vol. 1. Fluid Mechanics, 2nd edn. John Wiley, New York

Brebbia CA (1980) The boundary element method for engineers. Pentech Press, London

Bush MB (1987) Boundary element simulation of polymer extrusion processes. Engng Analysis 4:7–14

Bush MB, Phan-Thien N (1985) Three dimensional viscous flows with a free surface: flow out of a long square die. J Non-Newt Fluid Mech 18:211–218

Bush MB, Tanner RI (1983) Numerical solution of viscous flows using integral equation methods. Int J Num Meth Fluids 3:71–92

Bush MB, Milthorpe JF, Tanner RI (1984) Finite element and boundary element methods for extrusion computations. J Non-Newt Fluid Mech 16:37–51

Bush MB, Tanner RI, Phan-Thien N (1985) A boundary element investigation of extrudate swell. J Non-Newt Fluid Mech 18:143–162

Coleman CJ (1981) A contour integral formulation of plane creeping Newtonian flow. Q J Mech Appl Math 34:453–464

Coleman CJ (1984) On the use of boundary integral methods in the analysis of non-Newtonian fluid flow. J Non-Newt Fluid Mech 16:347–355

Crochet MJ, Keunings R (1982) J Non-Newt Fluid Mech 10:339–356

Cruse TA (1969) Numerical solution in three-dimensional elastostatics. Int J Solids Struct 5:1265–1274

Cruse TA, Snow DW, Wilson RB (1974) Numerical solutions in axisymmetric elasticity. Comput Struct 7:445–451

Danson DJ (1981) Boundary element methods, Brebbia CA (ed), Springer-Verlag, Berlin

Ferry JD (1980) Viscoelastic properties of polymers, 3rd edn. John Wiley, New York

Karabin ME, Smelser RE (1989) The boundary element in three-dimensional forging. In: Thompson EG, Wood RD, Zienkiewicz OC, Samuelsson A (eds) Numiform 89. Balkema, Rotterdam, pp 555–562

Kelmanson MA (1983) An integral equation method for the solution of singular slow flow problems. J Comput Phys 51:307–324

Kermanidis T (1975) A numerical solution for axially symmetrical elasticity problems. Int J Solids Struct 11:493–500

Ladyzhenskaya OA (1963) The mathematical theory of viscous incompressible flow. Gordon and Breach, New York

Larson RG (1988) Constitutive equations for polymer melts and solutions. Butterworths, Boston

Laun HM, Munstedt H (1978) Elongational behaviour of a low density polyethylene melt. Rheol Acta 17:415–425

Leonov AI (1976) Nonequilibrium thermodynamics and rheology of viscoelastic polymer media. Rheol Acta 15:85–98

Luo X-L, Tanner RI (1986) A streamline element scheme for solving viscoelastic flow problems. Part II: integral constitutive models. J Non-Newt Fluid Mech 22:61–89

Luo X-L, Tanner RI (1988) Finite element simulation of long and short circular die extrusion experiments using integral models, Int J Num Meth Engng 25:9–22

Meissner J (1975) Basic parameters, melt rheology, processing and end use properties of three similar low density polyethylene samples. Pure Appl Chem 42:553–612

Michaeli W (1984) Extrusion die design. Hanser, Munich

Onishi K, Kuroki T, Tanaka M (1984) An application of boundary element method to incompressible laminar viscous flows. Engng Analysis 1:122–127

Onishi K, Kuroki T, Tanaka M (1985) Topics in boundary element research, vol. 2. Brebbia CA (ed), Springer-Verlag, Berlin

Phan-Thien N (1988) Influence of wall slip on extrudate swell: a boundary element investigation. J Non-Newt Fluid Mech 26:327–340

Phan-Thien N, Tanner RI (1977) A new constitutive equation derived from network theory. J Non-Newt Fluid Mech 2:353–365

Phuoc HB, Tanner RI (1980) Thermally induced extrudate swell. J Fluid Mech 98:253–271

Rizzo FJ (1967) An integral equation approach to boundary value problems in classical elastostatics, QJ Appl Math 25:83–95

Skerget P, Alujevic A (1985) The solution of the Navier–Stokes equations in terms of vorticity–velocity variables by the boundary element method. Z Angew Math Mech 65:T245–T248

Sugeng F, Phan-Thien N, Tanner RI (1987) A study of non-isothermal non-Newtonian extrudate swell by a mixed boundary element and finite element method. J Rheol 31:37–58

Sugeng F, Phan-Thien N, Tanner RI (1988) A boundary element investigation of the pressure-hole effect. J Rheol 32:215–234

Tanner RI (1988) Engineering rheology, revised ed. Oxford University Press

Tasaka N, Onishi K (1985) Boundary integral equation formulations for steady Navier–Stokes equations using the Stokes fundamental solutions. Engng Analysis 2:128–132

Tran-Cong T, Phan-Thien N (1988a) Three dimensional study of extrusion processes by boundary element method. Part I. An implementation of high order elements and some Newtonian results. Rheol Acta 27:21–30

Tran-Cong T, Phan-Thien N (1988b) Three dimensional study of extrusion processes by boundary element method. Part II. Some non-Newtonian results. Rheol Acta 27:639–648

Walters K (1975) Rheometry. Chapman and Hall, London

Wu JC, Rizk YM (1979) Lecture notes in physics 90. Springer-Verlag, Berlin Heidelberg New York

Youngren GK, Acrivos A (1975) Stokes flow past a particle of arbitrary shape: a numerical method of solution. J Fluid Mech 69:377–403

Zheng R, Tanner RI (1988) A numerical analysis of calendering. J Non-Newt Fluid Mech 28:149–170

7 The Rigid-Plastic Finite-Element Method for Simulation of Deformation Processing

J.E. Jackson, Jr and M.S. Ramesh

7.1 Introduction

The use of finite-element methods (FEM) for metal-forming problems originated in the 1960s. One of the first approaches to the problem was the elastic-plastic finite-element method, developed by Marcal and King (1967). Yamada et al. (1968) and Lee and Kobayashi (1970, 1971) used FEM to solve a variety of problems in elastoplasticity, such as flat punch indentation, upsetting of solid cylinders, extrusion, and expansion of a hole in a plate. Relatively successful small-strain analysis of the above processes was achieved.

Several authors have published papers outlining the state-of-the-art in analysis. These include the work of Johnson and Sowerby (1978), Kobayashi (1979), Mahrenholtz (1982) and Haque et al. (1987).

Shabaik (1978, 1981, 1982) points out the distinctions between the various constitutive formulations used to simulate the deformations of metals. As shown in Fig. 7.1, the following four categories can be defined:

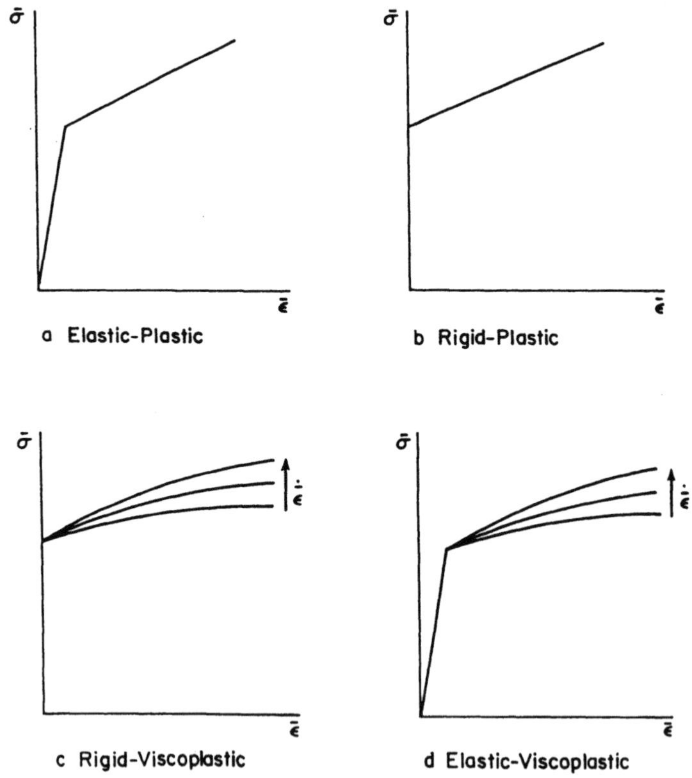

Fig. 7.1. Different types of material nonlinearities.

1. Rigid-plastic
2. Rigid-viscoplastic
3. Elastic-plastic
4. Elastic-viscoplastic

In reviewing the literature and also the available finite-element codes, it is apparent that two basic methods for solution of forging problems have evolved. These are (1) methods based on governing equations written with respect to a Lagrangian reference frame, and (2) those written with respect to an Eulerian reference frame. In general, finite-element codes assuming elastic-plastic and elastic-viscoplastic constitutive relations are formulated with respect to Lagrangian coordinate frames. Examples are MARC (MARC nd), ADINA (Bathe 1975), ABAQUS (1985), and NIKE2D (Hallquist 1986). Typically, these codes use the Updated Lagrangian Jaumann formulation described by Bathe (1982).

In contrast, the use of a rigid-plastic or rigid-viscoplastic constitutive relation often leads to a code based on an Eulerian reference frame. The following explanation for the two paths followed in the development of finite-element deformation processing codes is offered.

Since the Eulerian reference frame corresponds to a control volume approach, conservation of mass is not automatically satisfied within the equations of motion. Normally, a separate conservation of mass equation would be required. However, if the material is considered to be incompressible, the conservation of mass equation can be satisfied within the equations of motion through the use of a Lagrange multiplier or a penalty function.

In a Lagrangian reference frame, conservation of mass is automatically enforced. Thus, there is no additional conservation of mass equation necessary for an elastic-plastic or elastic-viscoplastic material model formulated within a Lagrangian reference frame.

In comparing the two basic approaches to deformation processing analysis, note that the Lagrangian approach making use of an elastic-plastic/viscoplastic constitutive relation allows calculation of residual stresses not available from an Eulerian rigid-plastic/viscoplastic approach. However, the stress and strain tensors used to formulate the rigid-plastic/viscoplastic method are simpler and, therefore, easier to code. Moreover, the relatively small elastic deformations are ignored in a rigid-plastic/viscoplastic analysis, thus allowing a less stringent convergence criterion for the nonlinear simultaneous equations. As a result, the rigid-plastic/viscoplastic codes tend to require much less computational expense. It should be noted that the two paths for forging analysis described above have become less distinct in recent years. For example, in order to describe powder material forming, methods have been developed which allow the inclusion of compressibility into an Eulerian-based formulation.

7.2 History of the Rigid-Plastic Finite-Element Method

The purpose of the preceding section has been to show how the rigid-plastic FEM fits into the total body of work on deformation processing simulation. Now the history of the rigid-plastic FEM will be outlined.

Lee and Kobayashi (1973) developed a "general matrix method" based on a variational principle for rigid-plastic deformation, the condition of incompressibility being applied using the Lagrange multiplier technique. The method proved to be very well suited to large deformation plasticity problems and was applied to solid cylinder compression, plane stress bore expansion, and flange drawing. Kobayashi (1977) applied an almost identical matrix method to analyse the cold-heading process, steady-state bar drawing and ring compression. In his paper, Kobayashi introduced a method to analyse processes in which only part of the body is deforming plastically. That is, a rigid zone is also present. Shah and Kobayashi (1974) presented detailed instructions on a method to handle rigid inclusions and also developed numerical methods to detect and model neutral points on die/workpiece boundaries. A detailed analysis of rigid-plastic ring compression was presented by Chen and Kobayashi (1978). Additional information on treatment of friction boundaries was provided by Hartley et al. (1980). Some additional applications of rigid-plastic FEM have been presented by Rebelo et al. (1982), Dung and Mahrenholtz (1982) and Tang and Kobayashi (1982a).

The above papers assume isothermal conditions. However, in many cases the coupling between nonuniform workpiece temperature distributions and the workpiece deformation is significant. Rebelo and Kobayashi (1980) developed an approach for such a coupled solution. Tang and Kobayashi (1982b) and Tang et al. (1982) have solved shell-nosing problems involving coupled heat transfer and deformation.

A particularly interesting study as to the applicability of the rigid-plastic FEM was discussed at the IUTAM Symposium held in 1978 at Tutzing, Germany. Kudo and Matsubara (1979) presented the results of 14 research groups using various codes for upsetting of solid cylinders. Comparisons were made for extent of yielded area, side profile, upsetting load, and folding prediction. The various analyses were categorized into two groups. One group used an elastic-plastic FEM (Prandtl–Reuss constitutive relation) and the second group used rigid-plastic FEM (Levy–Mises constitutive relation). Results were presented for perfectly plastic and linearly strain-hardening materials. Almost no difference was found between the elastic-plastic and rigid-plastic analyses. However, significant differences were found between the individual research groups. These differences were attributed to the number and shape of elements used, local increment, convergence criteria and method of incompressibility enforcement.

Since rigid-viscoplastic FEM is merely an extension of the older rigid-plastic FEM, many of the recent numerical advances, while done using a rigid-viscoplastic code, apply equally well for rigid-plastic applications.

7.3 Theory

7.3.1 Rate Form of Virtual Work

To formulate the rigid-plastic finite-element method, we begin with the following rate form of the virtual work equations (Chen and Han 1988):

$$\int_A T_i \dot{u}_i dA = \int_V \sigma_{ij} \dot{\epsilon}_{ij} dV \tag{7.1}$$

Here, body forces are neglected, T_i represents the surface traction components, u_i are velocity components and σ_{ij} is the Cauchy stress tensor. The rate-of-deformation tensor $\dot{\epsilon}_{ij}$ is defined by

$$\dot{\epsilon}_{ij} = \frac{1}{2}\left(\frac{\partial \dot{u}_i}{\partial x_j} + \frac{\partial \dot{u}_j}{\partial x_i}\right) \tag{7.2}$$

7.3.2 Flow Rule

In the flow theory of plasticity, increments of strain are related to stresses through a flow rule of the form

$$\dot{\epsilon}_{ij} = \gamma \frac{\partial G}{\partial \sigma_{ij}} \tag{7.3}$$

where

$$G = G(\sigma_{ij}, \epsilon_{ij}) \tag{7.4}$$

is a plastic potential function and

$$\gamma = \gamma[F(\sigma_{ij}, \epsilon_{ij})] \tag{7.5}$$

In the above note that the scalar γ is a function of the yield criterion F.

If $F < 0$ the material is rigid, while for $F > 0$ the material has yielded and has a nonzero strain rate. Assuming "associated flow", that is, the plastic potential and the yield function coincide, allows us to write $G = F$, which implies

$$\dot{\epsilon}_{ij} = \gamma \frac{\partial F}{\partial \sigma_{ij}} \tag{7.6}$$

For an isotropic, work-hardening material, the von Mises yield criterion F may be defined

$$F = \sqrt{J_2} - k$$

where J_2 is the second invariant of the deviatoric stress tensor and k is a parameter dependent upon the work-hardening characteristics of the material. By comparison with a uniaxial tension test it may be shown that

$$k = \frac{Y}{\sqrt{3}} \tag{7.7}$$

where Y is the yield stress in uniaxial tension. Evaluating the derivative in Eq. (7.7) gives

$$\frac{\partial F}{\partial \sigma_{ij}} = \frac{1}{2\sqrt{J_2}} \frac{\partial}{\partial \sigma_{ij}} [J_2] = \frac{1}{2\sqrt{J_2}} \frac{\partial}{\partial \sigma_{ij}} (\tfrac{1}{2} s_{ij} s_{ij}) = \frac{s_{ij}}{2\sqrt{J_2}} \tag{7.8}$$

In Eq. (7.8) the identity

$$J_2 = \tfrac{1}{2} s_{ij} s_{ij} \tag{7.9}$$

has been used, where

$$\sigma_{ij} = s_{ij} + \tfrac{1}{3} s_{kk}\, \delta_{ij} \tag{7.10}$$

Note that s_{ij} denotes the stress deviator tensor and δ_{ij} is the Kronecker delta.

Using the result of Eqs (7.6) and (7.8), the flow rule now may be written as

$$\dot{\epsilon}_{ij} = \frac{\gamma s_{ij}}{2\sqrt{J_2}} \tag{7.11}$$

Now square both sides of Eq. (7.11) to get

$$\dot{\epsilon}_{ij}\,\dot{\epsilon}_{ij} = \gamma^2 \frac{s_{ij}\,s_{ij}}{4J_2} = \gamma^2 \frac{2J_2}{4J_2} = \frac{1}{2}\gamma^2 \tag{7.12}$$

For a von Mises material model the *effective stress* $\bar{\sigma}$ can be shown to be

$$\bar{\sigma} = \sqrt{3J_2} = \sqrt{\tfrac{3}{2} s_{ij}\,s_{ij}} \tag{7.13}$$

while the *effective strain rate* $\dot{\bar{\epsilon}}$ may be written

$$\dot{\bar{\epsilon}} = \sqrt{\tfrac{2}{3} \dot{\epsilon}_{ij}\,\dot{\epsilon}_{ij}} \tag{7.14}$$

Using the above definitions for effective stress and strain rate yields the following relation for the scalar γ:

$$\gamma = \sqrt{2\,\dot{\epsilon}_{ij}\,\dot{\epsilon}_{ij}} = \sqrt{3}\,\dot{\bar{\epsilon}} \tag{7.15}$$

Therefore,

$$\dot{\epsilon}_{ij} = \frac{\sqrt{3}\,\dot{\bar{\epsilon}}\,s_{ij}}{2\sqrt{J_2}} = \frac{3}{2}\frac{\dot{\bar{\epsilon}}}{\bar{\sigma}} s_{ij} \tag{7.16}$$

or

$$s_{ij} = \frac{2}{3}\frac{\bar{\sigma}}{\dot{\bar{\epsilon}}} \dot{\epsilon}_{ij} \tag{7.17}$$

which is the *Levy–Mises equation*.

For a rigid-plastic work-hardening material, the effective stress may be written in the form

$$\bar{\sigma} = Y = Y_0 (1 + \beta\,\bar{\epsilon}) \tag{7.18}$$

where Y_0 is the initial yield stress and β is the strain-hardening coefficient of the material. Fig. 7.2 illustrates rigid-viscoplastic and rigid-plastic constitutive laws.

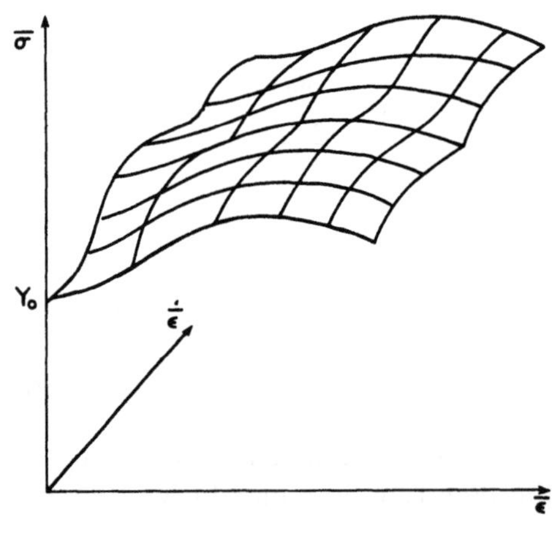

Fig. 7.2. The rigid-viscoplastic (strain rate dependent) and rigid-plastic constitutive laws.

In the former case stress is dependent upon both strain and strain rate, and is illustrated in the figure by a surface. The rigid-plastic case is described by a single line and is not dependent on strain rate.

7.3.3 Variational Formulation

The rate-of-work expression (7.1) may be rearranged to

$$\phi = \int_V s_{ij}\,\dot{\epsilon}_{ij}\,dV + \int_V \lambda\,\dot{\epsilon}_{ii}\,dV - \int_A T_i \dot{u}_i dA \tag{7.19}$$

where λ is the hydrostatic stress. The second term of the equation serves to enforce conservation of mass, in this case the incompressibility requirement. In the third term the traction T_i will be the friction force per unit area at the die/workpiece interface. To obtain the equilibrium equations, the functional (7.19) is made stationary with respect to the kinematically admissible velocity fields which satisfy compatibility and the velocity boundary conditions. Thus,

$$\delta\phi = 0 = \int_V s_{ij}\delta\dot{\epsilon}_{ij}dV + \int_V \lambda\delta\dot{\epsilon}_{ii}dV + \int_V \dot{\epsilon}_{ij}\delta\lambda dV - \int_A T_i\delta\dot{u}_i dA \tag{7.20}$$

As an alternative to the Lagrange multiplier approach, the incompressibility condition may be enforced by means of a penalty function. First, Eq. (7.1) is rewritten as

$$\int_A T_i\,\dot{u}_i dA = \int_V [s_{ij} - p\;\delta_{ij}]\dot{\epsilon}_{ij}\,dV \tag{7.21}$$

now set

$$p = -\,\alpha\frac{\partial \dot{u}_i}{\partial x_i} \tag{7.22}$$

where α is the penalty parameter and may be interpreted as a bulk stiffness.

After taking the variation of the functional the equilibrium equations may be expressed

$$\delta\phi = 0 = \int_V \left[s_{ij}\dot{u}_{ij} + \alpha\frac{\partial \dot{u}_i}{\partial x_i}\,\delta_{ij} \right] \delta\dot{\epsilon}_{ij}dV - \int_A T_i\dot{u}_i dA \tag{7.23}$$

The penalty parameter α is chosen to be a relatively large number, which serves to enforce the incompressibility condition.

In comparing the Lagrange multiplier and penalty function methods for enforcing incompressibility, note that the former results in more unknowns to be calculated. For example, for a four-node quadrilateral finite element in a plane strain or axisymmetric analysis, the element degrees of freedom will consist of eight nodal point velocities and the Lagrange multiplier λ. For the penalty function formulation, only the eight velocity degrees of freedom will be present; however, the penalty parameter sometimes leads to numerical ill-conditioning. Both formulations have been used to analyse deformation processing applications.

7.4 Finite-Element Implementation

7.4.1 Governing Equations

To obtain the elemental contributions, the governing equations need to be written in matrix form. Consider first the incompressibility condition

$$\dot{\epsilon}_{ii} = \delta_{ij}\dot{\epsilon}_{ij} = \{C\}^{\mathrm{T}}\{\dot{\epsilon}\} \tag{7.24}$$

where $\{C\}$ is the vector notation for the Kronecker delta, and

$$\{C\}^{\mathrm{T}} = [1, 1, 1, 0] \,. \tag{7.25}$$

For an axisymmetric analysis the strain-rate vector is defined as

$$\{\dot{\epsilon}\} = \begin{bmatrix} \dot{\epsilon}_r \\ \dot{\epsilon}_z \\ \dot{\epsilon}_\theta \\ \dot{\gamma}_{rz} \end{bmatrix} = \begin{bmatrix} \dfrac{\partial}{\partial r} & 0 \\ 0 & \dfrac{\partial}{\partial z} \\ \dfrac{1}{r} & 0 \\ \dfrac{\partial}{\partial z} & \dfrac{\partial}{\partial r} \end{bmatrix} \begin{bmatrix} \dot{u}_r \\ \dot{u}_z \end{bmatrix} \tag{7.26}$$

while for plane strain analysis

$$\{\dot{\epsilon}\} = \begin{bmatrix} \dot{\epsilon}_x \\ \dot{\epsilon}_y \\ \dot{\gamma}_{xy} \end{bmatrix} = \begin{bmatrix} \dfrac{\partial}{\partial x} & 0 \\ 0 & \dfrac{\partial}{\partial y} \\ \dfrac{\partial}{\partial y} & \dfrac{\partial}{\partial x} \end{bmatrix} \begin{bmatrix} \dot{u}_x \\ \dot{u}_y \end{bmatrix} \tag{7.27}$$

In matrix form the Levy–Mises equation may be expressed

$$\{s\} = \frac{\bar{\sigma}}{\dot{\bar{\epsilon}}}\,[D]\,\{\dot{\epsilon}\} \tag{7.28}$$

For axisymmetric and plane strain problems, respectively, the deviatoric components of Cauchy stress in matrix form are

$$\{s\}^{\mathrm{T}} = [s_r, s_z, s_\theta, s_{rz}] \tag{7.29}$$

and

$$\{s\}^{\mathrm{T}} = [s_x, s_y, s_{xy}] \tag{7.30}$$

The constitutive matrix $[D]$ for axisymmetric systems is

$$[D] = \begin{bmatrix} \frac{2}{3} & 0 & 0 & 0 \\ 0 & \frac{2}{3} & 0 & 0 \\ 0 & 0 & \frac{1}{3} & 0 \\ 0 & 0 & 0 & \frac{1}{3} \end{bmatrix} \tag{7.31}$$

where for plane strain the corresponding result is

$$[D] = \begin{bmatrix} \frac{2}{3} & 0 & 0 \\ 0 & \frac{2}{3} & 0 \\ 0 & 0 & \frac{1}{3} \end{bmatrix} \tag{7.32}$$

The continuous velocity vector must be replaced with nodal point values. For a four-node quadilateral this relationship is

$$\begin{bmatrix} \dot{u}_r \\ \dot{u}_z \end{bmatrix} = \begin{bmatrix} N_1 & 0 & N_2 & 0 & N_3 & 0 & N_4 & 0 \\ 0 & N_1 & 0 & N_2 & 0 & N_3 & 0 & N_4 \end{bmatrix} \begin{bmatrix} \dot{u}_1 \\ \dot{u}_2 \\ \dot{u}_3 \\ \dot{u}_4 \\ \dot{u}_5 \\ \dot{u}_6 \\ \dot{u}_7 \\ \dot{u}_8 \end{bmatrix} = [N]\{U\} \tag{7.33}$$

The odd-numbered and even-numbered velocities correspond to the r and z components, respectively. The plane strain discretization is similar to the above. The N_i are the finite-element shape functions, which for an isoparametric quadrilateral element are

$$N_1 = \frac{(1-s)(1-t)}{4} \qquad N_2 = \frac{(1+s)(1-t)}{4}$$
$$N_3 = \frac{(1+s)(1+t)}{4} \qquad N_4 = \frac{(1-s)(1+t)}{4} \tag{7.34}$$

where s and t represent a natural coordinate system.

It is also necessary to discretize the strain-rate versus velocity relations. Thus,

$$\{\dot{\epsilon}\} = [B]\{\dot{U}\} \tag{7.35}$$

where

$$[B] = \begin{bmatrix} Y_1 & 0 & Y_2 & 0 & Y_3 & 0 & Y_4 & 0 \\ 0 & X_1 & 0 & X_2 & 0 & X_3 & 0 & X_4 \\ G_1 & 0 & G_2 & 0 & G_3 & 0 & G_4 & 0 \\ X_1 & Y_1 & X_2 & Y_2 & X_3 & Y_3 & X_4 & Y_4 \end{bmatrix} \tag{7.36}$$

The terms in the $[B]$ matrix are

$$Y_1 = [z_{24} - z_{34}s - z_{23}t]/[8J]$$
$$Y_2 = [z_{31} + z_{34}s + z_{14}t]/[8J]$$
$$Y_3 = [z_{24} + z_{12}s - z_{14}t]/[8J]$$
$$Y_4 = [z_{13} - z_{12}s + z_{23}t]/[8J]$$

$$X_1 = -[r_{24} - r_{34}s - r_{23}t]/[8J]$$
$$X_2 = -[r_{31} + r_{34}s + r_{14}t]/[8J]$$
$$X_3 = -[r_{42} + r_{12}s - r_{14}t]/[8J]$$
$$X_4 = -[r_{13} - r_{12}s + r_{23}t]/[8J]$$
$$G_i = \frac{N_i}{r} \tag{7.37}$$

where

$$J = \frac{\partial r}{\partial s}\frac{\partial z}{\partial t} - \frac{\partial r}{\partial t}\frac{\partial z}{\partial s}$$
$$r_{ij} = r_i - r_j \tag{7.38}$$
$$z_{ij} = z_i - z_j$$

or,

$$|J| = \tfrac{1}{8}[r_{13}\,z_{24} - r_{24}\,z_{13}] + s\,[r_{34}\,z_{12} - r_{12}\,z_{34}] + t\,[r_{23}\,z_{14} - r_{14}\,z_{23}] \tag{7.39}$$

Fig. 7.3 illustrates the isoparametric quadilateral element. For plane strain, the derivation will proceed similar to the above except that the third row of $[B]$ should be deleted.

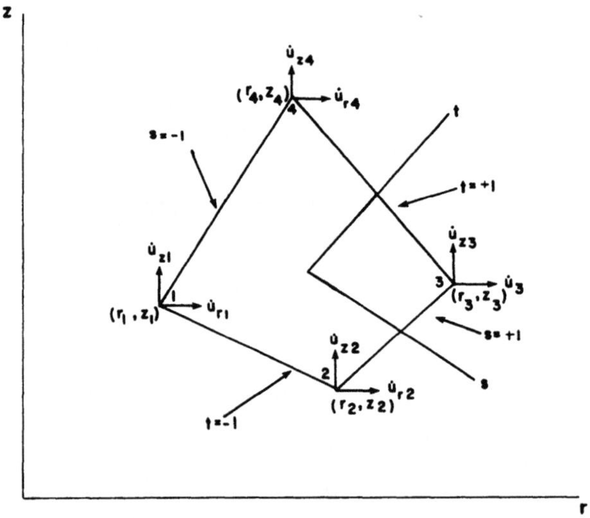

Fig. 7.3. Isoparametric quadrilateral element and natural coordinate system, showing velocity components at the nodes.

In matrix form Eq. (7.21) may be restated

$$\delta\phi = 0 = \int_V \delta\{[B]\{\dot{U}\}\}\{s\}dV + \int_A \lambda\delta\{[B]\{\dot{U}\}\}^{\mathrm{T}}[C]dV + \int_V \{[B]\{\dot{U}\}\}^{\mathrm{T}}[C]\,\delta\lambda dV - \int_A \delta\,\{[N]\{\dot{U}\}\}^{\mathrm{T}}\{T\}dA \tag{7.40}$$

Now incorporating the Levy–Mises equation gives

$$\delta\phi = 0 = \frac{\bar{\sigma}}{\dot{\bar{\epsilon}}}\int_V \delta[\dot{U}]^T[B]^T[D][B]\{\dot{U}\}dV$$

$$+ \lambda\int_V \delta[\dot{U}]^T[B]^T[C]dV + \delta\lambda\int_V [\dot{U}]^T[B]^T[C]dV - \int_A \delta[\dot{U}]^T[N]^T\{T\}dA \quad (7.41)$$

Finally, the following set of elemental equations is obtained:

$$\frac{\bar{\sigma}}{\dot{\bar{\epsilon}}}\int_V [K]\{\dot{U}\}dV + \lambda\int_V [B]^T[C]dV - \int_A [N]^T\{T\}dA = 0 \quad (7.42)$$

and

$$\int_V \{\dot{U}\}^T[B]^T[C]dV = 0$$

where

$$[K] = [B]^T[D][B] \quad (7.43)$$

The elemental contributions given by Eqs (7.43) and (7.44) represent nine simultaneous nonlinear equations. After the usual finite-element assembly procedure, the set of global unknowns (velocities and Lagrange multipliers) must be calculated iteratively. Standard or modified Newton–Raphson methods or a quasi-Newton method may be employed. Details of these solution methods are outlined by Chen and Han (1988) and by Bathe (1982).

Iteration continues until the nonlinear simultaneous equations converge. Various convergence criteria have been used. One which works well for many deformation processing applications is to require that

$$\frac{\left(\sum_{i=1}^{n} \Delta \dot{U}_i^2\right)^{1/2}}{\left(\sum_{i=1}^{n} \dot{U}_i^2\right)^{1/2}} < E \quad (7.44)$$

The convergence criterion E is problem dependent; however, values between 0.0001 and 0.001 are often effective.

7.5 Special Numerical Features of Deformation Processing Codes

7.5.1 Generation of Initial Guess Velocities

Solution procedures for nonlinear systems, such as the Newton–Raphson method, require an estimate of the nodal velocities to be supplied for the first time step. If this initial guess is too far off from the actual solution, the iterations may not converge. For simple geometries, such as cylinders and rings, the upper bound method may be used to provide the initial guess velocities. For more complicated geometries, the following procedure based on the work of Lyness et al. (1974) may be used. The procedures in the finite-element codes ALPID (Oh 1982) and CFORM (Jackson et al. 1988a) are based on this idea.

The initial guess velocities are generated by solving the system of nonlinear algebraic equations obtained in the deformation formulation by direct iteration, as follows:

$$\begin{bmatrix} \frac{\bar{\sigma}}{\dot{\bar{\epsilon}}} \int_V [K]\,dV & \{Q\} \\ \{Q\}^{\mathrm{T}} & 0 \end{bmatrix} \begin{bmatrix} \{\dot{U}\} \\ \lambda \end{bmatrix} = \begin{bmatrix} \{F\} \\ 0 \end{bmatrix} \tag{7.45}$$

with

$$\{Q\} = \int_V [B]^{\mathrm{T}} \{C\}\,dV, \quad \{F\} = \int_A [N]^{\mathrm{T}} \{T\}\,dA \tag{7.46}$$

7.5.2 Treatment of Rigid Inclusions

During forging opeiations, certain portions of the workpiece become rigid or almost nondeforming. A rigid region is characterized by its very low value of effective strain rate $\dot{\bar{\epsilon}}$, compared to other regions of the workpiece. Since $\dot{\bar{\epsilon}}$ appears in the denominator of the stiffness matrix, the system of Eq. (7.42) becomes very ill-conditioned when the rigid regions are included in the analysis. This leads to divergence in the Newton–Raphson iterations. This numerical problem is overcome by considering an effective strain-rate "offset" ($\dot{\bar{\epsilon}}_0$). If any element has a calculated $\dot{\bar{\epsilon}}$ less than $\dot{\bar{\epsilon}}_0$, the corresponding element $\dot{\bar{\epsilon}}$ is reset to $\dot{\bar{\epsilon}}_0$. The offset is calculated as a percentage of the average of the element effective strain rates. The value of $\dot{\bar{\epsilon}}_0$ chosen is dependent on the forging operation being analysed. For some applications (Gangjee 1987, Todt 1990) a value of $\dot{\bar{\epsilon}}_0$ two orders of magnitude less than the average effective strain rate works well. Below the offset strain rate, the effective stress is linearly varied to zero, as shown in Fig. 7.4. The offset strain-rate value is chosen because of numerical reasons only, and does not reflect the actual material behaviour.

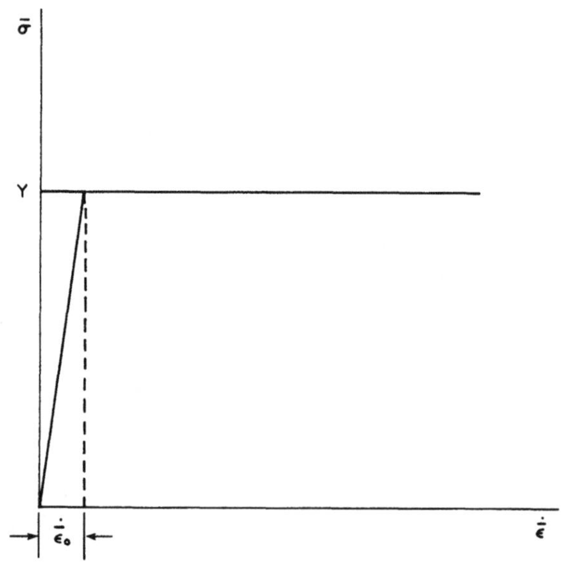

Fig. 7.4. Stress–strain rate curve for rigid-plastic materials modified to handle rigid inclusions.

7.5.3 Enforcement of Contact Boundary Conditions

The tool–workpiece interface is considered as a mixed boundary, with the velocity being specified in the direction of the die movement and traction being imposed in the form of friction tangential to the two surfaces. The friction boundary conditions will be discussed in the next section. In applying the velocity boundary condition, the following must be considered.

First, the die boundary must be suitably modelled. One way to do this is to break the die profile into segments. For straight segments, only the end point coordinates need be specified. For curved segments, the coordinates of a sufficient number of points on the segment must be specified, so that the segment may be represented by a smooth curve fit (for example, using cubic splines).

Second, as the workpiece deforms and folding occurs, parts of its boundary come into or lose contact with the die surface. A criterion must be established to decide when any nodes in the finite-element model of the workpiece are considered to be in contact with the die. The criterion must take into consideration both the vertical and horizontal distance of the node from the die. If a node is within a specified tolerance distance of the die, it is considered to be in contact with the die.

The calculations at a time step sometimes result in a node being positioned inside the boundary of a die. This may be avoided by choosing a small enough time step. The force at the die–workpiece interface is used to decide whether an element stays in contact with the die or not. The die stays in contact as long as the force is compressive (negative).

For flat dies moving vertically down, the velocity boundary condition for node i in contact with the die is

$$\dot{u}_{zi} = V_{\text{die}} \tag{7.47}$$

which states that the axial (vertical) velocity of node i is equal to the die velocity V_{die}.

For curved dies, the velocity boundary condition is given by

$$\dot{u}_{ni} = V_{\text{die}} \bullet \hat{n}_i \tag{7.48}$$

This states that the normal component of nodal velocity, $\dot{u}_{ni}$, is equal to the normal component of the die velocity. Here, $\hat{n}_i$ is the unit normal to the die at the point of contact. Since the boundary conditions are in terms of normal and tangential velocities, appropriate coordinate transformations (from radial–axial direction to normal–tangential direction) of the velocities of the boundary nodes are performed. Thus, the vector of incremental velocities $\{\Delta\dot{U}\}$ will contain some normal and tangential terms and some axial and radial terms.

The traction boundary conditions enter the last term of Eq. (7.42) as

$$\{F\} = \int_A [N]^{\mathrm{T}} \{T\}\, dA \tag{7.49}$$

Matrix $[N]$ is as defined in Eqs (7.33) and (7.34), and the traction stress vector is

$$\{T\} = \begin{Bmatrix} T_r \\ T_z \end{Bmatrix} \tag{7.50}$$

The evaluation of the integral in Eq. (7.49) is outlined in books on the finite-element method, such as that by Bathe (1982).

7.5.4 Friction Boundary Conditions

Friction and lubrication critically influence the metal-forming process. In most cases, low friction is desirable, because it reduces forming loads and stresses, and results in a more homogeneous deformation. In some processes, such as sheet metal forming, however, friction is needed to control metal flow. The lubrication system has to be carefully specified to achieve required levels of friction. Wilson (1979) has described four main regimes of lubrication in the die–workpiece interface, depending on the thickness of lubricant film entrapped in the interface. Which regime exists at any particular stage of a metal-forming process may depend on changes in lubricant and workpiece properties, speed of operation, surface roughness, temperature and geometry of die and workpiece. Different lubrication regimes may exist at different locations simultaneously, and change from one regime to another as the operation proceeds. A fundamental understanding of the mechanisms of friction and lubrication is critical both for the accurate computer simulation of metal-forming processes, and for understanding the limitations of different friction models used.

The magnitude of the frictional stresses at the die–workpiece interface has been commonly characterized in two ways. The first method of characterizing interface friction is by the use of a constant coefficient of friction (Coulomb friction),

$$\mu = \tau_f / p \tag{7.51}$$

where τ_f is the friction stress tangent to the surface and p is the stress normal to the surface (contact pressure). This is generally valid when the interface load is carried by a thin film of lubricant at the asperity peaks.

The second method of calculating interface friction assumes that the interface zone may be represented by a material of constant shear strength τ_f, which is assumed to be a constant fraction m (called the friction factor) of the shear strength k of the deforming body:

$$\tau_f = mk \tag{7.52}$$

The use of this model (also called the Tresca model) removes the necessity of measuring the contact pressure p. This model is often used in modelling forging processes. Determining appropriate values of m for specific applications is fairly difficult. Use may be made of experimental results such as those by Male and Cockcroft (1964–65), who used measurements of inner diameters of compressed rings to measure m, or those by Gunasekara and Mahadeva (1988), who used an extrusion forging test to predict m. The frictional stress acts in a direction opposite to the direction of relative motion between the workpiece and the tool.

If the traction is due to friction, then in Eq. (7.50),

$$T_z = 0$$

$$\text{and } T_r = -\mu p$$

$$\text{or } T_r = -mk \tag{7.53}$$

as the case may be. The negative sign in both the equations indicates that the friction force acts against the direction of relative motion between the workpiece and tool.

There is an inherent drawback in directly applying the frictional force to the interface nodes as outlined above. That is the fact that the direction of the frictional force is velocity dependent. A specific example of this is the "neutral point problem", which is discussed in the next section. In general, in order to apply the friction force directly, it is necessary to have an estimate for the velocity field in the workpiece. The response of the workpiece to the frictional forces is then determined iteratively until a converged velocity field is obtained.

As an alternative to applying friction forces directly, various workers have used special friction elements at the interface to simulate the friction conditions. The friction layer technique described by Hartley et al. (1979, 1988) is one such method. It does not require any prior knowledge of interfacial velocity. The stiffness matrix of each friction element is multipled by a function of the interface shear factor to provide different levels of interface restraint.

7.5.5 Treatment of Neutral Points

In processes such as ring compression, forging operations with complex die shapes and rolling, there sometimes exists a point along the tool–workpiece interface where the velocity of the workpiece relative to the tool velocity becomes zero. This point is called the neutral point. The relative velocity and hence the frictional stress changes direction at the neutral point. Since the location of the neutral point is not known a priori, special numerical techniques have to be used when applying the friction boundary conditions, whereby the direction of the frictional force is automatically chosen depending on the relative slip at the tool–workpiece interface.

Matsumoto et al. (1977) chose to model the friction stress as a step discontinuity. In other words, the frictional stress instantly changes direction at the neutral point, which is identified based on an initial guess for the velocity field. Fig. 7.5 illustrates this assumption. The model results in the last term of the functional, Eq. (7.19), being modified as follows:

$$\phi_1 = -\int_A T_i \dot{u}_i \, dA = -\int_{A_1} (-mk)\, V_R \, dA_1 + \int_{A_2} (-mk)\, V_R \, dA_2 \qquad (7.54)$$

where V_R is the relative velocity between workpiece and tool, A_1 is the portion of A on which V_R is positive and A_2 is the portion of A on which V_R is negative. A detailed derivation of the resulting expressions in the case of axisymmetric geometries is given by Matsumoto et al. (1977), while that for plane-strain geometry is given by Jackson et al. (1988b).

To improve convergence characteristics of the nonlinear equation solution, Chen and Kobayashi (1978) proposed a modified form of the friction stress term as

$$T_r = -\left[mk \left(\frac{2}{\pi}\right) \tan^{-1}\left(\frac{V_R}{a\,|V_{\text{die}}|}\right)\right] \qquad (7.55)$$

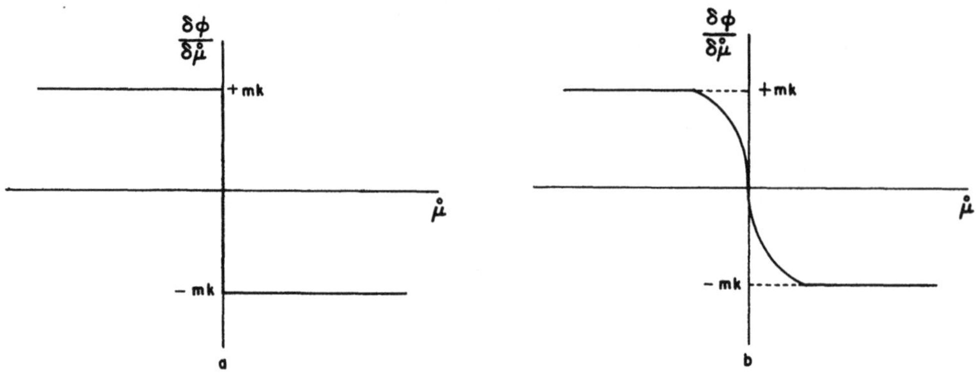

Fig. 7.5. Variation of frictional stress near a neutral point. **a** Matsumoto formulation; **b** Chen formulation.

where V_R is the relative velocity between workpiece and tool and $a = 10^{-5}$. The frictional contribution to the functional, Eq. (7.19), in this case is

$$\phi_1 = -\int_A \left[\int_0^{V_R} T_r \, dV_R \right] dA \tag{7.56}$$

The terms resulting from closed form evaluation of the frictional contribution to the governing equations are given by Chen and Kobayashi (1978) for axisymmetric geometry and by Jackson et al. (1988b) for plane strain geometry. Fig. 7.5 shows the distribution of the frictional stress obtained by this method. Note that the stress takes a magnitude equal to mk in regions away from a neutral point. Near a neutral point, there is a smooth transition from $+mk$ to $-mk$.

The closed form integration of the terms in the model proposed by Chen is feasible but very complex. An alternative is numerical integration via Gaussian quadrature of the form

$$I = mk \left(\frac{2}{\pi} \right) \sum_i F(s_i) \, \alpha_i \, |J|_i \tag{7.57}$$

where s_i is in terms of the natural coordinate system tangent to the side of the element where friction occurs. Though numerical integration is simpler, any loss in accuracy in adopting it needs to be carefully considered.

Results from Jackson et al. (1988b) indicate that the method of Chen is computationally more efficient, but the method due to Matsumoto has a larger domain of convergence. Hence, a reasonable solution strategy is to use the Matsumoto method for the first time step (for example, in conjunction with the special starting procedure) and the Chen method for all subsequent time steps.

7.5.6 Remeshing

During computer simulation of large-strain plastic deformation processes, the initial finite-element mesh becomes progressively distorted. The accuracy of the solution decreases with increasing distortion, until eventually a converged solution is no longer possible. Remeshing is the process of alleviating mesh distortions at suitable stages during the simulation, with a view to improving solution accuracy and convergence. The remeshing process is explained in terms of three major steps.

The first step is the detection of a need for remeshing. This is done by judging the severity of mesh distortion through geometric criteria. One such criterion is based on the included angles in an element. Element distortion becomes unacceptable as an included angle approaches 180°. Other criteria include the "element ratio" (Hageman 1987) (defined in terms of the lengths of the sides and diagonals of a quadrilateral element), the "distortion parameter" (Nicolas and Citipitioglu 1977) (calculated as a function of the determinant of the element Jacobian matrix and area) and the "strain gradient" (Cheng 1988) (which compares the strains in an element to that in its adjacent elements.

The second step in the remeshing process consists of rezoning the mesh to achieve more regular (less distorted) elements. The new mesh is created either using the old mesh as reference, or completely independent of the old mesh. Special consideration must be given to the boundary nodes and elements, because the boundary of the workpiece must remain unchanged during the rezoning process. Schemes discussed in the literature include the "linear relocation scheme" (Im and Kobayashi 1986), the "equipotential scheme" (Larenson 1980) and the isoparametric scheme" (Larenson 1980). A combination of these schemes often leads to more satisfactory results.

The last step in the remeshing process is to transfer or interpolate the calculated values from the old mesh to the new mesh, followed by iterations to obtain a converged solution using the new mesh. The general procedure is to identify the old element within which a new node lies. Using the old element as a reference, the nodal properties of the new node are obtained using interpolation schemes such as nodal averaging, bilinear interpolation, cubic interpolation and least squares interpolation.

All three of the above steps are discussed in greater detail by York (1988).

7.5.7 Examples: Isothermal Forging

This section presents examples of simulations of isothermal forging operations. The simulations were performed with the CFORM finite-element package (Jackson et al. 1988a). Fig. 7.6 shows the results of a simulation of the axisymmetric ring compression problem. A similar problem was solved by Matsumoto et al. (1977). The analysis was done with a standard ring specimen having an outer diameter to inner diameter to height ratio of 6:3:2. Due to symmetry, only one quarter of the cross-section was modelled. The material selected was Al 1100F with a rigid-plastic constitutive law. The die was assumed to move vertically down with a speed of approximately 25 mm s^{-1}.

The second example is a simulation of an axisymmetric centre punch forging process, shown in Fig. 7.7. The material in this case was SAE 52100 steel. The punch was modelled as moving vertically down with variable speed. Due to symmetry, only half the workpiece was modelled.

Fig. 7.8 illustrates the plane-strain forging of an H-section beam of constant cross section from a rectangular mild steel billet, with the die moving vertically down at 25 mm s^{-1}. Again, from symmetry considerations, only one quarter of the billet needed to be analysed.

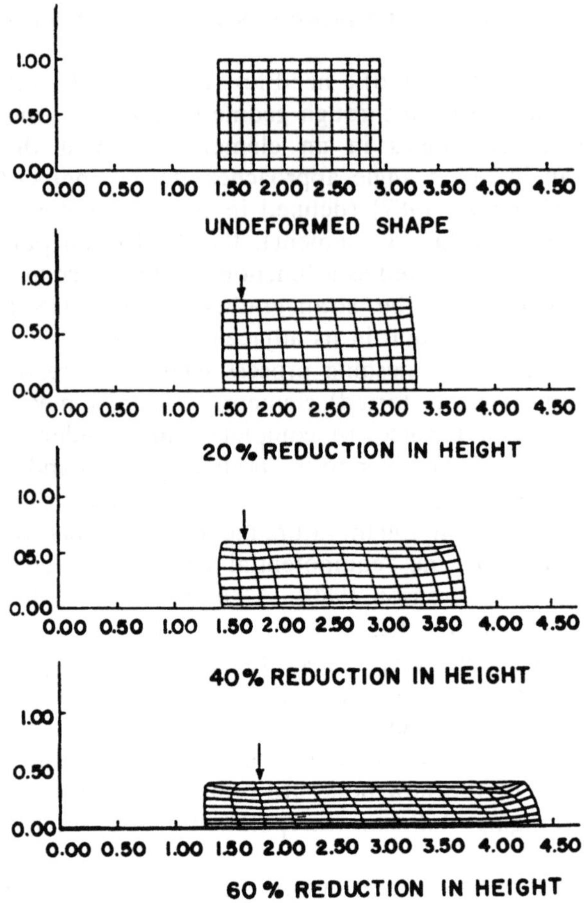

Fig. 7.6. Axisymmetric ring compression problem, arrow indicates location of neutral point.

7.6 Coupled Thermal and Deformation Processing

7.6.1 Introduction

During a metal-forming operation, a large amount of the mechanical work is converted into heat through large plastic deformations and through friction forces between tool and workpiece. If this heat generated is significant, and/or if the workpiece is initially at a much different temperature from the surroundings or the tools, heat transfer becomes an important consideration. This is because most materials have flow stresses that are temperature dependent, in addition to the previously discussed dependence on strain and strain rate (thermoviscoplastic behaviour). By coupling the previously discussed deformation model with a thermal analysis, it is possible to obtain a more realistic simulation of the metal-forming process.

Many researchers have determined the velocity field, either experimentally or by simulation, and then used the velocity field to determine the internal heat generation. The velocity and temperature fields are not coupled in these analyses. Examples are the works by Johnson and Kudo (1960) as applied to extrusion and Tay et al. (1974) as applied to machining. Another method, due to Bishop (1956),

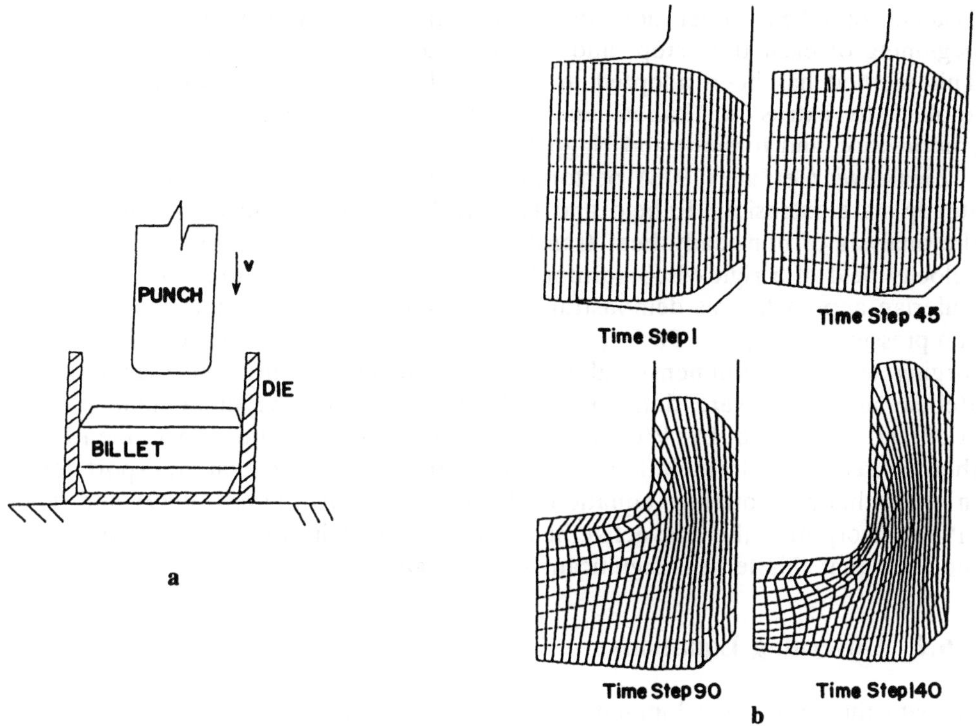

Fig. 7.7. Axisymmetric centre punch forging simulation. **a** Centre punch forging schematic; **b** forging simulation.

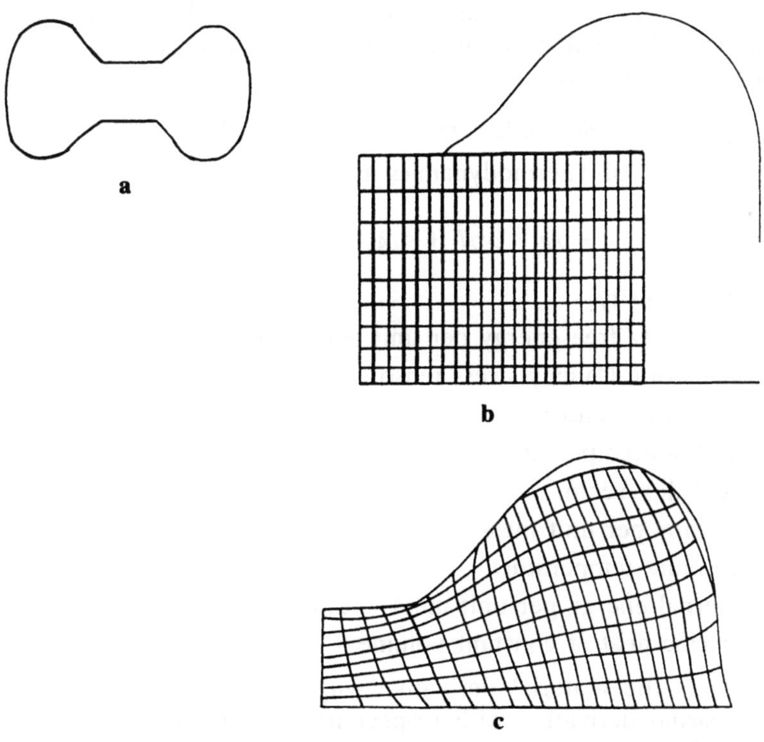

Fig. 7.8. Simulation of plane-strain H-section. **a** Required final cross-section; **b** initial billet; **c** deformed billet.

assumes that heat generation and transportation occur instantaneously at the beginning of each time step, and that conduction occurs during the time step. This method has been adopted in analyses by Altan and Kobayashi (1968) as applied to extrusion; Lahoti and Altan (1975) as applied to axisymmetric compression and torsion; and Nagpal et al. (1977) applied to forging.

Zienkiewicz et al. (1978) performed a coupled thermal analysis applied to steady state extrusion. Rebelo and Kobayashi (1980) extended this method for the analysis of nonsteady heat transfer coupling the temperature field with the velocity field of a rigid-viscothermoplastic material. The method, which used the Eulerian approach, was demonstrated using solid cylinder compression and ring compression examples. Park (1982) extended this method to consider block forging using three-dimensional solid finite elements. The above analyses by Rebelo and Park were done on modified versions of the finite-element code ALPIDT developed by Wu and Oh (1985). None of the above mentioned thermoviscoplastic formulations considered the convective heat transport terms in their thermal coupling equations. Formulations by Lamm (1989) and Todt (1990) incorporate the convective heat transport term in the analysis. Their work indicates that this term is significant in some analyses.

7.6.2 Governing Equations

As with the earlier development for the deformation analysis, the equations for the thermal analysis of the workpiece and the die are developed in Eulerian coordinates. They are derived from consideration of the First and Second Laws of Thermodynamics.

Using the Principles of Conservation of Mass, Linear Momentum and Angular Momentum with the definition of Helmholtz free energy, the First Law of Thermodynamics is written as

$$\rho\dot{\psi} = s_{ij}\dot{\epsilon}_{ij} + \rho h - q_{i,i} - \rho\eta\dot{T} - \rho T\dot{\eta} \tag{7.58}$$

and the Second Law of Thermodynamics as

$$-\rho\dot{\psi} - \rho\eta\dot{T} + s_{ij}\dot{\epsilon}_{ij} - \frac{1}{T}q_i T_{,i} \geq 0 \tag{7.59}$$

where ρ Mass density

ψ Helmholtz free energy per unit mass $= \mu - T\eta$

s_{ij} Deviatoric Cauchy stress tensor

$\dot{\epsilon}_{ij}$ Strain-rate tensor

h Heat source per unit mass

q_i Heat flux vector

η Entropy per unit mass

T Absolute temperature

μ Internal energy per unit mass $= C_v T$

C_v Specific heat at constant volume

A dot represents the material derivative with respect to time, and a comma represents a partial derivative with respect to a coordinate. A repeated index implies summation.

The first three terms of Eq. (7.59) correspond to the dissipation of mechanical energy into heat (internal dissipation). Truesdell and Noll (1965) postulate that the internal dissipation must be greater than or equal to zero, so that

$$s_{ij}\,\dot{\epsilon}_{ij} \geq \rho\eta\dot{T} + \rho\dot{\psi} \tag{7.60}$$

This implies that only some fraction ν of the mechanical power is dissipated into heat. Gordon (1955) and Greenfield and Bever (1956) found that the mechanical power *not* dissipated into heat was in the range of 0.05 to 0.15 of the total mechanical power $s_{ij}\,\dot{\epsilon}_{ij}$. The Second Law may thus be represented as an equality.

Incorportion of this form of the Second Law into the energy expression, Eq. (7.58), together with a few simplifications leads to the following,

$$\rho\,C_\nu\frac{\partial T}{\partial t} + \rho\,C_\nu\,\dot{u}_i\frac{\partial T}{\partial x_i} - k\,T_{,i} - \nu\,\bar{\sigma}\,\dot{\bar{\epsilon}} = 0 \tag{7.61}$$

Here, it is assumed that the specific heat at constant volume C_ν is constant. Further, heat q_i has been defined using the Fourier Law of Heat Conduction (assuming constant thermal conductivity, k) as

$$q_i = -kT_{,i} \tag{7.62}$$

The mechanical power term $s_{ij}\,\dot{\epsilon}_{ij}$ has been rewritten in terms of the effective stress and equivalent strain rate as $\bar{\sigma}\,\dot{\bar{\epsilon}}$. Also, the internal heat generation term ρh has been dropped, because the only internal heat source considered here is that due to internal plastic deformations, which is given by $\nu\,\bar{\sigma}\,\dot{\epsilon}$. Eq. (7.61) together with appropriate boundary conditions constitute the governing equations for heat transfer in the workpiece.

The governing equations for heat transfer in the die are the same, except that, since the die is rigid, there is no plastic deformation, that is, $\nu = 0$.

7.6.3 Finite-Element Discretization of the Heat Equation

The temperature T at any point in a finite element may be expressed in terms of the corresponding nodal point values as

$$T = [N]\,\{T\} \tag{7.63}$$

where $[N]$ is the matrix of shape functions, defined in Eq. (7.34) for a linear, four-noded isoparametric element, and $\{T\}$ is the vector of nodal temperatures.

The Galerkin Finite Element Method (GFEM) may be utilized for discretization of the heat equation, Eq. (7.61). The GFEM requires that the integral over the global problem domain of the weighted residual should vanish. The weighting functions are taken to be the shape functions N_i, and the residual is defined to be error incurred when the temperature T in Eq. (7.61) is replaced by the approximation in Eq. (7.63). Thus, the GFEM requires that

$$\int_V [N]^{\mathrm{T}}\left[\rho\,C_\nu\,[N]\,\{\dot{T}\} + \rho\,C_\nu\,\dot{u}_i\frac{\partial[N]}{\partial x_i}\{T\} - q_{i,i} - \nu\,\bar{\sigma}\,\dot{\bar{\epsilon}}\right]dV = 0 \tag{7.64}$$

After application of the Green–Gauss theorem to the integral of the third term, and using Eq. (7.62), the above equation becomes,

$$\int_A [N]^{\mathrm{T}} q_i n_i \, dA + \int_V k \frac{\partial [N]^{\mathrm{T}}}{\partial x_i} \frac{\partial [N]}{\partial x_i} \{T\} \, \alpha \, dV$$

$$+ \int_V \left[\rho \, C_v \, [N]^{\mathrm{T}} [N] \{\dot{T}\} + \rho \, C_v \, \dot{u}_i \, [N]^{\mathrm{T}} \frac{\partial [N]}{\partial x_i} \{T\} - [N]^{\mathrm{T}} \, \nu \, \bar{\sigma} \, \dot{\bar{\epsilon}} \right] dV = 0 \quad (7.65)$$

For axisymmetric problems, $x_1 = r$ and $x_2 = \theta$. For plane-strain problems, $x_1 = x$ and $x_2 = y$.

Conservation of energy at the massless surface of the body at any given instant may be expressed as follows (see Fig. 7.9),

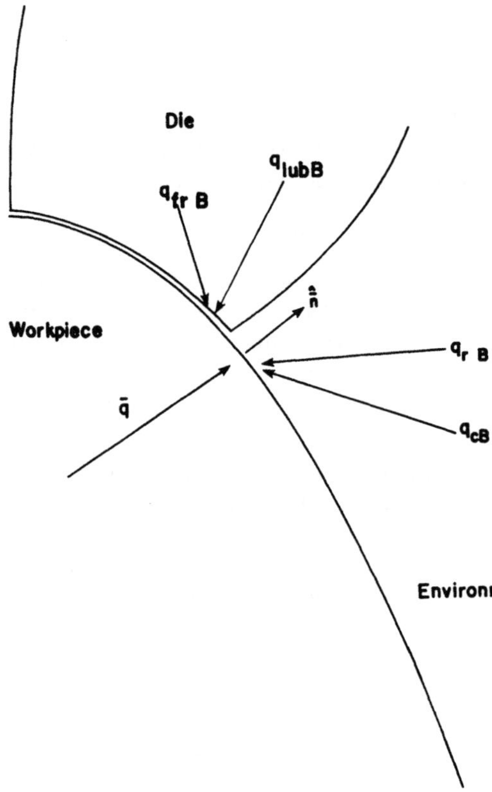

Fig. 7.9. Conservation of energy at massless material surface.

$$q_i n_i = - [q_{\mathrm{fr\,B}} + q_{\mathrm{r\,B}} + q_{\mathrm{c\,B}} + q_{\mathrm{lub\,B}}] \quad (7.66)$$

where

$q_{\mathrm{fr\,B}}$ Heat flux due to friction $= |T_{\mathrm{r}}| \, |V_{\mathrm{R}}|$

$q_{\mathrm{r\,B}}$ Heat flux due to radiation $= \epsilon\sigma \, [T_{\mathrm{c}}^4 - T^4]$

$q_{\mathrm{c\,B}}$ Heat flux due to free convection $=$ h $[T_{\mathrm{c}} - T]$

$q_{\mathrm{lub\,B}}$ Heat flux due to convection through lubricant $= h_{\mathrm{lub}} \, [T_{\mathrm{D}} - T]$.

Here

T_{r} Friction force at die–workpiece interface

V_R Relative velocity of workpiece with respect to die
ϵ Emissivity of the material surface
σ Stefan–Boltzman constant
h Free convection coefficient
h_{lub} Convection coefficient for the lubricant
T_c Temperature of environment
T_D Time-dependent die temperature
T Time-dependent surface temperature of workpiece = $[N]\ \{T\}$.

Substituting Eq. (7.66) in (7.65) and rearranging,

$$[C]\{\dot{T}\} + [\mathbf{K}]\{T\} = \{f\} \tag{7.67}$$

where

$$[C] = \int_V \rho\, C_v\, [N]^T [N]\, d\,V \tag{7.68}$$

$$[\mathbf{K}] = \int_V \left[k\, \frac{\partial [N]^T}{\partial x_i} \frac{\partial [N]}{\partial x_i} + \rho\, C_v\, \dot{u}_i\, [N]^T \frac{\partial [N]}{\partial X_i} \right] dV + \int_{A_t} h_{lub}\, [N]^T [N]\, dA_t$$
$$+ \int_{A_f} [h + ([N]\{T\})^3]\, [N]^T [N]\, dA_f \tag{7.69}$$

$$\{f\} = \int_V [N]^T \nu\, \bar{\sigma}\, \dot{\bar{\epsilon}}\, dV + \int_{A_t} [N]^T \left[\frac{1}{2} |T_r|\, |V_R| + h_{lub}\, T_D \right] dA_t$$
$$+ \int_{A_f} [N]^T [\sigma\, \epsilon\, T_e^4 + h\, T_c]\, dA_f \tag{7.70}$$

Here, A_t is the portion of the element in contact with a die and A_f is the portion of the element in the global boundary not in contact with a die. The factor 1/2 enters the second term on the right-hand side of Eq. (7.70) because it is assumed that the heat generated due to friction at the die–workpiece interface is evenly distributed between workpiece and die. Note that the nonlinearity of the stiffness matrix [**K**] results solely from including radiation heat flux as a boundary condition (the last integral in Eq. (7.69)). Also, [**K**] is nonsymmetric because of the convective term arising from the material time derivative of the internal energy (the second term of the volume integral in Eq. (7.69)). The capacitance matrix $[C]$ is always symmetric.

7.6.4 Time Discretization of the Heat Equation

As pointed out earlier, the system of equations obtained from the spatial discretization of the heat equation is nonsymmetric. This drawback may be overcome by choosing a suitable time discretization scheme, such that the resultant system of algebraic equations is symmetric.

One such scheme is the least squares scheme, as presented by Zienkiewicz (1986). The discretization in time is done using linear, one-dimensional shape functions (see Fig. 7.10)

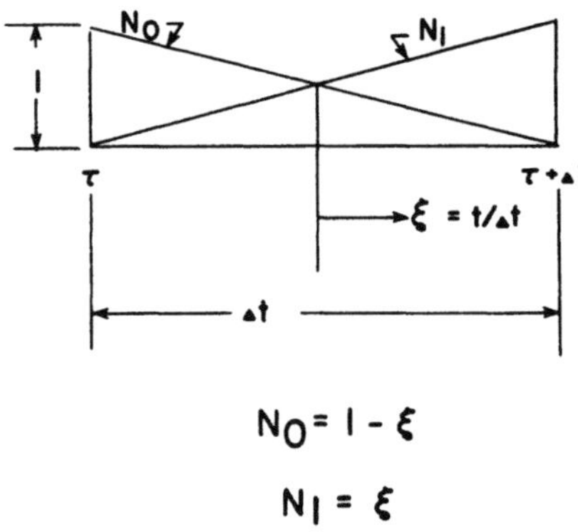

Fig. 7.10. Two-node finite time element and associated linear shape functions.

$$N_0 = 1 - \xi \,;\, N_1 = \xi, \text{ with } \xi = t/\Delta t \tag{7.71}$$

The solution process from start to finish involves a series of such time elements. The temperature $\{T\}$, heat capacitance matrix $[C]$, stiffness matrix $[\mathbf{K}]$ and the column vector $\{f\}$ are all assumed to vary linearly in time as

$$\{T\} = N_0 \{T\}_0 + N_1 \{T\}_1 \tag{7.72}$$

$$[C] = N_0 [C]_0 + N_1 [C]_1 \tag{7.73}$$

$$[\mathbf{K}] = N_0 [\mathbf{K}]_0 + N_1 [\mathbf{K}]_1 \tag{7.74}$$

$$\{f\} = N_0 \{f\}_0 + N_1 \{f\}_1 \tag{7.75}$$

Also

$$\{\dot{T}\} = \dot{N}_0 \{T\}_0 + \dot{N}_1 \{T\}_1 = -\frac{1}{\Delta t}\{T\}_0 + \frac{1}{\Delta t}\{T\}_1 \tag{7.76}$$

Here, the only unknown is the nodal set of temperatures $\{T\}_1$.

The residual is defined as

$$\begin{aligned}\{R\} &= [C]\{\dot{T}\} + [\mathbf{K}]\{T\} - \{f\} \\ &= [(N_0 [C]_0 + N_1 [C]_1)(\dot{N}_0 \{T\}_0 - \dot{N}_0 \{T\}_1) + (N_0 [\mathbf{K}]_0 + N_1 [\mathbf{K}]_1) \times \\ &\quad (N_0 \{T\}_0 + N_1 \{T\}_1) - (N_0 \{f\}_0 + N_1 \{f\}_1)]\end{aligned} \tag{7.77}$$

when the results of Eqs (7.72)–(7.76) are used. The least squares weighted residual method requires that

$$\varphi = \int_0^1 \{R\}^{\mathrm{T}} \{R\}\, d\xi \tag{7.78}$$

be minimized with respect to the unknown temperatures $\{T\}_1$.

Upon carrying out the required operations, the resultant system of equations is

$$[K]^{\text{eff}} \{T\}_1 = \{f\}^{\text{eff}} \tag{7.79}$$

where

$$[K]^{\text{eff}} = \frac{1}{3\Delta t}([C]_0^{\mathrm{T}}[C] + [C]_1^{\mathrm{T}}[C]_1) + \frac{\Delta t}{30}[\mathbf{K}]_0^{\mathrm{T}}[\mathbf{K}]_0$$

$$+ \frac{\Delta t}{5}[\mathbf{K}]_1^{\mathrm{T}}[\mathbf{K}]_1 + \frac{1}{6\Delta t}([C]_1^{\mathrm{T}}[C]_0 + [C]_0^{\mathrm{T}}[C]_1) + \tfrac{1}{12}([\mathbf{K}]_0^{\mathrm{T}}[C]_0$$

$$+ [C]_0^{\mathrm{T}}[\mathbf{K}]_0 + [\mathbf{K}]_0^{\mathrm{T}}[C]_1 + [C]_1^{\mathrm{T}}[\mathbf{K}]_0 + [\mathbf{K}]_1^{\mathrm{T}}[C]_0$$

$$+ [C]_0^{\mathrm{T}}[\mathbf{K}]_1) + \tfrac{1}{4}([\mathbf{K}]_1^{\mathrm{T}}[C]_1 + [C]_1^{\mathrm{T}}[\mathbf{K}]_1) + \frac{\Delta t}{20}([\mathbf{K}]_1^{\mathrm{T}}[\mathbf{K}]_0$$

$$+ [\mathbf{K}]_0^{\mathrm{T}}[\mathbf{K}]_1) \tag{7.80}$$

$$\{f\}^{\text{eff}} = \left[\frac{1}{6\Delta t}([C]_1^{\mathrm{T}}[C]_0 + [C]_0^{\mathrm{T}}[C]_1) + \tfrac{1}{12}([\mathbf{K}]_0^{\mathrm{T}}[C]_0 + [\mathbf{K}]_1^{\mathrm{T}}[C]_0\right.$$

$$+ [\mathbf{K}]_0^{\mathrm{T}}[C]_1 - [C]_0^{\mathrm{T}}[\mathbf{K}]_1 - [C]_1^{\mathrm{T}}[\mathbf{K}]_0 - [C]_1^{\mathrm{T}}[\mathbf{K}]_1)$$

$$+ \tfrac{1}{4}([\mathbf{K}]_1^{\mathrm{T}}[C]_1 - [C]_0^{\mathrm{T}}[\mathbf{K}]_0) + \frac{1}{3\Delta t}([C]_0^{\mathrm{T}}[C]_0$$

$$+ [C]_1^{\mathrm{T}}[C]_1) - \frac{\Delta t}{20}([\mathbf{K}]_0^{\mathrm{T}}[\mathbf{K}]_0 + [\mathbf{K}]_1^{\mathrm{T}}[\mathbf{K}]_1) - \frac{\Delta t}{30}([\mathbf{K}]_1^{\mathrm{T}}[\mathbf{K}]_0$$

$$\left. + [\mathbf{K}]_0^{\mathrm{T}}[\mathbf{K}]_1)\right]\{T\}_0 + \left[\tfrac{1}{3}[C]_0^{\mathrm{T}} + \tfrac{1}{6}[C]_1^{\mathrm{T}} + \frac{\Delta t}{12}([\mathbf{K}]_0^{\mathrm{T}}\right.$$

$$\left. + [\mathbf{K}]_1^{\mathrm{T}})\right]\{f\}_0 + \left[\tfrac{1}{6}[C]_0^{\mathrm{T}} + \tfrac{1}{3}[C]_1^{\mathrm{T}} + \frac{\Delta t}{12}[\mathbf{K}]_0^{\mathrm{T}}\right.$$

$$\left. + \frac{\Delta t}{4}[\mathbf{K}]_1^{\mathrm{T}}\right]\{f\}_1 \tag{7.81}$$

If small displacements are incurred in time Δt, then the matrices $[C]$ and $[\mathbf{K}]$ may be assumed to be constant over the time element. In this case, only $\{T\}$ and $\{f\}$ need be discretized as per Eqs (7.72), (7.75) and (7.76). In this case,

$$[K]^{\text{eff}} = \frac{1}{\Delta t}[C]_1^{\mathrm{T}}[C]_1 + \tfrac{1}{2}([\mathbf{K}]_1^{\mathrm{T}}[C]_1 + [C]_1^{\mathrm{T}}[\mathbf{K}]_1)$$

$$+ \frac{\Delta t}{3}[\mathbf{K}]_1^{\mathrm{T}}[\mathbf{K}]_1 \tag{7.82}$$

$$\{f\}^{\text{eff}} = \left[\frac{1}{\Delta t}[C]_1^{\mathrm{T}}[C]_1 + \tfrac{1}{2}([\mathbf{K}]_1^{\mathrm{T}}[C]_1 - [C]_1^{\mathrm{T}}[\mathbf{K}]_1)\right.$$

$$\left. - \frac{\Delta t}{6}[\mathbf{K}]_1^{\mathrm{T}}[\mathbf{K}]_1\right]\{T\}_0 + \left[\tfrac{1}{2}[C]_1^{\mathrm{T}} + \frac{\Delta t}{6}[\mathbf{K}]_1^{\mathrm{T}}\right]\{f\}_0$$

$$+ \left[\tfrac{1}{2}[C]_1^{\mathrm{T}} + \frac{\Delta t}{3}[\mathbf{K}]_1^{\mathrm{T}}\right]\{f\}_1 \tag{7.83}$$

Since the system of Eq. (7.79) is nonlinear in $\{T\}$, although symmetric, the solution must be calculated iteratively, using a suitable convergence criterion such as one based on the Euclidean norm.

The developments in this section and the previous one may also be used to analyse heat transfer in the die. The major difference is that the die, unlike the workpiece, is rigid. This leads to simplifications in the previously developed equations. One, $\nu = 0$ in Eq. (7.61) and all subsequent equations. Two, the heat capacitance matrix $[C]$ is constant over time Δt. Hence, this matrix need not be discretized in time using Eq. (7.73).

The choice of the time integration scheme to be used is based upon the conditioning and stability of the problem being simulated and computer time used by the method. Generally, the heat transfer problem for the workpiece is more complicated than that for the die. Zienkiewicz (1986) demonstrates that the manipulation of the time increment Δt can influence the existence of nodal temperature oscillations in the solution. There exists a critical time step increment (the inverse of the largest eigenvalue associated with the assembled system of elements), below which the solutions will not give rise to oscillating temperature fields. Oscillating temperature fields are less likely to be obtained for the workpiece than for the die. This is because the plastic deformation term $\nu \bar{\sigma} \dot{\bar{\epsilon}}$ tends to damp the artificial oscillations (Todt 1990).

7.6.5 Combined Mechanical and Thermal Analysis

The coupling between the deformation analysis and the heat analysis of the workpiece is given through the material constitutive relationship for the workpiece. This is because for a rigid-thermoviscoplastic material, the flow stress is a function of effective strain, effective strain rate and temperature. Further, the problem of thermal analysis consists of two parts, the thermal analysis of the workpiece and that of the die. The two are connected through the boundary conditions imposed at the die workpiece interface (see Eq. (7.66)).

One approach to solving the combined problem is to alternate between the deformation and thermal analyses iteratively until both solutions converge. The deformation analysis gives the nodal velocities $\dot{U}$ (from which strain and strain rate may be calculated); the thermal solution of the workpiece yields the temperature distribution T^{W} in the workpiece; the thermal analysis of the die yields the temperature distribution T^{D} in the die. Assuming a converged solution has been obtained for all the above quantities $\dot{U}_i$, T_i^{W} and T_i^{D} at some time t_i (the subscripts referring to the time step), the next step is to obtain converged solutions for the velocities and temperatures at time step $t_{i+1} = t_i + \Delta t$. The steps necessary to do this are as follows:

1. The finite-element mesh of the workpiece is updated based on velocities $\dot{U}_i$ and the time step Δt, to give the mesh configuration at time t_{i+1}. The effective strains are then calculated for the new configuration.
2. The converged velocities and temperatures for time step t_i are used as initial guesses for the solutions at time step t_{i+1}, that is,

$$
\begin{aligned}
{}^{0}\dot{U}_{i+1} &= \dot{U}_i \\
{}^{0}T_{i+1}^{W} &= T_i^{W} \\
{}^{0}T_{i+1}^{D} &= T_i^{D}
\end{aligned}
$$

   Here, the superscript on the left is the iteration count for the iterative cycle between deformation and thermal analyses.

3. For $j = 0, 1, 2 \ldots$ until convergence,
   (a) Based on ${}^{j}T^{W}_{i+1}$, perform deformation analysis on workpiece to converge to ${}^{j+1}\dot{U}_{i+1}$.
   (b) Based on ${}^{j+1}\dot{U}_{i+1}$ and ${}^{j}T^{D}_{i+1}$, perform thermal analysis of workpiece to converge to ${}^{j+1}T^{W}_{i+1}$.
   (c) Based on ${}^{j+1}T^{W}_{i+1}$, perform thermal analysis of die to converge to ${}^{j+1}T^{D}_{i+1}$.
   (d) Check for convergence, that is, check to see if ${}^{j+1}\dot{U}_{i+1}$, ${}^{j+1}T^{W}_{i+1}$, and ${}^{j+1}T^{D}_{i+1}$ differ appreciably from ${}^{j}\dot{U}_{i+1}$, ${}^{j}T^{W}_{i+1}$ and ${}^{j}T^{D}_{i+1}$, respectively. If yes, convergence is not satisfied. In this case, update j by one and return to step 3. If convergence is satisfied, then update i by one and return to step 1.

The steps are illustrated in Fig. 7.11. It must be kept in mind that in each of the steps (a), (b) and (c) above, a nonlinear system of algebraic equations has to be solved iteratively.

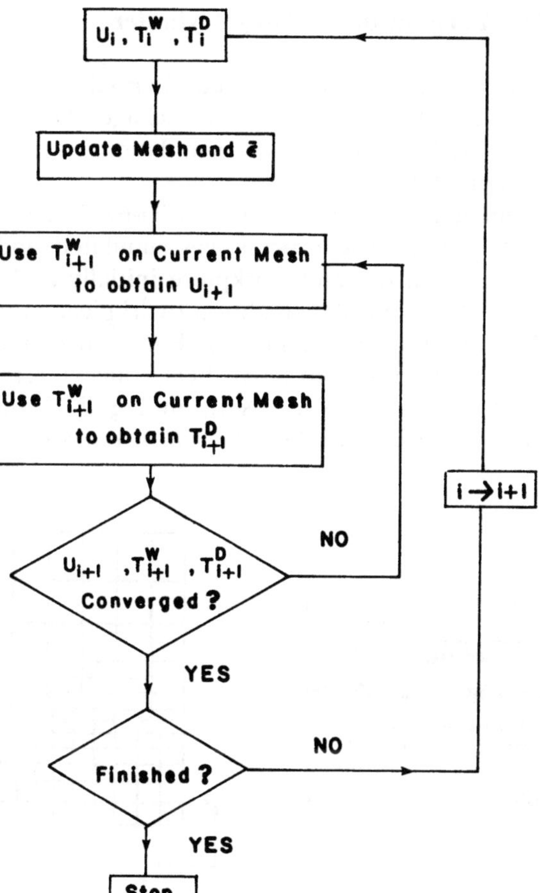

Fig. 7.11. Flow chart illustrating coupled thermal and deformation analysis.

7.6.6 Convective Heat Transport

In the equation governing heat transfer, Eq. (7.61),

$$\rho C_v \frac{\partial T}{\partial t} + \rho C_v \dot{u}_i \frac{\partial T}{\partial x_i} - k\, T_{,i} - \nu \bar{\sigma} \dot{\bar{\epsilon}} = 0 \tag{7.61}$$

the second term (arising from the Eulerian description of the material derivate of temperature T) is the convective heat transport term. It deserves special mention because it results in a nonsymmetric system of equations after partial discretization. To get around this problem, many researchers have ignored the term in their formulation. An alternative is to include the term, and use a suitable time discretization scheme that makes the equation symmetric, as discussed in the previous section. In their numerical studies, Lamm (1989) and Todt (1990) have noted a significant difference between results obtained with the convection term included and those where the term is neglected. Further studies, including experimental work, should help clarify the significance of the convective heat transport term, and also the validity of neglecting or including it in the analysis.

7.6.7 Examples: Forging of Ti Alloy Cylinder

This section presents examples of coupled thermal and deformation analyses performed with the CFORM finite-element package (Jackson et al. 1988a). The parameters used to define the forging simulations were patterned from the example given by Park (1982).

The process modelled is compression of a Ti-6242-Si alloy cylinder between dies as shown in Fig. 7.12. The process was simulated under two conditions – isothermal condition (both die and workpiece initially at 1171.9 K; die velocity 8.467×10^{-2} mm s^{-1}) and hot-die condition (workpiece initially at 1171.9 K; die initially at 644 K; die speed 1.27 mm s^{-1}). For each condition, two runs were made, one run neglecting the convective heat transfer term, one including the convective heat transfer term. The results in Fig. 7.13 show the temperature distribution in the workpiece and die for each of the four cases.

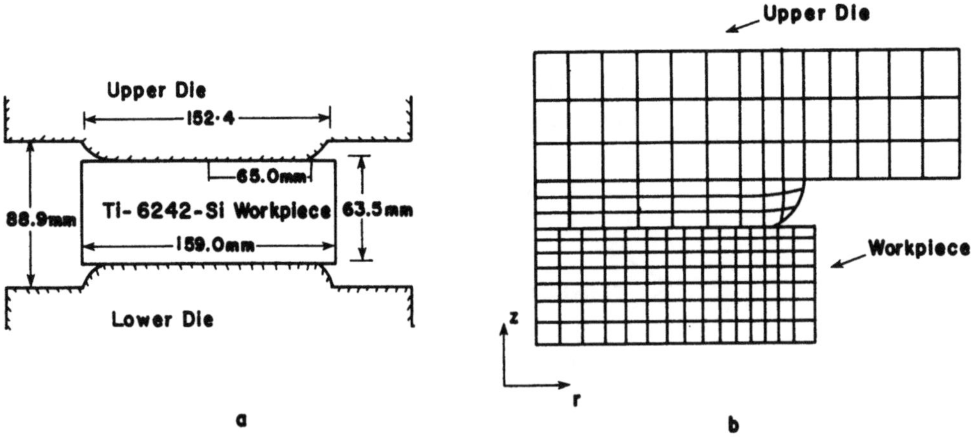

Fig. 7.12. Coupled thermal and deformation analysis of cylinder compression. **a** Cross-sectional view of workpiece and die geometries; **b** finite-element models of workpiece and dies.

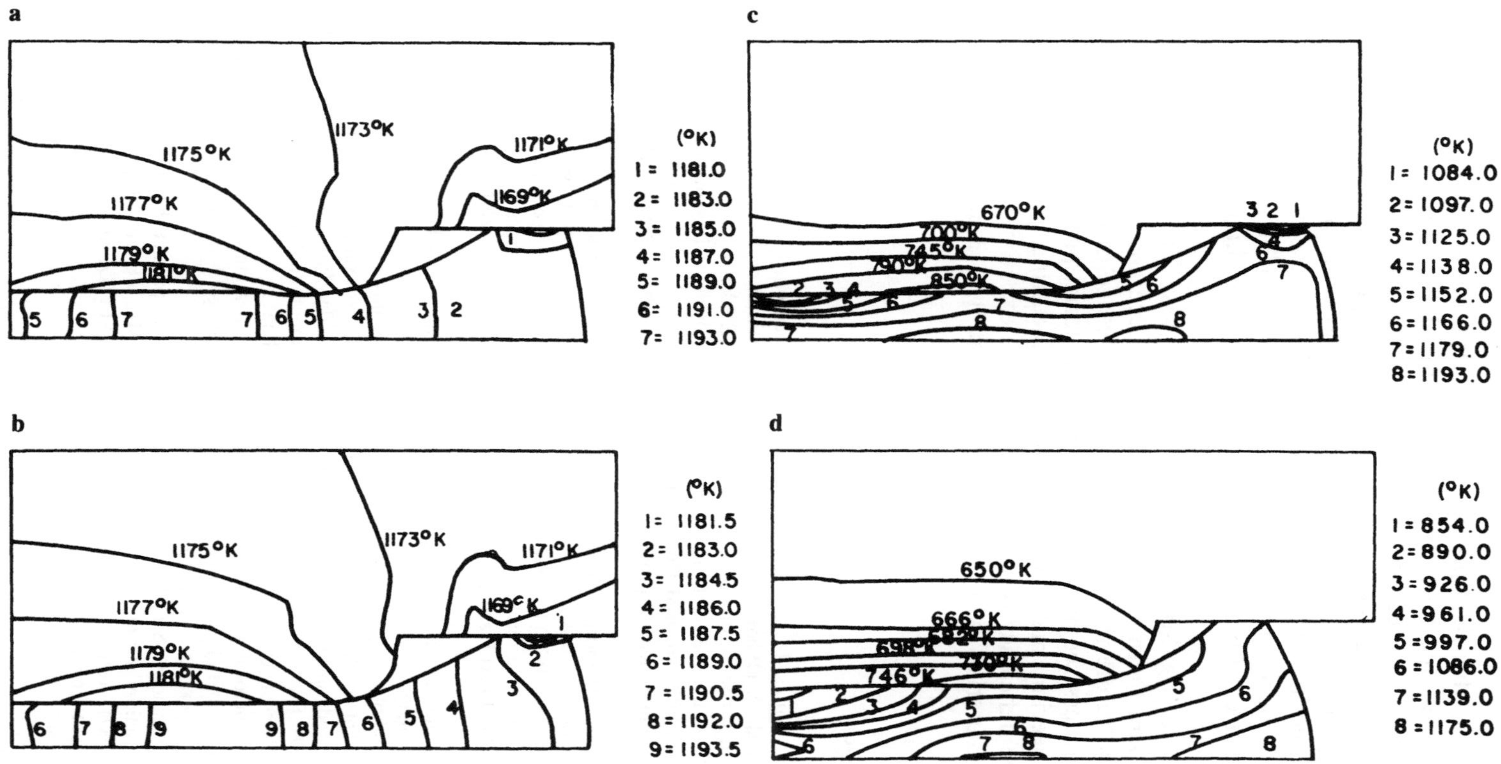

Fig. 7.13 Cylinder compression, temperature distribution results for Ti-6242-Si workpiece and die as predicted by CFORM. **a**,**b** 70% compression, isothermal conditions; **c** 70% compression, hot-die conditions; **d** 58% compression, hot-die conditions. **b** and **d** include convective heat transport.

7.7 Conclusion

In this chapter, the rigid-plastic finite-element method, as applied to metal-forming processes, has been discussed in some detail. After the basic governing equations were presented, it was shown how finite-element methodology could be implemented to allow development of computer codes for simulation of deformation processing. A number of special problems arise during development of numerical methods to realize such a capability. Resolution of several of these difficulties has been outlined. The successful methodology has been illustrated through several examples of deformation processing simulation.

It has been shown how the basic rigid-plastic analysis can be extended to rigid-viscoplastic and rigid-thermoviscoplastic analyses. The process of combined deformation and thermal analysis of the workpiece and die has been outlined.

The biggest hurdle in the way of large-scale usage of numerical simulations in the design process seems to be the lack of material property datasets for many of the engineering materials used in manufacturing processes. The way to overcome the deficiency is to perform extensive experiments on various common engineering materials with a view to formulating a reliable material property database.

References

ABAQUS user's manual (1985). Hibbitt, Karlsson and Sorenson, Providence, Rhode Island

Altan T, Kobayashi S (1968) A numerical method for estimating the temperature distributions in extrusion through conical dies. J Engng Ind Trans ASME 90:107

Bathe KJ (1975) A finite element program for automatic dynamic incremental non-linear analysis. Report 82448-2, MIT, (revised December 1978)

Bathe KJ (1982) Finite element procedures in engineering analysis. Prentice-Hall, Englewood Cliffs, New Jersey

Bishop JFW (1956) An approximation method for determining the temperatures reached in steady-state motion problems of plane plastic strain. Q J Mech Appl Math 9:236

Chen CC, Kobayashi S (1978) Rigid plastic finite element analysis of ring compression. In: Applications of numerical methods to forming processes. ASME Applied Mechanics Division, vol 28, pp 163–174

Chen WF, Han DJ (1988) Plasticity for structural engineers. Springer-Verlag, Berlin Heidelberg New York

Cheng J (1988) Automatic adaptive remeshing for finite element simulation of forming processes. Int J Num Meth Engng, 26:1–18

Dung NL, Mahrenholtz O (1982) Progress in the analysis of unsteady metal forming processes using the finite element method. In: Pittman JFT, Wood RD, Alexander JM, Zienkiewicz OC (eds) Numerical methods in industrial forming processes. Pineridge Press, Swansea, pp 187–196

Gangjee T (1987) Friction and lubrication in finite element analysis of forging processes. PhD dissertation, Clemson University, Clemson, South Carolina

Gordon P (1955) Microcalorimetric investigation of recrystallization of copper. J Metals 7:1043–1052

Greenfield P, Bever MB (1956) Evolution of energy stored by gold–silver alloy cold-worked at −95°C and at room temperature. Acta Metall 4:433–439

Gunasekara JS, Mahadeva S (1988) An alternative method to predict the effects of friction in metal forming. In: Friction and material characterizations. ASME MD vol 10, pp 55–62

Hageman LJ (1987) Automatic adaptive remeshing in EALPID, an advanced forging simulation program. Comput Engng 2:93–97

Hallquist JO (1986) NIKE2D – A vectorized implicit finite deformation finite element code for analyzing the static and dynamic response of 2-D solids with interactive reasoning and graphics. Lawrence Livermore National Laboratory, Livermore, California

Haque I, Jackson JE Jr, Gangjee T, Raikar A (1987) Empirical and finite element approaches to forging die design: a state-of-the-art survey. J Mater Shaping Technol 5:23–33

Hartley P, Sturgess CEN, Rowe GW (1979) Friction in finite-element analyses of metal forming processes. Int J Mech Sci 21:301–311

Hartley P, Sturgess CEN, Rowe GW (1980) Influence of friction on the prediction of forces, pressure distributions and properties in upset forging. Int J Mech Sci 22:743–753

Hartley P, Pillinger I, Sturgess CEN (1988) Modelling frictional effects in finite-element simulation of metal forming. In: Friction and material characterizations. ASME MD vol 10, pp 91–96

Im YT, Kobayashi S (1986) Analysis of axisymmetric forgings of porous materials by the finite element method. Adv Manuf Proc 1:473–499

Jackson JE Jr, Haque I, Gangjee T (1988a) The CFORM finite element code for material forming analysis. In: Wang SY, Hackett RM, DeLeeuw SL, Smith AM (eds) Proceedings of the fourteenth conference on theoretical and applied mechanics, School of Engineering, University of Mississippi, USA, pp 567–574

Jackson JE Jr, Haque I, Gangjee T, Ramesh M (1988b) Some numerical aspects of frictional modeling in material forming processes. In: Friction and material characterizations. ASME MD vol 10, pp 39–43

Johnson W, Kudo H (1960) The use of upper bound solutions for the determination of temperature distribution in fast hot-rolling and axisymmetric extrusion processes. Int J Mech Sci, vol 1, p 175

Johnson W, Sowerby R (1978) Metal forming processes: analysis and technology. In: Applications of numerical methods to forming processes. ASME (Applied Mechanics Division), vol 28, pp 1–13

Kobayashi S (1977) Rigid plastic finite element analysis of axisymmetric metal forming processes. In: Numerical modeling of manufacturing processes. ASME PVP-PB-025, New York, pp 49–65

Kobayashi (1979) The mechanics of plastic deformation in metal forming. In: Proceedings of the US National Congress of Applied Mechanics, West Period, North Hollywood, California pp 115–146

Kudo H, Matsubara S (1979) Joint examination project of validity of various numerical methods for the analysis of metal forming processes. In: Metal forming plasticity. IUTAM symposium, 1978, Tutzing, Germany, Springer-Verlag, Berlin Heidelberg New York, pp 378–423

Lahoti G, Altan T (1975) Predictions of temperature distributions in axisymmetric compression and torsion. J Engng Mat Technol Trans ASME 97:113

Lamm BE (1989) Simulating heat transfer in metal forming. MS thesis, Clemson University, Clemson, South Carolina

Larenson W (1980) Grid generation tools for the finite element analyst. In: Conaway JH (ed.) Proceedings first chataqua on finite element modeling, Wallace Press, pp 119–136

Lee CH, Kobayashi S (1970) Elastoplastic analysis of plane-strain and axisymmetric flat punch indentation by the finite element method. Int J Mech Sci 12:349–370

Lee CH, Kobayashi S (1971) Analysis of axisymmetric upsetting and plane-strain side-pressing of solid cylinders by the finite element method. J Engng Ind Trans ASME Ser B 93(2):445

Lee CH, Kobayashi S (1973) New solutions to rigid-plastic deformation problems using a matrix method. J Engng Ind Trans ASME 95:865–873

Lyness JF, Owen DRJ, Zienkiewicz OC (1974) Finite element analysis of the steady flow of non-Newtonian fluids through parallel sided conduits. In: Oden JT, Zienkiewicz OC, Gallagher RH, Taylor C (eds) Finite Element Methods in Flow Problems. UAH Press, Huntsville, AL, pp 489–503

Mahrenholtz O (1982) Different finite element approaches to large plastic deformations. Comput Meth Appl Mech Engng 33:454–468

Male AT, Cockcroft MG (1964–65) A method for the determination of the coefficient of friction of metals under conditions of bulk plastic deformation. J Inst Metals 93:38–46

Marc (nd) MARC General purpose finite element program. vol. A. User information manual. MARC Analysis Research Corporation, Palo Alto, CA

Marcal PV, King IP (1967) Elastic plastic analysis of two dimensional stress systems by the finite element method. Int J Mech Sci 9:143

Matsumoto H, Oh SI, Kobayashi S (1977) A note on the matrix method for rigid-plastic finite element analysis of ring compression. In: Proceedings of the 18th MTDR Conference, London, September 1977, pp 3–9

Nagpal V, Lahoti G, Altan T (1977) A numerical method for simultaneous prediction of metal flow and temperatures in upset forging of rings. ASME paper no. 77-WA/PROD-35

Nicolas VT, Citipitioglu E (1977) A general isoparametric finite element program SDRC SUPERB. Comput Struct 7:303–313

Oh SI (1982) Finite element analysis of metal forming processes with arbitrarily shaped dies. Int J Mech Sci 24(8):479–493

Park JJ (1982) Application of the finite element method to metal forming problems. PhD dissertation, University of California, Berkeley

Rebelo N, Kobayashi S (1980) A coupled analysis of viscoplastic deformation and heat transfer – I and II. Int J Mech Sci 22:699–718

Rebelo N, Rydstad H, Schroder G (1982) Simulation of material flow in closed-die forging by model techniques and rigid-plastic finite element method. In: Pittman JFT, Wood RD, Alexander JM, Zienkiewicz OC (eds) Numerical Methods in Industrial Forming Processes. Pineridge Press, Swansea, pp 237–246

Shabaik AH (1978) Analysis of forming processes: experimental and numerical methods. In: Applications of numerical methods to forming processes. ASME (Applied Mechanics Division) vol 28, pp 15–25

Shabaik AH (1981) Mechanics of plastic deformation in metal forming processes: experimental and numerical methods. In: Shawki GSA, Metwalli SM (eds) Current advances in mechanical design and production. Pergamon Press, pp 269–288

Shabaik AH (1982) Plane strain extrusion and forging. In: Pittman JFT, Wood RD, Alexander JM, Zienkiewicz OC (eds) Numerical methods in industrial forming processes. Pineridge Press, Swansea, pp 321–331

Shah SN, Kobayashi (1974) Rigid-plastic analysis of cold heading by the matrix method. In: Proceedings of the 15th MTDR Conference, Birmingham, pp 603–610

Tang MC, Kobayashi S (1982a) An investigation of the shell nosing process by the finite element method, part 1: nosing at room temperature. J Engng Ind Trans ASME 104(3):305–311

Tang MC, Kobayashi S (1982b) An investigation of the shell nosing process by the finite element method, part 2: nosing at elevated temperatures (hot nosing). J Engng Ind Trans ASME 104(3):312–318

Tang MC, Hsu M, Kobayashi S (1982) Analysis of shell nosing – process mechanics and preform design. Int J Mach Tool Des Res 22(14):293–307

Tay AO, Stevenson MG, Davis GV (1974) Using the finite element method to determine temperature distributions in orthogonal machining. Proc Inst Mech Engrs 188(55/74):627

Todt M (1990) Simulating die-workpiece heat transfer during metal forming processes, MS thesis, Clemson University, Clemson, South Carolina

Truesdell C, Noll W (1965) The nonlinear field theories of mechanics. In: Flugge S (ed.) Encyclopedia of physics, Springer-Verlag, Berlin Heidelberg New York 3/3 Sect. 79

Wilson WRD (1979) Friction and lubrication in bulk metal-forming processes. J Appl Metalwork 1(1):7–19

Wu WT, Oh SI (1985) ALPIDT: a general purpose FEM code for simulation of nonisothermal forming processes. In: NAMRC conference, Berkeley, California

Yamada Y, Yoshimura N, Sakurai T (1968) Plastic stress-strain matrix and its application for the solution of elastic-plastic problems by the finite element method. Int J Mech Sci 10:343–354

York II AR (1988) A study of finite element remeshing procedures for large strain problems with specific application to the CFORM-FEM computer code, MS Thesis, Clemson University, Clemson, South Carolina

Zienkiewicz OC (1986) The finite element method, 3rd edn. McGraw-Hill Book Company, New York

Zienkiewicz OC, Oñate E, Heinnrich JC (1978) Plastic flow in metal forming. I: coupled thermal behavior in extrusion. II: thin sheet forming. In: Applications of numerical methods to forming processes. ASME (Applied Mechanics Division), vol 28, pp 107–120

8 The Viscoplastic Approach for the Finite-Element Modelling of Metal-Forming Processes

J.-L. Chenot and M. Bellet

8.1 Introduction

There is a growing interest in industry in the use of numerical models for the simulation of industrial forming processes. These models allow the prediction of different parameters, which are of extreme importance for the design of the optimized forming sequence:

- Final shape of the part
- Loads on the machine and the tools
- Possibility of defects on the surface, or in the part
- Local deformations in the part
- Temperature in the tool and in the part
- Tool wear
- Evolution of the physical structure of the part after the deformation and temperature history

If the material properties and the friction behaviour are known, the practician is able to simulate many trials of forming, and select the best one with respect to industrial constraints.

There is a large variety of computer programs available which may appear increasingly sophisticated, but many have in common a basic approximation, namely the neglect of the elastic effects, and consequently make use of a rigid plastic, or a viscoplastic constitutive equation, the problem being formulated in terms of the velocity field.

As soon as large deformations are involved the neglect of the elastic effects does not affect significantly the flow or the stress distribution during the forming step. However, one should be aware that some problems cannot be solved by this approach, among which are: calculation of residual stresses for cold forming after unloading, or for hot forming after unloading and cooling, and the spring-back effect which may be significant in some cases. However, some authors have proposed the use of an elastoplastic or elastoviscoplastic approach only during the unloading step (Lee and Dawson 1988).

While the elastoplastic or elastoviscoplastic approaches seem to be very dependent conceptually on the old displacement formulation for elastic solids, the plastic or viscoplastic approximation, often called the flow formulation, is closer to fluid mechanics concepts. An eventual unification between the two approaches seems possible.

The first finite-element approach of viscoplastic flow (Cornfield and Johnson, 1973) made use of a modified elastic computer code, with a variable modulus, and approximate incompressibility. Then more systematic formulations were presented by Lee and Kobayashi (1973) and Zienkiewicz and Godbole (1974).

Among the first two-dimensional applications are

- Forging of dense material (Oh et al. 1978)
- Forging of porous material (Shima and Mori 1978)
- Deformation of bi-material (Perrot and Chenot 1981)

Other works are referenced in the reviews by Kobayashi (1982) and Zienkiewicz (1984). Then these works were extended to more complex geometries with a three-dimensional approach:

- Steady-state approach of rolling (Chenot 1983)
- Incremental forging of simple parts (Park and Kobayashi 1984, Mori et al. 1984)
- Incremental forging of complex parts (Surdon and Chenot 1987, Cescutti et al. 1987)
- Forging of porous metals (Chenot 1989)

8.2 Constitutive Equations for a Viscoplastic Material

8.2.1 The Viscoplastic Potential Approach

Most of the more common constitutive laws for idealizing real materials have been obtained from experimental approaches. But we can show that a large number of these laws may be expressed in terms of a viscoplastic potential. The frame of the thermodynamic approach has been described in Freudenthal and Geiringer (1958) and we shall only summarize the most important results for viscoplastic materials.

We suppose here that there exists a viscoplastic potential ϕ, which is a function of the strain rate tensor $\dot{\boldsymbol{\epsilon}}$, such that the constitutive equation may be written as

$$\boldsymbol{\sigma} = \frac{\partial \phi(\dot{\boldsymbol{\epsilon}})}{\partial \dot{\boldsymbol{\epsilon}}} \tag{8.1}$$

which is equivalent to the following form expressed in terms of the components:

$$\sigma_{ij} = \frac{\partial \phi(\dot{\boldsymbol{\epsilon}})}{\partial \dot{\epsilon}_{ij}} \tag{8.2}$$

Regarding the incompressible material case, we see that ϕ must be independent of the rate of compressibility:

$$\dot{\theta} = \text{tr}\,(\dot{\boldsymbol{\epsilon}}) = \text{div}(\mathbf{v}) \tag{8.3}$$

where $\mathbf{v}$ is the velocity field. The trace of a tensor is defined in the usual way, in cartesian coordinates:

$$\text{tr}\,(\dot{\boldsymbol{\epsilon}}) = \sum_i \dot{\epsilon}_{ii} \tag{8.4}$$

such that, for incompressible materials it is easy to show that we obtain in fact the stress deviator in 8.1:

$$\mathbf{s} = \frac{\partial \phi(\dot{\boldsymbol{\epsilon}})}{\partial \dot{\boldsymbol{\epsilon}}} \tag{8.5}$$

For isotropic materials ϕ must be a function only of the invariants of the strain rate tensor: the first invariant is (8.3), the second invariant is related to

$$\dot{\bar{\epsilon}} = \left(\frac{2}{3}\sum_{ij} \dot{\epsilon}_{ij}^2\right)^{1/2} \tag{8.6}$$

and the third invariant may be taken as the determinant of the strain rate tensor, but it is of less common use, and will be ignored in the following. For isotropic incompressible materials, the most common form is

$$\mathbf{s} = \frac{\partial\phi(\dot{\bar{\epsilon}})}{\partial\dot{\boldsymbol{\epsilon}}} = \frac{2}{3}\frac{1}{\dot{\bar{\epsilon}}}\frac{\partial\phi(\dot{\bar{\epsilon}})}{\partial\dot{\bar{\epsilon}}}\dot{\boldsymbol{\epsilon}} \tag{8.7}$$

and

$$\dot{\theta} = 0 \tag{8.8}$$

Let us introduce now the second invariant of the stress tensor

$$\bar{\sigma} = (\tfrac{3}{2}\,\mathbf{s}{:}\mathbf{s})^{1/2} \tag{8.9}$$

The notation ":" indicates the doubly contracted product, namely

$$\mathbf{s}{:}\mathbf{s} = \sum_{ij} s_{ij}s_{ij} \tag{8.10}$$

Then we obtain from (8.7) the monodimensional constitutive equation

$$\bar{\sigma} = \frac{\partial\phi(\dot{\bar{\epsilon}})}{\partial\dot{\bar{\epsilon}}} \tag{8.11}$$

We can also transform Eq. (8.7) into the viscous form

$$\mathbf{s} = 2\eta(\dot{\bar{\epsilon}})\dot{\boldsymbol{\epsilon}} \tag{8.12}$$

where the pseudo viscosity parameter is

$$\eta(\dot{\bar{\epsilon}}) = \frac{1}{3\dot{\bar{\epsilon}}}\frac{\partial\phi(\dot{\bar{\epsilon}})}{\partial\dot{\bar{\epsilon}}} \tag{8.13}$$

Conversely, any constitutive law which takes the general form given by (8.12) and (8.13) corresponds to a viscoplastic potential

$$\phi(\dot{\bar{\epsilon}}) = \int_0^{\dot{\bar{\epsilon}}} 3\eta(\lambda)\lambda d\lambda \tag{8.14}$$

In the general case, a complementary or dual, viscoplastic potential can be introduced as a function of the stress tensor $\phi'(\boldsymbol{\sigma})$, such that

$$\dot{\boldsymbol{\epsilon}} = \frac{\partial\phi'(\boldsymbol{\sigma})}{\partial\boldsymbol{\sigma}} \tag{8.15}$$

or with the index notation

$$\dot{\epsilon}_{ij} = \frac{\partial\phi'(\boldsymbol{\sigma})}{\partial\sigma_{ij}} \tag{8.16}$$

For the incompressible isotropic material and when ϕ' is only a function of the equivalent stress,

$$\dot{\boldsymbol{\epsilon}} = \frac{3}{2}\frac{\partial\phi'(\bar{\sigma})}{\partial\bar{\sigma}}\frac{\mathbf{s}}{\bar{\sigma}} \tag{8.17}$$

which gives the monodimensional equation

$$\dot{\bar{\epsilon}} = \frac{\partial\phi'(\bar{\sigma})}{\partial\bar{\sigma}} \tag{8.18}$$

The rate of dissipated energy takes the following form in the general case:

$$\dot{w} = \boldsymbol{\sigma}:\dot{\boldsymbol{\epsilon}} = \phi(\dot{\boldsymbol{\epsilon}}) + \phi'(\boldsymbol{\sigma}) \tag{8.19}$$

Returning again to the incompressible isotropic material (where the third invariant is not considered), by use of (8.11) and (8.8) we obtain

$$\dot{w} = \bar{\sigma}\,\dot{\bar{\epsilon}} = \phi(\dot{\bar{\epsilon}}) + \phi'(\bar{\sigma}) \tag{8.20}$$

If we suppose that (8.11) is invertible (at least theoretically), and g is the inverse of $\partial\phi/\partial\dot{\bar{\epsilon}}$, we get

$$\phi'(\bar{\sigma}) = \bar{\sigma}\,g(\bar{\sigma}) - \phi(g(\bar{\sigma})) \tag{8.21}$$

which allows us to deduce ϕ' from the potential ϕ; the reverse is also possible.

8.2.2 Examples of Viscoplastic Behaviour

8.2.2.1 The Norton–Hoff Law

The Norton–Hoff law was introduced for the monoaxial creep analysis by Norton (1929), and extended to three-dimensional analysis by Hoff (1954). It is generally written

$$\mathbf{s} = 2K(\sqrt{3}\,\dot{\bar{\epsilon}})^{m-1}\,\dot{\boldsymbol{\epsilon}} \tag{8.22}$$

and the monodimensional expression is

$$\bar{\sigma} = K(\sqrt{3})^{m+1}\,\dot{\bar{\epsilon}}^{m} \tag{8.23}$$

We observe that

- $m = 1$ corresponds to the Newtonian fluid with a viscosity $\eta = K$.
- $m = 0$ is the plastic flow rule for a material obeying the von Mises criterion with a yield stress $\sigma_0 = \sqrt{3}K$.
- $0 < m < 1$ is a first approximation for hot metals (or molten polymers): for more common metals m lies between 0.1 and 0.2, but for superplastic materials we have $0.5 \leq m \leq 0.7$.

The situation is depicted in Fig. 8.1 with the two limit cases.

Now the first viscoplastic potential is obviously

$$\phi(\dot{\bar{\epsilon}}) = \frac{K}{m+1}(\sqrt{3}\,\dot{\bar{\epsilon}})^{m+1} \tag{8.24}$$

and the second one

$$\phi(\bar{\sigma}) = \frac{m}{m+1}K\left(\frac{\bar{\sigma}}{K\sqrt{3}}\right)^{\frac{1}{m}+1} \tag{8.25}$$

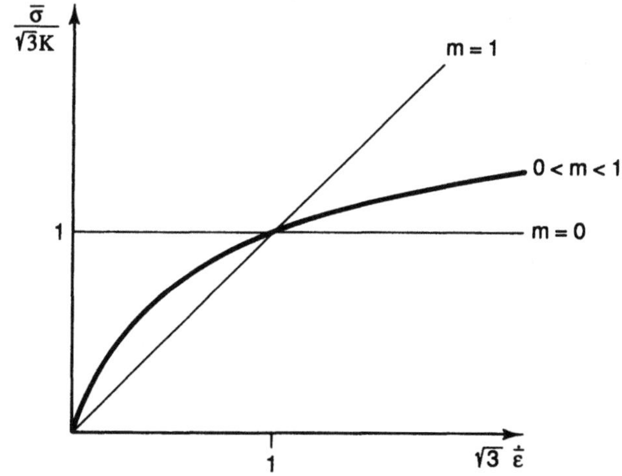

Fig. 8.1. One-dimensional viscoplastic behaviour.

8.2.2.2 The Sellars–Tegart Law

In one dimension, the Sellars–Tegart Law is written (Sellars and Tegart, 1972)

$$\dot{\bar{\epsilon}} = A(\sinh(\alpha\bar{\sigma}))^{\frac{1}{m}} \tag{8.26}$$

or

$$\bar{\sigma} = \frac{1}{\alpha}\sinh^{-1}\left(\frac{\dot{\bar{\epsilon}}}{A}\right)^{m} \tag{8.27}$$

and using (8.7) and (8.11) we obtain

$$\mathbf{s} = \frac{2}{3}\frac{1}{\alpha}\sinh^{-1}\left(\frac{\dot{\bar{\epsilon}}}{A}\right)^{m}\frac{1}{\dot{\bar{\epsilon}}}\dot{\boldsymbol{\epsilon}} \tag{8.28}$$

It is well known that the Norton–Hoff law may be seen as an approximation of the Sellars–Tegart constitutive equation when $\dot{\bar{\epsilon}}/A \ll 1$; it is easily verified in (8.26) by writing

$$\dot{\bar{\epsilon}} \approx A\left(\alpha\bar{\sigma}\right)^{\frac{1}{m}} \tag{8.29}$$

8.2.2.3 Regularization of the Constitutive Law

We shall see in Sect. 8.4 that the numerical calculation introduces the second derivative of ϕ, which tends to infinity if $\dot{\bar{\epsilon}}$ tends to 0, for both the Norton–Hoff law and the Sellars–Tegart law.

A regularized Norton–Hoff law can be written

$$\mathbf{s} = 2K(\sqrt{3})^{m-1}\left(\dot{\bar{\epsilon}}_0^2 + \dot{\bar{\epsilon}}^2\right)^{\frac{m-1}{2}}\dot{\boldsymbol{\epsilon}} \tag{8.30}$$

in which $\dot{\bar{\epsilon}}_0$ is a small constant such that if $\dot{\bar{\epsilon}} \gg \dot{\bar{\epsilon}}_0$, (8.30) is nearly identical to (8.22) and if $\dot{\bar{\epsilon}} \ll \dot{\bar{\epsilon}}_0$ the constitutive law tends to become purely Newtonian.

This form allows us to express the viscoplastic potential in its regularized form:

$$\phi(\dot{\varepsilon}) = \frac{K}{m+1} (\sqrt{3})^{m+1} \left(\dot{\varepsilon}_0^2 + \dot{\varepsilon}^2\right)^{\frac{m+1}{2}} \tag{8.31}$$

The same method applies to Eq. (8.28) where $\dot{\varepsilon}$ may be replaced by $(\dot{\varepsilon}_0^2 + \dot{\varepsilon}^2)^{1/2}$.

8.2.3 The Compressible Viscoplastic Material

For a porous material the relative density ρ_r is no longer constant so that the first invariants for the stress and strain rate tensors must be introduced. Following the work on compressible plasticity (Shima and Oyane 1976) we define the equivalent strain rate and the equivalent stress:

$$\dot{\varepsilon} = \frac{1}{\rho_r} \left(\frac{2}{3c} \dot{\mathbf{e}}:\dot{\mathbf{e}} + \frac{1}{9f} (\mathrm{tr}(\dot{\boldsymbol{\varepsilon}}))^2 \right)^{1/2} \tag{8.32}$$

$$\bar{\sigma} = \left(\frac{3c}{2} \mathbf{s}:\mathbf{s} + f (\mathrm{tr}(\boldsymbol{\sigma}))^2 \right)^{1/2} \tag{8.33}$$

where $\dot{\mathbf{e}}$ is the deviatoric strain rate tensor.

The c and f parameters in (8.32) and (8.33) are functions of the relative density ρ_r, which must give the usual form when it tends to unity, so we have

$$c(1) = 1 \text{ and } f(1) = 0 \tag{8.34}$$

These parameters are obtained by experiments carried out on samples of various density and their variation is shown in Fig. 8.2.

The generalization of this approach to compressible viscoplastic material has been studied extensively in Abouaf (1985) and Abouaf et al. (1988). If a

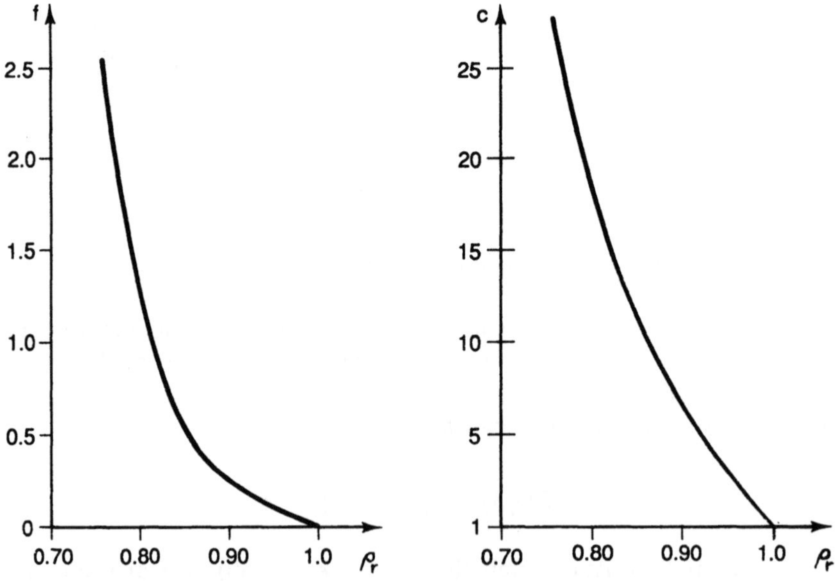

Fig. 8.2. Variations of the f and c parameters for a porous material.

viscoplastic potential $\phi(\dot{\bar\epsilon})$ is known for a given dense material, the potential for the same material in a nondense state will be $\phi(\rho_r\dot{\bar\epsilon})$, where $\dot{\bar\epsilon}$ is defined by (8.32). The constitutive equation is immediately derived:

$$\sigma = \frac{1}{\rho_r\dot{\bar\epsilon}}\frac{\partial\phi(\rho_r\dot{\bar\epsilon})}{\partial(\rho_r\dot{\bar\epsilon})}\left[\frac{2}{3c}\dot{\mathbf{e}} + \frac{1}{9f}(\mathrm{tr}(\dot{\boldsymbol{\epsilon}}))\,\mathbf{I}\right] \tag{8.35}$$

here **I** is the unit tensor with

$$\mathrm{I}_{ij} = 1 \text{ if } i=j$$
$$\mathrm{I}_{ij} = 0 \text{ if } i\neq j \tag{8.36}$$

In (8.35) we have introduced two parameters K and m which are supposed to be independent of the relative density ρ_r (but are functions of temperature T in general), and the c and f functions which depend on ρ_r and T.

From (8.35) we can express the rate of density evolution:

$$\frac{\dot\rho_r}{\rho_r} = -\mathrm{tr}(\dot{\boldsymbol{\epsilon}}) = 9\rho_r\, p\, \dot{\bar\epsilon}\left(\frac{\partial\phi(\rho_r\dot{\bar\epsilon})}{\partial(\rho_r\dot{\bar\epsilon})}\right) \tag{8.37}$$

8.2.4 The Anisotropic Material

Following the approach of Hill (1950) the equivalent stress, in the axes of orthotropy, is given by

$$\bar\sigma = \{\tfrac{1}{2}(F(\sigma_{yy}-\sigma_{zz})^2 + G(\sigma_{zz}-\sigma_{xx})^2 + H(\sigma_{xx}-\sigma_{yy})^2 + 2L\sigma_{yz}^2$$
$$+ 2M\sigma_{zx}^2 + 2N\sigma_{xy}^2)\}^{1/2} \tag{8.38}$$

It is easy to show that the equivalent strain rate can be defined by

$$\dot{\bar\epsilon} = \{\tfrac{2}{3}(F'\dot\epsilon_{xx}^2 + G'\dot\epsilon_{yy}^2 + H'\dot\epsilon_{zz}^2 + 2L'\dot\epsilon_{yz}^2 + 2M'\dot\epsilon_{zx}^2$$
$$+ 2N'\dot\epsilon_{xy}^2)\}^{1/2} \tag{8.39}$$

where

$$F' = \frac{3F}{FG+GH+HF} \qquad G' = \frac{3G}{FG+GH+HF} \qquad H' = \frac{3H}{FG+GH+HF} \tag{8.40}$$

$$L' = \frac{3}{L} \qquad M' = \frac{3}{M} \qquad N' = \frac{3}{N} \tag{8.41}$$

and the anisotropic viscoplastic law can be defined with a generalized Norton–Hoff law according to Eq. (8.23), or with the viscoplastic potential (8.25). Now the deviatoric stress tensor will be

$$s_{xx} = 2\eta F'\dot\epsilon_{xx} \qquad s_{yy} = 2\eta G'\dot\epsilon_{yy} \qquad s_{zz} = 2\eta H'\dot\epsilon_{zz}$$
$$s_{xy} = 2\eta N'\dot\epsilon_{xy} \qquad s_{yz} = 2\eta L'\dot\epsilon_{yz} \qquad s_{zx} = 2\eta M'\dot\epsilon_{zx} \tag{8.42}$$

with $\eta = K(\sqrt{3}\,\dot{\bar\epsilon})^{m-1}$ and we have again the incompressibility requirement (8.3).

If these equations are to be written in arbitrary axes, we must introduce two fourth-rank tensors

- Tensor **A** such that the equivalent stress takes the form

$$\bar{\sigma} = (\tfrac{1}{2}\,\boldsymbol{\sigma}\mathbf{A}\boldsymbol{\sigma})^{1/2} \tag{8.43}$$

- Tensor **C** for calculating the equivalent strain rate:

$$\dot{\bar{\epsilon}} = (\tfrac{2}{3}\,\dot{\boldsymbol{\epsilon}}\mathbf{C}\dot{\boldsymbol{\epsilon}})^{1/2} \tag{8.44}$$

The tensor **A** can be expressed in terms of the F, G, H, L, M, N coefficients and the rotation matrix which allows transformation of the orthotropy axes to the laboratory reference system. But to our knowledge there does not exist a macroscopic law to describe properly the evolution of **A**. One of the first attempts was by Dawson and Mathur (1989) for cold forming with a complex microscopic model.

8.2.5 Influence of Temperature and Internal Parameters

8.2.5.1 Temperature

Apart from very specific forming conditions where the process is very slow and may be considered as isothermal, the temperature distribution plays an important role, and the viscoplastic potential must be temperature dependent. The most common law is multiplicative and is similar to the influence of an activation energy, so that we shall write the new potential as

$$\phi_T(\dot{\boldsymbol{\epsilon}},T) = \phi(\dot{\boldsymbol{\epsilon}}) \exp\left(\frac{\alpha}{T}\right) \tag{8.45}$$

Another point of view which is generally adopted for the Norton–Hoff law, is to consider that the consistency in (8.24) is a function of temperature:

$$K = K_0 \exp\left(\frac{\alpha}{T}\right) \tag{8.46}$$

In this case the two approaches are identical but it has been shown experimentally that the rate sensitivity index m may also depend on temperature:

$$m = m_0 + m_1\, T \tag{8.47}$$

and in this case Eq. (8.45) does not apply.

8.2.5.2 Work Hardening

Another important physical phenomenon is work hardening, which is present in hot forming processes when the strain rate is large enough. It has been recognized, for example, that during a single pass of hot rolling, work hardening plays a significant part, but the delay between two passes allows the hardening due to the first sequence to be cancelled when the second one starts. The isotropic strain hardening may also be introduced in the plastic potential with a multiplicative form:

$$\phi_{\bar{\epsilon}}(\dot{\boldsymbol{\epsilon}},\bar{\epsilon}) = \phi(\dot{\boldsymbol{\epsilon}})\, h(\bar{\epsilon}) \tag{8.48}$$

where $\bar{\epsilon}$ is the equivalent strain which is defined for a material point by

$$\bar{\epsilon} = \int \dot{\bar{\epsilon}} dt \tag{8.49}$$

Different forms for the h function are used, for example

$$h(\bar{\epsilon}) = h_0 \bar{\epsilon}^n \quad \text{or} \quad h(\bar{\epsilon}) = h_0 + h_1 \bar{\epsilon}^n \tag{8.50}$$

It is generally admitted that work hardening has little effect on the m coefficient, except for superplastic materials in which a decrease in m can be found as $\bar{\epsilon}$ increases (Ghosh and Hamilton 1979).

8.2.5.3 *Damage*

Modelling of damage may be introduced with the compressible material approach, provided that the constitutive law is defined for a density decrease. In many instances a more phenomenological formulation is adopted, and makes use of a cumulated damage parameter D, which varies between 0 (no damage) and 1 (failure).

Several forms have been presented in the literature; only the Latham and Cockcroft approximation will be presented here (with some modification). For a given material let us define the integral

$$C = \int_0^{\bar{\epsilon}_r} \sigma_T \dot{\bar{\epsilon}} \, dt \tag{8.51}$$

where $\bar{\epsilon}_r$ is the maximum deformation the material can exhibit before fracture, for a given path of deformation, and σ_T is the maximum principal stress corresponding to the current stress tensor if it is positive, or zero if it is negative.

The approximation is that the C parameter depends only on the material, and not on the stress–strain history. If we keep this first approximation, we can define the damage by the simple differential equation

$$\dot{D} = \sigma_T \dot{\bar{\epsilon}} / C \tag{8.52}$$

The effect of damage on the constitutive law may be introduced simply by putting

$$\phi_D(\dot{\boldsymbol{\epsilon}}, D) = \phi(\dot{\boldsymbol{\epsilon}})(1 - D) \tag{8.53}$$

For the Norton–Hoff law this is again equivalent to considering that the consistency is a function of the damage parameter:

$$K = K_0(1 - D) \tag{8.54}$$

We observe that when $K = 0$ (i.e. $D = 1$) in a given subdomain the material may exhibit arbitrary deformation with a vanishing stress.

8.2.5.4 *Metallurgical Structure*

Many metallurgical parameters have been recognized to play an important role in mechanical behaviour, and to undergo important variations when the material is subjected to a complex stress–strain and temperature history. Today there is a trend in the field of metallurgy to characterize individual physical parameters in a really quantitative way, and to propose mathematical laws for the prediction of their evolution and their effect in constitutive laws. For the sake of illustration the typical effect of a metallurgical parameter G (which can be the mean grain size) is presented according to Smelser et al. (1986). The time evolution of G is governed by

$$\frac{dG}{dt} = C_1 \exp(C_2 Z^p)\, \dot{\bar{\epsilon}} \left(1 - \frac{G}{G^*}\right) \tag{8.55}$$

where the Zener–Hollomon parameter is

$$Z = \dot{\bar{\epsilon}} \exp\left(\frac{\alpha}{T}\right) \tag{8.56}$$

and

$$G^* = C_3 Z^l \tag{8.57}$$

Now the Norton–Hoff equation is simply updated by

$$K = K_G \left(\frac{G}{G_0}\right)^\gamma \tag{8.58}$$

if G_0 is a reference mean grain size parameter.

8.2.6 Friction Behaviour

We suppose that there exists an interface between a rigid body (tool) and a viscoplastic domain (the deformed part), which is represented by a surface of contact. The situation is shown in Fig. 8.3 in the vicinity of a material point **M** lying at the interface.

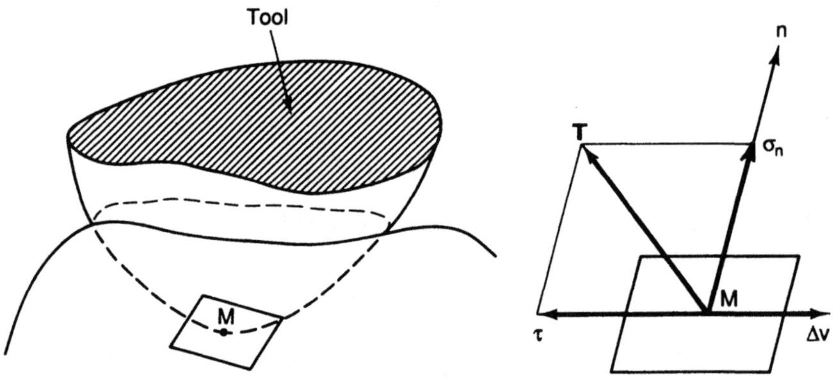

Fig. 8.3. Contact with friction.

The interaction between the tool and the part corresponds to a stress vector **T** which is the resultant of

- $\sigma_n = (\boldsymbol{\sigma}\mathbf{n})\mathbf{n}$, the normal stress which must be negative (compression) for the contact to occur
- $\boldsymbol{\tau}$, the shear stress which is due to friction

- $\mathbf{n}$, the normal to the surface
- $\Delta\mathbf{v} = \mathbf{v} - \mathbf{v}^0$, the velocity difference between a material point of the part, and a coincident material point of the tool

For a general friction law the shear stress is a function of many factors: velocity difference, normal stress, temperature, local roughness of the part and of the tool, properties of the lubricant, etc. Here we shall restrict ourselves to isotropic friction which can be derived from a friction potential ϕ_f depending on the norm $|\Delta\mathbf{v}|$ of the velocity difference, such that the stress will take the simple form (Moreau 1970)

$$\boldsymbol{\tau} = -\frac{\partial\phi_f(|\Delta\mathbf{v}|)}{\partial|\Delta\mathbf{v}|}\frac{\Delta\mathbf{v}}{|\Delta\mathbf{v}|} \tag{8.59}$$

The Norton friction law is analogous to the Norton–Hoff material law and the associated potential is

$$\phi_f(|\Delta\mathbf{v}|) = \frac{a}{q+1}|\Delta\mathbf{v}|^{q+1} \tag{8.60}$$

which gives immediately

$$\boldsymbol{\tau} = -a\,|\Delta\mathbf{v}|^{q-1}\Delta\mathbf{v} \tag{8.61}$$

and it can be shown that this form is obtained if one considers a thin intermediate layer of a material with a rate sensitivity index equal to q. In many instances it has proved more convenient to write

$$\boldsymbol{\tau} = -\alpha K\,|\Delta\mathbf{v}|^{m-1}\Delta\mathbf{v} \tag{8.62}$$

where K and m correspond to the material of the part. The relationship between $|\boldsymbol{\tau}|$ and $|\Delta\mathbf{v}|$ is quite similar to the curves in Fig. 8.1 and if $m = 0$ the Tresca limit is obtained for $0 \leq \alpha \leq 1$.

By analogy with the constitutive equation (8.30), the friction law can be regularized such that (8.62) is replaced by

$$\boldsymbol{\tau} = -\alpha K\left(\Delta\mathbf{v}^2 + \Delta\mathbf{v}_0^2\right)^{\frac{q-1}{2}}\Delta\mathbf{v} \tag{8.63}$$

8.3 The Thermomechanical Formulation

8.3.1 The Mechanical Formulation

We consider a domain Ω representing the part which undergoes viscoplastic deformation. The boundary $\partial\Omega$ can be divided into subsets:

- $\partial\Omega^f$ where the part is in contact with the tool(s)
- $\partial\Omega^T$ for which surface traction may be prescribed: T^d or 0 if it is a free surface

If we choose the velocity field as the main unknown, the problem to solve in the general case is the following:

- In Ω:

$$\boldsymbol{\sigma} = \frac{\partial \phi(\dot{\boldsymbol{\epsilon}})}{\partial \dot{\boldsymbol{\epsilon}}} \tag{8.64}$$

$$\text{div}\boldsymbol{\sigma} + \rho \mathbf{g} = 0 \tag{8.65}$$

(the inertia contribution is neglected here).

- On $\partial\Omega^{\mathrm{f}}$

$$\Delta \mathbf{v} \cdot \mathbf{n} = 0 \tag{8.66}$$

$$\boldsymbol{\tau} = -\frac{\partial \phi_{\mathrm{f}}(\Delta \mathbf{v})}{\partial \Delta \mathbf{v}} \tag{8.67}$$

- On $\partial\Omega^{\mathrm{T}}$

$$\boldsymbol{\sigma}\mathbf{n} = \mathbf{T}^d \tag{8.68}$$

It can be shown that this problem is equivalent to finding the stationary value of the functional

$$\Phi(\mathbf{v}) = \int_{\Omega} \phi(\dot{\boldsymbol{\epsilon}})\, dV - \int_{\Omega} \rho \mathbf{g} \cdot \mathbf{v}\, dV + \int_{\partial\Omega^{\mathrm{f}}} \phi_{\mathrm{f}}(\Delta \mathbf{v})\, dS - \int_{\partial\Omega^{\mathrm{T}}} \mathbf{T}^d \cdot \mathbf{v}\, dS \tag{8.69}$$

$$\text{with } \Delta \mathbf{v} \cdot \mathbf{n} = 0 \text{ on } \partial\Omega^{\mathrm{f}} \tag{8.70}$$

In general this functional admits a minimum for the solution.

An equivalent formulation arises from the virtual work principle (or may be interpreted as the Gâteau differential of the functional). We set for any virtual velocity $\mathbf{v}^*$

$$\int_{\Omega} \frac{\partial \phi(\dot{\boldsymbol{\epsilon}})}{\partial \dot{\boldsymbol{\epsilon}}} : \dot{\boldsymbol{\epsilon}}^*\, dV - \int_{\Omega} \rho \mathbf{g} \cdot \mathbf{v}^*\, dV + \int_{\partial\Omega^{\mathrm{f}}} \frac{\partial \phi_{\mathrm{f}}(\Delta \mathbf{v})}{\partial \Delta \mathbf{v}} \mathbf{v}^*\, dS - \int_{\partial\Omega^{\mathrm{T}}} \mathbf{T}^d \cdot \mathbf{v}^*\, dS = 0 \tag{8.71}$$

$$\text{with } \Delta \mathbf{v} \cdot \mathbf{n} = 0 \text{ on } \partial\Omega^{\mathrm{f}} \tag{8.72}$$

8.3.2 The Isotropic Incompressible Material

With the same notations as in Sect. 8.3.1, the functional will be almost identical:

$$\Phi(\mathbf{v}) = \int_{\Omega} \phi(\dot{\bar{\varepsilon}})\, dV - \int_{\Omega} \rho \mathbf{g} \cdot \mathbf{v}\, dV + \int_{\partial\Omega^{\mathrm{f}}} \phi_{\mathrm{f}}(|\Delta \mathbf{v}|)\, dS - \int_{\partial\Omega^{\mathrm{T}}} \mathbf{T}^d \cdot \mathbf{v}\, dS \tag{8.73}$$

$$\text{with div } \mathbf{v} = 0 \text{ in } \Omega \text{ and } \Delta \mathbf{v} \cdot \mathbf{n} = 0 \text{ on } \partial\Omega^{\mathrm{f}} \tag{8.74}$$

The mathematical constraint due to the incompressibility requirement is not easy to cope with directly. Two methods are most frequently proposed.

The Lagrange Multiplier Method. The functional (8.73) is transformed into

$$\Phi_p(\mathbf{v}) = \Phi(\mathbf{v}) - \int_\Omega p \operatorname{div} \mathbf{v} \, dV \tag{8.75}$$

where it is shown that the Lagrange multiplier can be identified with the hydrostatic pressure, such that it is a mixed method, and the solution is obtained by

$$\mathrm{Max}_p \mathrm{Min}_{\mathbf{v}} \, \Phi_p(\mathbf{v},p) \tag{8.76}$$

(and p must be prescribed at one point at least). The major drawback of this method is that the number of unknown functions is increased.

The Penalty Method. The initial problem is "slightly" modified by allowing a small rate of compressibility, such that the functional (8.73) is replaced by

$$\Phi_r(\mathbf{v}) = \Phi(\mathbf{v}) + \frac{r}{2}\int_\Omega (\operatorname{div} \mathbf{v})^2 \, dV \tag{8.77}$$

where r is the penalty coefficient, positive and as large as possible.

The constitutive equation appears to be perturbated and (8.77) implies that

$$p = -r \operatorname{div} \mathbf{v} \tag{8.78}$$

which possesses some analogy with Eq. (8.37) for the compressible material. On the other hand it can be observed that, if r is large enough, the perturbation with respect to the Lagrange multiplier method is actually negligible.

Today one of the most common functionals is Norton's, with a penalty approach, which can be expressed by

$$\Phi(\mathbf{v}) = \int_\Omega \frac{K}{m+1}(\sqrt{3}\,\dot{\bar{\epsilon}})^{m+1} dV - \int_\Omega \rho \mathbf{g} \cdot \mathbf{v} \, dV + \int_\Omega \frac{rK}{2}(\operatorname{div}(\mathbf{v}))^2 \, dV$$

$$+ \int_{\partial\Omega^f} \frac{\alpha K}{q+1} |\Delta \mathbf{v}|^{q+1} \, dS - \int_{\partial\Omega^T} \mathbf{T}^d \cdot \mathbf{v} \, dS \tag{8.79}$$

Note that K has been introduced in the penalty term, but this may be insufficient to avoid problems of units, as the different contributions are not homogeneous.

8.3.3 The Free Surface Problem for Stationary Processes

We suppose that the boundary conditions are not contact conditions on the entire surface $\partial\Omega$ of the domain Ω: it is generally the case in forming problems where the material is allowed to flow at least in one direction. If one neglects the atmospheric pressure on a free surface, the following conditions must be imposed for a stationary process:

$$\mathbf{v} \cdot \mathbf{n} = 0 \tag{8.80}$$

$$\mathbf{T} = \boldsymbol{\sigma}\,\mathbf{n} = 0 \tag{8.81}$$

These conditions are not "classical" in the sense that four scalar conditions are imposed instead of three for a three-dimensional problem. One can view the supplementary condition as the condition which allows us to determine the shape of the free surface. For example if we impose $\mathbf{T} = \mathbf{0}$ on a "trial" free surface, the actual one will be obtained by the condition $\mathbf{v} \cdot \mathbf{n} = 0$. Several formulations can be used:

The Stream Line Method. The free surface is determined by the family of curves defined by the equations

$$\frac{dx}{v_x} = \frac{dy}{v_y} = \frac{dz}{v_z} \tag{8.82}$$

The Optimum Control Method. We find the free surface as the surface S which minimizes the following functional:

$$J(S) = \int_S (\mathbf{v} \cdot \mathbf{n})^2 \, dS \tag{8.83}$$

The Weak Form. Here the free surface will verify the equation

$$\int_S (\mathbf{v} \cdot \mathbf{n}) \, w \, dS = 0 \tag{8.84}$$

for any function w belonging to a suitable subspace.

8.3.4 Time-Dependent Processes

When acceleration is neglected, the very simple form of the viscoplastic constitutive equation, where no time derivative of the stress tensor is present, allows us to consider only the evolution of the material point for which the equation is obviously

$$\frac{d\mathbf{x}}{dt} = \mathbf{v} \tag{8.85}$$

If dynamic problems are to be solved we must introduce the acceleration $\boldsymbol{\gamma}$ in the virtual work expression (8.72) which is transformed into

$$\int_\Omega \frac{\partial \phi(\dot{\boldsymbol{\epsilon}})}{\partial \dot{\boldsymbol{\epsilon}}} : \dot{\boldsymbol{\epsilon}}^* \, dV + \int_\Omega \rho \boldsymbol{\gamma} \cdot \mathbf{v}^* \, dV - \int_\Omega \rho \mathbf{g} \cdot \mathbf{v}^* \, dV +$$

$$\int_{\partial\Omega^{\mathrm{f}}} \frac{\partial \phi_{\mathrm{f}}(\Delta \mathbf{v})}{\partial \Delta \mathbf{v}} \cdot \mathbf{v}^* \, dS - \int_{\partial\Omega^{\mathrm{T}}} \mathbf{T}^d \cdot \mathbf{v}^* \, dS = 0 \tag{8.86}$$

and the equations of evolution are again (8.85) to which we must add

$$\frac{d\mathbf{v}}{dt} = \boldsymbol{\gamma} \tag{8.87}$$

8.3.5 The Membrane Formulation

The previous general concepts for viscoplasticity have been presented in the context of bulk forming simulation. In the case of the hot forming of thin sheets for instance, the analysis remains fully valid but the discretization and the numerical resolution of the equations lead to noneconomical models. The membrane formulation has now been widely used to shorten those computation costs (Kobayashi and Kim 1978, Wang 1984, Wood et al. 1985, Bellet and Chenot 1989). We will recall here the main features of the membrane formulation and show how the previous viscoplasticity equations have to be adapted.

8.3.5.1 Membrane Mechanics

The membrane formulation consists in considering the thin deformable sheet as a geometric surface. Hence, the material points of this surface can be identified by two parameters named material coordinates and denoted θ^1 and θ^2. For instance, if the sheet is initially plane and parallel to the xy plane, θ^1 and θ^2 can be chosen as initial x and y coordinates of the material points in this undeformed state. These two material coordinates, that we can also see as Lagrangian or curvilinear coordinates, remain constant throughout the forming process, and there is a bijective relation at any instant t between the spatial position vector $\mathbf{X}$ and θ^i coordinates:

$$\mathbf{X} = \mathbf{X}(\theta^i, t) \tag{8.88}$$

This allows for the definition of a local tangent basis to the deformed configuration of the sheet at time t. The two vectors $\mathbf{g}_1$ and $\mathbf{g}_2$ given by

$$\mathbf{g}_i = \frac{\partial \mathbf{X}}{\partial \theta^i} \tag{8.89}$$

are tangential to the sheet surface at each material point. A natural tangential basis can then be defined by adding to the previous vectors their unit normal:

$$\mathbf{g}_3 = \frac{1}{|\mathbf{g}_1 \wedge \mathbf{g}_2|} \mathbf{g}_1 \wedge \mathbf{g}_2 \tag{8.90}$$

The metric tensor is then defined by its covariant and contravariant components:

$$g_{ij} = \mathbf{g}_i \, \mathbf{g}_j \tag{8.91}$$

$$g^{ij} = \mathbf{g}^i \, \mathbf{g}^j \tag{8.92}$$

with reciprocal vectors $\mathbf{g}^i$ given by

$$\mathbf{g}^i \, \mathbf{g}_j = \delta^i_j \tag{8.93}$$

The stress state of the deformed sheet can be expressed through the Cauchy stress tensor:

$$\boldsymbol{\sigma} = \sigma^{ij} \, \mathbf{g}_i \otimes \mathbf{g}_j \tag{8.94}$$

where σ^{ij} are the contravariant components of the stress tensor within the current tangential basis.

The membrane formulation consists basically of neglecting the dissipated power of the out-of-plane stresses in front of in-plane stresses. Hence, for any virtual velocity field $\mathbf{v}^*$, and the associated strain rate tensor $\dot{\boldsymbol{\epsilon}}^*$, we have

$$\boldsymbol{\sigma} : \dot{\boldsymbol{\epsilon}}^* = \sum_{ij} \sigma^{ij} \dot{\epsilon}_{ij} = \sigma^{11} \dot{\epsilon}^*_{11} + \sigma^{22} \dot{\epsilon}^*_{22} + 2\sigma^{12} \dot{\epsilon}^*_{12}$$

$$= \sum_{ij} \sigma^{ij} \mathbf{v}^*_{i/j} = \sigma^{11} \mathbf{v}^*_{1/1} + \sigma^{22} \mathbf{v}^*_{2/2} + \sigma^{12} \mathbf{v}^*_{1/2}$$

$$+ \sigma^{21} \mathbf{v}^*_{2/1} \tag{8.95}$$

where $/j$ denotes the covariant derivation with respect to θ^j. Clearly, this is equivalent to assuming a plane stress state throughout the sheet. Hence, the stress tensor description is reduced to three components.

8.3.5.2 Viscoplastic Potential

Now the second invariant of $\dot{\boldsymbol{\epsilon}}$ becomes

$$\dot{\bar{\epsilon}} = \left(\frac{2}{3} \sum_{ij} \dot{\epsilon}^{ij} \dot{\epsilon}_{ij}\right)^{1/2} \tag{8.96}$$

and the tensorial viscoplastic flow rule is in covariant coordinates:

$$s_{ij} = \frac{2}{3} \frac{1}{\dot{\bar{\epsilon}}} \frac{\partial \phi(\dot{\bar{\epsilon}})}{\partial \dot{\bar{\epsilon}}} \dot{\epsilon}_{ij} \tag{8.97}$$

that is for the Norton–Hoff potential:

$$s_{ij} = 2K(\sqrt{3}\, \dot{\bar{\epsilon}})^{m-1} \dot{\epsilon}_{ij} \tag{8.98}$$

The contravariant components are obtained by

$$s^{ij} = \sum_{kl} g^{ik} g^{jl} s_{kl} \tag{8.99}$$

and the rate of dissipated energy is

$$\dot{\mathbf{w}} = \mathbf{s}:\dot{\boldsymbol{\epsilon}} = \sum_{ij} s^{ij} \dot{\epsilon}_{ij} = \sum_{ij} s_{ij} \dot{\epsilon}^{ij} = \bar{\sigma}\, \dot{\bar{\epsilon}} \tag{8.100}$$

where

$$\bar{\sigma} = \left(\frac{3}{2} \sum_{ij} s^{ij} s_{ij}\right)^{1/2} \tag{8.101}$$

8.3.5.3 The Viscoplastic Functional

The membrane approximation induces some differences with respect to the general approach given in Sect. 8.3.1. Let Ω denote the surface domain representing the thin sheet. Volume integrals in the viscoplastic functional will be replaced by surface integrals since

$$dV = h\, dS \tag{8.102}$$

where h is the local thickness. In the same way, friction stresses now apply on regions Ω^{f} of the surface Ω and not on its boundary. A frequent case in hot sheet forming is the role of a pressure difference between both sides of the sheet, as a deformation source (thermoforming process). The load vector is normal to the surface and given by $\mathbf{df} = \pm\, p\, \mathbf{g}_3\, dS$. At the boundary $\partial\Omega$ of the surface Ω, kinematic conditions can be prescribed or, if not, lineic forces $\mathbf{T}^d$ can be prescribed on a part $\partial\Omega^{\mathrm{T}}$ of the boundary. Finally, the viscoplastic functional is expressed as

$$\Phi(\mathbf{v}) = \int_{\Omega} \phi(\dot{\bar{\epsilon}})\, h dS - \int_{\Omega} \rho \mathbf{g} \cdot \mathbf{v}\, h dS - \int_{\Omega^p} \pm\, p \mathbf{g}_3 \cdot \mathbf{v}\, dS$$
$$+ \int_{\Omega^{\mathrm{f}}} \phi_{\mathrm{f}}(|\Delta \mathbf{v}|)\, dS - \int_{\partial\Omega^{\mathrm{T}}} \mathbf{T}^d \cdot \mathbf{v}\, dl \tag{8.103}$$

8.3.5.4 *The Incompressibility Relation*

In the tangential basis, we have

$$\mathrm{tr}\,(\dot{\boldsymbol{\epsilon}}) = \sum_{ij} \delta^{ij}\, \dot{\epsilon}_{ij} = \sum_{ij} \delta_{ij}\, \dot{\epsilon}^{ij} \tag{8.104}$$

so that the thickness deformation can be easily computed as a function of the in-plane terms:

$$\frac{\dot{h}}{h} = \dot{\epsilon}_{33} = -\sum_{ij}^{2} \delta^{ij}\, \dot{\epsilon}_{ij} \tag{8.105}$$

A more convenient and equivalent alternative can be used by considering the relation

$$dS = \sqrt{g}\, d\theta^1\, d\theta^2 \tag{8.106}$$

where g denotes the determinant of the metric tensor $\mathbf{g}$. The incompressibility can then be expressed by

$$h\, dS = \text{constant} \tag{8.107}$$

and so, between two deformed configurations, we have

$$h\sqrt{g} = h'\sqrt{g'} \tag{8.108}$$

In other words the sheet thickness can always be easily computed after the membrane deformations are known, in order to fulfil the incompressibility condition. The consequence is that no additional term is needed in the expression of the viscoplastic functional.

8.3.5.5 *Friction Stresses*

The usual practice in sheet metal forming is to consider a Coulomb friction model, rather than boundary layer models such as the Norton model presented in Sect. 8.2.6 or the Tresca model. We shall use here a regularized Coulomb model, the shear stress being given by

$$\boldsymbol{\tau} = -\mu\,\sigma_n\,(\Delta\mathbf{v}^2 + \Delta\mathbf{v}_{0}^2)^{-1/2}\,\Delta\mathbf{v} \tag{8.109}$$

where σ_n is the normal stress and μ the friction coefficient. As σ_n depends on the velocity field, the friction potential does not exist any more. The equilibrium equation issued from the principle of virtual work has to be solved with iterative contact estimation and σ_n calculation (Massoni 1987, Germain et al. 1989). It should be noted however, that in the case of the thermoforming process, σ_n is equal to the forming pressure which presses the sheet against the tool surface.

The associated friction potential is, in this case:

$$\phi_f(|\Delta\mathbf{v}|) = \mu\,p\,(|\Delta\mathbf{v}|^2 + \Delta\mathbf{v}_{0}^2)^{1/2} \tag{8.110}$$

8.3.6 The Thermomechanical Approach

The general heat equation is written in the domain Ω:

$$\rho c \frac{dT}{dt} = \operatorname{div}(k\,\mathbf{grad}T) + \dot{q} \tag{8.111}$$

where ρ Material density

c Heat capacity

k Conduction coefficient

$\dot{q}$ Internal rate of heat production due to viscoplastic deformation; it is generally admitted that it is nearly equal to the rate of plastic energy given by (8.19), i.e.: $\dot{q} = r\,\dot{w}$ where $0.9 \leq r \leq 1$.

On the boundary of Ω several conditions may appear:

- On the free surface we have convection and radiation such that the heat flux is

$$-k\,\mathbf{grad}(T)\cdot\mathbf{n} = \epsilon_r\,\sigma_r\,(T^4 - T_c^4) + h_c\,(T - T_c) \tag{8.112}$$

where ϵ_r Surface emissivity

σ_r Stefan constant

h_c Convection coefficient

- On the surface in contact with the tool we can impose also a heat flux which is the resultant dissipation:

$$-k\,\mathbf{grad}(T)\cdot\mathbf{n} = h_{cd}\,(T - T_c) + \frac{b}{b + b_T}\,\dot{w}_f \tag{8.113}$$

where h_{cd} Conduction coefficient between the tool and the part

b (resp. b_t) Effusivity of the material of the part (resp. the tool) which controls the division of the heat fluxes

$\dot{w}_f$ Rate of friction dissipation energy

As an example the formula (8.62) gives

$$\dot{w}_f = \alpha K\,|\Delta\mathbf{v}|^{q+1} \tag{8.114}$$

The thermal coupling with the material flow is due to the variation of the coefficients in the viscoplastic law as a function of temperature (see Sect. 8.2.5.1).

8.4 The Steady-State Problem

8.4.1 Finite-Element Discretization of the Mechanical Problem

The domain Ω is discretized into finite elements in the usual way (see Hughes 1987, Zienkiewicz and Taylor 1989). Elements are defined by the vector coordinate $\mathbf{X}_n$ of each node n. The unknown velocity field $\mathbf{v}$ is discretized in terms of the interpolation functions:

$$\mathbf{v} = \sum_n \mathbf{V}^n N_n(\mathbf{x}) \tag{8.115}$$

where $\mathbf{V}^n$ Velocity vector at node n, with components V_i^n

N_n Global interpolation function which takes the value 1 at node n, and 0 for any other node (more general finite elements will not be considered here for simplicity)

The shape function N_n is generally defined in each element Ω_e, with the isoparametric element formalism. Local coordinates x are introduced and reference shape functions N_ν^0 are defined on a reference element Ω_0. A bijective mapping is then introduced between Ω_e and Ω_0 by

$$\mathbf{x} = \sum_{\nu=1}^{\nu_m} X^{n_e(e,\nu)} N_\nu^0(\xi) \tag{8.116}$$

$n_e(e,\nu)$ allows us to calculate the global number of the νth node of element Ω_e. Now the global shape function can be defined in element Ω_e:

$$N_{n_e(e,\nu)} = N_\nu^0(\xi) \tag{8.117}$$

The strain rate tensor can be evaluated from Eq. (8.115)

$$\dot{\epsilon}_{ij} = \frac{1}{2} \sum_n \left(V_i^n \frac{\partial N_n(\mathbf{x})}{\partial x_j} + V_j^n \frac{\partial N_n(\mathbf{x})}{\partial x_i} \right) \tag{8.118}$$

and the classical $\mathbf{B}$ operator is defined from (8.118)

$$\dot{\epsilon}_{ij} = \sum_{nk} B_{ijnk} V_k^n \tag{8.119}$$

or in a more compact form

$$\dot{\boldsymbol{\epsilon}} = \sum_n \mathbf{B}_n \mathbf{V}^n \tag{8.120}$$

At this stage it is worth noting that operator $\mathbf{B}$ is a function only of vectors $\mathbf{X}_n$.

The general expression of the functional gives rise to the discretized form

$$\Phi(\mathbf{V}) = \int_\Omega \Phi\left(\sum_{nk} B_{ijnk} V_k^n\right) dV - \int_\Omega \rho \sum_{nk} g_k V_k^n N_n \, dV$$

$$+\int_{\partial\Omega^f}\Phi_f\left(\sum_n \mathbf{V}^n N_n - \mathbf{v}^0\right) dS - \int_{\partial\Omega^T}\sum_{nk} T_k^d V_k^n N_n\, dS \tag{8.121}$$

By differentiating with respect to each unknown and equating to zero, we obtain the set of nonlinear equations

$$\mathbf{R}(\mathbf{V}) = \mathbf{0} \tag{8.122}$$

with components

$$R_k^n(\mathbf{V}) = \int_\Omega \sum_{ij}\frac{\partial\phi(\dot{\epsilon})}{\partial\dot{\epsilon}_{ij}} B_{ijnk}\, dV - \int_\Omega \rho g_k N_n\, dV + \int_{\partial\Omega^f}\frac{\partial\phi_f(\Delta\mathbf{v})}{\partial V_k^n}\, dS$$

$$-\int_{\partial\Omega^T} T_k^d N_n\, dS = 0 \tag{8.123}$$

The system (8.122) is solved by the Newton–Raphson method (Ortega and Rheinbolt 1970). Starting from an initial velocity field $\mathbf{V}^{(r)}$, at iteration r we set

$$\mathbf{V}^{(r+1)} = \mathbf{V}^{(r)} + \Delta\mathbf{V}^{(r)} \tag{8.124}$$

Putting (8.124) into (8.122) and linearizing leads to

$$\mathbf{R}(\mathbf{V}^{(r+1)}) = \mathbf{R}(\mathbf{v}^{(r)}) + \frac{\partial\mathbf{R}}{\partial\mathbf{V}}(\mathbf{V}^{(r)})\cdot\Delta\,\mathbf{V}^{(r)} \tag{8.125}$$

and we have the new iterate

$$\mathbf{V}^{(r+1)} = \mathbf{V}^{(r)} - \theta(\mathbf{H}^{(r)})^{-1}\mathbf{R}(\mathbf{V}^{(r)}) \tag{8.126}$$

where θ is a sub-incrementation parameter which is chosen such that

$$|\mathbf{R}(\mathbf{V}^{(r+1)})| \le |\mathbf{R}(\mathbf{V}^{(r)})| \tag{8.127}$$

and the symmetric matrix

$$\mathbf{H} = \frac{\partial\mathbf{R}}{\partial\mathbf{V}} \tag{8.128}$$

can be obtained by differentiating (8.123):

$$H_{kk'}^{nn'} = \int_\Omega \sum_{iji'j'}\frac{\partial^2\phi}{\partial\dot{\epsilon}_{ij}\,\partial\dot{\epsilon}_{i'j'}}(\mathbf{V})\, B_{ijnk}\, B_{i'j'n'k'}\, dV$$

$$+\int_{\partial\Omega^f}\frac{\partial^2\phi_f}{\partial V_k^n\,\partial V_{k'}^{n'}}(\Delta\,\mathbf{V})\, N_n\, N_{n'}\, dS \tag{8.129}$$

8.4.1.1 *The Isotropic Incompressible Material*

We shall examine the Norton–Hoff case with a penalty approach. The derivative of the functional takes the following form:

$$R_k^n = 2\int_\Omega K(\sqrt{3}\,\dot{\bar{\epsilon}})^{m-1}\sum_{ij}\dot{\epsilon}_{ij}\, B_{ijnk}\, dV - \int_\Omega \rho g_k N_n\, dV$$

$$+\int_{\partial\Omega^{\mathrm{f}}} \alpha K|\Delta \mathbf{v}|^{q-1}\, \Delta V_k\, N_n\, dS - \int_{\partial\Omega^{\mathrm{T}}} T_k^d\, N_n\, dS$$

$$+\int_{\Omega} \rho K \operatorname{div}(\mathbf{v}) \sum_i B_{iink}\, dV = 0 \tag{8.130}$$

and the hessian **H** is

$$H_{kk'}^{nn'} = 2\int_{\Omega} K\Big[(\sqrt{3}\,\dot{\bar{\epsilon}})^{m-1} \sum_{ij} B_{ijnk}\, B_{ijn'k'} - 2\,(m-1)(\sqrt{3}\,\dot{\bar{\epsilon}})^{m-3}$$

$$\times \sum_{iji'j'} \dot{\epsilon}_{ij}\, B_{ijnk}\, \dot{\epsilon}_{i'j'}\, B_{i'j'n'k'}\Big]\, dV + \int_{\partial\Omega^{\mathrm{f}}} \alpha K\, [|\Delta \mathbf{v}|^{q-1}\, N_n\, N_{n'}\, \delta_{kk'}$$

$$-\,(1-p)\, |\Delta \mathbf{v}|^{q-3}\, \Delta V_k\, \Delta V_{k'}\, N_n\, N_{n'}]\, dS$$

$$+\int_{\Omega} rK \sum_{ii'} B_{iink}\, B_{i'i'n'k'}\, dV \tag{8.131}$$

The last term in Eq. (8.131), which takes into account the incompressibility by the penalty method, must be considered with a special attention. The element must possess a sufficient number of degrees of freedom, and the penalty contribution to the hessian must be integrated with a reduced scheme (Zienkiewicz and Godbole 1974, Hughes and Liu 1979).

8.4.2 Stress Calculation

The stress tensor (or the stress deviatoric tensor when the material is incompressible) can be expressed from the velocity field by differentiation. But the resultant stress field is discontinuous at the element interfaces. The most popular method is local smoothing, followed by averaging the different values obtained for a given node (see Hinton and Campbell 1974); a global smoothing procedure is also described in the same reference. Let $\boldsymbol{\sigma}$ be the (discontinuous) stress field obtained from the velocity field, and $\tilde{\boldsymbol{\sigma}}$ the smoothed stress field defined by its nodal values. $\tilde{\boldsymbol{\sigma}}$ is calculated by minimizing the quadratic least-squares expression

$$Q(\tilde{\boldsymbol{\sigma}}) = \int_{\Omega} (\boldsymbol{\sigma} - \tilde{\boldsymbol{\sigma}})^2\, dV \tag{8.132}$$

In fact (8.132) applies separately for each component $\tilde{\sigma}_{ij}$ of $\tilde{\boldsymbol{\sigma}}$.

A more accurate method has been proposed for three-dimensional analysis by Bertrand-Corsini (1987). The numerical derivatives of the velocity field are obtained by a generalized finite-difference scheme (Liszka and Orkisz 1980) which allows use of unstructured meshes and provides a first-order scheme for the first derivatives (the error is second-order).

8.4.2.1 The Incompressible Material

This topic deserves special treatment: it has been observed that the pressure field obtained either from the Lagrange formulation, or from the penalty equation is not accurate and may exhibit spurious modes. A new method was first described in Neyret et al. (1982) for two-dimensional problems and extended to three-dimensional analysis in Bertrand-Corsini et al. (1986). The numerical differentiation of the velocity field gives the deviatoric stress tensor **s** and the pressure field p should satisfy the equilibrium equation

$$\mathbf{grad}(p) = \mathbf{div}(\mathbf{s}) \tag{8.133}$$

This equation is approximated in a least-squares sense by minimizing with respect to p the following function:

$$F(p) = \sum_{e,i,n} \left[\int_{\Omega_e} \left(\frac{\partial p}{\partial x_i} - \sum_j \frac{\partial s_{ij}}{\partial x_j} \right) N_n \, dV \right]^2 \tag{8.134}$$

Here p is defined by the nodal values on the same mesh and with the same shape functions as for the velocity field $\mathbf{v}$.

8.4.3 Free Surface Calculation

The stream-line method has been proposed by Tanner et al. (1975) for polymer processing, and by many other authors. It is relatively easy to implement provided that the mesh corresponds (at least approximately) to stream lines. Eq. (8.82) is discretized into

$$\frac{\Delta x}{v_x} = \frac{\Delta y}{v_y} = \frac{\Delta z}{v_z} = \Delta t \tag{8.135}$$

If the flow is roughly along the z direction, Δz can be imposed and allows calculation of Δx and Δy from a given velocity field, the domain is then updated and the velocity field is recalculated until convergence occurs.

The optimal control approach is more difficult to implement but it has proved to be effective for the prediction of spread in hot rolling; see Bern (1987), David (1987), Bertrand-Corsini et al. (1988). The whole procedure consists of defining the unknown free surface by one parameter corresponding to each node on this surface. The functional (8.83) can be minimized with respect to these parameters for a velocity field solution of the viscoplastic general problem, and an approximate Newton–Raphson procedure is used.

The weak formulation is illustrated in Karagiannis (1989).

8.4.4 Thermal Coupling and Strain Hardening

The procedure outlined here is quite similar to that of Pittman and Nakazawa (1984). The Euler form of the heat equation for stationary processes is

$$\rho c \mathbf{v} \cdot \mathbf{grad}(T) = \mathrm{div}(k\, \mathbf{grad}(T)) + \dot{Q} \tag{8.136}$$

The temperature field is discretized with the same elements:

$$T = \sum_n T^n N_n \tag{8.137}$$

and the Galerkin method allows us to write

$$\forall n' \int_{\Omega} \rho c \mathbf{v} \cdot \mathbf{grad}(T)\, N_{n'}\, dV + \int_{\Omega} k \mathbf{grad}(T) \cdot \mathbf{grad}(N_{n'})\, dV$$

$$- \int_{\Omega} \dot{Q} N_{n'}\, dv - \int_{\partial\Omega} \varphi N_{n'}\, dS = 0 \tag{8.138}$$

In fact this formulation is not accurate enough as shown by Hughes and Brooks (1982), and in the first integral in (8.138), the weighting function $N_{n'}$, must be changed into

$$\tilde{N}_{n'} = N_{n'} + \sum_{i} \frac{\partial N_{n'}}{\partial x_i} \alpha_i \tag{8.139}$$

The extra term depends on the velocity field and is given explicitly by Pittman and Nakazawa (1984). If we combine the modified heat equation with the mechanical equation, the whole system may be solved by the Newton–Raphson method with $\mathbf{v}$ and T as unknowns. If we add the strain hardening effect the equivalent strain obeys the equation:

$$\mathbf{v} \cdot \mathbf{grad}\bar{\epsilon} = \dot{\bar{\epsilon}} \tag{8.140}$$

which is simpler than (8.138) but may be treated in the same way, if $\bar{\epsilon}$ is discretized according to

$$\bar{\epsilon} = \sum_{n} \bar{\epsilon}^{n} N_n \tag{8.141}$$

and the final problem has the coupled unknowns $\mathbf{v}$, T and $\bar{\epsilon}$.

In fact if the coupling is not too "strong" it is possible to solve alternately

- The velocity problem with given T and $\bar{\epsilon}$
- The temperature problem with given $\mathbf{v}$ and $\bar{\epsilon}$
- The equivalent strain problem with given $\mathbf{v}$

The procedure is iterated until convergence.

For nearly adiabatic processes a simplified calculation is possible if the mesh is built on the stream lines. The method is described in David (1987) and may be summarized by

$$T^{n'} - T^{n} = \frac{1}{\rho c} (\dot{Q}^{n'} + \dot{Q}^{n}) \frac{|\mathbf{X}^{n'} - \mathbf{X}^{n}|}{|\mathbf{V}^{n'} + \mathbf{V}^{n}|} \tag{8.142}$$

if the nodes n and n' are consecutive on the same stream line.

The same method applies to the calculation of equivalent strain.

8.4.5 Error Estimate and Mesh Optimization

There is an increasing effort to introduce reliability concepts in software, and the a-posteriori error estimates seem to be a very promising way when they are associated with adaptive remeshing to decrease the error where it is too high.

The basic concepts of error estimate were introduced into the finite-element calculation of viscoplastic flow by Zienkiewicz et al. (1988). This method is based on the energy norm of the difference between the calculated, discontinuous, strain rate tensor, and the smoothed one in the sense outlined in Sect. 8.4.2. The mean error can then be approximated by

$$|d|^2 = \frac{1}{V(\Omega)} \int_{\Omega} \eta(\dot{\bar{\epsilon}})\,(\dot{\boldsymbol{\epsilon}} - \tilde{\dot{\boldsymbol{\epsilon}}}):(\dot{\boldsymbol{\epsilon}} - \tilde{\dot{\boldsymbol{\epsilon}}})\,dV \tag{8.143}$$

where $V(\Omega)$ Volume of the domain
$\eta(\dot{\bar{\epsilon}})$ Generalized viscosity defined by (8.13)

A mean error may also be defined in each element by

$$|d|_e^2 = \frac{1}{V(\Omega_e)} \int_{\Omega_e} \eta(\dot{\bar{\epsilon}})\,(\dot{\boldsymbol{\epsilon}} - \tilde{\dot{\boldsymbol{\epsilon}}}):(\dot{\boldsymbol{\epsilon}} - \tilde{\dot{\boldsymbol{\epsilon}}})\,dV \tag{8.144}$$

and this measure is the basis for a local refinement method which is described in some detail in Huang (1989). The generalization to thermocoupled analysis is summarized in Zienkiewicz (1989).

Another method of remeshing has been proposed by Cescutti and Chenot (1987) which minimizes a functional of the position of the nodes, the expression of which in two dimensions is

$$\pi(\mathbf{X}) = \sum_e \int_{\Omega_e} \left|\frac{\partial \mathbf{x}}{\partial \xi}\right|^2 \frac{1}{J_e^2}\,dV + \lambda \sum_e W_e \left(\int_{\Omega_e} dV\right)^2 \tag{8.145}$$

where J_e Jacobian of the transformation defined for each element
W_e Weighting factor

The first term in Eq. (8.145) is a smoothing contribution and the second one introduces a kind of adaptation as the W_e factor may depend on the error estimate. The optimum mesh is obtained by solving the set of equations

$$\frac{\partial \pi(\mathbf{X})}{\partial \mathbf{X}} = 0 \tag{8.146}$$

This method has been generalized and applied to three-dimensional problems by Cescutti et al. (1987).

8.4.6 Applications of Stationary Forming Processes

8.4.6.1 Continuous Casting (Dalin and Chenot 1988)

One of the most crucial problems in continuous casting of slabs is the control of bulging of the wider side of the slab. The bulging is responsible for the deformation of the solidified shell near the interface with the liquid, which can produce internal cracks. Bulging is caused by the ferrostatic pressure on the relatively thin solidified shell between the support rolls as shown in Fig. 8.4.

The model takes into account the ferrostatic pressure, the gravity forces and a given heterogeneity of temperature. The shape of the free surface is iteratively calculated by the stream-line method outlined in Sect. 8.4.3, and a special treatment for enforcing the contact with the downstream support roll; see Dalin (1987). The results during the iteration process are represented in Fig. 8.5.

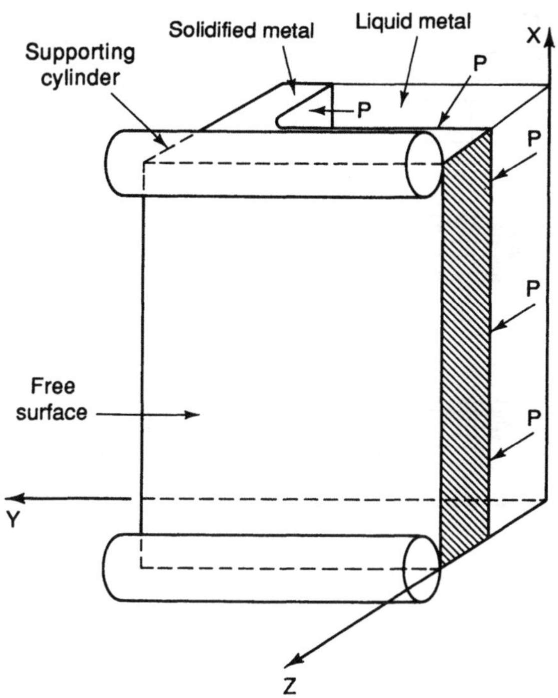

Fig. 8.4. Schematic representation of continuous casting.

8.4.6.2 Hot Shape Rolling

A three-dimensional finite-element model for numerical modelling of hot shape rolling has been reported in Bertrand-Corsini et al. (1989). The computer code is stationary, takes into account the influence of strain hardening and temperature distribution, and allows the prediction of the thermal and the mechanical parameters for a multipass rolling sequence, including spread.

For example the influence of waiting time between two passes was modelled for a sequence of blooming passes on a reversible mill. The constitutive equation is

$$\bar{\sigma} = \sqrt{3}\,(\sqrt{3}\;\dot{\bar{\epsilon}})^{0.18}\left(\frac{5.23\times 10^5}{T} - 400\right)\exp\left(\frac{-0.02}{0.09+\bar{\epsilon}}\right) \tag{8.147}$$

and the friction law is

$$\tau = 80\,|\Delta \mathbf{v}|^{0.05} \tag{8.148}$$

The thermal parameters are

- Conductivity 30 W m^{-1} K^{-1}
- Specific heat 5.7×10^6 J K^{-1} m^{-3}
- Interface conductance 5000 W K^{-1} m^{-2}

and for interstand computations

- Emissivity 0.7
- Convective heat transfer coefficient $h = 25$ W K^{-1} m^{-2}

A typical mesh is given in Fig. 8.6.

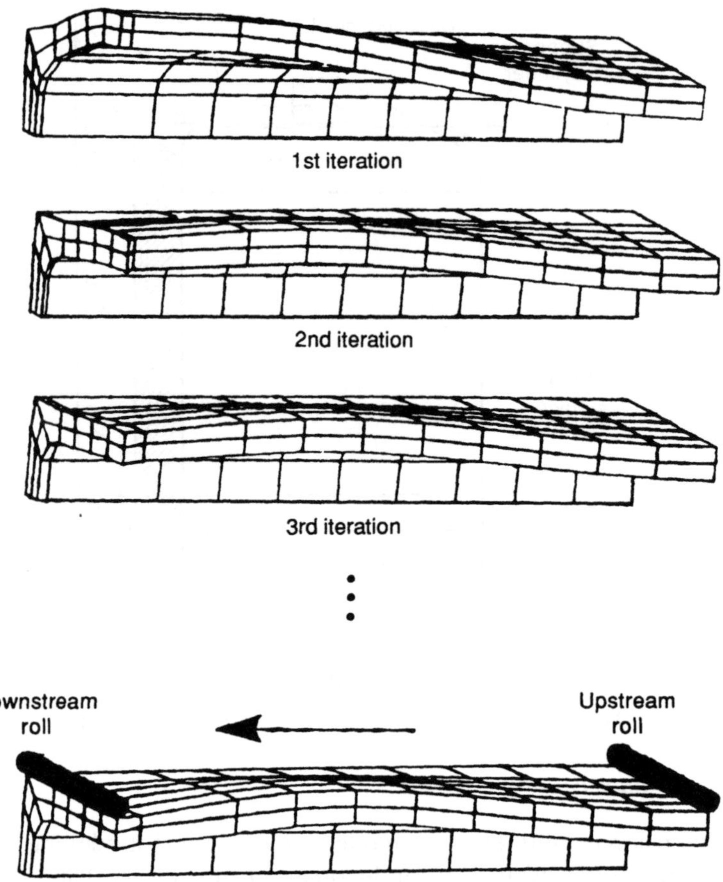

Fig. 8.5. Deformed mesh for successive iterations in calculating bulging in continuous casting (Dalin and Chenot 1988).

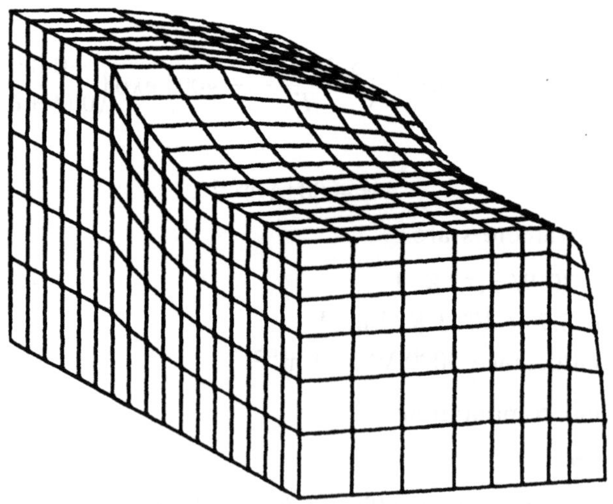

Fig. 8.6. Modelling of a blooming sequence: typical mesh.

The isotemperature map is represented in Fig. 8.7a for the final pass. The time evolution of maximum temperature is pictured in Fig. 8.7b showing the succession of heating due to deformation and cooling due to the waiting time between two passes.

8.5 Incremental Processes

8.5.1 Time Integration for the Quasistatic Approach

Many studies have been devoted to time integration for general constitutive equations with inertia effects, see for example Geradin et al. (1983) for implicit schemes, but relatively little work has been done on purely viscoplastic behaviour.

The following parameters are introduced at time t:

- Domain Ω_t, and its surface in contact with the tools $\partial\Omega_t^{\mathrm{f}}$
- Current position of a material point x_t and its velocity v_t
- Strain rate tensor $\dot{\boldsymbol{\epsilon}}_t$ and stress tensor $\boldsymbol{\sigma}_t$.

Three integration schemes can be discussed to solve Eq. (8.85).

8.5.1.1 The One-Level Explicit Scheme

We suppose the domain Ω_t is known at time t, such that the minimization of the functional on Ω_t gives the velocity field. For any node n we can write

$$\mathbf{X}_{t+\Delta t}^{n} = \mathbf{X}_{t}^{n} + \Delta t\, \mathbf{V}_{t}^{n} \tag{8.149}$$

This allows us to update the domain Ω_t into $\Omega_{t+\Delta t}$, and to continue the calculation. Due to its simplicity this scheme is obviously the most popular.

8.5.1.2 The One-Level Implicit Scheme

Eq. (8.149) is replaced by

$$\mathbf{X}_{t+\Delta t}^{n} = \mathbf{X}_{t}^{n} + \Delta t \mathbf{V}_{t+\Delta t}^{n} \tag{8.150}$$

and the velocity field is calculated by the minimization of the functional on the unknown domain $\Omega_{t+\Delta t}$. More precisely the nonlinear Eq. (8.123) is transformed into

$$R_k^n(\mathbf{V}_{t+\Delta t}) = \int_{\Omega_{t+\Delta t}} \sum_{ij} \frac{\partial \phi(\dot{\epsilon}_{t+\Delta t})}{\partial \dot{\epsilon}_{ij}} B_{ijnk}(\mathbf{X}_{t+\Delta t})\, dV - \int_{\Omega_{t+\Delta t}} g_k N_n\, dV$$

$$+ \int_{\partial\Omega_{t+\Delta t}^{\mathrm{f}}} \frac{\partial \phi_{\mathrm{f}}(\Delta \mathbf{v}_{t+\Delta t})}{\partial V_k^n} N_n\, dS - \int_{\partial\Omega_{t+\Delta t}^{\mathrm{T}}} T_{k\,(t+\Delta t)}^{d} N_n\, dS = 0 \tag{8.151}$$

(with an additional penalty term if the material is incompressible).

If an exact Newton–Raphson method is implemented the derivatives of (8.151) with respect to $\mathbf{V}_{t+\Delta t}$ produce several extra terms as compared to (8.124). If the evolution of the geometric configuration is not too important, the extra terms can be neglected.

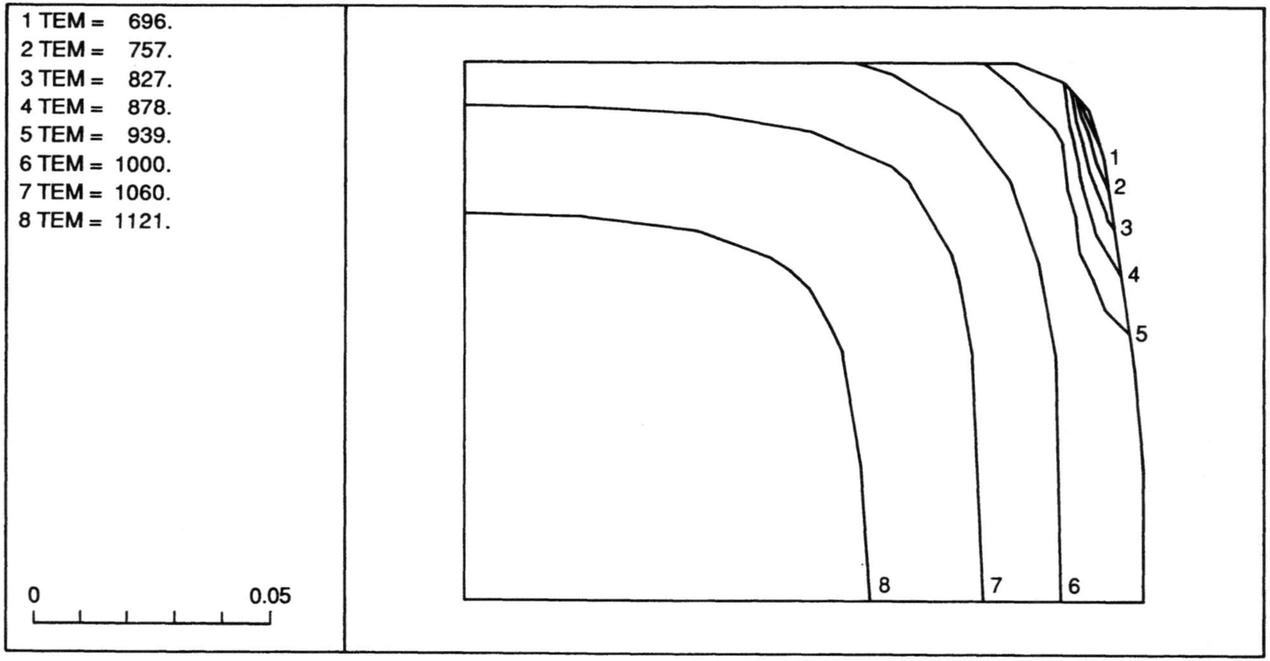

Fig. 8.7a. Thermal results for a blooming sequence. Typical isotemperature in a cross-section.

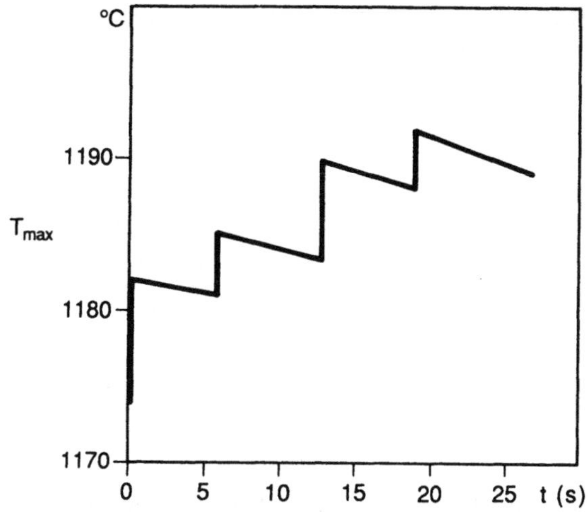

Fig. 8.7b. Thermal results for a blooming sequence. Evolution of the maximum of core temperature as a function of time.

8.5.1.3 The Two-Level Implicit Scheme

We suppose that, at time t, Ω_t and v_t are known, so that the new node coordinates can be expressed as

$$\mathbf{X}^n_{t+\Delta t} = \mathbf{X}^n_t + \tfrac{1}{2}\Delta t\,(\mathbf{V}^n_t + \mathbf{V}^n_{t+\Delta t}) \qquad (8.152)$$

and the equations are again written on the final configuration as in (8.151). This method was tested in two-dimensions (Bohatier and Chenot 1985) and in three dimensions (Surdon and Chenot 1987), and was proved to be more accurate on numerical examples of forging.

8.5.2 The Contact Problem

Here we shall follow the presentation of Chenot (1987) and Montmitonnet et al. (1988).

We suppose that the surface of the tool is described by the function g such that

- $g(x) = 0$ is the surface of the tool
- $g(x) < 0$ is the interior of the tool

For simplicity we shall suppose that the tool is motionless.

8.5.2.1 The One-Level Explicit Scheme

Three items must be analysed during one increment for a node-wise contact:

1. The node is on the free surface at the beginning of the increment, and penetrates inside the tool at the end of the increment:

$$g(\mathbf{X}^m_t) > 0 \quad \text{and} \quad g(\mathbf{X}^m_{t+\Delta t}) \le 0 \qquad (8.153)$$

We can reproject orthogonally the node with coordinates $\mathbf{X}^m_{t+\Delta t}$ on the surface of the tool, or reduce the time increment such that we have exactly

$$g(\mathbf{X}^m_t + \Delta t^* \, \mathbf{V}^m_t) = 0 \qquad (8.154)$$

and choose the minimum of the Δt^* for all the nodes coming into contact. The second method is more accurate but may increase significantly the number of time increments, mainly in three-dimensional analysis.

2. The node is already on the contact surface at the beginning of the increment; we impose:

$$\Delta \mathbf{V}^m \cdot \mathbf{n} = 0 \qquad (8.155)$$

and we observe that the normal stress is compressive:

$$\sigma_n = (\boldsymbol{\sigma}\,\mathbf{n}) \cdot \mathbf{n} < 0 \qquad (8.156)$$

Then the material point must remain in contact with the tool, and if due to the curvature of the tool, at the end of increment the node exhibits a small departure from the tool surface, it must be reprojected.

3. For a node on the contact surface, for which the normal stress is no longer negative, we must allow a loss of contact.

8.5.2.2 The Two-Level Implicit Scheme

Contact Establishment. If the trial node $\mathbf{X}^m_{t+\Delta t}$ obtained by (8.152) lies inside the tool, the time increment is split into two terms:

- Δt_1 during which there is no contact and which is calculated according to

$$g(\mathbf{X}^m_t + \Delta t_1 \, \mathbf{V}^m_t) = 0 \qquad (8.157)$$

- $\Delta t_2 = \Delta t - \Delta t_1$ corresponding to contact such that, at the end of the increment we must verify

$$g(\mathbf{X}^m_t + \Delta t_1 \, \mathbf{V}^m_t + \Delta t_2 \, \mathbf{V}^m_{t+\Delta t}) = 0 \qquad (8.158)$$

which allows us to impose a scalar condition on $\mathbf{V}^m_{t+\Delta t}$. It is interesting to observe that if the tool surface is not planar in the vicinity of the contact the condition (8.158) is not exactly equivalent to

$$\mathbf{V}^m_{t+\Delta t} \cdot \mathbf{n} = 0 \qquad (8.159)$$

Sliding Contact. If $g(\mathbf{X}^m_t) = 0$ and the normal stresses at the beginning and at the end of increment are compressive, the contact occurs during the whole increment and the discrete condition of contact will be

$$g(\mathbf{X}^m_t + \tfrac{1}{2}\Delta t(\mathbf{V}^m_t + \mathbf{V}^m_{t+\Delta t})) = 0 \qquad (8.160)$$

which is again not equivalent to (8.159).

Loss of Contact. If at the end of an increment, with a contact hypothesis, the normal stress is positive, the condition (8.160) must be released, such that loss of contact may occur.

A simplified version of this two-step contact algorithm has been reported in Surdon and Chenot (1987).

8.5.3 Introduction of Dynamic Effects

The functional approach is no longer valid and the equilibrium equation with the inertia term must be expressed by the virtual work principle, as in Sect. 8.3.4. The two-step implicit scheme described in Sect. 8.5.2.2 can be shown to be consistent with a linear time interpolation of the velocity field during one increment, such that γ is piecewise constant during each increment $[t, t+\Delta t]$:

$$\gamma = \frac{\mathbf{V}_{t+\Delta t} - \mathbf{V}_t}{\Delta t} \tag{8.161}$$

One can thus write the virtual work principle (8.86) at the beginning or at the end of the increment, or preferably at the mid-increment corresponding to $t + \frac{1}{2}\Delta t$.

Another approach has been proposed by Arminjon and Chenot (1987) in which the acceleration is continuous, and is a linear function of time during each increment.

8.5.4 Thermal Coupling

Among the first incremental thermocoupled analyses we can cite the two-dimensional approaches (Rebelo and Kobayashi 1980, Argyris and Doltsinis 1981, and Kobayashi 1984) and the extension to three-dimensional analysis (Cescutti et al. 1987, Soyris et al. 1988). Using the Galerkin method (Comini et al. 1974) the heat equation is discretized into

$$\mathbf{C}\dot{\mathbf{T}} + \mathbf{H}\mathbf{T} - \mathbf{Q} = 0 \tag{8.162}$$

C Heat capacity matrix
H Conduction matrix
Q Thermal loading vector, including the heat production due to viscoplastic deformation and friction at the boundary

A two-step second-order scheme (see Hogge 1981) is used in which the quantities appearing in (8.162) are approximated by

$$\mathbf{T} = a\,\mathbf{T}_{t-\Delta t} + (\tfrac{3}{2} - 2a - g)\,\mathbf{T}_t + (a - \tfrac{1}{2} + g)\,\mathbf{T}_{t+\Delta t} \tag{8.163}$$

$$\dot{\mathbf{T}} = \frac{1-2g}{\Delta t}(\mathbf{T}_t - \mathbf{T}_{t-\Delta t}) + \frac{g}{\Delta t}(\mathbf{T}_{t+\Delta t} - \mathbf{T}_t) \tag{8.164}$$

$$\mathbf{C} = (\tfrac{1}{2} - g)\,\mathbf{C}_{t-\Delta t} + (\tfrac{1}{2} + g)\,\mathbf{C}_t \tag{8.165}$$

and the same for **H** and **Q**.

For stability and precision reasons, the Dupont scheme seems to be preferred ($a = 0.25$ and $g = 1$).

Now the coupling with the mechanical problem may be

- Exact with a Newton–Raphson procedure on the unknowns **V** and **T**, or with an iterative process which involves succesive calculation of **V** and **T** separately but with an update of the coupling conditions;
- Approximate with calculation of the velocity field followed by the calculation of temperature, without iteration within one time step.

8.5.5 Dynamic Remeshing or Complete Remeshing

The adaptive remeshing method of Zienkiewicz and Huang (1989) can be applied to transient forming processes. But too frequent remeshing may be time consuming and the interpolation of some variables from an old grid to a new one may introduce errors. A complementary method can help to avoid many of the intermediate remeshing steps, based on the Arbitrary Lagrangian Eulerian method proposed by Donea (1983). The concept of optimal dynamic remeshing is presented in Cescutti and Chenot (1987) and illustrated in Cescutti et al. (1988). The basic idea is to introduce a (numerical) velocity $\mathbf{V}^{(M)}$ of the mesh which is different from the actual kinematic velocity of the material **V**. If the initial mesh is optimum in the sense of Sect. 8.4.5 and we want it to remain optimum, we must differentiate (8.146) with respect to time to obtain

$$\frac{\partial^2 \pi(\mathbf{X})}{\partial \mathbf{X}^2} \mathbf{V}^{(M)} = 0 \tag{8.166}$$

The condition on the boundary for a node m is

$$(\mathbf{V}^{m\,(M)} - \mathbf{V}^m) \cdot \mathbf{n} = 0 \tag{8.167}$$

Now a physical parameter P on the mesh must be convected according to

$$P^m_{t+\Delta t} = P^m_t + ((\dot{P}^m_t + \mathbf{grad}(P^m_t) \cdot (\mathbf{V}^m_t - \mathbf{V}^{m\,(M)}_t))\,\Delta t \tag{8.168}$$

(for an explicit scheme).

8.5.6 The Membrane Approximation

As already seen in Sect. 8.3.5, sheet forming simulation problems can be reduced to membrane deformation problems. The spatial representation is then only two-dimensional for three-dimensional problems: the finite-element discretization is made of linear or quadratic triangles or quadrangles. However the spatial velocity field is three-dimensional and we have the choice between considering its components in the natural tangent basis of $\mathbf{g}_i$ vectors or in the fixed cartesian frame. This last option is the more commonly used and the problem is then to relate the covariant or contravariant components of $\dot{\boldsymbol{\epsilon}}$, which are used in the functional evaluation (Eq. (8.103)) and is derivatives, to the cartesian components of the velocity of the nodes.

Using the relation of Sect. 8.4.1, this can be achieved in the following way:

$$v_{i/j} = \frac{\partial \mathbf{v}}{\partial \theta^j} \cdot \mathbf{g}_i = \sum_n \frac{\partial N_n}{\partial \theta^j} \mathbf{V}^n \cdot \mathbf{g}_i \tag{8.169}$$

Denoting $\mathbf{I}^k$ as the unit vectors of the cartesian frame,

$$v_{i/j} = \sum_{nk} \frac{\partial N_n}{\partial \theta^j} \mathbf{g}_i \cdot \mathbf{I}^k V_k^n \tag{8.170}$$

Since the covariant components of the strain rate tensor are given by

$$\dot{\epsilon}_{ij} = \tfrac{1}{2}(v_{i/j} + v_{j/i}) \tag{8.171}$$

The expression of the **B** operator of Eqs (8.119) and (8.120) is slightly modified with respect to the usual case. Now we have

$$B_{ijnk} = \frac{1}{2}\left(\frac{\partial N_n}{\partial \theta^j} \mathbf{g}_i \cdot \mathbf{I}^k + \frac{\partial N_n}{\partial \theta^i} \mathbf{g}_j \cdot \mathbf{I}^k\right) \tag{8.172}$$

with

$$\mathbf{g}_i = \frac{\partial \mathbf{x}}{\partial \theta^i} = \sum_n \frac{\partial N_n}{\partial \theta^i} \mathbf{X}^n \tag{8.173}$$

It is worth noting that the derivatives $\partial N_n/\partial \theta^j$ are constant and can be calculated only once at the beginning of the simulation.

If we now come back to the functional expression (8.103), the discretized form is given by

$$\begin{aligned}\Phi(\mathbf{V}) = &\int_\Omega \phi\Big(\sum_{nk} B_{ijnk} V_k^n\Big) h\sqrt{g}\, d\theta^1 d\theta^2 - \int_\Omega \rho \sum_{nk} (g_k V_k^n N_n)\, h\sqrt{g}\, d\theta^1 d\theta^2 \\ &+ \int_{\Omega^f} \phi_f\Big(\sum_n N_n \mathbf{V}^n - \mathbf{V}^0\Big) \sqrt{g}\, d\theta^1 d\theta^2 - \int_{\Omega^p} \pm p \sum_{nk} g_{3_k} V_k^n N_n \sqrt{g}\, d\theta^1 d\theta^2 \\ &- \int_{\partial\Omega^T} \sum_{nk} T_k^d V_k^n N_n \sqrt{g}\, d\theta^1 d\theta^2 \end{aligned} \tag{8.174}$$

the minimization of which is performed in the same way as already explained in Sect. 8.4.1.

8.5.7 Applications

8.5.7.1 Superplastic Forming of Thin Sheet Metals

The superplastic forming process has been developed during the last 20 years in the aerospace industry. This technique takes advantage of the high stability of the viscoplastic deformation in some specific alloys called "superplastic". This stability is explained by high values of the strain rate sensitivity coefficient m, in a certain range of high temperatures and low strain rates. Moreover, the flow stress is considerably lower than usual and very large deformation (up to $\bar{\epsilon} \approx 2.5$) can be achieved by simple pressure forming. Actual industrial applications concern specific titanium or aluminium alloys (Stephen 1987). In the case of titanium alloys, this technique can be associated with the diffusion-bonding process to produce complex structural components in a single operation (SPF/DB process).

The basic forming operation is the bulging of thin sheets by a one-sided gas pressure in a shaped mould. The two main results that are to be supplied by a numerical simulation are the final thickness profile, in order to fit in-flight structural requirements, and the pressure-time history, in order to form the part correctly.

Obviously the thickness profile depends on the geometry and on the alloy rheology (especially m coefficient), and also on the lubrication conditions. It has been shown that a Coulomb friction model is appropriate to model stresses at sheet/tool interfaces (Bellet 1988). The regularized law (8.110) is used:

$$\boldsymbol{\tau} = -\mu p\,(\mathbf{v}^2 + \mathbf{v}_0^2)^{-1/2}\,\mathbf{v} \tag{8.175}$$

since the normal stress is equal to the forming pressure and the tools are fixed.

The other main objective of the simulation is to determine the pressure curve $p(t)$ to be applied to the sheet. Fig. 8.8 illustrates schematically superplastic behaviour. There is a strain-rate range in which m is maximum: the superplastic domain. If the sheet is formed outside this domain, necking will occur. The minimization of the forming time (for productivity and quality reasons) then leads naturally to prescribe the upper bound of the superplastic domain as the maximum strain rate in the sheet, at any instant t.

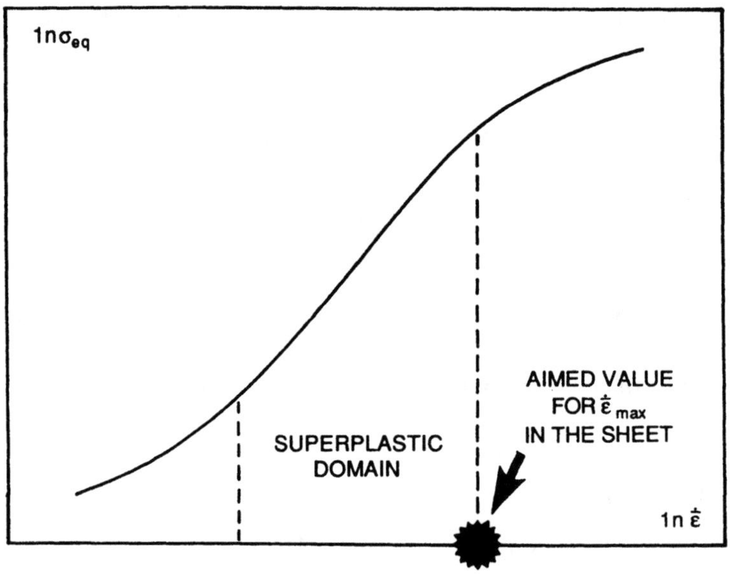

Fig. 8.8. Typical behaviour law of a superplastic alloy, and basic idea of the strain rate control (Bellet and Chenot 1989).

The strain-rate regulation can be achieved by successive incremental corrections of p that are computed according to the evolution of the maximum strain rate during the previous time steps (Bellet 1988). The constraint can also be entered as an additional equation in the global nonlinear system: the solution for pressure is then performed simultaneously (Wargadipura et al. 1989).

Hemisphere Bulging. The forming geometry is given in Fig. 8.9. The material is Ti-6Al-4V alloy. Although the initial thickness is rather high (8 mm), membrane analysis is still valid in most of the part. The computation was run using the

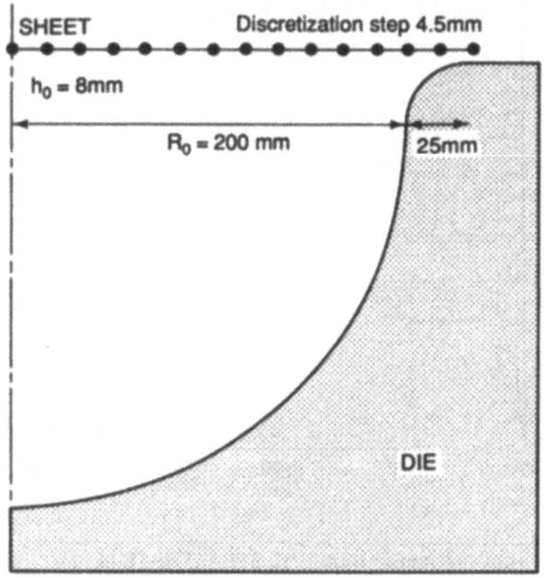

Fig. 8.9. Hemisphere bulging (Bellet and Chenot 1989).

SUPFORM2 model, with two-noded one-dimensional elements, according to the theoretical model presented in Sect. 8.3.5 and 8.5.6.

Fig. 8.10 shows the computed pressure curve and the associated regulation of the maximum strain rate around 3×10^{-4} per second. As regards the final thickness profile, the comparison between computed and industrial experimental results is shown in Fig. 8.11. The agreement is quite good.

Three-Dimensional SPF/DB Component. The studied component comprises four sheets of Ti-6-4 alloy, with an internal vertical rib, which is formed by the two core sheets, and divides the circular part in two (Fig. 8.12). Only the deformation of internal sheets has been studied, as it is the most critical. By reason of symmetry, only an upper quarter of the system has been considered.

The FEM mesh comprises 1027 nodes and 1944 triangular elements. The friction conditions are the following: as the sheet in question is to contact the external sheet without any lubrication (to allow diffusion bonding), the contact is considered as sticking, except in the rib region. To form the vertical rib, the internal sheet comes into contact with itself and this can be simulated by a sliding contact on a vertical fictitious wall. The initial thickness of the core sheets is 0.74 mm. The comparisons between the computation and the experiments have been done at two different forming times 1500 and 2500 s. Fig. 8.13 shows the successive computed deformed meshes which are in fairly good agreement with the experiment. Fifty-four increments were necessary to get the final configuration.

As regards the thickness distribution, several cross-sections of the part have been analysed (Fig. 8.12) and generally, the agreement with the experiment is quite good (Fig. 8.14).

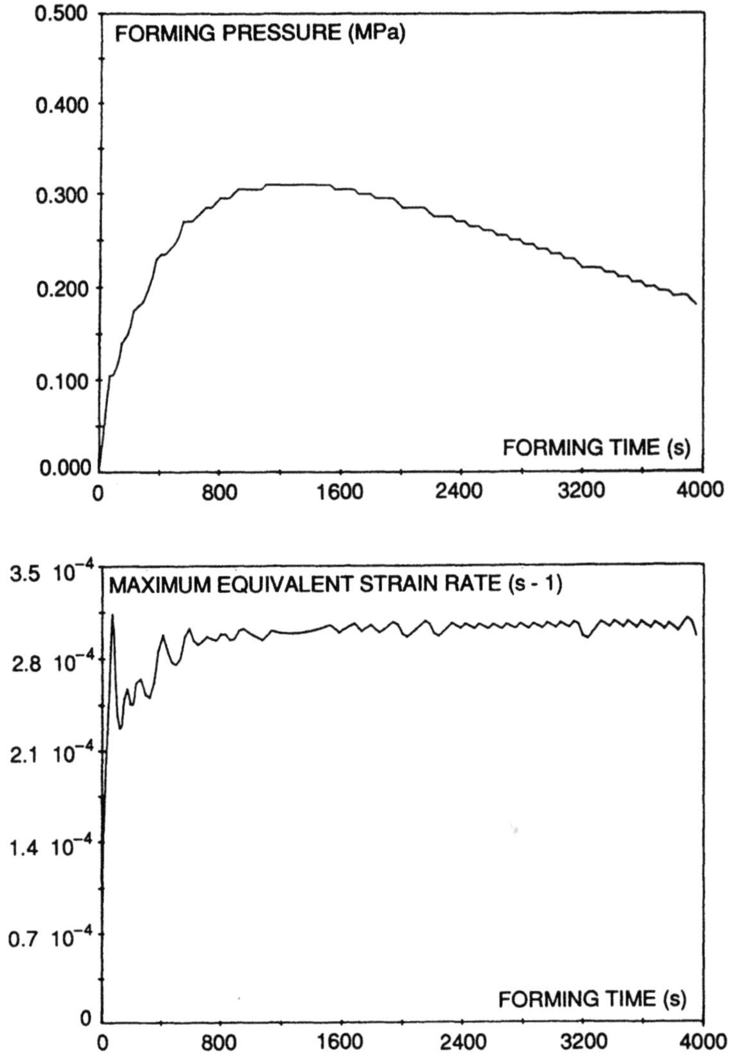

Fig. 8.10. Optimum pressure curve and associated strain-rate control (Bellet and Chenot 1989).

8.5.7.2 Forging

Numerous research groups have worked in recent years on numerical modelling, with a viscoplastic approach, of the two-dimensional (or axisymmetrical) forging process. Most of the features described in Sect. 8.5.7 have been tested with laboratory codes. However there are relatively few industrial codes able to take into account very complex geometries with user-friendly pre-processors, especially for the meshing and remeshing steps.

For the three-dimensional problem the development seems to be at an earlier stage, and relatively few works have been published on complex geometries. An industrial application on the forging of a steel connecting rod will be briefly summarized here according to Soyris et al. (1988). The preform is obtained by side rolling and forging at a temperature of about 1200°C. The numerical results have been obtained with the FORGE3 code which incorporates several important features (see Chenot, 1989):

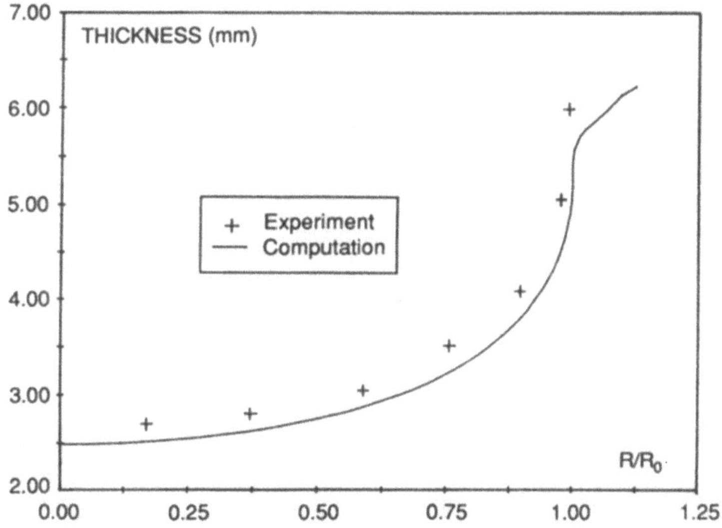

Fig. 8.11. Computed and experimental thickness profiles (Bellet and Chenot 1989).

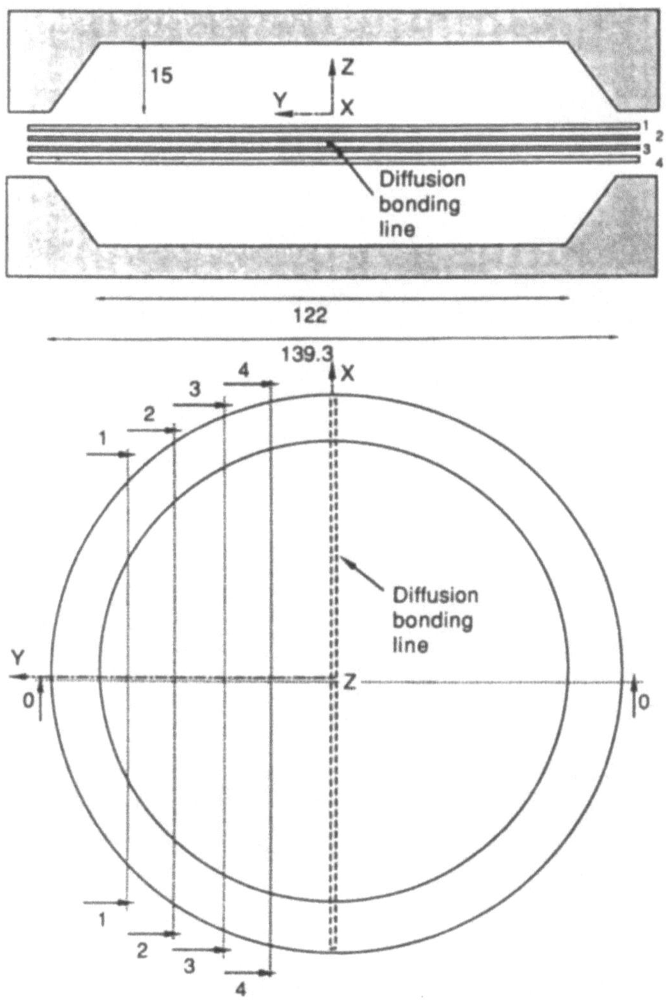

Fig. 8.12. Four-sheet SPF/DB component studied (Bellet and Chenot 1989).

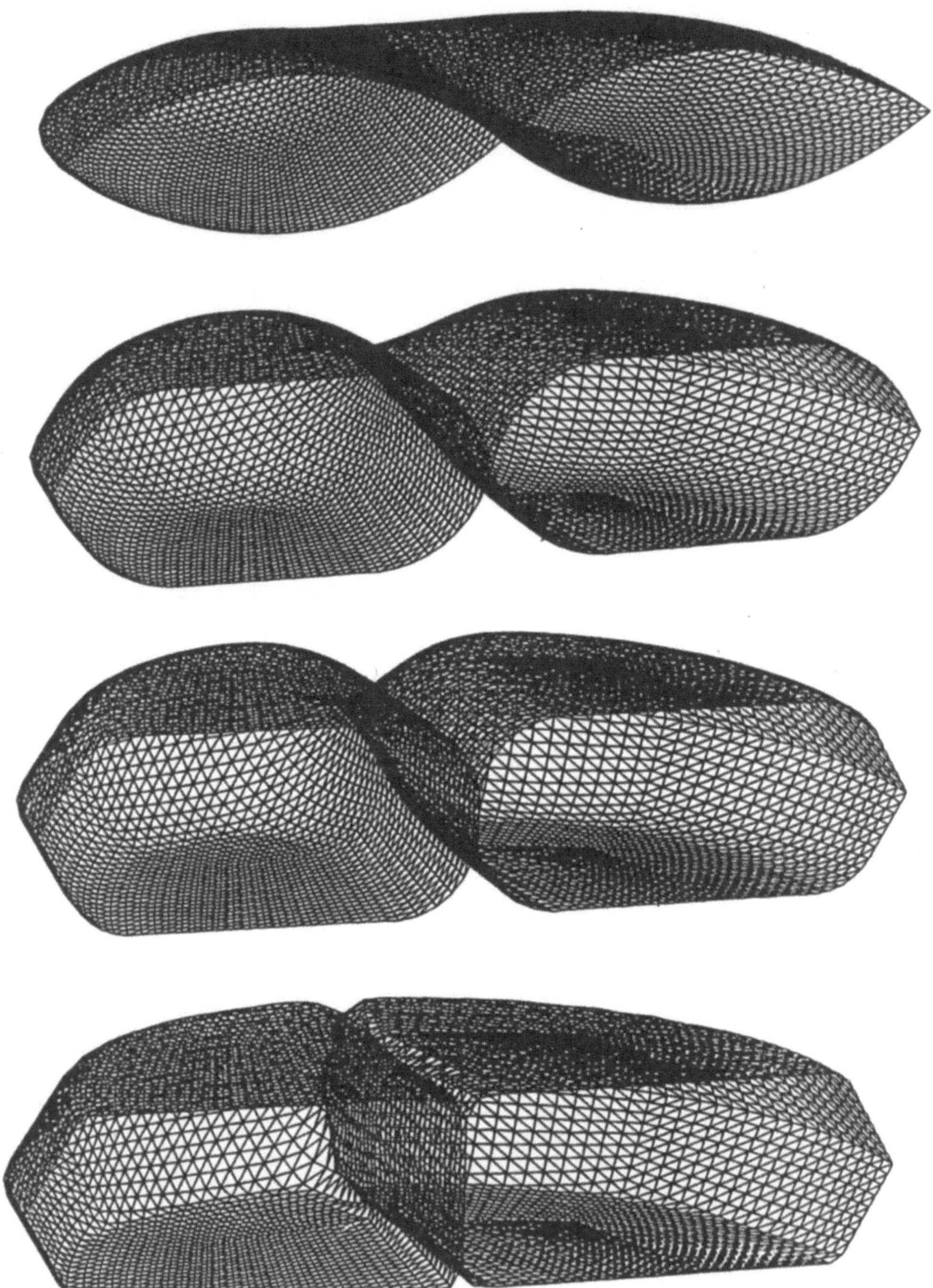

Fig. 8.13. Successive deformed meshes (Bellet and Chenot 1989).

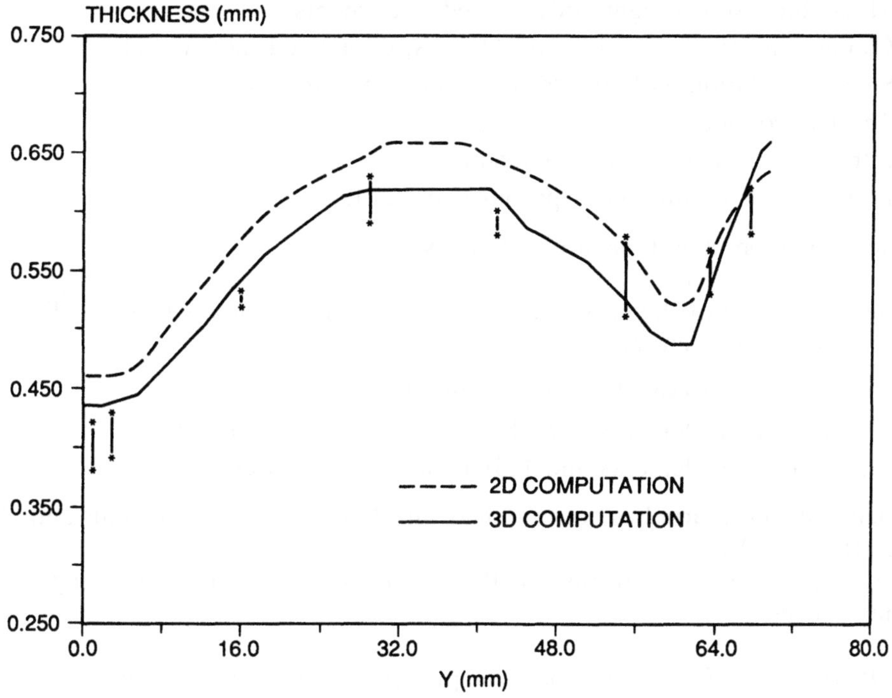

a

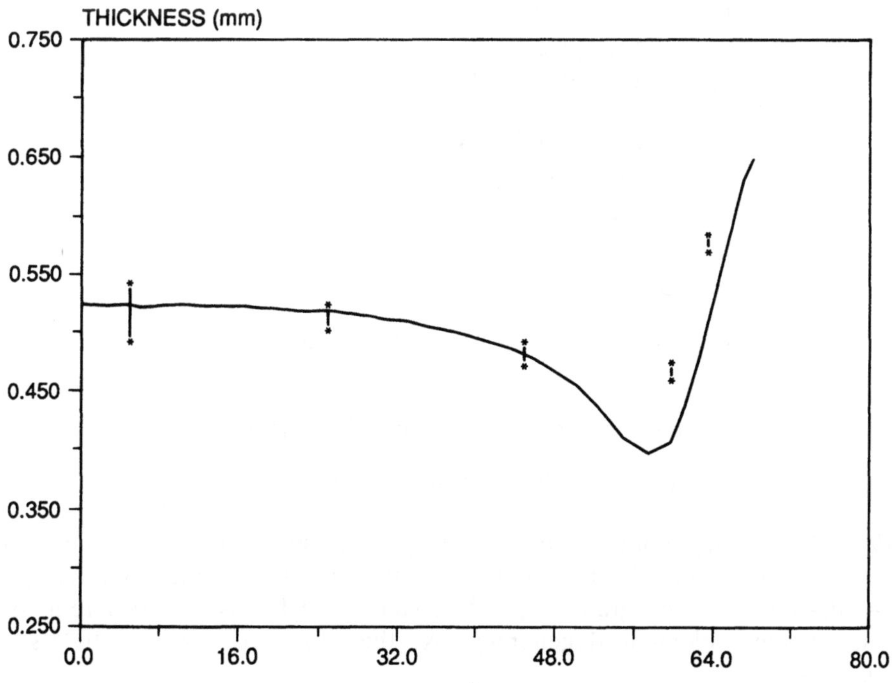

b

Fig. 8.14. Computed thickness profile in two cross-sections (0 and 4). Vertical segments are for the experimental results (Bellet and Chenot 1989).

- Three-dimensional eight-node hexaedral elements
- Contact and friction with complex tools, discretized into triangles
- Stress calculation with an efficient smoothing procedure
- Thermal coupling
- Introduction of physical parameters
- Possibility of simulation of porous material flow

The simulation conditions are as follows.

Meshes. Because of the symmetry (vertical plane) only one half of the dies and the material have been meshed.

- Upper die and lower die are different (Fig. 8.15a)
- 2070 triangular elements and 1112 nodes for the half upper die
- 2053 triangular elements and 1101 nodes for the lower die

The half-rolled preform has been meshed with 1860 eight-node elements and 2583 nodes (Fig. 8.15b).

Fig. 8.15 shows the preform and the dies in position at the beginning of the forging process.

The Parameter Values. The material properties and the boundary conditions have been chosen as representative of those in metal forming.

The mechanical parameters are

$$
\begin{aligned}
K &= K_0 \exp(b/T) \\
K_0 &= 1.55 \times 10^6 \text{ Pa s}^{-m} \\
b &= 4.2 \times 10^3 \text{ K}^{-1} \\
\rho &= 7.8 \times 10^3 \text{ kg m}^{-3} \\
c &= 7.0 \times 10^2 \text{ J kg}^{-1} \text{ K}^{-1} \\
k &= 2.3 \times 10^1 \text{ W m}^{-1} \text{ K}^{-1} \\
b &= 1.12 \times 10^4 \text{ kg s}^{-5/2} \text{ K}^{-1} \\
T_d &= 120°C \\
b_d &= 1.24 \times 10^4 \text{ kg}^{-1} \text{ s}^{-5/2} \text{ K}^{-1} \\
T_a &= 20°C \\
\epsilon &= 0.7 \\
h_{cd} &= 6 \times 10^3 \text{ W m}^{-2} \text{ K}^{-1}, \quad H_c = 6 \text{ W m}^{-2} \text{ K}^{-1}
\end{aligned}
$$

The initial temperature is uniform: $T = 1200°\text{C}$

The Dupont scheme has been used: $a = 0.25$, $g = 1$

Results. At the beginning of the forging process simulation, the time step is $\Delta t = 1.75 \times 10^{-3}$ s and the height between the two dies at the reference point is $h = 40.4$ mm. The program stopped at step 203 (that is for 70% reduction) because of some element degeneracy. At this stage $\Delta t = 7.28 \times 10^{-4}$ s and $h = 16.7$ mm.

Fig. 8.16 shows the deformed material at 44%, 60% and 70% reduction.

At the end of the forging, the equivalent strain isovalues in the symmetry plane are high in the head of the connecting rod and negligible elsewhere.

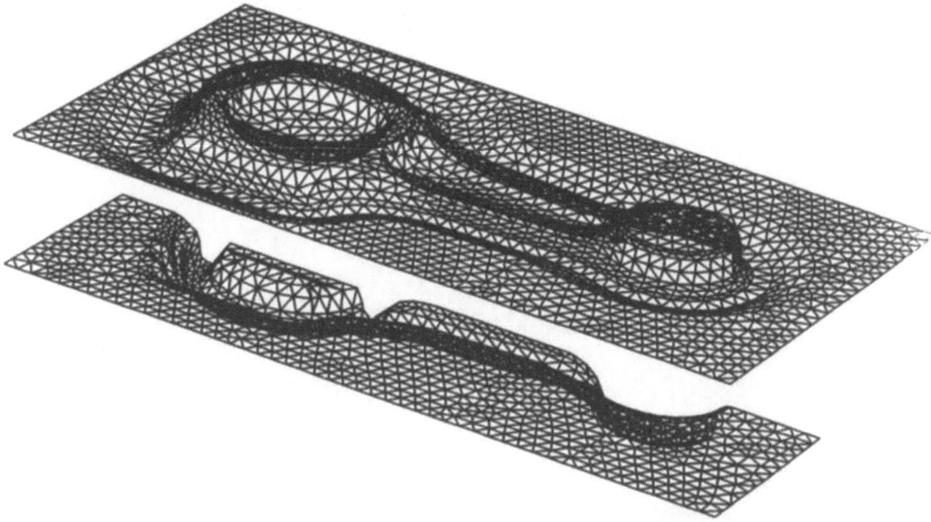

a

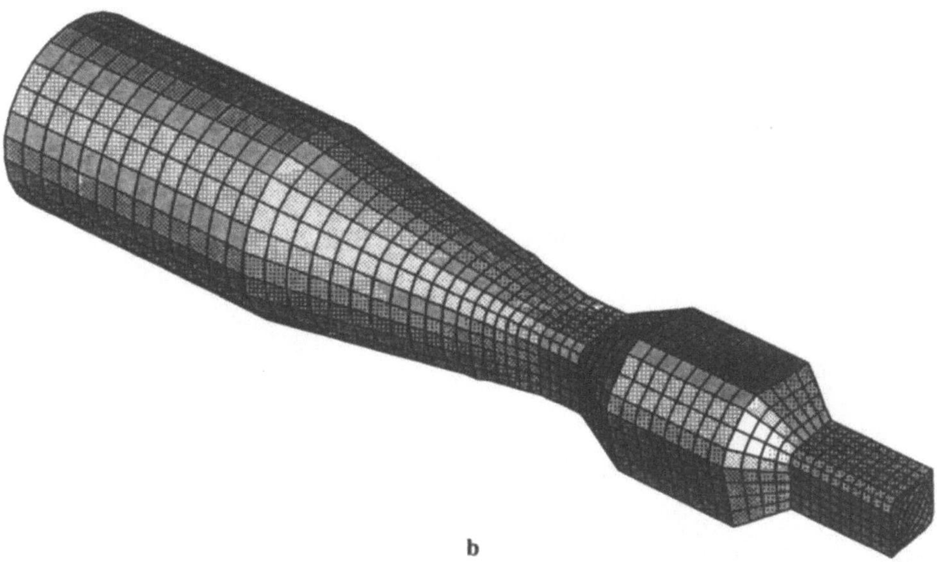

b

Fig. 8.15. Forging of a connecting rod. **a** Mesh of the dies; **b** mesh of the preform.

At 70% reduction, the temperature isovalues in the symmetry plane give an almost uniform temperature in the volume; the coolest parts are in the vicinity of the contact surfaces.

8.6 Conclusion

The basic concepts for finite-element modelling of viscoplastic flow have been described in some detail for the practical application to the simulation of either hot or cold deformation processes, provided that the elastic effects can be neglected. After the first historical studies, the topics in progress in different

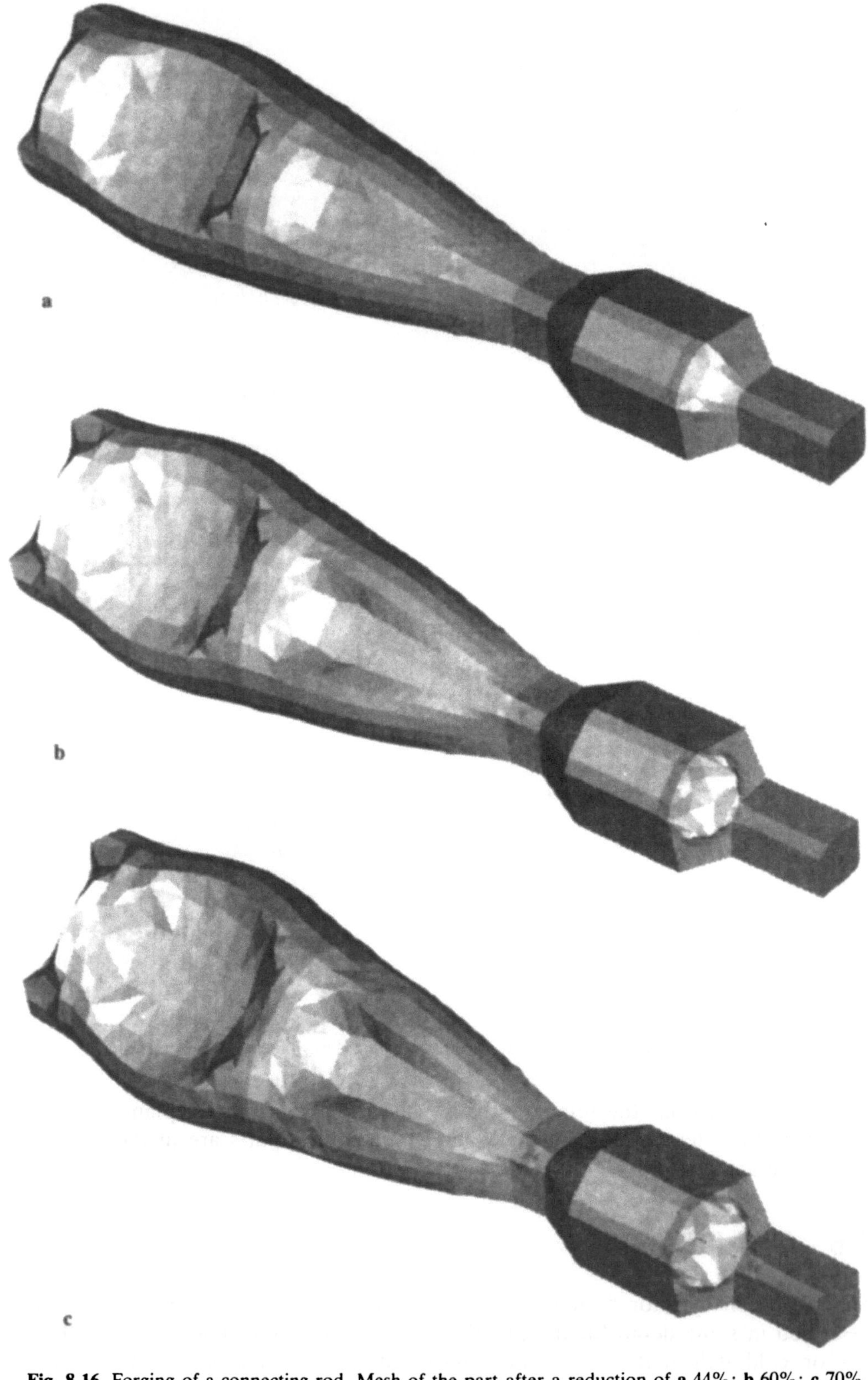

Fig. 8.16. Forging of a connecting rod. Mesh of the part after a reduction of **a** 44%; **b** 60%; **c** 70%.

laboratories are mainly oriented towards more accurate time integration and contact analysis, provision of free surface in stationary processes, error estimates and adaptivity, and the complete mechanical coupling with a more realistic physical description. In the future we can foresee developments in the fields of:

- Generalization of the approaches to three-dimensional problems
- Introduction of new methods of resolution to decrease the computer time and the memory space
- Development of industrial codes
- Use of computer codes for a better formulation of constitutive laws and an improved accuracy in physical parameter identification

References

Abouaf M (1985) Modelling of compaction of sintered metallic powder. Continuum mechanics approach. Thesis, Grenoble University (in French)

Abouaf M, Chenot JL, Raisson G, Bauduin P (1988) Finite element simulation of hot isostatic pressing of metal powders, Int J Num Meth Engng 25:191–212

Argyris J, Doltsinis JS (1981) On the natural formulation and analysis of large deformation coupled thermomechanical problems. Comput Meth Appl Mech Engng 25:195–253

Arminjon M, Chenot JL (1987) An implicit formulation for finite element analysis of dynamic plastic deformation in 2D-geometry. In: Pande GN, Middleton J (eds) Numerical techniques for engineering analysis and design. Martinus Nijhoff, Dordrecht, pp T2/1–T2/10

Bellet M (1988) Numerical modelling of sheet superplastic forming. Thesis, Ecole des Mines de Paris, Sophia-Antipolis (in French)

Bellet M, Chenot JL (1989) Numerical modelling of thin sheet superplastic forming. In: Thomson EG, Wood RD, Zienkiewicz OC, Samuelson A (eds) Numerical methods in industrial forming processes. AA Balkema, Rotterdam, pp 401–406

Bern A (1987) Contribution to the finite element modelling of free surfaces for stationary processes. Thesis, Ecole des Mines de Paris, Sophia-Antipolis (in French)

Bertrand-Corsini C (1987) Finite element modelling of hot shape rolling. Thesis, Ecole des Mines de Paris, Sophia-Antipolis (in French)

Betrand-Corsini C, David C, Chenot JL, Buessler P (1986) Stresses calculation in finite element analysis of three-dimensional hot shape rolling. In: Mattiasson K, Samuelson A, Wood RD, Zienkiewicz OC (eds) Numerical methods in industrial forming processes. AA Balkema, Rotterdam, pp 207–212

Bertrand-Corsini C, David C, Bern A, Montimitonnet P, Chenot JL, Bussler P, Fau F (1988) A 3D thermomechanical analysis of steady flows in hot forming. Application to hot flat rolling and hot shape rolling. In: Chenot JL, Oñate E (eds) Modelling of metal forming processes. Kluwer, Dordrecht, pp 271–280

Bertrand-Corsini C, Montmitonnet P, Chenot JL, Buessler P (1989) A three-dimensional analysis of hot rolling with a steady state thermomechanical approach. In: Thompson EG, Wood RD, Zienkiewicz OC, Samuelson A (eds) Numerical methods in industrial forming processes. AA Balkema, Rotterdam, pp 303–308

Bohatier C, Chenot JL (1985) Finite element formulations for non-steady-state large viscoplastic deformation. Int J Num Meth Engng 21:1697–1708

Cescutti JP, Chenot JL (1987) A geometrical continuous remeshing procedure for application to finite element calculation of non-steady state forming processes. In: Pande GN, Middleton J (eds) Numerical techniques for engineering analysis and design. Martinus Nijhoff, Dordrecht, pp S17/1–S17/10

Cescutti JP, Soyris N, Surdon G, Chenot JL (1987) Thermomechanical finite element calculations of three-dimensional hot forging with remeshing. In: Lange K (ed) Advanced technology of plasticity. Springer-Verlag, Berlin, pp 1051–1058

Cescutti JP, Wey E, Chenot JL, Mosser PE (1988) Finite element calculations of hot forging with continuous remeshing. In: Chenot JL, Oñate E (eds) Modelling of metal forming processes. Kluwer, Dordrecht, pp 207–216

Chenot JL (1983) A study of spread in hot rolling 3D finite element calculation. Report Greco CNRS (Grandes deformations et endommagement) no 63/1982 (in French)

Chenot JL (1987) Finite element calculation of unilateral contact with friction in nonsteady-state processes. In: Pande GN, Middleton J (eds) Numerical techniques for engineering, analysis and design. Martinus Nijhoff, Dordrecht, pp T1–T10

Chenot JL (1989) Finite element simulation of powder forming. In: Thomson EG, Wood RD, Zienkiewicz OC, Samuelson A, Numerical methods in industrial forming processes. AA Balkema, Rotterdam, pp 11–16

Comini G, Del Guidice S, Lewis RW, Zienkiewicz OC (1974) Finite element solution of non linear heat conduction problems with special reference to phase change. Int J Num Meth Engng 8:613–624

Cornfield GC, Johnson RH (1973) Theoretical predictions of plastic flow in hot rolling including the effect of various temperature distributions. J Iron Steel Inst 211:567–573

Dalin JB (1987) Modélisation numerique de la couleé continue. Thèse de Doctorat, Ecole des Mines de Paris, Sophia Antipolis (in French)

Dalin JB, Chenot JL (1988) Finite element computation of bulging in continuously cast steel with a visco-plastic model. Int J Num Meth Engng 25:147

David C (1987) 3D finite element modelling of a viscoplastic material flow applied to rolling. Thesis, Ecole des Mines de Paris, Sophia-Antipolis (in French)

Dawson PR, Mathur KK (1989) A numerical formulation for anisotropy in metal forming analysis. In: Thompson EG, Wood RD, Zienkiewicz OC, Samuelson A (eds) Numerical methods in industrial forming processes. AA Balkema, Rotterdam, pp 535–542

Donea J (1983) Arbitrary Lagrangian-Eulerian finite element method. In Belytschko T, Hughes TJR (eds) Computational methods for transient analysis. Elsevier, Amsterdam, pp 473–516

Freudenthal AM, Geiringer H (1958) The mathematical theories of inelastic continuum. In: Flügge S (ed) Handbuch der Physik band VI, Elasticity and plasticity. Springer-Verlag, Berlin, pp 229–313

Geradin M, Hogge M, Idelson S (1983) Implicit finite element method. In: Belytschko T, Hughes TJR (eds) Computational methods for transient analysis. Elsevier, Amsterdam, pp 417–472

Germain Y, Chung K, Wagoner RH (1989) A rigid viscoplastic finite element program for sheet metal forming analysis. Int J Mech Sci 31:1–24

Ghosh AK, Hamilton CH (1979) Mechanical behaviour and hardening characteristics of a superplastic Ti-GAI-4V alloy. Met Trans A 10:699

Hill R (1950) The mathematical theory of plasticity. Clarendon Press, Oxford

Hinton E, Campbell JS (1974) Local and global smoothing of discontinuous finite element functions using a least square method. Int J Numer Meth Engng 8:461–480

Hoff NJ (1954) Approximate analysis of structures in the presence of moderately large steps deformation. Quart Appl Math 2(1):49

Hogge MA (1981) A comparison of two and three level integration schemes for non-linear heat conduction. In: Lewis RW, Morgan K, Zienkiewicz OC, Wood RD, Alexander JM (eds) Numerical methods in heat transfer. John Wiley, Chichester, pp 75–90

Huang GC (1989) Error estimates and adaptive remeshing in finite element analysis of forming processes. PhD thesis, University College of Swansea, Swansea

Hughes TJR (1987) The finite element method. Linear static and dynamic finite element analysis. Prentice-Hall, Englewood Cliffs, New Jersey

Hughes TJR, Brooks AN (1982) A theoretical framework for Petrov–Galerkin methods with discontinuous weight functions: application to the streamline procedure. In: Gallagher RH et al. (eds) Finite elements in fluids, vol 4. John Wiley, Chichester.

Hughes TJR, Liu WK (1979) Finite element analysis of incompressible viscous fluid by the penalty function formulation. J Comp Phys 30(1):60

Karagiannis A (1989) Modelling of single component and bicomponent extrusion flows, thesis, MacMaster University

Kobayashi S, Kim JH (1978) Deformation analysis of axisymmetric sheet metal forming processes by the rigid-plastic finite element method. In: Mechanics of sheet metal forming. Plenum Press, New York, p 341

Kobayashi S (1982) A review on the finite element method and metal forming process modelling. J Appl Metal Working 2:163–169

Kobayashi S (1984) Thermoviscoplastic-analysis of metal forming problems by the finite element method. In: Pittman JFT, Zienkiewicz OC, Wood RD, Alexander JM (eds) Numerical analysis of forming processes. John Wiley, Chichester, pp 45–69

Lee CH, Kobayashi S (1973) New solutions to rigid plastic deformation problems using a matrix

method. J Engng Ind Trans ASME 95:865–873
Lee YS, Dawson PR (1988) Obtaining residual stresses in metal forming using a viscoplastic analysis for loading. In: Nakazawa S, William K, Rebelo N (eds) Advances in inelastic analysis. The American Society of Mechanical Engineers, New York, pp 283–298
Liska T, Orkisz J (1980) The finite difference method at arbitrary irregular grids and its application in applied mechanics. Comput Struct 11:83–95
Massoni E (1987) Finite element modelling of the deep-drawing process. Thesis, Ecole des Mines de Paris, Sophia-Antipolis (in French)
Montmitonnet P, Germain Y, Chenot JL (1988) Application of the finite element method to contact, friction and lubrication in metal forming. J Theor Appl Mech 7(suppl 1): 193–208.
Moreau JJ (1970) On plastic and viscoplastic friction laws. CR Acad Sci 271 (in French)
Mori K, Osakada K, Fukuda M (1984) Simulation of three dimensional deformation in metal forming by the rigid-plastic finite element method. In: Japan Society for Technology of Plasticity (ed) Advanced Technology of Plasticity, pp 1009–1014
Neyret B, Agassant JF, Chenot JL (1982) Finite element calculation and experimental investigation of two-dimensional molten polymer flows. In: Pittman JFT, Wood RD, Alexander JM, Zienkiewicz OC (eds) Numerical methods in industrial forming processes. Pineridge Press, Swansea, pp 535–546
Norton FH (1929) Creep of steel at high temperature. McGraw-Hill, New York
Oh SI, Rebelo N, Kobayashi S (1978) Finite element formulation for the analysis of plastic deformation of rate-sensitive materials in metal forming. In: Lippman H (ed) Metal forming plasticity. Springer-Verlag, Berlin, pp 273–291
Ortega JM, Rheinbolt WC (1970) Iterative solution of non-linear equations in several variables. Academic Press, New York
Park JJ, Kobayashi S (1984) Three dimensional finite element analysis of block compression. Int J Mech Sci 26(3):165–176
Perrot J, Chenot JL (1981) Finite element calculation of the evolution of an inclusion in a uniformly strained matrix. In: Hultz J, Lemaitre J (eds) Physical non-linearities in structural analysis, Springer-Verlag, Berlin pp 27–30
Pittman JFT, Nakazawa S (1984) Finite element analysis of polymer processing operations. In: Pittman JFT, Zienkiewicz OC, Wood RD, Alexander JM (eds) Numerical analysis of forming processes, John Wiley, Chichester, pp 165–218
Rebelo N, Kobayashi S (1980) A coupled analysis of viscoplastic deformation and heat transfer. Theoretical considerations, applications. Int J Mech Sci 22:699–718
Sellars CM, McTeggart G (1972) Hot workability. Int Met Rev 17:1–24
Shima S, Mori K (1978) Analysis of metal forming by the rigid-plastic finite element method based on plasticity theory for porous materials. In: Lippman H (ed) Metal forming plasticity. Springer-Verlag, Berlin, pp 305–317
Shima S, Oyane M (1976) Plasticity theory for porous metals. Int J Mech Sci 18:285–291
Smelser RE, Richmond O, Thompson EG (1986) A numerical study of the effects of die profile on extrusion. In: Mattiasson K, Samuelson A, Wood RD, Zienkiewicz OC (eds) Numerical methods in industrial forming processes. AA Balkema, Rotterdam, pp 305–312
Soyris N, Cescutti JP, Coupez T, Brachotte G, Chenot JL (1988) Three dimensional finite element calculations of the forging of a connecting rod. In: Chenot JL, Oñate E (eds) Modelling of metal forming processes. Kluwer, Dordrecht, pp 227–236
Stephen D (1987) Designing for superplastic alloys. Nato-AGARD lecture series no 154, Superplastic, paper no. 7
Surdon G, Chenot JL (1987) Finite element calculation of three-dimensional hot forging, Int J Num Meth Engng 24:2107–2117
Tanner RI, Nickell RE, Bilger RW (1975) Finite element methods for the solution for some incompressible non-Newtonian fluid mechanics problems with free surfaces. Comp Meth Appl Mech Engng 6:155–174
Wang NM (1984) A rigid-plastic rate-sensitive finite element method for modelling sheet metal forming processes. In: Pittman JFT, Zienkiewicz OC, Wood DR, Alexander JM (eds) Numerical analysis of forming processes. John Wiley, New York, p 117
Wargadipura AHS, Bonet J, Wood RD (1989) Finite element simulation of superplastic thin sheet forming with bounding. In: Proceedings 2nd international conference on computational plasticity, Barcelona, p 849
Wood RD, Mattiasson K, Honnor ME, Zienkiewicz OC (1985) Viscous flow and solid mechanics approaches to the analysis of thin sheet forming. In: Computer modelling of sheet metal forming

processes. The Metallurgical Society, London, p 121
Zienkiewicz OC (1984) Flow formulation for numerical solution of forming processes. In: Pittman JFT, Zienkiewicz OC, Wood RD, Alexander JM (eds) Numerical analysis of forming processes. John Wiley, Chichester, pp 1–44
Zienkiewicz OC, Huang GC (1989) Adaptive modelling of transient coupled metal forming processes. In: Thompson EG, Wood RD, Zienkiewicz OC, Samuelson A (eds) Numerical methods in industrial forming processes. AA Balkema, Rotterdam, pp 3–10
Zienkiewicz OC, Godbole PN (1974) Flow of plastic and viscoplastic solids with special reference to extrusion and forming problems. Int J Num Meth Engng 8:3–16
Zienkiewicz OC, Taylor RL (1989) The finite element method, 4th edn. McGraw-Hill
Zienkiewicz OC, Liu YC, Huang GC (1988) Error estimation and adaptivity in flow formulation for forming problems. Int J Num Meth Engng 24:23–42

9 The Elastic-Plastic Finite-Element Method

I. Pillinger

9.1 Introduction

The rigid-plastic and viscoplastic finite-element techniques described in the previous chapters are useful approaches to the modelling of metal deformation when the elastic component of strain may reasonably be ignored. This is often the case, for example, for hot working conditions. In other situations, it is important to take into account the elastic as well as the plastic deformation of the material, and it is vital to do so if the unloading behaviour is to be predicted.

Finite-element (FE) methods were first developed to analyse the elastic deformation of structures and so it is not surprising that some of the earliest FE studies of metal forming adopted a similar approach, but using an elastic-plastic constitutive relationship instead of the elastic Hookean one (Marcal and King 1967, Yamada et al. 1968, Zienkiewicz et al. 1969).

Plastic deformation is of course nonlinear, and so most studies of nonsteady-state processes use an incremental formulation in which the relationship between, say, force and displacement is assumed to be piecewise linear over each small step of the deformation. This strategy is reasonably successful for processes involving small amounts of total deformation, but two fundamental problems with this simple approach soon become apparent. The first problem is practical, in that in order to achieve the desired level of accuracy, very small increments of deformation have to be used, and this leads to very extensive computer run times for the analysis of anything other than small deformation problems. The second difficulty concerns the validity of the elastic-plastic formulation itself when applied to situations involving finite (as opposed to infinitesimal elastic) deformation (Lee 1976). In particular, for finite deformation it is not sufficient just to divide the deformation process up into small increments, it is essential to use the correct definitions of stress rate and strain rate (or, equivalently, stress increment and strain increment) for each quasilinear step. The so-called finite-strain elastic-plastic formulations are therefore derived from the theory of deforming continuous solids, such as that set out in Chap. 2.

In providing a solution to the second problem associated with infinitesimal-strain elastic-plastic formulations, it turns out that, to a great extent, the first problem is also solved. Finite-strain programs can typically use increment sizes which are at least an order of magnitude greater than those to which infinitesimal-strain programs are restricted. Indeed, in most metalforming simulations, the maximum increment size is limited by the nonlinearity of the boundary conditions (i.e. by the changing contact between the workpiece and the tools) rather than by the nonlinearity of the governing equations.

Elastic-plastic FE models come in many different forms. For example, some adopt an Eulerian formulation and some use a Lagrangian one. In some implementations, an explicit stiffness expression is obtained and in others the incremental changes in the state parameters are calculated implicitly from the differential equations describing the material behaviour. This chapter describes just one particular variation, namely an implicit finite-strain elastic-plastic updated-

Lagrangian formulation. Many aspects of this approach are of course shared by other FE models, such as rigid-plastic and viscoplastic, so only the distinguishing features of the elastic-plastic method will be described in detail.

9.2 Incremental Methodology

The basic approach adopted by the formulation described in this chapter is to divide the deformation process up into small increments, typically of around 1% nominal strain. The increments need not be all the same size, and it may be advantageous for the user of the FE program to specify a variation of increment size during the process. The size of increments may also be modified by the program itself (see Sect. 9.5.1).

The basic methodology is as follows. Assuming the values of all the state variables are known at the start of a given increment, construct a set of equations based on these state variables which can be solved in order to find the values of all the state variables at the end of the increment. This is then the starting point for the next increment, and so on.

Because the governing equations are based upon a reference state of the workpiece (at the start of the increment) this approach is called *Lagrangian*. Since a new reference state is used for each increment, it is called *updated Lagrangian*.

In terms of the entities defined in Chap. 2, the initial configuration Γ coincides with the start of an increment and the "current" configuration γ is at some stage during the increment. Adopting the same notation as in Chap. 2, let a point P of the workpiece have Cartesian coordinates X^i and let this material at this point be deformed to a position P' in the γ configuration which has Cartesian coordinates x^i. Define a convected or embedded axis system which deforms with the material so that in configuration γ, the coordinates of P' in this axis system are the same as the coordinates of P in the Cartesian system, i.e. X^i. Let ξ^i be the coordinates of P' in the convected axis system. Note that as configuration γ approaches Γ, the convected axis system tends to the Cartesian one. If $\mathbf{e}_i$ is the orthonormal basis of the Cartesian axis system, then the basis $\mathbf{g}_i$ corresponding to the convected axis system may be defined as in Eq. (2.9):

$$\mathbf{g}_i = \frac{\delta x^j}{\delta \xi^i}\mathbf{e}_j \tag{9.1}$$

or

$$\mathbf{g}_i = \frac{\delta x^j}{\delta X^i}\mathbf{e}_j$$

$$= x^{j;i}\,\mathbf{e}_j \tag{9.2}$$

since, by definition, the convected coordinates of P' are the same as the Cartesian coordinates of P, and using the notation introduced in Sect. 2.6.1.

It is necessary, also, to define how infinitesimal area elements deform. Let an infinitesimal area element be defined at point P in the reference configuration Γ by the two nonparallel vectors $\mathbf{V}^1$ and $\mathbf{V}^2$ (Fig. 9.1) so that:

$$\mathbf{V}^1 \times \mathbf{V}^2 = \mathbf{N}\,dA \tag{9.3}$$

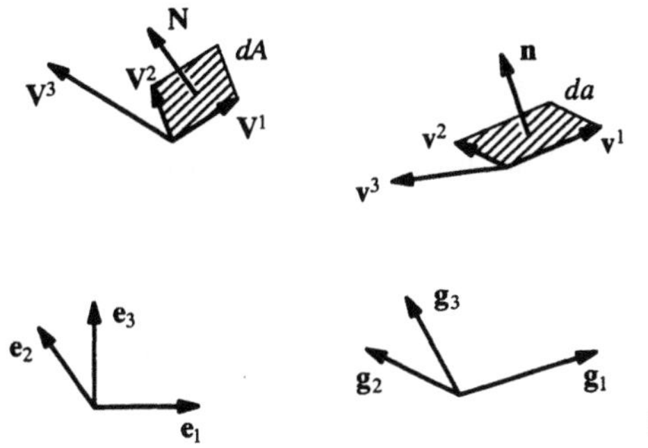

Fig. 9.1. Undeformed and deformed area elements.

where **N** is the unit vector normal to the area and dA is its size. This of course defines an area element in the shape of a parallelogram, but the argument that follows is quite general as any arbitrary shape may be decomposed into an appropriate number of parallelograms or triangles (i.e. half parallelograms).

Suppose in the current configuration γ, this area has deformed into another parallelogram, this time situated at P' and defined by the bounding vectors $\mathbf{v}^1$ and $\mathbf{v}^2$ (Fig. 9.1):

$$\mathbf{v}^1 \times \mathbf{v}^2 = \mathbf{n}\, da \tag{9.4}$$

in which **n** is the unit normal to the deformed area element and da is its area. The vectors $\mathbf{v}^1$ and $\mathbf{v}^2$ are of course obtained by applying to $\mathbf{V}^1$ and $\mathbf{V}^2$ the same deformation transformation that mapped P to P'.

Let $\mathbf{V}^3$ be an *arbitrary* vector in the reference configuration that deforms to $\mathbf{v}^3$ in the current configuration (Fig. 9.1). Then write

$$\mathbf{V}^i = V^{ij}\, \mathbf{e}_j \tag{9.5}$$

and

$$\begin{aligned} \mathbf{v}^i &= v^{ij}\, \mathbf{g}_j \\ &= V^{ij}\, \mathbf{g}_j \end{aligned} \tag{9.6}$$

by the properties of convected coordinates. Taking the scalar product of $\mathbf{v}^3$ and the scaled normal to the deformed area element, and using Eq. (9.4) gives

$$\mathbf{n}\, da \cdot \mathbf{v}^3 = (\mathbf{v}^1 \times \mathbf{v}^2) \cdot \mathbf{v}^3 \tag{9.7}$$

But from Eq. (9.6), using the properties of a vector triple product:

$$\begin{aligned} \mathbf{n}\, da \cdot \mathbf{v}^3 &= (V^{1i}\, \mathbf{g}_i \times V^{2j}\, \mathbf{g}_j) \cdot V^{3k}\, \mathbf{g}_k \\ &= \det(V^{ij})(\mathbf{g}_1 \times \mathbf{g}_2) \cdot \mathbf{g}_3 \end{aligned} \tag{9.8}$$

in which det() indicates the determinant of the enclosed matrix. Substituting Eq. (9.2) gives:

$$\begin{aligned} \mathbf{n}\, da \cdot \mathbf{v}^3 &= \det(V^{ij})(x^{k;1}\, \mathbf{e}_k \times x^{m;2}\, \mathbf{e}_m) \cdot x^{n;3}\, \mathbf{e}_n \\ &= \det(V^{ij}) \det(x^{m;n})(\mathbf{e}_1 \times \mathbf{e}_2) \cdot \mathbf{e}_3 \\ &= \det(x^{m;n})(V^{1i}\, \mathbf{e}_i \times V^{2j}\, \mathbf{e}_j) \cdot V^{3k}\, \mathbf{e}_k \end{aligned} \tag{9.9}$$

using the property of triple products again. Thus

$$\mathbf{n}\,da \cdot \mathbf{v}^3 = J\,(\mathbf{V}^1 \times \mathbf{V}^2) \cdot \mathbf{V}^3$$
$$= J\,\mathbf{N}\,dA \cdot \mathbf{V}^3 \tag{9.10}$$

from Eq. (9.3) and using the symbol J for the determinant for the Jacobian transformation $x^{i;j}$. If n_{0i} are Cartesian components of $\mathbf{N}$ then

$$\mathbf{n}\,da \cdot \mathbf{v}^3 = J\,n_{0i}\,V^{3i}\,dA \tag{9.11}$$

Using Eq. (9.6), the left-hand side of Eq. (9.11) may also be expressed in terms of n_i, the Cartesian components of $\mathbf{n}$

$$\mathbf{n}\,da \cdot \mathbf{v}^3 = n_j\,\mathbf{e}_j\,da \cdot V^{3i}\,\mathbf{g}_i \tag{9.12}$$

But from Eq. (9.2)

$$\mathbf{e}_j \cdot \mathbf{g}_i = x^{j;i} \tag{9.13}$$

so

$$\mathbf{n}\,da \cdot \mathbf{v}^3 = n_j\,x^{j;i}\,da\,V^{3i} \tag{9.14}$$

Comparing Eqs (9.11) and (9.14), and noting that $\mathbf{V}^3$ was an *arbitrary* vector leads to the result

$$J\,n_{0i}\,dA = x_{j;i}\,n_j\,da \tag{9.15}$$

Note, subscripts have been used exclusively in Eq. (9.15) since all components are expressed with respect to the orthonormal Cartesian basis.

9.3 Discretization

The fundamental principle of the FE method is that it conceptually divides up the workpiece into a number of disjoint regions, or elements. Each element is associated with a number of points, or nodes, which are usually, but not exclusively, on the outside of the element. Adjoining elements share nodes that are on their common interface.

The value of any state variable at a point within an element may be obtained by interpolation between the values of that state variable at the nodes of the element. In particular, material displacement and body or surface forces may be obtained in this way so that the general description of continuum deformation involving an infinite number of degrees of freedom reduces to an expression of the form

$$\mathbf{f} = \mathbf{K}\,\mathbf{d} \tag{9.16}$$

where $\mathbf{f}$ and $\mathbf{d}$ are finite-length vectors of the components of force and displacement, respectively, at all the nodes of the workpiece, and $\mathbf{K}$ is a matrix known as the *stiffness matrix*.

Eq. (9.16) is the usual expression derived for linear problems. The corresponding incremental expression used for nonlinear problems such as metal forming is

$$\Delta\mathbf{f} = \mathbf{K}\,\Delta\mathbf{d} \tag{9.17}$$

Within a given element, the interpolation is defined by a set of functions N_I of position, one function for each node of the element. Then, for example, the change in displacement $\Delta u_i = x_i - X_i$ at P is given in terms of the changes in displacement Δd_{Ii} of all the nodes of the element by

$$\Delta u_i = N_I(X_j)\, \Delta d_{Ii} \tag{9.18}$$

in which summation is carried out for the repeated upper-case subscript over all the nodes of the element.

The definition of the interpolation functions depends upon the type of element being used. Usually, they are defined in terms of local element coordinates. These are coordinates in an axis system with its origin at the element centroid, such that the corners of the element have coordinates +1 or −1 (Fig. 9.2). In the local element axis system, the coordinates of point P are χ_i. For three-dimensional eight-noded linear hexahedral elements, for example, the interpolation function for the node with local coordinates (J, K, L) is given by

$$N_{(JKL)}(\chi_i) = \tfrac{1}{8} JKL\,(\chi_1 + J)(\chi_2 + K)(\chi_3 + L), \quad J,K,L = \pm 1 \tag{9.19}$$

It can be seen that this function attains the value one at the node with local coordinates (J, K, L) and is zero at all the other nodes.

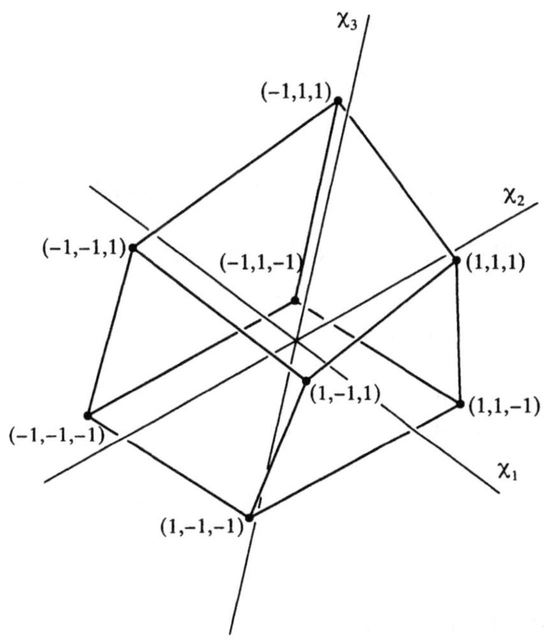

Fig. 9.2. Local axis system for a hexahedral element.

9.4 Elastic-Plastic Formulation

9.4.1 Principle of Virtual Work

The starting point for the derivation of the elastic-plastic stiffness equations for a given element is the Principle of Virtual Work which states that for equilibrium,

$$\delta W_{\mathrm{I}} - \delta W_{\mathrm{E}} = 0 \tag{9.20}$$

in which δW_{I} is the increment of internal (deformation) energy within a region caused by an arbitrary variation in internal displacement and δW_{E} is the increment of external work caused by the corresponding displacement of the boundaries of the region. From Eq. (2.196), for an updated-Lagrangian approach:

$$\delta W_{\mathrm{I}} = \int_V P_{ij}\, \delta(u_{j;i})\, dV \tag{9.21}$$

in which P_{ij} is the *nominal* or *Lagrange* stress, $\delta(u_{j;i})$ is the virtual displacement gradient, and the integration is carried out over the reference volume V of the element. As before, it is permitted to replace the superscripts used in Chap. 2 by subscripts as only Cartesian components will be used in this chapter.

If forces f_{Im} act on the nodes of the element, and $\delta(d_{Im})$ are the virtual nodal displacements, the variation in external work is simply given by

$$\delta W_{\mathrm{E}} = f_{Im}\, \delta(d_{Im}) \tag{9.22}$$

Combining Eqs (9.21) and (9.22):

$$f_{Im}\, \delta(d_{Im}) = \int_V P_{ij}\, \delta(u_{j;i})\, dV \tag{9.23}$$

The principle of virtual work will also apply at the end of the increment, a time Δt after the start, so

$$(f_{Im} + \dot{f}_{Im}\, \Delta t)\delta(d_{Im}) = \int_V (P_{ij} + \dot{P}_{ij}\, \Delta t)\, \delta(u_{j;i})\, dV \tag{9.24}$$

Subtracting Eq. (9.23) from (9.24) and eliminating Δt gives

$$\delta(d_{Im})\dot{f}_{Im} = \int_V \delta(u_{j;i})\, \dot{P}_{ij}\, dV \tag{9.25}$$

But since virtual displacements are arbitrary, Eq. (9.25) must also be true for arbitrary variations in displacement *rate*:

$$\delta(\dot{d}_{Im})\dot{f}_{Im} = \int_V \delta(\dot{u}_{j;i})\, \dot{P}_{ij}\, dV \tag{9.26}$$

9.4.2 Constitutive Relationship

The relationship between stress rate and strain rate may be written as

$$\dot{\tau}_{ij} = D_{ijkl}\dot{\epsilon}_{kl} \tag{9.27}$$

in which the strain rate $\dot{\epsilon}_{kl}$ is defined by

$$\dot{\epsilon}_{ij} = \tfrac{1}{2}\,(\dot{u}_{i;j} + \dot{u}_{j;i}) \tag{9.28}$$

As explained in Chap. 3, for an isotropic elastic-plastic material, the constitutive expression is derived from the Prandtl–Reuss flow rule and von Mises' yield criterion. From Eq. (3.35) it can be seen that

$$D_{ijkl} = 2G\left(\delta_{ik}\delta_{jl} + \delta_{ij}\delta_{kl}\left(\frac{\nu}{1-2\nu}\right) - \frac{3\,\sigma'_{ij}\sigma'_{kl}}{2\bar{\sigma}\left(1+\dfrac{Y'}{3G}\right)}\right) \tag{9.29}$$

in which G is the rigidity modulus ($= E/2(1+\nu)$, where E is Young's modulus), ν is Poisson's ratio, $\bar{\sigma}$ is the generalized or equivalent stress defined by

$$\bar{\sigma}^2 = \tfrac{3}{2}\,\sigma'_{ij}\sigma'_{ij} \tag{9.30}$$

Y' is the slope of the generalized stress versus plastic strain curve, σ'_{ij} is the deviatoric stress (Sect. 1.1.2.2) and δ_{ij} is the Kronecker delta.

In order to maintain the objectivity of the constitutive relationship, $\dot{\tau}_{ij}$ is interpreted to be the *corotational* or *Jaumann* rate of the stress obtained by multiplying the Cauchy stress σ_{ij} by the determinant of the deformation gradient J. The scaling factor J is applied to the Cauchy stress because, physically, J is the ratio of deformed to undeformed infinitesimal volumes, and this allows for the fact that when the constitutive expression is eventually used in Eq. (9.26), the integration is carried out over the reference (undeformed) volume, not the current (deformed) volume on which Cauchy stress is defined.

The corotational components of stress rate are simply the components of the ordinary stress rate expressed in a Cartesian coordinate system that *rotates* with the deforming material. This coordinate system is not of course the convected one, since the latter is, in general, not Cartesian.

In mathematical terms, since $x_{i:j}$, the deformation gradient, is a real, nonsingular transformation, it may be decomposed uniquely into a rotational transformation r_{ij} and a symmetric deformation q_{ij}:

$$x_{i:j} = r_{ik}q_{kj} \tag{9.31}$$

Differentiating this with respect to time, and evaluating in the reference configuration when $r_{ij} = q_{ij} = \delta_{ij}$:

$$\begin{aligned}\dot{x}_{i:j} = \dot{u}_{i:j} &= \dot{r}_{ik}\,q_{kj} + r_{ik}\,\dot{q}_{kj}\\ &= \dot{r}_{ij} + \dot{q}_{ij}\end{aligned} \tag{9.32}$$

Since $\dot{q}_{ij}$ is symmetric:

$$\dot{r}_{ij} = \tfrac{1}{2}\,(\dot{u}_{i:j} - \dot{u}_{j:i}) = \langle \dot{u}_{i:j}\rangle \tag{9.33}$$

where the angle brackets denote the skew-symmetric part of the enclosed matrix.

τ_{ij} are defined to be the components of $J\sigma_{ij}$ in the Cartesian axis systems rotated by r_{ij} with respect to the reference axis system:

$$J\sigma_{i:j} = \tau_{mn}\,r_{im}\,r_{jn} \tag{9.34}$$

Differentiating Eq. (9.34) with respect to time, and once again evaluating in the reference configuration:

$$\begin{aligned}\frac{d(J\sigma_{ij})}{dt} &= \dot{\tau}_{mn}r_{im}r_{jn} + \tau_{mn}\dot{r}_{im}r_{jn} + \tau_{mn}r_{im}\dot{r}_{jn}\\ &= \dot{\tau}_{ij} + \sigma_{kj}\dot{r}_{ik} + \sigma_{ik}\dot{r}_{jk}\end{aligned} \tag{9.35}$$

since all definitions of stress are the same in the reference configuration. From Eq. (9.33)

$$\frac{d(J\sigma_{ij})}{dt} = \dot{\tau}_{ij} + \sigma_{kj}\langle \dot{u}_{i;k}\rangle + \sigma_{ik}\langle \dot{u}_{j;k}\rangle \tag{9.36}$$

9.4.3 Relationship Between Cauchy and Nominal Stress

Replacing the superscripts in Eq. (2.80) by subscripts for the orthonormal basis $\mathbf{e}_i$ produces the following definition of Cauchy stress:

$$\mathbf{t}_i = \sigma_{ij}\,\mathbf{e}_j \tag{9.37}$$

Similarly, from Eq. (2.92)

$$\mathbf{T}_i = P_{ij}\,\mathbf{e}_j \tag{9.38}$$

in which (Eqs (2.76) and (2.89))

$$\mathbf{t}_i\,da_i = \mathbf{t}\,da = \mathbf{T}\,dA = \mathbf{T}_k\,dA_k \tag{9.39}$$

is the force acting on a given infinitesimal area and (Eqs (2.78) and (2.90))

$$da_i = n_i\,da \tag{9.40}$$

$$dA_k = n_{0k}\,dA \tag{9.41}$$

But from Eq. (9.15)

$$n_{0k}\,dA = \frac{1}{J}x_{i;k}\,n_i\,da \tag{9.42}$$

Combining Eqs (9.39), (9.41) and (9.42)

$$\mathbf{T}_k\,dA_k = \mathbf{T}_k\,\frac{1}{J}x_{i;k}\,n_i\,da \tag{9.43}$$

But from Eqs (9.39) and (9.40)

$$\mathbf{T}_k\,dA_k = \mathbf{t}_i\,da_i = \mathbf{t}_i\,n_i\,da \tag{9.44}$$

Comparing Eqs (9.43) and (9.44) gives

$$J\,\mathbf{t}_i = \mathbf{T}_k\,x_{i;k} \tag{9.45}$$

and substitution for $\mathbf{t}_i$ and $\mathbf{T}_k$ from Eqs (9.37) and (9.38) respectively leads to the result

$$J\,\sigma_{ij}\,\mathbf{e}_j = P_{kj}\,\mathbf{e}_j\,x_{i;k} \tag{9.46}$$

or

$$J\,\sigma_{ij} = P_{kj}\,x_{i;k} \tag{9.47}$$

9.4.4 Nominal Stress Rate

Differentiating Eq. (9.47) with respect to time in the reference configuration and combining with Eq. (9.36) gives

$$\dot{P}_{ij} + \sigma_{kj}\,\dot{u}_{i;k} = \dot{\tau}_{ij} + \sigma_{kj}\langle \dot{u}_{i;k}\rangle + \sigma_{ik}\langle \dot{u}_{j;k}\rangle \tag{9.48}$$

Thus

$$\begin{aligned}\dot{P}_{ij} &= \dot{\tau}_{ij} - \sigma_{kj}\,\dot{\epsilon}_{ik} - \sigma_{ik}\,\dot{\epsilon}_{jk} + \sigma_{ik}\,\dot{u}_{j;k}\\ &= \dot{\tau}_{ij} - 2\sigma_{ik}\,\dot{\epsilon}_{kj} + \sigma_{ik}\,\dot{u}_{j;k}\end{aligned} \tag{9.49}$$

since both σ_{ij} and $\dot{\epsilon}_{ij}$ are symmetric. Substituting for $\dot{\tau}_{ij}$ from Eq. (9.27) gives

$$\dot{P}_{ij} = (D_{ijkl} - 2\sigma_{ik}\delta_{jl})\dot{\epsilon}_{kl} + \sigma_{ik}\,\dot{u}_{j;k} \tag{9.50}$$

9.4.5 Governing Rate Equations

Using the expression for P_{ij} given in Eq. (9.50) in the virtual work equation (9.26) results in

$$\delta(\dot{d}_{Im})\dot{f}_{Im} = \int_V \delta(\dot{u}_{j;i})[(D_{ijkl} - 2\sigma_{ik}\delta_{jl})\dot{\epsilon}_{kl} + \sigma_{ik}\dot{u}_{j;k}]\,dV \tag{9.51}$$

or, since σ_{ij}, $\dot{\epsilon}_{ij}$ and D_{ijkl} are all symmetric:

$$\delta(\dot{d}_{Im})\dot{f}_{Im} = \int_V [\delta(\dot{\epsilon}_{ij})(D_{ijkl} - 2\sigma_{ik}\delta_{jl})\dot{\epsilon}_{kl} + \delta(\dot{u}_{j;i})\sigma_{ik}\dot{u}_{j;k}]\,dV \tag{9.52}$$

9.4.6 Incremental Equations

Multiplying Eq. (9.52) by Δt^2 and using incremental expressions in place of the products of rate and time increment produces the incremental form of the virtual work equations:

$$\begin{aligned}\delta(\Delta d_{Im}) = \int_V &[\delta(\Delta\epsilon_{ij})(D_{ijkl} - 2\sigma_{ik}\delta_{jl})\Delta\epsilon_{kl}\\ &+ \delta(\Delta u_{j;i})\sigma_{ik}\Delta u_{j;k}]\,dV\end{aligned} \tag{9.53}$$

With the exception of $\Delta\epsilon_{ij}$, the definition of all the incremental quantities in Eq. (9.53) is quite straightforward. The definition of $\Delta\epsilon_{ij}$ needs some care. The simplest approach, which defines it to be the symmetric part of $\Delta u_{i;j}$ (the infinitesimal definition) is *not* satisfactory since this does not define an objective measure. Moreover, whatever expression is adopted has to be a linear function of Δu_{ij} or else it will not be possible to obtain a stiffness expression in the form of Eq. (9.17).

One possible definition of $\Delta\epsilon_{ij}$ is called the *linearized corotational* (LCR) expression for strain increment (Pillinger et al. 1986):

$$\begin{aligned}\Delta\epsilon_{ij} = \tfrac{1}{2}[r_{ki}\,\Delta u_{k;j} + r_{kj}\,\Delta u_{k;i} - \tfrac{1}{2}(\langle r_{ik}\rangle\langle \Delta u_{j;k}\rangle\\ + \langle r_{jk}\rangle\langle \Delta u_{i;k}\rangle)]\end{aligned} \tag{9.54}$$

The value of the rotational matrix r_{ij} is estimated during the previous increment (see Sect. 9.5.2) so that Eq. (9.54) is, indeed, a linear function of $\Delta u_{i;j}$. It can be seen from this equation that for rigid-body rotation when $\Delta u_{i;j} = r_{ij} - \delta_{ij}$, then

$\Delta u_{i;j}$ is approximately skew-symmetric (providing the incremental angles of rotation are not too large) and so the right-hand side of Eq. (9.54) vanishes. Furthermore, Eq. (9.54) reduces to the symmetric part of $\Delta u_{i;j}$ when there is no rotation.

9.4.7 Effect of Plastic Incompressibility

Plastic incompressibility is an implicit assumption of the Prandtl–Reuss flow rule used to derive the constitutive matrix D_{ijkl} in Eq. (9.29). As a consequence, solving the elastic-plastic stiffness equations tends to enforce plastic incompressibility at every point of an element.

In a continuum, the condition of volume constancy will introduce one constraint for every yielded particle, each of which has three independent components of displacement. There is therefore one volume constraint for every three degrees of freedom.

Solving the standard elastic-plastic equations however generally imposes a far higher level of constraint on the FE mesh than this. For example, with three-dimensional eight-noded linear elements, seven volume constraints are imposed for each element. Since, for extensive meshes, the number of nodes is approximately equal to the number of elements, this means that there are approximately seven volume constraints for every three degrees of freedom. The implicit imposition of volume constancy at every point of an element clearly leads to a very overconstrained deformation which can, in severe cases, produce totally incorrect predictions of workpiece deformation.

The solution to this problem is to relax the condition that volume must be constant at every point of an element and instead merely to require that the *total* volume of an element be constant during plastic deformation. (This does not of course prevent it from having an *elastic* change in volume.)

There are various ways of achieving this objective. One way is to perform underintegration of the volume–strain components when numerically integrating the stiffness expressions. Another method, called *element dilatation* (Nagtegaal et al. 1974) will be described here.

If the deviatoric strain increment $\Delta\epsilon'_{ij}$ is defined by

$$\Delta\epsilon'_{ij} = \Delta\epsilon_{ij} - \tfrac{1}{3}\,\delta_{ij}\Delta\epsilon_{pp} \tag{9.55}$$

then from Eq. (9.29) it can be shown that

$$\delta(\Delta\epsilon_{ij})D_{ijkl}\Delta\epsilon_{kl} = \delta(\Delta\epsilon'_{ij})D_{ijkl}\Delta\epsilon'_{kl} + \kappa\,\delta(\Delta\epsilon_{pp})\Delta\epsilon_{qq} \tag{9.56}$$

in which κ is the elastic bulk modulus $E/3(1-2\nu)$. The last term in Eq. (9.56) represents the contribution of the volume strain to the element stiffness.

The element dilatation method defines the volume strain increment to be an independent function $\Delta\phi$, called the dilatation increment, of position. The nature of the dilatation function depends on the type of element, i.e. on how many volume constraints need to be imposed per element to correspond with the level of constraint in a continuous solid. For three-dimensional eight-noded linear elements it can be seen that one volume constraint is required per element, so in this case $\Delta\phi$ is constant throughout the element. The single volume constraint is then the condition

$$\int_V \Delta\epsilon_{jj}\, dV = \int_V \Delta\phi\, dV$$
$$= V\,\Delta\phi \tag{9.57}$$

or

$$\Delta\phi = \frac{1}{V}\int_V \Delta\epsilon_{jj}\, dV \tag{9.58}$$

Using element dilatation, Eq. (9.53) becomes

$$\delta(\Delta d_{Im})\Delta f_{Im} = \int_V [\delta(\Delta\epsilon_{ij})(D_{ijkl} - 2\sigma_{ik}\delta_{jl})\Delta\epsilon_{kl} + \delta(\Delta u_{j;i})\sigma_{ik}\Delta u_{j;k}$$
$$- \kappa\delta(\Delta\epsilon_{ii})\Delta\epsilon_{jj} + \kappa\delta(\Delta\phi)\Delta\phi]\, dV \tag{9.59}$$

If, for sake of convenience, it is assumed that three-dimensional eight-noded linear elements are being used, substituting for $\Delta\phi$ from Eq. (9.58) into Eq. (9.59) leads to

$$\delta(\Delta d_{Im})\Delta f_{Im} = \int_V [\delta(\Delta\epsilon_{ij})(D_{ijkl} - 2\sigma_{ik}\delta_{jl})\Delta\epsilon_{kl} + \delta(\Delta u_{j;i})\sigma_{ik}\Delta u_{j;k}$$
$$- \kappa\delta(\Delta\epsilon_{ii})\Delta\epsilon_{jj}]\, dV + \frac{\kappa}{V}\left(\int_V \delta(\Delta\epsilon_{ii})dV\right)\left(\int_V \Delta\epsilon_{jj}dV\right) \tag{9.60}$$

9.4.8 Interpolation

Using the interpolation function defined in Eq. (9.18) and differentiating with respect to the reference coordinates:

$$\Delta u_{i;j} = N_{I;j}\Delta d_{Ii} \tag{9.61}$$

and from the definition of LCR strain in Eq. (9.54):

$$\Delta\epsilon_{ij} = B_{ijIm}\Delta d_{Im} \tag{9.62}$$

in which

$$B_{ijIm} = \tfrac{1}{2}(B^*_{ijIm} + B^*_{jiIm}) \tag{9.63}$$

and

$$B^*_{ijIm} = r_{mi}N_{I;j} - \tfrac{1}{4}\langle r_{ik}\rangle(\delta_{jm}N_{I;k} - \delta_{km}N_{I;j}) \tag{9.64}$$

Substituting for $\Delta u_{i;j}$ and $\Delta\epsilon_{ij}$ (Eqs (9.61) and (9.62)) into Eq. (9.60) produces

$$\delta(\Delta d_{Im})\Delta f_{Im} = \delta(\Delta d_{Im})\int_V [B_{ikIm}(D_{ijkl} - 2\delta_{jl}\sigma_{ik})B_{klJn}$$
$$+ N_{I;i}\delta_{mn}\sigma_{ik}N_{J;k} - \kappa B_{iiIm}B_{jjJn}]dV\,\Delta d_{Jn}$$
$$+ \delta(\Delta d_{Im})\left[\frac{\kappa}{V}\left(\int_V B_{iiIm}\, dV\right)\left(\int_V B_{jjJn}dV\right)\right]\Delta d_{Jn} \tag{9.65}$$

Cancelling the arbitrary variations $\delta(\Delta d_{Im})$ in Eq. (9.65) produces the element stiffness equations

$$\Delta f_{Im} = (K^{\epsilon}_{ImJn} + K^{\sigma}_{ImJn} + K^{\phi}_{ImJn})\,\Delta d_{Jn} \tag{9.66}$$

where

$$K^{\epsilon}_{ImJn} = \int_V B_{ijIm}\, D_{ijkl}\, B_{klJn}\, dV \tag{9.67}$$

is the element *deformation stiffness matrix*

$$K^{\sigma}_{ImJn} = \int_V (N_{I;i}\delta_{mn}\sigma_{ik}N_{J;k} - 2B_{ijIm}\delta_{jl}\sigma_{ik}B_{klJn})\, dV \tag{9.68}$$

is the element *stress-increment correction matrix* and

$$K^{\phi}_{ImJn} = \kappa\left[\frac{1}{V}\left(\int_V B_{iiIm}\, dV\right)\left(\int_V B_{jjJn}\, dV\right) - \int_V B_{iiIm}B_{jjJn}\, dV\right] \tag{9.69}$$

is the element *dilatation correction matrix*.

9.4.9 Calculation of Global Stiffness Matrix

The volume integrals in Eqs (9.67)–(9.69) are evaluated by numerical integration. This is usually performed in the local element coordinate system after multiplying all the integrands by the determinant of the Jacobian mapping of local to reference Cartesian coordinates. Integration in local element coordinates considerably simplifies the calculation procedure since, as has been seen in Sect. 9.3, the interpolation functions are usually expressed in terms of local element coordinates of position. The integrands in Eqs (9.67)–(9.69) involve derivatives of the shape functions with respect to reference Cartesian coordinates, but these may easily be evaluated at the sample points of the numerical quadrature using the chain rule.

9.5 FE Solution

Once the stiffness matrices have been calculated for all the elements, these may be assembled into the global matrix **K** of Eq. (9.17). This may then be solved to obtain the incremental nodal displacements corresponding to the applied incremental nodal forces, subject to the prevailing boundary conditions determined by the contact between the workpiece and the tools. Since the assembly and solution of the global stiffness matrix, and the imposition of frictional contact, are standard procedures common to many different types of FE approach, details will not be given here. For more information see Rowe et al. (1991).

9.5.1 Elastic-Plastic Transition

One detail of the solution of the elastic-plastic stiffness equations however is worth noting. It is a characteristic of the elastic-plastic constitutive relationship

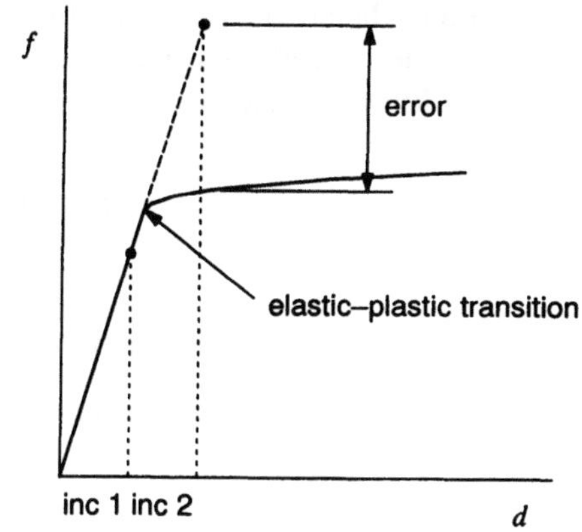

a

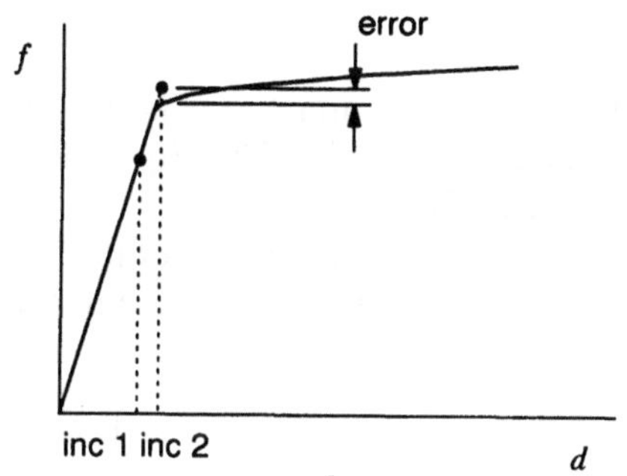

b

Fig. 9.3. Schematic representation of the effect of overestimating the element stiffness during elastic-plastic transition increments: **a** fixed increment size; **b** increment size adjusted to cause the element to yield just before the end of an increment.

that there is a sharp change in the effective modulus when the material begins to yield. Typically, the coefficients of D_{ijkl} will be found to decrease by several orders of magnitude when the material passes through the elastic-plastic transition. As a result, the stiffness of an element which is elastic at the start of an increment but which yields part way through the increment will be far too large when the element is supposed to be deforming plastically. To counteract this problem, it is prudent to have the FE program check for elements in this category and to have it reduce the size of increments in which yielding occurs (Yamada et al. 1968). While in theory it would be possible to adjust the increment size so that the elastic-plastic transition always occurs exactly at the end of an increment, in practice this would result in an unacceptably large number of tiny increments being performed as each of possibly several thousand elements goes through yield. Instead, it is more practical to accept a small amount of error and adjust the increment size so that elements yield at, or just before, the end of an increment (Fig. 9.3).

9.5.2 Calculation of Incremental Strain

At any point, the incremental displacement gradient may be calculated quite simply from the incremental nodal displacement using Eq. (9.61).

The calculation of incremental strain is also quite straightforward since it is not necessary to adopt the rather complicated LCR expression used earlier. It will be recalled that the expression for strain increment used in the stiffness formulation had to be a linear function of the incremental displacement gradients, but this is not a requirement once the displacement gradients are known. Instead, the strain increment is calculated to be

$$\Delta\epsilon_{ij} = \tfrac{1}{2}(x_{k;i}\, x_{k;j} - \delta_{ij}) \tag{9.70}$$

where

$$x_{i;j} = \Delta u_{i;j} + \delta_{ij} \tag{9.71}$$

From the decomposition of the deformation gradient into rotational and symmetric parts given by Eq. (9.31)

$$\begin{aligned} x_{k;i}\, x_{k;j} &= r_{km}\, q_{mi}\, r_{kn}\, q_{nj} \\ &= \delta_{mn}\, q_{mi}\, q_{nj} \\ &= q_{mi}\, q_{mj} \end{aligned} \tag{9.72}$$

since r_{ij} is an orthogonal matrix, i.e. $\mathbf{r}^{\mathrm{T}}\,\mathbf{r} = \mathbf{I}$. The symmetric matrix q_{ij} is simply the stretch tensor **U** defined in Chap. 2. The strain increment defined by Eq. (9.70) therefore corresponds to the Green strain described in Sect. 2.9.3.

In addition to the strain increment, it is also necessary to evaluate the rotational matrix r_{ij}, since this is required for the stiffness equations of the next increment. From Eq. (9.31)

$$r_{ij} = x_{i;k}\, q_{kj}^{-1} \tag{9.73}$$

From Eq. (9.72), q_{ij} is the square root of the matrix

$$h_{ij} = q_{mi}\, q_{mj} \tag{9.74}$$

The square root is most easily calculated using Newton's method i.e. the nth estimate of q_{ij} is given by

$$q_{ij}^{(n)} = \tfrac{1}{2}(q_{ij}^{(n-1)} + p_{ij}^{(n-1)}) \tag{9.75}$$

where

$$p_{ik}^{(n-1)}\, q_{kj}^{(n-1)} = h_{ij} \tag{9.76}$$

and

$$q_{ij}^{(0)} = \tfrac{1}{2}(h_{ij} + \delta_{ij}) \tag{9.77}$$

Since the components of $\Delta u_{i;j}$ are normally much less than one, the iterative procedure usually converges very rapidly.

9.5.3 Calculation of Incremental Deviatoric Stress

At the start of an increment, the material at a point within an element may either be deforming elastically, or it may be deforming plastically. In the latter case, as has been seen in Chap. 3, its stress state will lie on the initial yield locus. Similarly, at the end of the increment, the material at the deformed point may be deforming elastically or plastically. Fig. 9.4 shows the various possibilities,

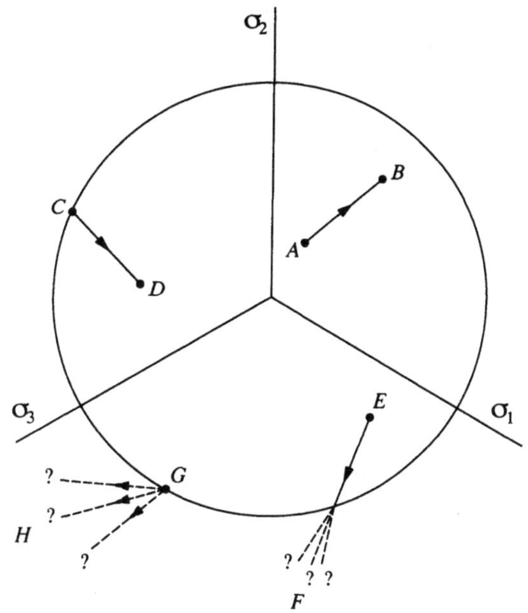

Fig. 9.4. Different types of stress increment represented on the π stress plane.

depicted, with respect to the von Mises yield locus, in the π stress plane. Note, the yield locus at the end of the increment is not shown. Both of the increments AB and CD are purely elastic. Increment EF has an initial elastic part and then an elastic-plastic part after the point in the increment at which the stress first satisfies the yield criterion. Increment GH begins at yield, so the whole increment is elastic-plastic.

It is clear from the above, that the first step in calculating incremental deviatoric stress is to evaluate the *hypothetical elastic* deviatoric stress increment $\Delta\sigma_{ij}^{e\prime}$ from the deviatoric strain increment using Hooke's law by *assuming* that the material does not yield:

$$\Delta\sigma_{ij}^{e\prime} = 2G\,\Delta\,\epsilon_{ij}^{\prime} \tag{9.78}$$

By adding this stress increment to the initial deviatoric stress $\sigma_{ij}^{\prime}$, it is possible to determine whether or not the final stress state lies inside the initial yield locus. If it does, the whole increment was, indeed, elastic and $\Delta\sigma_{ij}^{e\prime}$ is the actual change in deviatoric stress. (There is a complication here if the material undergoes strain, strain-rate or temperature softening, so that the final yield locus is *inside* the initial one. Consideration of this problem is beyond the scope of this chapter.)

If the stress state at the end of the increment lies outside the initial yield locus, then at least part of the increment was elastic-plastic. Consideration of the initial stress state will determine whether or not the material started with a purely elastic deformation. If it did, the calculated hypothetical elastic increment of deviatoric stress can be used to find out what proportion r of the strain increment $\Delta\epsilon_{ij}$ occurred in the elastic regime, and hence the corresponding elastic change in deviatoric stress $r\Delta\sigma_{ij}^{e\prime}$, and the stress state $\sigma_{ij}^{\prime} + r\Delta\sigma_{ij}^{e\prime}$ when the material first started to yield.

The only problem remaining, then, is to calculate the elastic-plastic increment in deviatoric stress corresponding to a deviatoric strain increment of $\Delta\epsilon'_{ij}$ when the material starts at yield with a stress of σ'_{ij}. It can be seen from the above that this statement of the problem includes the situation in which the increment has an initial elastic portion, providing it is understood that in this case, $\Delta\epsilon'_{ij}$ is replaced by $(1-r)\Delta\epsilon'_{ij}$, σ'_{ij} is replaced by $\sigma'_{ij} + r\Delta\sigma^{e\prime}_{ij}$ and $\Delta\sigma^{e\prime}_{ij}$ (when it is used in the following description) is replaced by $(1-r)\Delta\sigma^{e\prime}_{ij}$.

The simplest way to calculate the increment in deviatoric stress during the elastic-plastic part of the increment would be to use the elastic-plastic constitutive expression. However the elastic-plastic constitutive matrix D_{ijkl} (Eq. (9.29)) is derived from the Prandtl–Reuss flow rule which, strictly, is only applicable to infinitesimal increments of strain. If the incremental form of Eq. (9.27) is used to evaluate finite changes in stress, Nagtegaal and de Jong (1980) have shown that the FE solution obtained may be inaccurate or unstable.

The Prandtl–Reuss expression may be written as

$$\Delta\epsilon_{ij} = \frac{1}{E}((1+\nu)\Delta\sigma_{ij} - \nu\,\delta_{ij}\Delta\sigma_{kk}) + \Delta\lambda\,\sigma'_{ij} \tag{9.79}$$

where $\Delta\lambda$ is a constant of proportionality that depends on the strain increment, the state of stress at the start of the increment and the current amount of work hardening. The last term in Eq. (9.79) may be identified with the increment of plastic strain. Since we are concerned at the moment with deviatoric components only,

$$\Delta\epsilon'_{ij} = \frac{1}{2G}\Delta\sigma'_{ij} + \Delta\lambda\,\sigma'_{ij} \tag{9.80}$$

or

$$\begin{aligned}\Delta\sigma'_{ij} &= 2G\Delta\epsilon'_{ij} - 2G\Delta\lambda\,\sigma'_{ij}\\ &= \Delta\sigma^{e\prime}_{ij} - 2G\Delta\lambda\,\sigma'_{ij}\end{aligned} \tag{9.81}$$

Fig. 9.5 illustrates the problem of using the Prandtl–Reuss flow expression for finite increments of strain. For simplicity, it is assumed in this figure that there is no work hardening, i.e. $Y' = 0$. In this case it can easily be shown that the Prandtl–Reuss equations require that $\sigma'_{ij}\Delta\sigma'_{ij} = 0$, so the line AB is perpendicular to the initial stress vector σ'_{ij}. This means that the final stress state B must lie

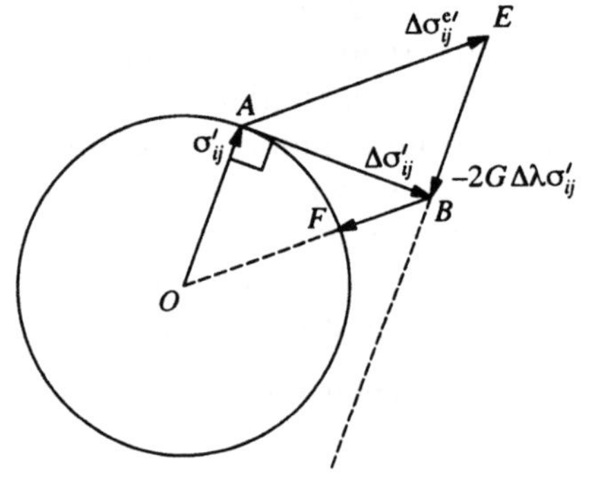

Fig. 9.5. Interpretation of the Prandtl–Reuss flow rule for finite increments of stress.

outside the yield locus, which is prohibited by the yield criterion. The larger the increment size, the worse this problem will be.

Extending line EB until it reaches the yield locus is not a solution, as it can be seen from the diagram that it is quite possible that EB produced does not intersect with the initial yield locus. Applying a correction to the stress state B to make the final stress state correspond to the nearest point (F) on the yield locus is a partial solution, but can introduce large errors into the FE model.

To find a better solution, it is necessary to look at the vector defined by the second term on the right-hand side of Eq. (9.81) in a little more detail. As can be seen from this equation, and from Fig. 9.5, this vector is parallel to the *initial* stress vector σ'_{ij}. This comes about because Prandtl–Reuss flow requires that the plastic strain rate is normal to the von Mises yield locus at the current stress state. For infinitesimal increments there is no ambiguity about this stress state, but for finite increments, a choice has to be made with regard to the stress state used: the one at the start of the increment, the one at the end, or some intermediate state.

As has been seen, the simple approach that uses normality at the start of the increment can lead to instability. In Sect. 3.8, a technique was described in which the hypothetical elastic increment of stress is added to the initial value and then the final stress state is evaluated as the nearest point on the yield locus, the so-called *radial-return* method. This is equivalent to assuming that the plastic strain increment is normal to the yield locus at the final stress state.

An alternative method, the *mean-normal* (Rice and Tracey 1973, Tracey 1976) makes the assumption that the plastic strain increment is parallel to the stress half way through the increment (with no work hardening this would mean that the plastic strain increment would be normal to the yield locus half way through the step, hence the name of the technique).

The value of stress half way through the increment will not actually be known until the stress at the end of the increment has been calculated. Fortunately, this does not matter because, as can be seen from Fig. 9.6, the mid-increment stress vector OC is parallel to the line OD, where D is the point in stress space corresponding to the stress:

$$\hat{\sigma}'_{ij} = \sigma'_{ij} + \tfrac{1}{2}\,\Delta\sigma^{e\prime}_{ij} \tag{9.82}$$

which *is* known since the hypothetical elastic increment of stress $\Delta\sigma^{e\prime}_{ij}$ has already been calculated.

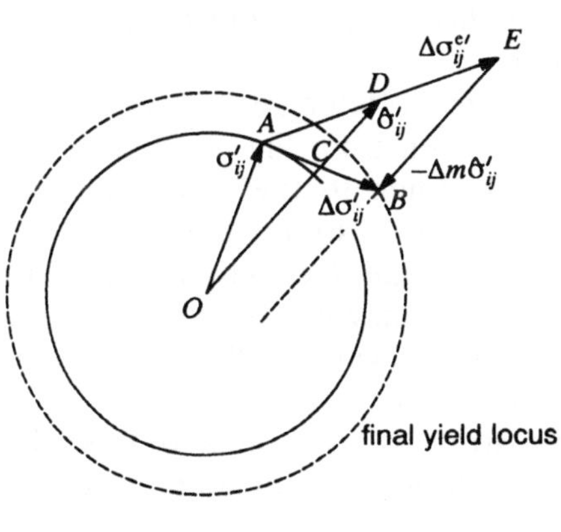

Fig. 9.6. Illustration of the mean-normal method for determining the change in stress during an elastic-plastic increment.

A modified form of the Prandtl–Reuss equations is therefore used:

$$\Delta\sigma'_{ij} = \Delta\sigma^{e\prime}_{ij} - \Delta m\, \hat{\sigma}'_{ij} \tag{9.83}$$

Δm is another proportionality factor which needs to be evaluated in order to determine the deviatoric stress increment. A value of Δm is chosen which makes the stress at the end of the increment satisfy yield. The von Mises yield criterion may be written as

$$\tfrac{3}{2}\sigma'_{ij}\sigma'_{ij} = Y^2(\bar{\epsilon}^{\mathrm{p}}, \dot{\bar{\epsilon}}, T) \tag{9.84}$$

in which Y is a function of generalized plastic strain $\bar{\epsilon}^{\mathrm{p}}$, plastic strain rate $\dot{\bar{\epsilon}}^{\mathrm{p}}$ and temperature T. The stress $\sigma'_{ij} + \Delta\sigma'_{ij}$ at the end of the increment must satisfy the yield criterion for the value of the yield function Y corresponding to a plastic strain of $\bar{\epsilon}^{\mathrm{p}} + \Delta\bar{\epsilon}^{\mathrm{p}}$, plastic strain rate $\dot{\bar{\epsilon}}^{\mathrm{p}}$ and final temperature T^{f} where

$$\Delta\bar{\epsilon}^{\mathrm{p}} = (\tfrac{2}{3}\Delta\epsilon^{\mathrm{p}}_{ij}\Delta\epsilon^{\mathrm{p}}_{ij})^{1/2} \tag{9.85}$$

and

$$\Delta\epsilon^{\mathrm{p}}_{ij} = \frac{\Delta m}{2G}\hat{\sigma}'_{ij} \tag{9.86}$$

from the definition of the modified Prandtl–Reuss equations.

From Eqs (9.83)–(9.86), the yield condition at the end of the increment is

$$P(\Delta m)^2 + Q\,\Delta m + R = Y^2\left(\bar{\epsilon}^{\mathrm{p}} + \frac{\sqrt{P}}{2G}\Delta m, \dot{\bar{\epsilon}}^{\mathrm{p}}, T^{\mathrm{f}}\right) \tag{9.87}$$

where

$$P = \tfrac{3}{2}\hat{\sigma}'_{ij}\hat{\sigma}'_{ij} \tag{9.88}$$

$$Q = -3\,\hat{\sigma}'_{ij}(\sigma'_{ij} + \Delta\sigma^{e\prime}_{ij}) \tag{9.89}$$

$$R = \tfrac{3}{2}(\sigma'_{ij} + \Delta\sigma^{e\prime}_{ij})(\sigma'_{ij} + \Delta\sigma^{e\prime}_{ij}) \tag{9.90}$$

The solution of Eq. (9.87) (i.e. the value of Δm at the intersection of the curves in Fig. 9.7) is carried out iteratively and is usually rapidly convergent. The iteration is computationally efficient since the coefficients P, Q and R need only to be evaluated once at the start of the iteration.

It should be noted that while strain or temperature softening is not specifically prohibited by the mean-normal method, it can be seen from Fig. 9.6 that if the yield locus at the end of the increment (corresponding to the final values of plastic strain and temperature and the average value of strain rate during the increment) is much smaller than the initial locus, the line through E parallel to OD may not intersect with the yield locus, and there may not be a solution for Δm. This is not a problem with the radial-return method which will always obtain a solution whatever material hardening or softening is exhibited.

Whichever method of calculating deviatoric stress is used, this will be the change in stress in the corotational reference frame. Moreover, if material rotation

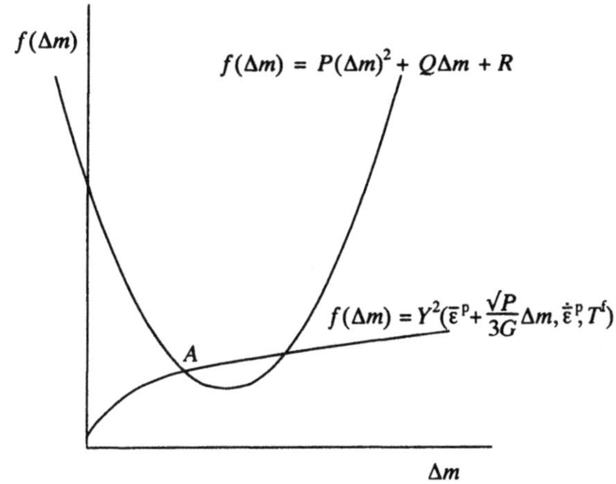

Fig. 9.7. Method of evaluating the proportionality factor Δm.

is present, the Cartesian reference components of stress will change even if the strain increment is zero, simply as a result of the usual rules for evaluating the components of a tensor in different reference frames. The final reference components of stress $\sigma_{ij}^{f\prime}$ at the end of the increment are therefore given by

$$\sigma_{ij}^{f\prime} = (\sigma'_{mn} + \Delta\sigma'_{mn})\, r_{mi}\, r_{nj} \tag{9.91}$$

9.5.4 Calculation of Incremental Hydrostatic Stress

A great deal of space has been devoted to the calculation of changes in deviatoric stress, and indeed, from the point of view of plastic behaviour, the deviatoric components of stress are the most important. However, in order for contact forces to be calculated, the *total* components of stress must also be known. Hence it is necesssary to evaluate the hydrostatic stress.

Since it is assumed as part of the elastic-plastic approach that plastic flow is incompressible, it follows that volume strain is purely elastic. In principle therefore, it should be possible to evaluate incremental hydrostatic stress by multiplying the incremental volume strain $\Delta\epsilon_{\kappa\kappa}$ by $\kappa/3$.

In practice, this approach usually leads to grossly inaccurate estimates of hydrostatic stress. This is because, in most situations, the elastic modulus tends to be several orders of magnitude greater than the plastic one. Small and unavoidable errors in incremental volume strain therefore lead to very much larger errors in incremental hydrostatic stress which can swamp the deviatoric components.

One solution to this problem is to obtain values of total (not incremental) hydrostatic stress which satisfy the principles of equilibrium for the current distribution of deviatoric stress (Alexander and Price 1977):

$$\frac{\delta\,\sigma'_{kk}}{\delta X_i} = -3\,\frac{\delta\,\sigma'_{ij}}{\delta X_j} \tag{9.92}$$

The distribution of deviatoric stress is known throughout the workpiece. Integration of Eq. (9.92) along a given line therefore allows the change in hydrostatic stress to be calculated between its end points. Since the hydrostatic stress at any free surface is known to be equal to the negative of the deviatoric stress normal to that surface, a number of starting points can usually be found for the integration, so that the hydrostatic stress can be determined anywhere within the workpiece. This procedure, of course, is not suitable for processes in which there may not be any free surface at which to start the calculation, such as completely closed-die forging. Another problem with this method is that the spatial gradients of deviatoric stress in Eq. (9.92) have to be obtained by numerical differentiation and so are very prone to numerical error. It is usually advisable, therefore, to apply some sort of smoothing to the raw deviatoric stress values before using this method of calculating hydrostatic stress.

9.6 Examples

A range of examples of the results of elastic-plastic FE analysis will be described in later chapters, but it is useful at this stage to briefly show the results of the modelling of two very different processes which illustrate the principal advantages of the elastic-plastic technique.

9.6.1 Multiple Cropping

Cropping is a process which is widely used in industry when moderately thin stock needs to be divided up into sections prior to some other operation, such as forging. The shape of the cropped edge is of great significance in these situations. In order to minimize material wastage and to allow subsequent forming operations to be performed accurately, the cropped face should, ideally, be flat and at right angles to the main axis of the workpiece. In practice, the cropped face is rarely flat or perpendicular to the workpiece axis, and the study of the parameters that affect the quality of the cropped face is clearly of great importance. The main factors influencing the nature of the cropped face are the way fracture is initiated and propagated across the workpiece during the cropping operation (dependent on material properties and the cropping geometry, e.g. die clearance) and, particularly for thin materials, elastic spring-back after the cropping is finished. The elastic-plastic FE method is therefore an ideal way of investigating the cropping process.

Various criteria for predicting when and where fracture initiates during forming processes have been studied by Hartley et al. (1989) using values of workpiece variables calculated by an elastic-plastic FE program, and excellent correlation with experiment was obtained with the generalized plastic work criterion.

Fig. 9.8a shows a section of the multiple cropping process in which a sheet is simultaneously cropped into a number of similar strips. Because of the symmetry involved, it is only necessary to model in the FE analysis one section of the sheet, and one of the FE meshes used is illustrated in Fig. 9.8b together with the tool surfaces used to impose the deformation.

The FE analysis modelled the cropping of 2.5 mm thick Duraloy 2014 aluminium alloy sheet under plane-strain conditions. Analyses were performed for two different heat treatments of this alloy. With the first treatment, the alloy was maintained at a temperature of 500°C for 2 h, quenched in water and then aged

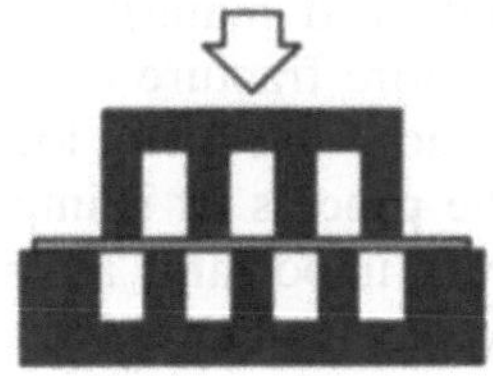

a

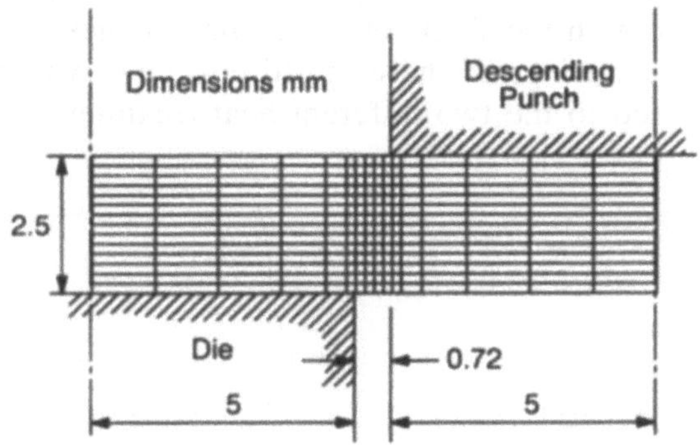

b

Fig. 9.8. **a** Schematic representation of the multiple-cropping process; **b** typical initial FE mesh used for the analysis of the process.

at room temperature for 4 days. The second treatment was a process of annealing. Fig. 9.9 shows the deformed FE meshes for the two treatments of the material at the stages in the deformation when the cropping fracture was observed to occur in the two cases. As expected, the annealed material was much more

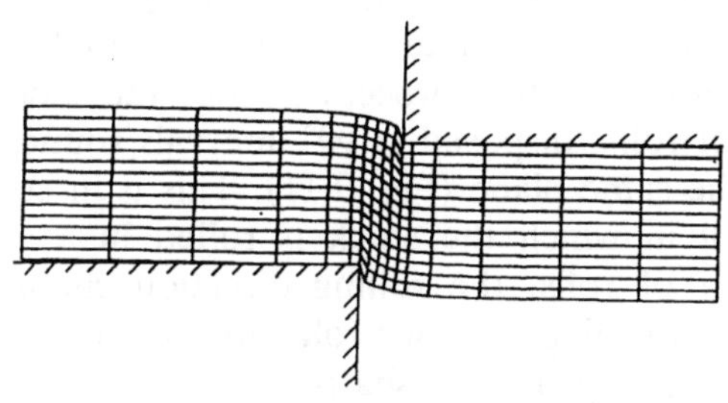

a

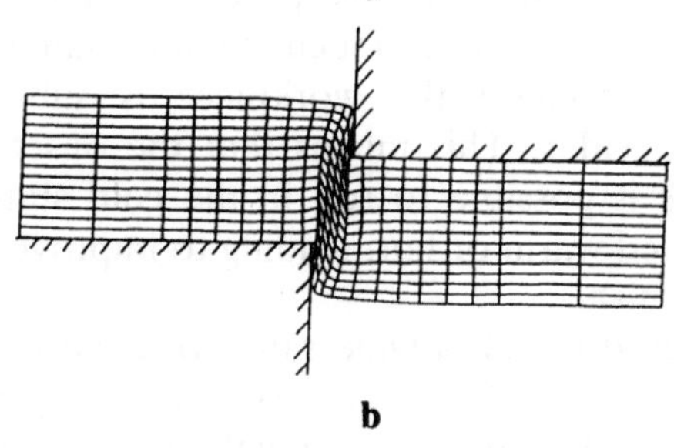

b

Fig. 9.9. Deformed FE meshes just before fracture occurs during the multiple-cropping process: **a** quenched and aged aluminium alloy; **b** annealed aluminium alloy.

ductile than the quenched and aged material, and underwent more extensive deformation during cropping before fracture occurred.

In both cases, the deformation was very localized and most of the sheet remained elastic throughout the process. It is important to note though, that the elastic deformation was still very important, as can be seen from the separation between the die and the workpiece in Fig. 9.9a.

Although the elastic-plastic program used for this analysis was not capable of simulating the propagation of the fracture across the sheet, it correctly predicted the site of fracture initiation observed experimentally. Moreover, the deformation pattern was consistent with the shape of the cropped edges, as can be seen from Fig. 9.10 which shows tracings of the cropped edges of experimental specimens that had been subjected to the two different heat treatments.

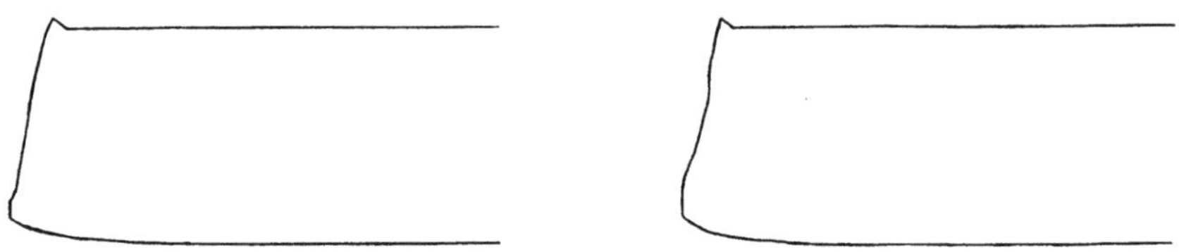

a **b**

Fig. 9.10. Edge profile of cropped experimental specimens: **a** quenched and aged aluminium alloy; **b** annealed aluminium alloy.

9.6.2 Four-Roll Pass Rolling

Rolling is another process in which elastic unloading can be very important in determining product quality, including final dimensions and residual stress distribution.

Four-roll pass rolling is a relatively new type of rolling process developed at the Anshan Institute of Iron and Steel Technology, People's Republic of China. The distinguishing feature of this process is that the stock material is rolled simultaneously by four rolls, disposed at 90° intervals round the workpiece, with all four roll axes in the same plane (Fig. 9.11). This is in contrast to other types of section rolling mill in which the stock is rolled in a series of passes each consisting of one pair of rolls, the rolling direction changing by 90° between passes. In four-roll pass rolling, all four rolls are driven from a single motor by means of a gear train linking the roll shafts.

The great advantage of the four-roll pass process is that the level of compressive hydrostatic stress in the roll gap is much higher than would be the case in traditional rolling mills because the workpiece is subjected to compressive deformation on all four sides. This means that the onset of ductile fracture is inhibited and so makes it possible to roll in the cold state sections that would normally have to be rolled warm or hot, leading to improved product dimensional tolerance and surface finish.

The process can be used to roll a wide range of continuous sections. Fig. 9.12 shows the central portion of a four-roll pass mill set up to roll square section from round mild-steel stock, with the original stock section superimposed. An elastic-plastic analysis of this process was carried out using the initial mesh in

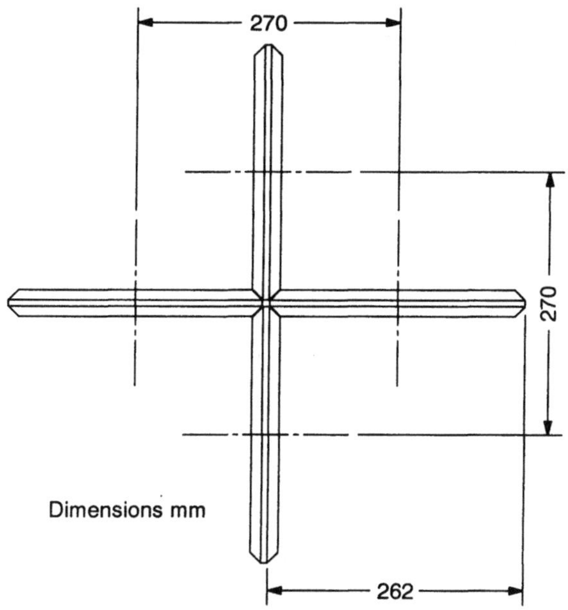

Fig. 9.11. Vertical section through a typical four-roll pass mill.

Fig. 9.13a. Due to the symmetry of the process, only one quarter of the workpiece was modelled. Fig. 9.13b shows the deformed mesh when steady-state conditions had been reached. The shading indicates the plastically deforming zone.

Fig. 9.14 shows more detailed information about the state of the workpiece on a longitudinal section through its axis. Fig. 9.14a indicates that the unloading of

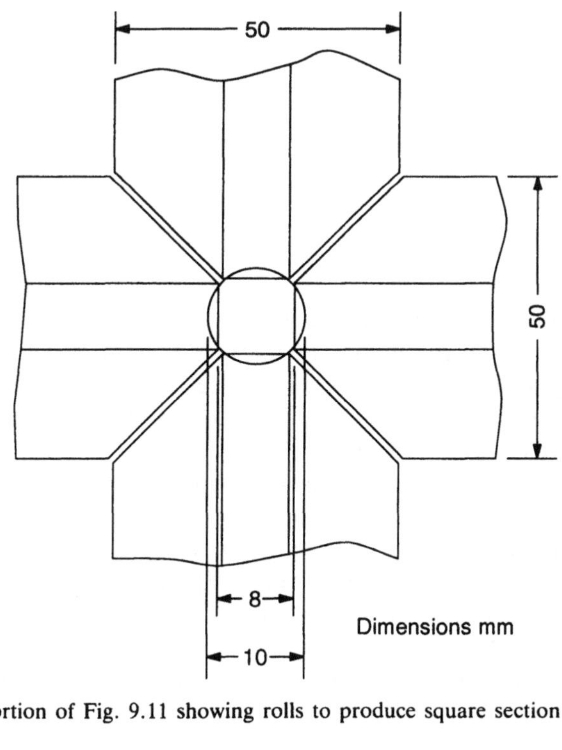

Fig. 9.12. Central portion of Fig. 9.11 showing rolls to produce square section from round stock.

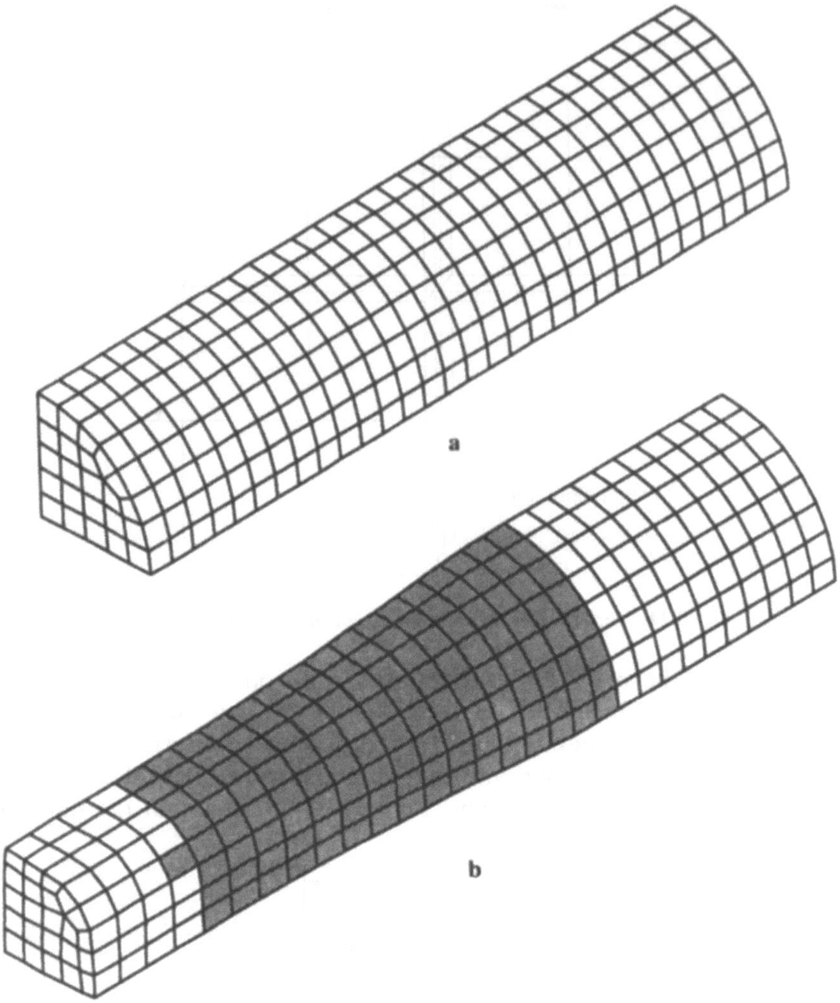

Fig. 9.13. **a** initial mesh used for the elastic-plastic FE analysis of the four-roll pass rolling of square section from round stock; **b** deformed mesh when steady-state conditions had been reached showing the plastically deforming region (shaded).

the specimen after exit from the roll gap begins at the centre and progresses towards the outside. The generalized stress contours in Fig. 9.14b show the process of unloading in more detail. The average generalized stress in the section after rolling, while well below yield, can be seen to be of the order of 100 to 200 MN m^{-2}. The generalized stress is considerably higher at the leading edges of the workpiece at the top and on the axis, but these effects are very localized and reflect the nature of the deformation before steady state is reached.

The variation of the vertical component of stress across the longitudinal section is illustrated in Fig. 9.14c. It can be seen that whereas this stress is compressive throughout most of the thickness of the workpiece when it is in the roll gap, after unloading the stress is mainly tensile. One interesting detail is that the maximum compressive vertical stress at the surface in contact with the rolls is at the entrance and at the exit of the roll gap. In between these two positions, the vertical stress is somewhat less compressive.

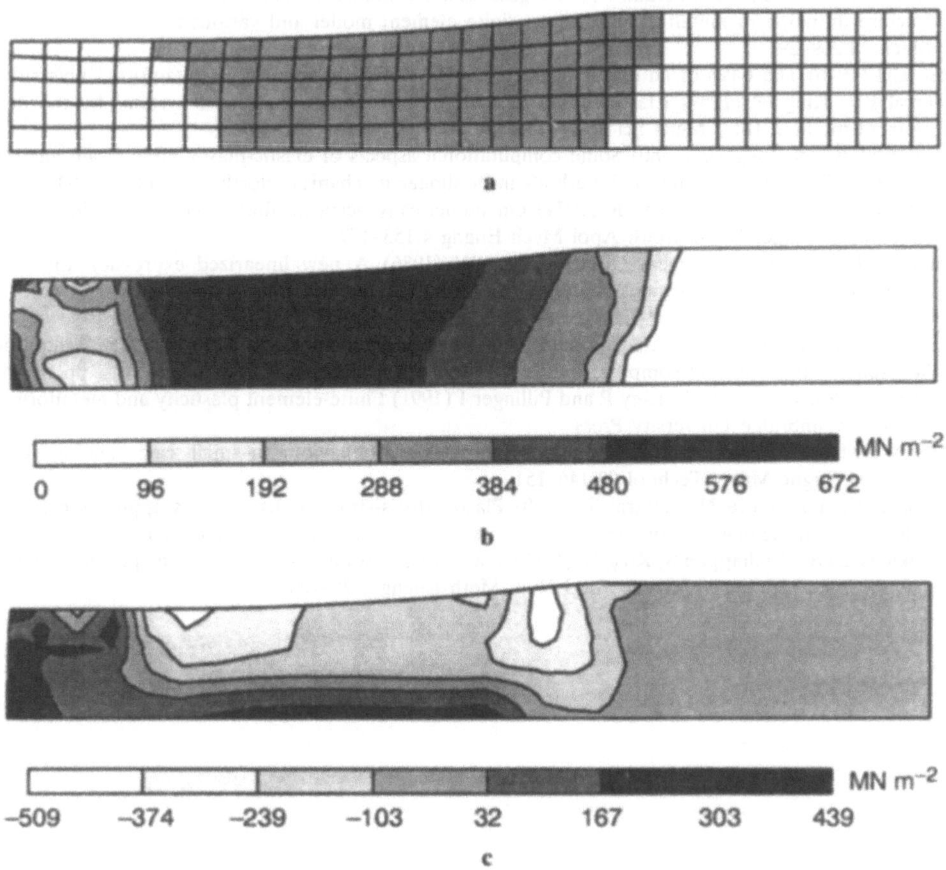

Fig. 9.14. Longitudinal centre-line sections through the deformed workpiece during the rolling of square section from round stock: **a** plastically deforming region (shaded); **b** distribution of generalized stress; **c** distribution of the normal, vertical component of stress.

9.7 Conclusions

This chapter has described the theoretical formulation for an elastic-plastic FE analysis of metal forming and has discussed some topics of the FE calculation that are of particular relevance to the elastic-plastic approach. Two metal-forming examples have been described which illustrate the main advantages of the elastic-plastic technique, namely its ability to consider deformation in the elastic region of the deforming workpiece and its ability to predict workpiece behaviour during unloading.

Acknowledgements. I should like to thank Mr Wen Shuwen for providing the results of his anlaysis of the four-roll pass rolling process.

References

Alexander JM, Price JWH (1977) Finite element analysis of hot metal forming. In: Alexander JM (ed) Proceedings 18th machine tool design research conference, Macmillan, pp 267–274

Hartley P, Clift SE, Salimi-Namin J, Sturgess CEN, Pillinger I (1989) The prediction of ductile fracture initiation in metalforming using a finite element model and various fracture criteria. Res Mech 28:269–293

Lee EH (1976) The basis of an elastic-plastic code. SUDAM report no 76-1, Stanford University

Marcal PV, King IP (1967) Elastic-plastic analysis of two-dimensional stress systems by the finite element method. Int J Mech Sci 9:143–155

Nagtegaal JC, de Jong JE (1980) Some computational aspects of elastic-plastic large strain analysis. In: Oden JT (ed) Computational methods in nonlinear mechanics, North-Holland, pp 303–339

Nagtegaal JC, Parks DM, Rice JR (1974) On numerically accurate finite element solutions in the fully plastic range. Comp Meth Appl Mech Engng 4:153–177

Pillinger I, Hartley P, Sturgess CEN, Rowe GW (1986) A new linearized expression for strain increment for the finite-element analysis of deformations involving finite rotation. Int J Mech Sci 28:253–262

Rice JR and Tracey DM (1973) Computational fracture mechanics. In: Robinson AR, Schnobrich WC (eds) Numerical and computer methods in structural mechanics, Academic Press, pp 585–623

Rowe GW, Sturgess CEN, Hartley P and Pillinger I (1991) Finite-element plasticity and metalforming analysis. Cambridge University Press

Tracey DM (1976) Finite element solutions for crack tip behaviour in small-scale yielding. Trans ASME J Engng Mater Technol 98:146–151

Yamada Y, Yoshimura N, Sakurai T (1968) Plastic stress-strain matrix and its application for the solution of elastic-plastic problems by the finite element method. Int J Mech Sci 10:343–354

Zienkiewicz OC, Valliappan S, King IP (1969) Elasto-plastic solutions of engineering problems "initial stress", finite element approach. Int J Num Meth Engng 1:75–100

Part III
Specialized Applications and Topics

10 Numerical Modelling of Forging

P. Hartley, F.R. Hall and S.M. Roberts

10.1 Introduction

Numerical modelling of forging combines the study of an ancient manual craft with the use of modern computer technology and mathematical modelling techniques. The hand forging craft did not become a major industrial process until the development of Nasmyth's steam hammer in 1840 allowed the production of heavy hot forgings on a regular basis. Since this date, advances in forging equipment have improved the efficiency and industrial viability of the process. Current attention is focused on the structure and quality of the forged product, and this is one area where mathematical modelling can make a real contribution.

Forging is often used as a major method of shaping large castings, to reduce the cross-section of the ingot or to improve the metallurgical structure and mechanical properties. Forging is also a means of producing discrete components in which the flow and grain structure can be controlled so that the forged components have good strength and toughness. A wide variety of workpiece shapes and sizes are processed in forging, ranging from 300 t steam turbine rotors, to aircraft landing gear, engine discs, gears, connecting rods, bolts and rivets. Recent data in the UK estimate that about 25% of all forgings are for the aerospace and automotive industries.

10.1.1 Forging Processes

The term forging covers a wide variety of metal-forming processes; these include open-die forging, cogging, closed-die forging, heading, upsetting, indenting, coining, flashless and near-net-shape forging. Other innovations in this field also include roll forging, rotary forging and isothermal forming. Most of the terms relate to the form of tool design or to the prescribed movement of the forming tool, although the latter refers to how the forming temperature is controlled. Processes may also be classified as hot forging, cold forging or, for intermediate temperatures, warm forging. Useful descriptions of forging processes can be found in Rowe (1977), De Garmo et al. (1984) and Kalpakjian (1989).

Open-die forging is the simplest process in which a workpiece is compressed between flat platens as in cogging or upsetting. Cogging usually involves an incremental process, whereas upsetting is completed in one operation. There may in some cases be a simple shaped cavity in the platens. In closed-die forging (sometimes called impression-die forging) the tools will contain more complex die cavities into which the material is formed. Some material will, however, flow outwards in the peripheral gap surrounding the workpiece, between the two dies, called the flash regions. The material in this region can play a significant role in providing a back-pressure as a resistance to flow. This helps to force the material into the die cavities. This final filling of the cavities is often the most difficult stage as high pressures are encountered and die failure is a danger.

Many forged components are produced in multistage processes where the initial sizing stages are open-die operations and the later finishing operations are of a closed-die nature. Forgings produced in this way will then need the flash removed in a further trimming operation. Machining operations are also common for finishing of the product. Both these post-forging operations are wasteful of material and time. In an attempt to eliminate flash and minimize subsequent machining, flashless forging has been developed (also called completely closed-die forging). In this process the workpiece is completely enclosed within the dies and no flash is formed. Accurate control of the billet volume and careful die design is essential. A billet with a low volume will result in under filling of the cavities whereas too high a volume will induce very high pressures, possible leading to die failure.

Precision forging (alternatively called near-net-shape or net-shape forging) is a further development which aims to minimize, and where possible, eliminate completely, any post-forging machining operations. Higher forces are necessary to obtain the precise tolerances required. Aluminium and magnesium alloys can be produced successfully with this process but for steels it is more difficult and special dies are required.

The open-die forging operation, often used as the first stage in a multistage process, is sometimes replaced with roll forging in which the cross-section of a bar is reduced or shaped by moving it into two or four shaped rolls. This is usually a quicker method of shaping the first preform and involves less costly equipment.

Rotary forging, and the closely related orbital forging, involves the incremental deformation of the workpiece. As the forging tool moves down compressing the workpiece it is also rotating. The centre of rotation may be fixed or may itself be moving to obtain a predetermined tool motion suitable for the component being produced. The process is particularly suitable for disc-shaped or conical parts. As only a small part of the workpiece is deformed at any time the maximum load can be greatly reduced although the total energy consumption per component is little affected.

10.1.2 Cold, Hot and Warm Forging

The dominant phenomenon of plastic deformation that occurs in metal forming is basically a process of shear in which atoms or planes of atoms slide over one another without any change in volume. It is strongly dependent on the rate of deformation and the temperature at which it occurs. Forging processes are often classified as either cold forging or hot forging. The definitions are very imprecise but cold working is said to occur where the temperature of deformation is below the recrystallization temperature, up to a maximum of approximately 0.3 T_m, where T_m is the melting temperature (in K). Hot forging occurs above the recrystallization temperature, where the working temperature is above 0.6 T_m. The intermediate temperature range is usually classed as warm working. As recrystallization is a kinetically activated process and a function of temperature, time at which the temperature is maintained, material composition and deformation, the boundaries between one regime and another cannot be clearly defined. As a very rough indication the recrystallization temperature for titanium is about 1200 K, for low carbon steel about 900 K and for some aluminium alloys is about 500 K (see Cottrell 1967 for a discussion on recrystallization).

Good dimensional accuracy and surface finish can be maintained in cold forging but higher forces than in hot forging are required (see Feldmann 1961 on cold forging). Deformation at high temperature will increase material ductility and hence reduce the dangers of cracking, but it also gives rise to problems of surface oxidation, and the grain structure is more difficult to control as it is dependent on a complex deformation and temperature history. Forging dies can be subject to thermal shock, particularly during initial use before a stable temperature has been reached. Chilling effects of a cool die in contact with a hot workpiece can be a problem where the cooled parts of the workpiece will be less ductile than those areas which are further away from the dies.

The development of warm forging and also isothermal forging processes has been encouraged as a means of reducing these problems. Warm forging is an attempt to take advantage of the more useful features of both hot and cold forging. The higher temperature compared to cold forging means increased material ductility and reduced loads. Compared to hot forging the process requires less energy per component, less scaling is evident and better dimensional control is obtained.

Isothermal forging (sometimes called hot-die forging) is a process in which the dies are heated to the same temperature as the workpiece. The temperature is controlled very carefully and the process is often performed in a controlled atmosphere. As the workpiece stays at a high temperature throughout the process it maintains a low strength and good ductility. This enables good dimensional accuracy to be achieved. The process has found useful application in the forging of titanium and superalloys.

10.1.3 Lubrication in Forging

An improvement in lubricant behaviour to reduce energy requirements due to friction is important. Of equal importance is the use of a lubricant as a parting agent to minimize, and eliminate if possible, pick-up of material from the workpiece onto the dies. Minimizing material pick-up will help to control the dimensional accuracy of the workpiece and extend the die life.

Variations in frictional stress at the interface of the die and workpiece will have a direct influence on the metal flow and homogeneity within the workpiece. In the simple example of cylinder upsetting, high interfacial friction results in greater free surface barrelling and hence the development of secondary tensile stresses on the circumference.

A major problem with the use of lubricants is in ensuring that they remain in place and function as desired. Lubrication thinning and breakdown is a common occurrence. It is difficult to give the precise cause of lubricant breakdown as the relative importance of different parameters varies from one process to another. It is also undesirable to maintain a lubricant which is too thick as a matt surface may result.

In cold forging, mineral oils and soaps such as sodium stearate, applied to the workpiece, are common lubricants which can usually result in friction coefficients of less than 0.1. For hot forging the use of graphite or molybdenum disulphide, applied to the dies, is more common, resulting in friction coefficients in the range 0.1 and 0.2. Lubricants are often still characterized by the coefficient of friction, although at the high pressures associated with metal forming this is not an appropriate measure. Of more relevance is the interface shear factor m, from the relationship, $\tau = mk$, where τ is the surface shear stress, k is the shear yield stress of the adjacent workpiece material.

At very high temperatures, say above 1300 K, compacted glass powder can be an effective lubricant and is often used in isothermal forming of titanium. In nonisothermal forming it is possible to use glass cloth as a thermal barrier to minimize thermal effects of die contact.

10.1.4 Finite-Element Models

The preceding sections have provided a very brief overview of forging and commented on some of the main problem areas of the process. Computer simulation of forging is a technique which involves the modelling of an infinite variety of shapes and materials under continually changing conditions of temperature and deformation rate. The structure of the materials themselves will change as the forging conditions change. Many of the existing finite-element codes must of necessity make a number of simplifying assumptions, but the development of more sophisticated and rigorous codes is continuing. Frictional boundary conditions in particular present a major problem. It is possible to incorporate friction models that give a reasonable performance where the lubricant is always present but there is still much work to be done to develop friction models that will allow for local variations in a lubricant behaviour due perhaps to lubricant breakdown, or to allow for changing thermal properties of a lubricant throughout a process.

10.2 Finite-Element Plasticity

This chapter is concerned with the application of a finite-strain elastic-plastic finite-element method to simulate forging processes. The formulation for the technique used here has been described in detail in Chap. 9, and further information is given in Rowe et al. (1991). The program, epfep3, is suitable for fully three-dimensional thermomechanical analyses but for the purposes of this chapter the examples are restricted to cold-forging analyses in plane strain. These will be described later.

As with the majority of metal-forming finite-element analyses, the first step is to represent the workpiece with a geometric model consisting of an assemblage of finite elements. Equations relating the distribution of forces and displacements of the model are established (these are the stiffness equations) and the boundary conditions, die movement and friction for example, are imposed. Due to the nonlinear behaviour of the material and the nonlinear variation of the geometric and boundary states, the stiffness equations must be solved in a series of small linear steps or increments.

The incremental stiffness equations for the elastic-plastic technique used here, described in Eqs (9.17) and (9.66), are solved by Gaussian elimination and back-substitution. In order to improve the accuracy of the solution, a second-order Runge–Kutta technique is adopted in which the finite-element equations are based upon the state variables half-way through the increment. These mid-increment values are first estimated by a preliminary solution based upon the state of the workpiece at the start of the increment. The full increment is then repeated using the new data obtained from the half-way stage of the initial calculation. The boundary conditions for the solution of the finite-element equations are obtained by the program automatically detecting any contact between external nodes of the mesh and the moving, or stationary, die surfaces.

In this program the die surfaces are modelled as sets of rigid primitive boundary surfaces, which may be made to rotate or translate by any required amount during a given increment. The primitive surfaces available at present are rectangular or annular planes, cones and cylinders, spheres and toroids. Die surfaces are represented by assembling whole or partial primitive surfaces in die groups. Each surface or group of surfaces will have associated properties such as friction or thermal conductivity. Any level of friction may be specified for the boundary surfaces. When the program detects that an external workpiece element face has come into contact with a boundary surface it will create, temporarily, a "friction element" attached to the workpiece element. The nodes of this element can be constrained and its properties modified so as to give a level of frictional restraint varying from zero friction for perfect lubrication, to sticking friction for complete lubricant breakdown, with any intermediate level as desired. As mentioned earlier, the appropriate friction law for metal-forming processes is $\tau = mk$, where m is termed the interface shear factor. The friction element properties and hence its stiffness equation is modified by a function of this interface shear factor. In practice, ring-tests are often used to estimate the value of m but it is difficult to determine precisely what this value should be. It becomes extremely difficult when problems such as lubricant breakdown occur.

The material properties required for an elastic-plastic analysis obviously include the elastic coefficients (Young's modulus, Poisson's ratio) in addition to the thermal parameters (conductivity, specific heat) and the yield stress function. The elastic properties are usually assumed to be constant but the thermal properties may be a function of the temperature which varies through the process and the yield stress may also be a function of the temperature as well as of strain and strain rate. Temperature changes are calculated in epfep3 after each mechanical step, although a large number of very small thermal steps is adopted to ensure a stable thermal solution. As the temperature calculations are based on a finite-difference type of approach, the computational overhead for the thermal steps is very small, of the order of 1% or 2% (Pillinger et al. 1987). The change in temperature in an increment is determined from the increase in energy of the element due to the work of plastic deformation, the increase in energy due to frictional work at external boundary faces and the change of energy due to flow of heat through the faces of the element. The present elastic-plastic model assumes isotropic material behaviour, although an anisotropic version is under development (Van Houtte et al. 1989).

10.3 Forging Applications

The numerical modelling of forging involves the representation of many complex phenomena. The examples in this section are selected to represent typical geometric shapes common to open and closed die forging to highlight problems largely associated wtih geometric changes in the workpiece model. To simplify the examples they are all two-dimensional plane strain. With the epfep3 program this means applying appropriate constraints to one layer of the three-dimensional elements to prevent any out-of-plane deformation. All the examples use commercially-pure aluminium as the workpiece model material and are assumed to be cold-forging operations at room temperature. In all cases a reasonably low friction is assumed on all the die surfaces, represented by $m = 0.1$. The simulations are conducted on one half of the workpiece cross-section only due to vertical symmetry about the centre line. The lower die is always assumed to be stationary and the upper die translates down towards the lower die in steps of 0.5% per increment. The movement is expressed as a percentage of the initial undeformed height of the workpiece.

10.3.1 Open-Die Forging

In this first example open-die forging is considered in which a circular bar of 200 mm diameter is compressed between two platens. Both platens are curved as shown in Fig. 10.1 which also illustrates the initial finite-element mesh. This provides a shape typical of turbine blade forging. Each curved die surface is represented by a segment of a rigid cylindrical primitive surface. The distorted meshes for this example are shown in Fig. 10.2 and clearly indicate how material is being compressed and moved away from the vertical centre-line of the workpiece. The displacment vectors in Fig. 10.3 illustrate this more clearly. The displacement vectors indicate the incremental displacement of each nodal point in the mesh. The material is shown to be closely following the curved die profiles.

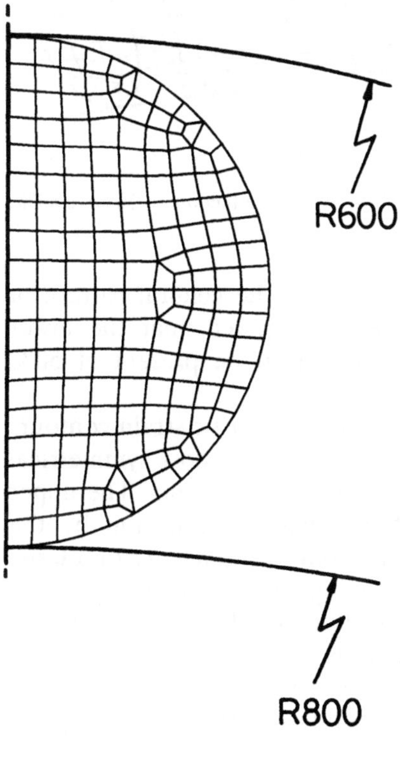

Fig. 10.1. Initial finite-element mesh and die profiles for open-die forging of a circular bar (dimensions in mm).

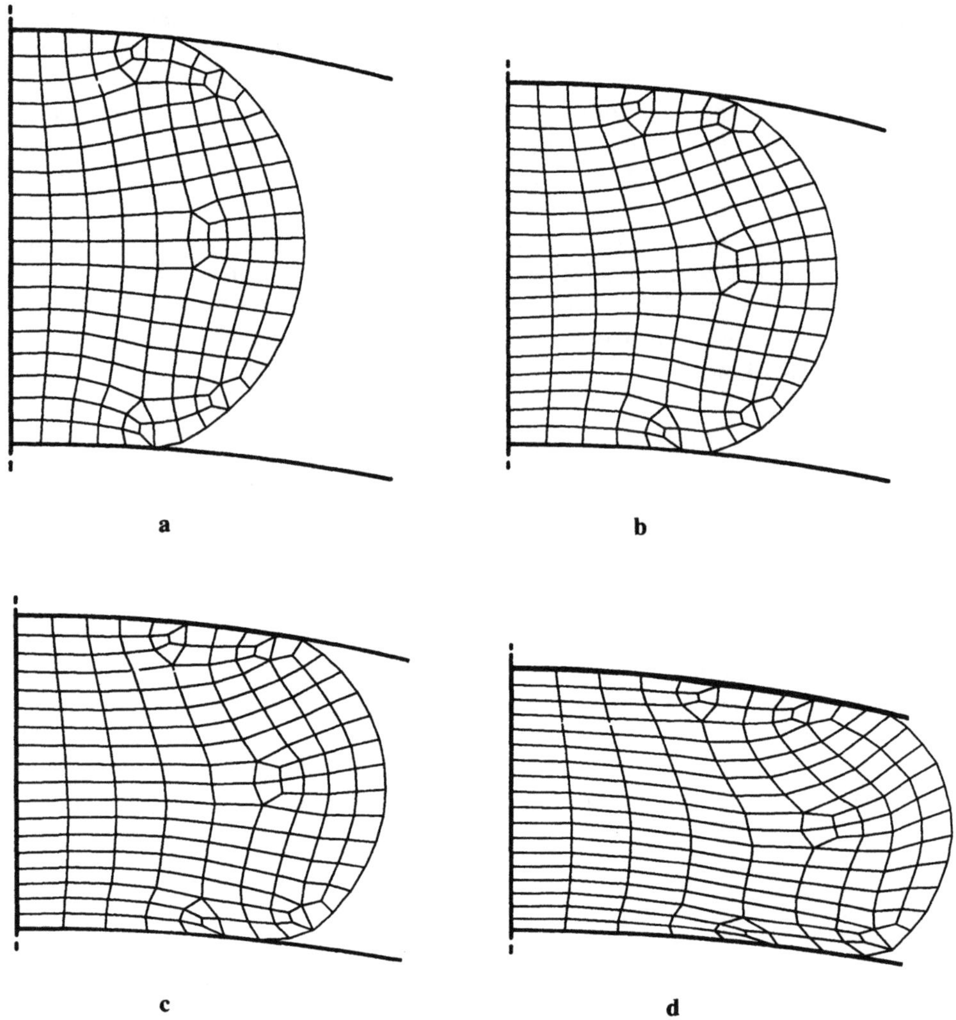

Fig. 10.2. Open-die forging of a circular bar: distorted meshes. **a** 20% reduction; **b** 30%; **c** 40%; **d** 50%.

As can be seen, there is not a great deal of difference in the general pattern of flow at 10% and at 50%, although a rigid zone is suggested near the lower contact area at 10% reduction which does not appear to be present at 50% reduction.

This phenomenon becomes more clear if the generalized plastic strain contours in Fig. 10.4 are examined. At 10% reduction the pattern is almost symmetrical with regions of low strain in both the upper and lower contact areas. The upper region of low deformation was obscured on the previous illustration by the displacement vectors. An interesting observation from the patterns of plastic strain is that the region of maximum strain is moving throughout the process. Initially this occurs in the centre of the workpiece, but as deformation proceeds the region of maximum strain is located more towards the lower platen. By 40% reduction this region has spread to be in contact with the lower platen and at 50% reduction the area of maximum strain can be found away from the workpiece

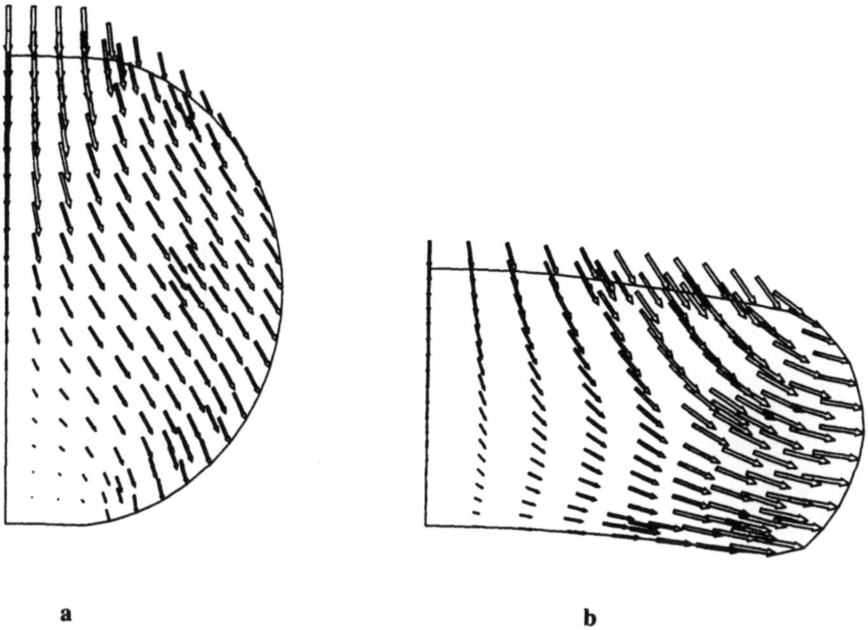

Fig. 10.3. Open-die forging of a circular bar: displacement vectors. **a** 10% reduction; **b** 50%.

centre line. Clearly the shape of the dies is causing the lower parts of the workpiece to be subjected to much greater rates of increase of plastic strain than is occurring in the upper regions.

10.3.2 Closed-Die Forging

A more complex closed-die forging operation is illustrated next. In this example a rectangular section workpiece is compressed between profile dies typical of forging processes. The mesh and profiles are shown in Fig. 10.5. In this case 11 rigid boundary surfaces were used to construct the die profiles.

The distorted meshes are shown in Fig. 10.6. Here it can be seen how the main cavities on each die surface are being filled and how the material is moving away from the workpiece centre towards the flash region. At 40% reduction the die cavities are almost completely filled but there are small areas near the sharp corners that are yet to fill. This corresponds well with experience which indicates that small areas such as this are normally the last ones to be filled properly. In the finite-element simulation, however, there is little point in continuing with this model any further as the coarse mesh will not reproduce the detailed filling very well. This is an obvious candidate for remeshing and an example of this will be given later. There is also evidence at point "a" in Fig. 10.6d that the extended side of one of the elements appears to be crossing into the die. Another indication that remeshing prior to this stage is advisable.

The flow patterns in Fig. 10.7 show a fascinating series of patterns changing significantly as the deformation proceeds. This is precisely the sort of information that is almost impossible to obtain experimentally. At 10% reduction most of the material is moving down with the upper die, albeit at different rates, with some

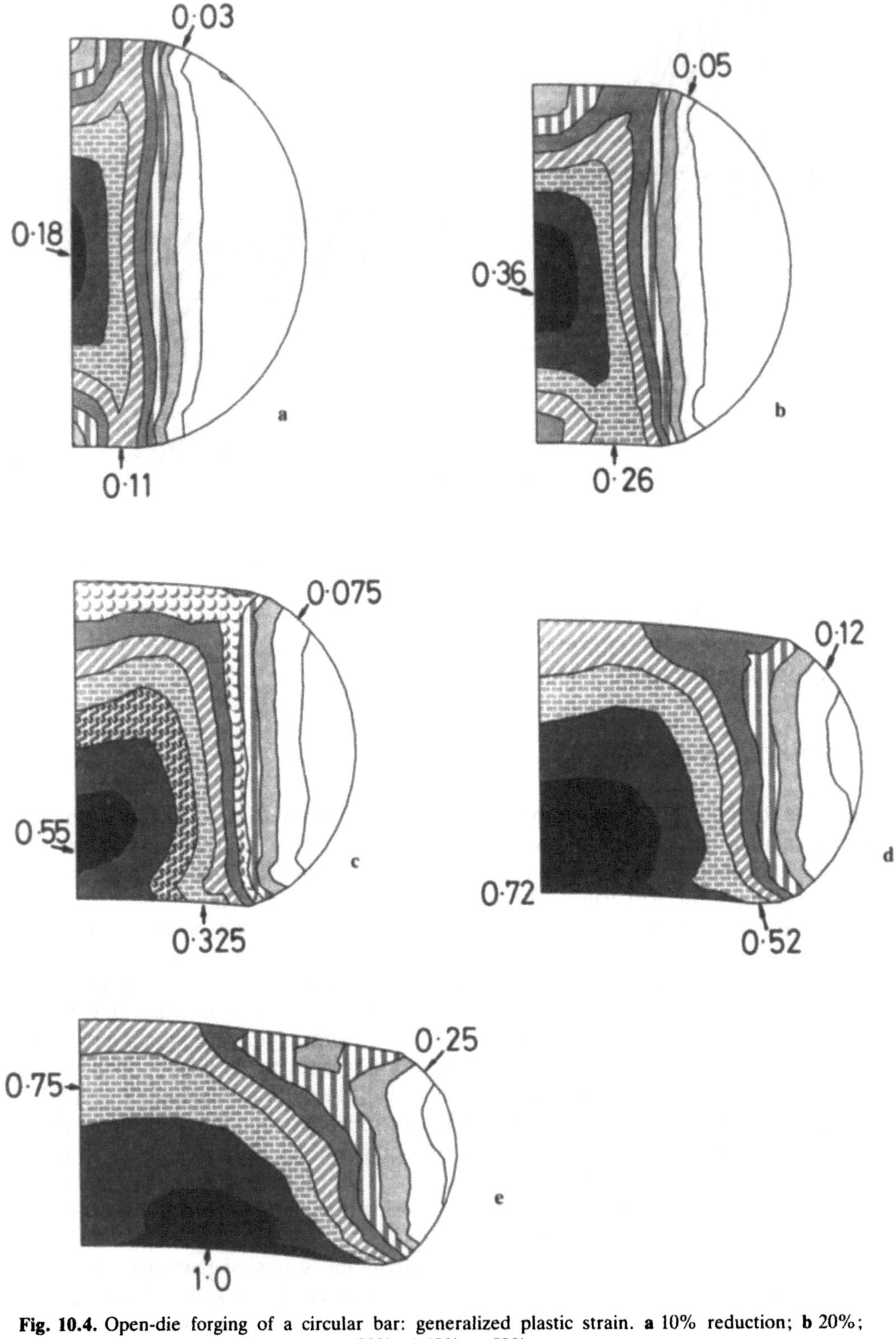

Fig. 10.4. Open-die forging of a circular bar: generalized plastic strain. **a** 10% reduction; **b** 20%; **c** 30%; **d** 40%; **e** 50%.

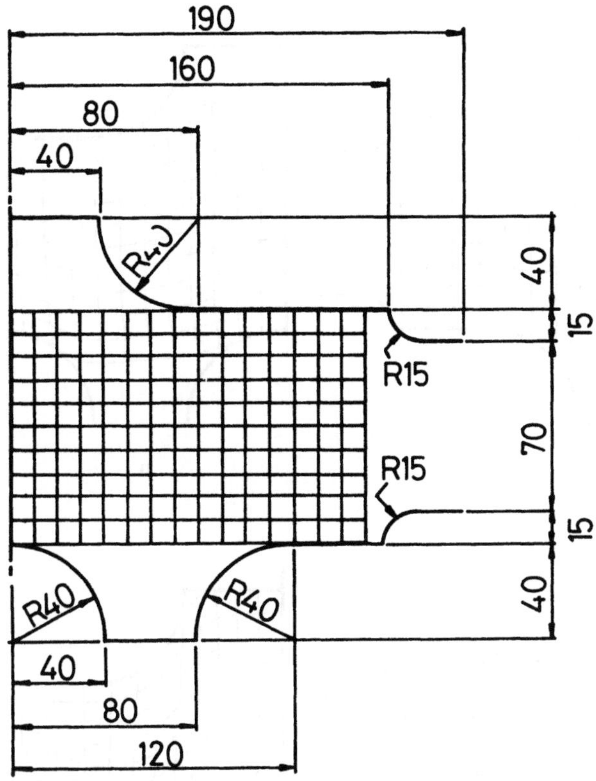

Fig. 10.5. Initial finite-element mesh and die profiles for closed-die forging of a horizontal rectangular bar (dimensions in mm). The mesh is 150 mm × 100 mm.

material moving towards the flash region. At 20% reduction a clear change can be seen as material on the left side of the model has changed direction and is now moving upwards into the upper die cavity opposite to the movement of the die. This is also clear at 30% reduction where more material is now moving into the upper die cavity. This is occurring simultaneously with material moving into the lower die cavity, the combined effect of which is to slow down the rate at which material is moving towards the flash gap. At 35% reduction we see the next major change in the flow pattern. This has been brought about principally because the material moving into the upper die cavity has now come into contact with the upper die and is being pushed down as the die descends. At 40% reduction there is small but uniform movement throughout the workpiece, except where the material is now being pushed more quickly into the flash region. In this example the flash area has not been used to provide a back pressure but as a volume into which excessive material may flow. We shall see later an example of completely closed-die forging where this is not possible. Fig. 10.8 shows the changing patterns of plastic strain for this example.

In spite of the difficulties in modelling the filling of the finished detail the mesh for this example has performed remarkably well. However, if we look at a similar case but this time with the long side of the workpiece model placed vertically instead of horizontally (see Fig. 10.9) then problems become apparent well before the die cavities can be filled.

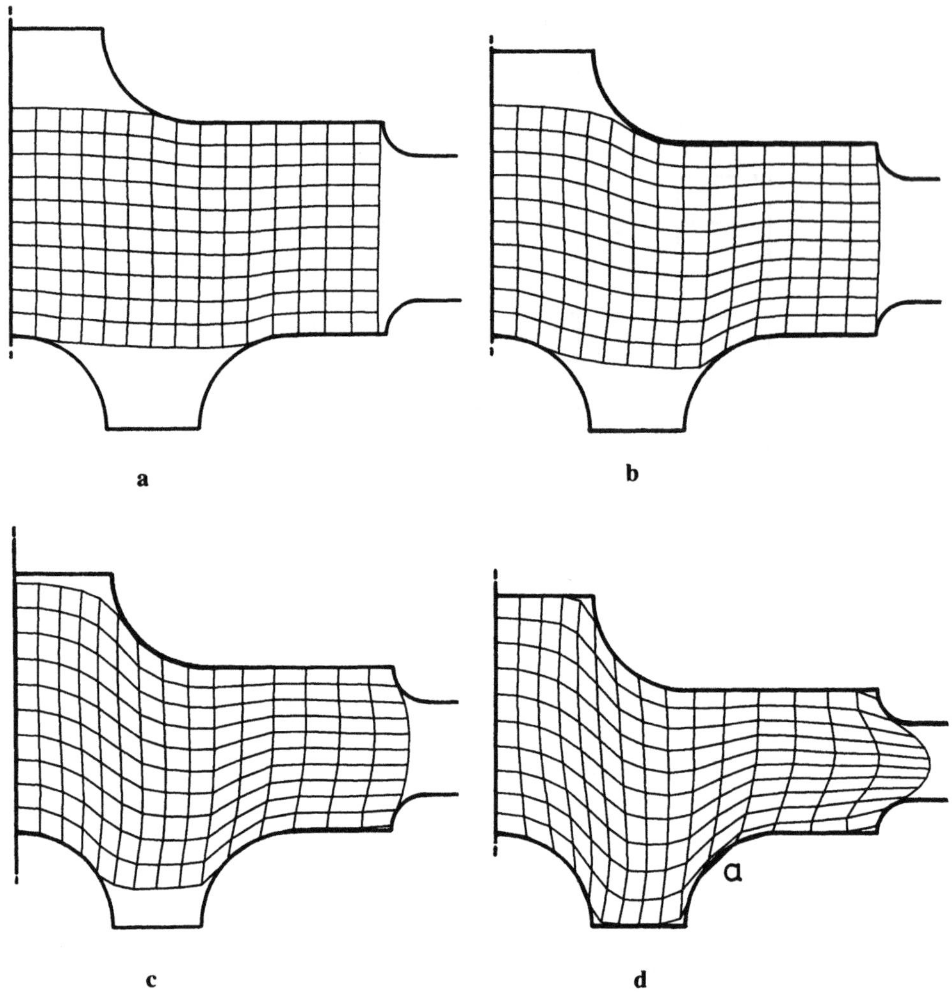

Fig. 10.6. Closed-die forging of a horizontal rectangular bar: distorted meshes. **a** 10% reduction; **b** 20%; **c** 30%; **d** 40%.

Fig. 10.10 shows the distorted meshes which behave well to 20% reduction, but at 30% reduction the element at the lower right-hand corner is becoming extended and at 40% reduction it is too distorted for the analysis to continue. Up to this level the flow patterns (Fig. 10.11) are clearly changing as before and although the analysis could not be completed the results do show that with the upright workpiece the upper-die cavity is filled well before the lower one, much more quickly than in the previous example. This indicates how process simulations can be used to assess various preform routes to optimize the process.

10.3.3 Completely Closed-Die Forging

Here the ability of the finite-element program to cope with completely closed-die forging is considered. The initial mesh is slightly smaller than in the first

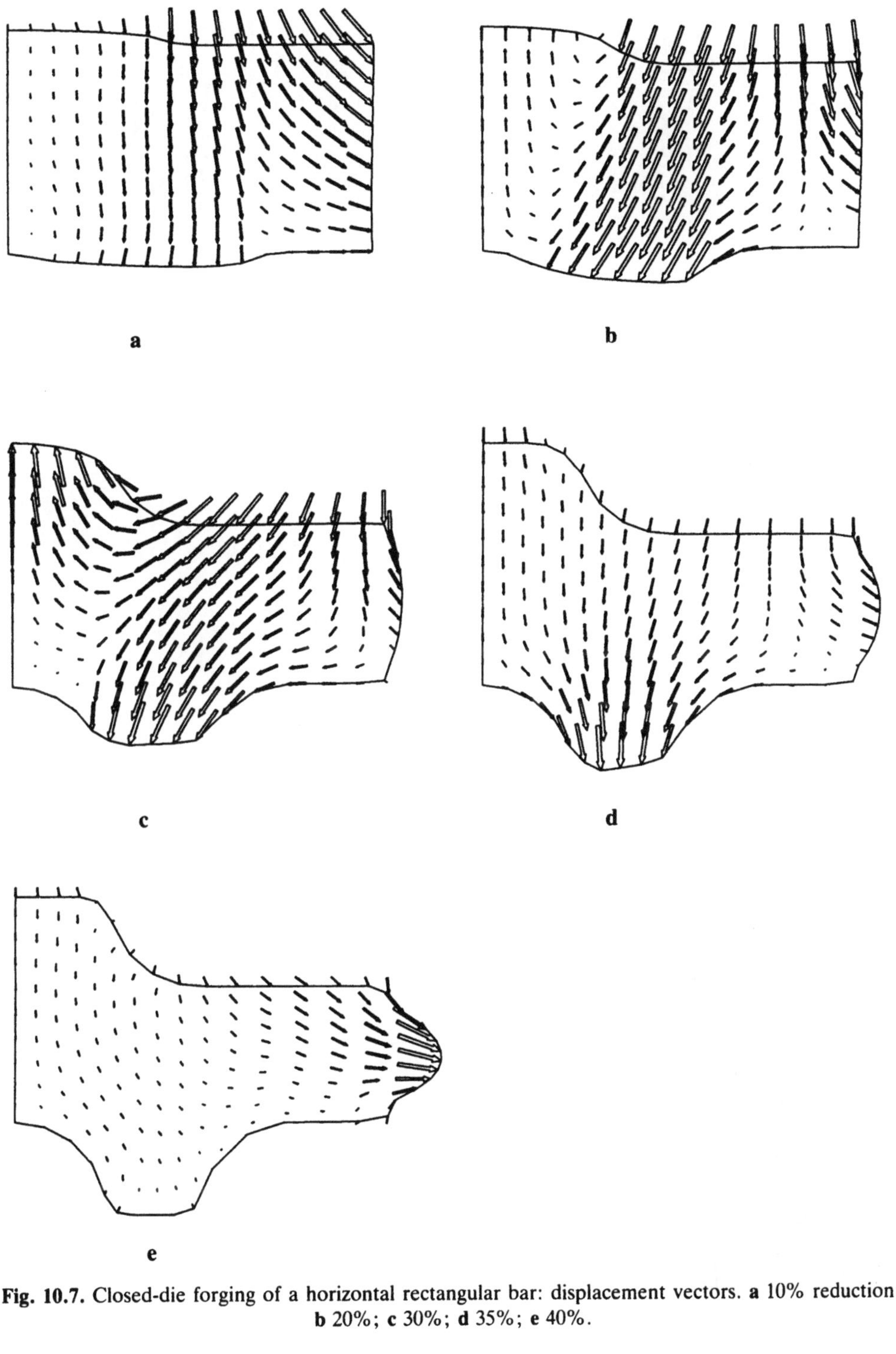

Fig. 10.7. Closed-die forging of a horizontal rectangular bar: displacement vectors. **a** 10% reduction; **b** 20%; **c** 30%; **d** 35%; **e** 40%.

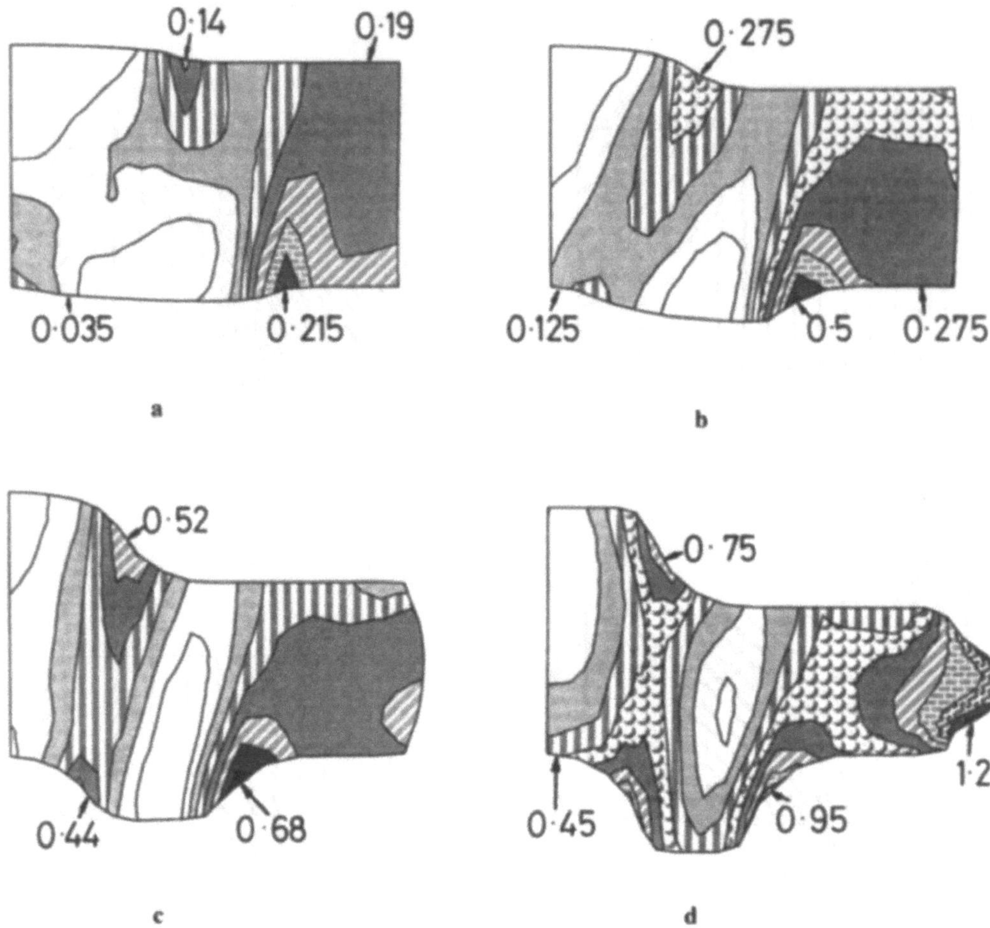

Fig. 10.8. Closed-die forging of a horizontal rectangular bar: generalized plastic strain. **a** 10% reduction; **b** 20%; **c** 30%; **d** 40%.

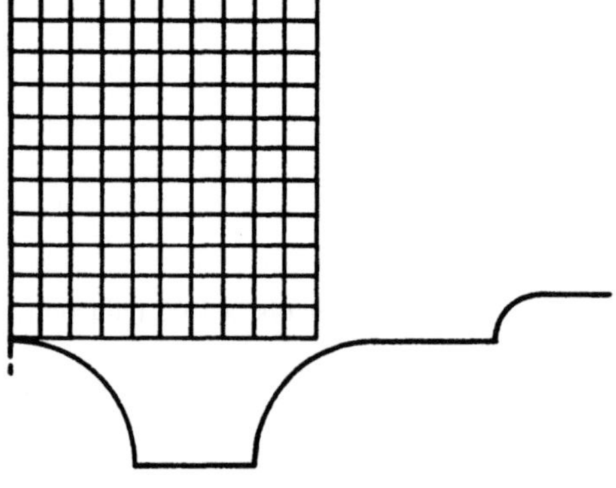

Fig. 10.9. Initial finite-element mesh and die profiles for closed-die forging of a vertical rectangular bar (dimensions in mm). The mesh is 100 mm × 150 mm. The die profiles are as in Fig. 10.5.

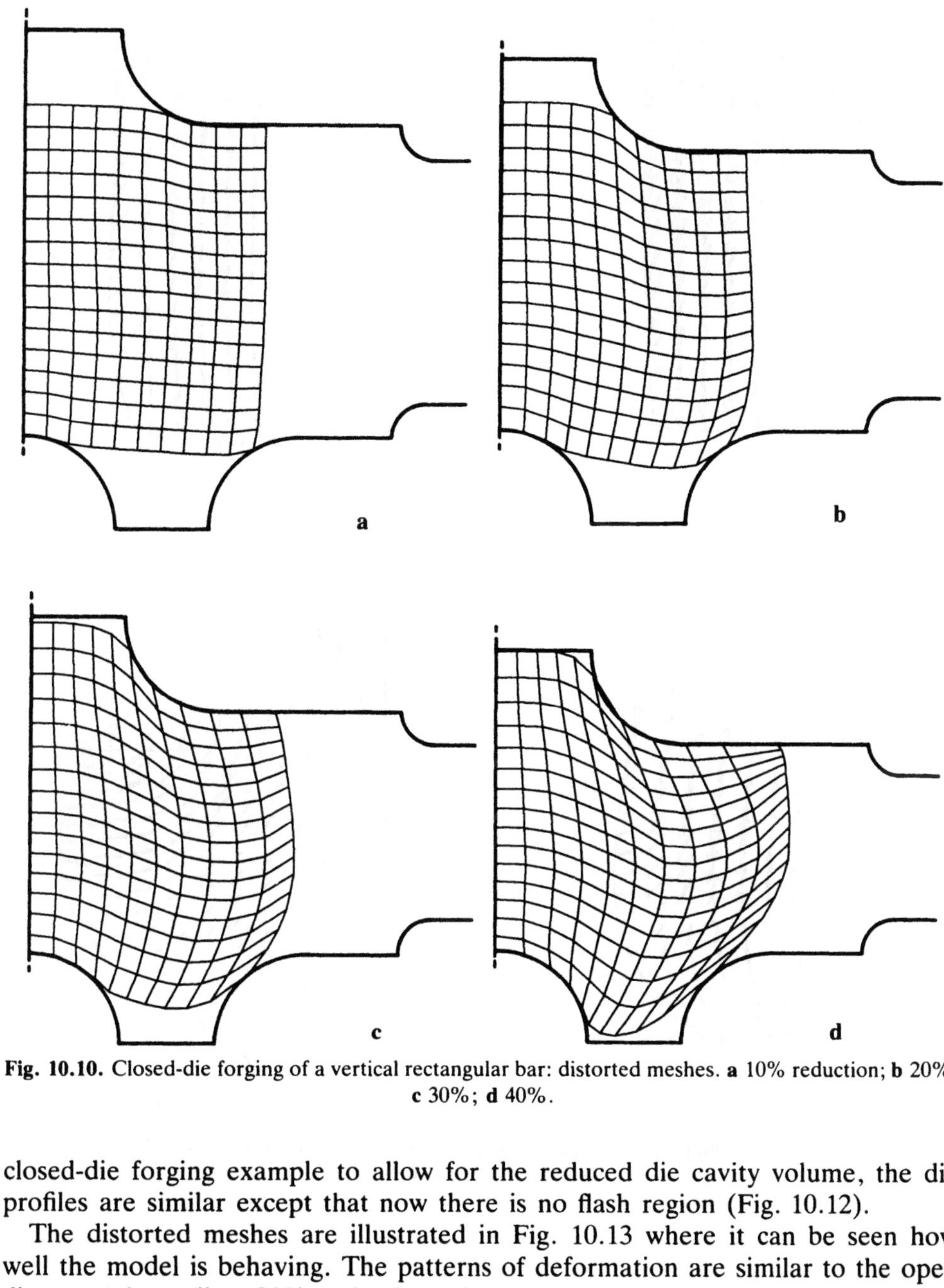

Fig. 10.10. Closed-die forging of a vertical rectangular bar: distorted meshes. **a** 10% reduction; **b** 20%; **c** 30%; **d** 40%.

closed-die forging example to allow for the reduced die cavity volume, the die profiles are similar except that now there is no flash region (Fig. 10.12).

The distorted meshes are illustrated in Fig. 10.13 where it can be seen how well the model is behaving. The patterns of deformation are similar to the open die example until at 30% reduction where the material moving to the right has come into contact with the die wall. For the purposes of this example the vertical die wall is considered to move down as part of the upper die profile and with the same frictional restraint as on all the other surfaces. At 40% reduction the die cavities are almost completely filled and the mesh appears to be behaving reasonably well. However, there may well be inaccuracies developing that are not obvious from the flow patterns.

At the final stage of deformation there are very few free surfaces left. As explained in Chap. 9, it is necessary to have free surfaces available, the more

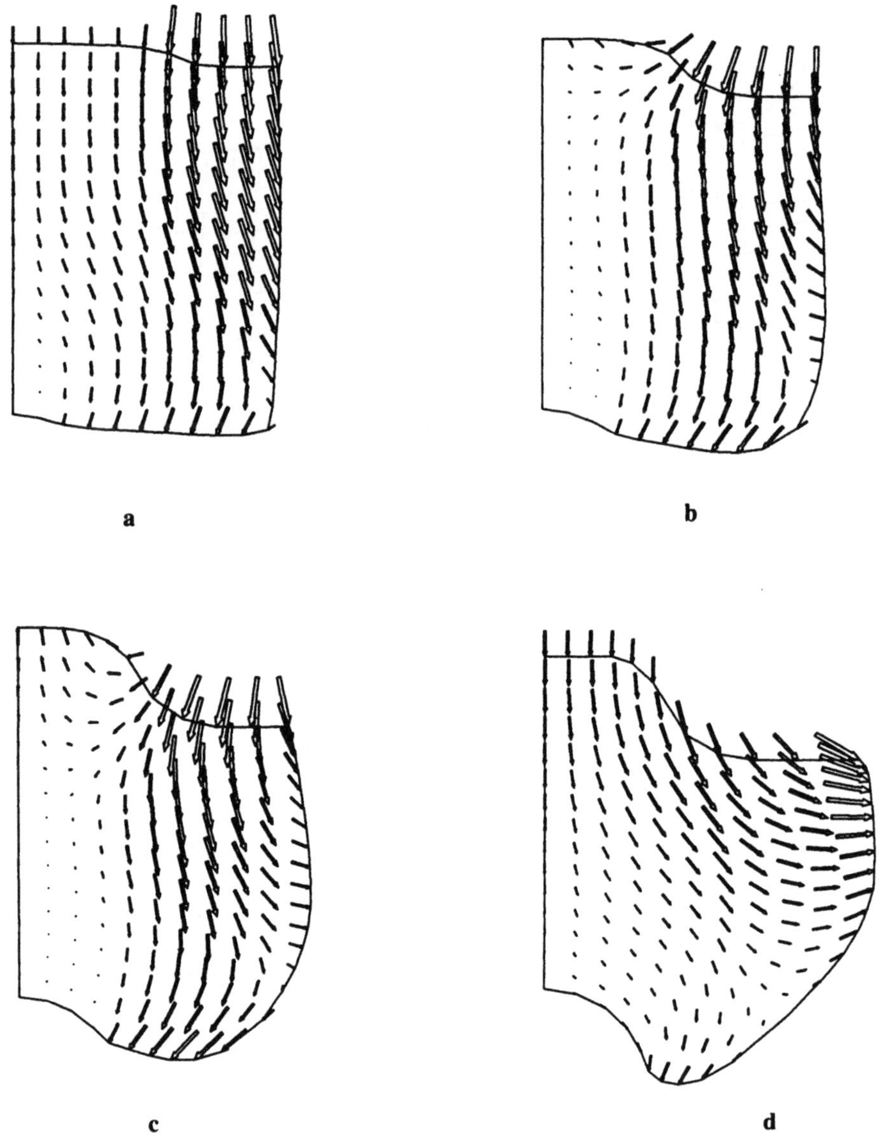

Fig. 10.11. Closed-die forging of a vertical rectangular bar: displacement vectors. **a** 10% reduction; **b** 20%; **c** 30%; **d** 40%.

the better, from which to initiate the calculation of the distribution of deviatoric stress gradients and hence accurately calculate the distribution of hydrostatic stress throughout the workpiece. In this example the lack of free surfaces will possibly result in the development of erroneous stress components although the generalized stress may still be reasonable. There is also a danger of over estimating the elastic volume change where the model is so heavily constrained. The flow diagrams in Fig. 10.14 show the now familiar series of changing patterns. The comparatively large vectors in Figs 10.14e and 10.14f indicate the last areas to be filled as the finite-element mesh tries to completely fill all the cavities.

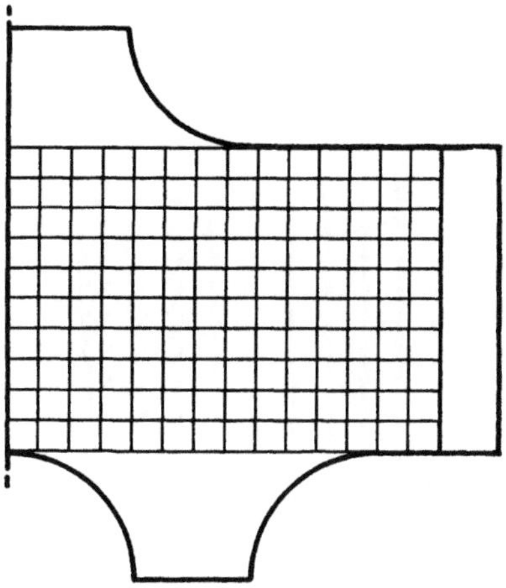

Fig. 10.12. Initial finite-element mesh and die profiles for completely closed-die forging of a horizontal rectangular bar (dimensions in mm). The mesh is 140 mm × 100 mm.

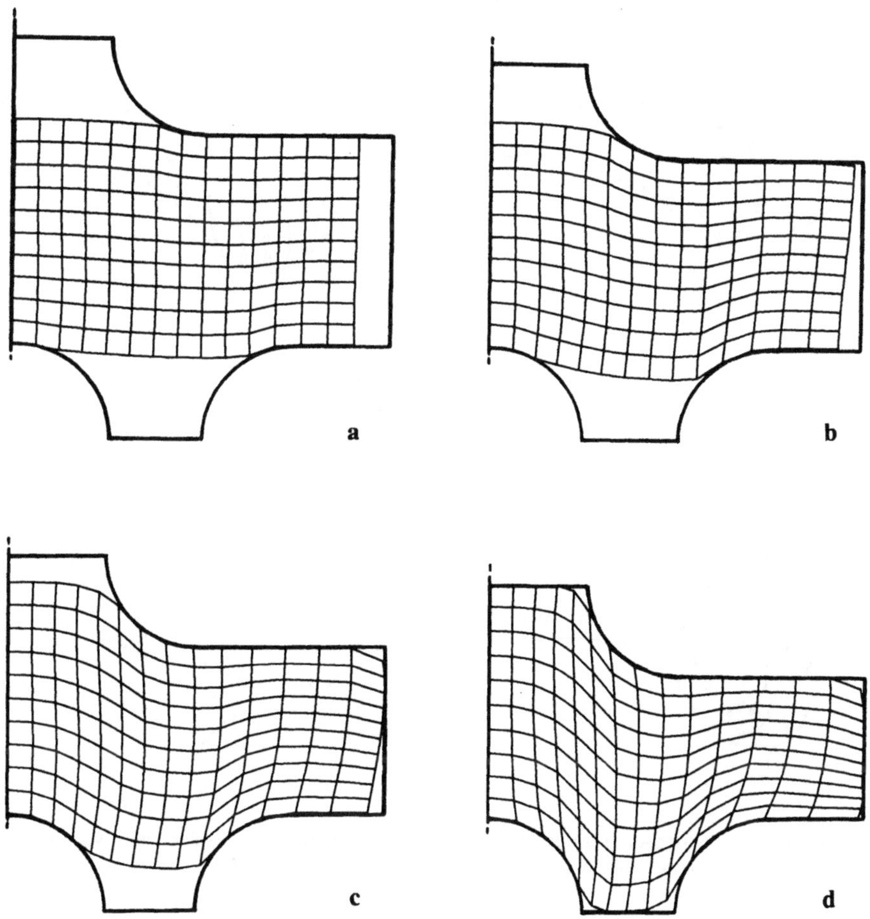

Fig. 10.13. Completely closed-die forging of a horizontal rectangular bar: distorted meshes. **a** 10% reduction; **b** 20%; **c** 30%; **d** 40%.

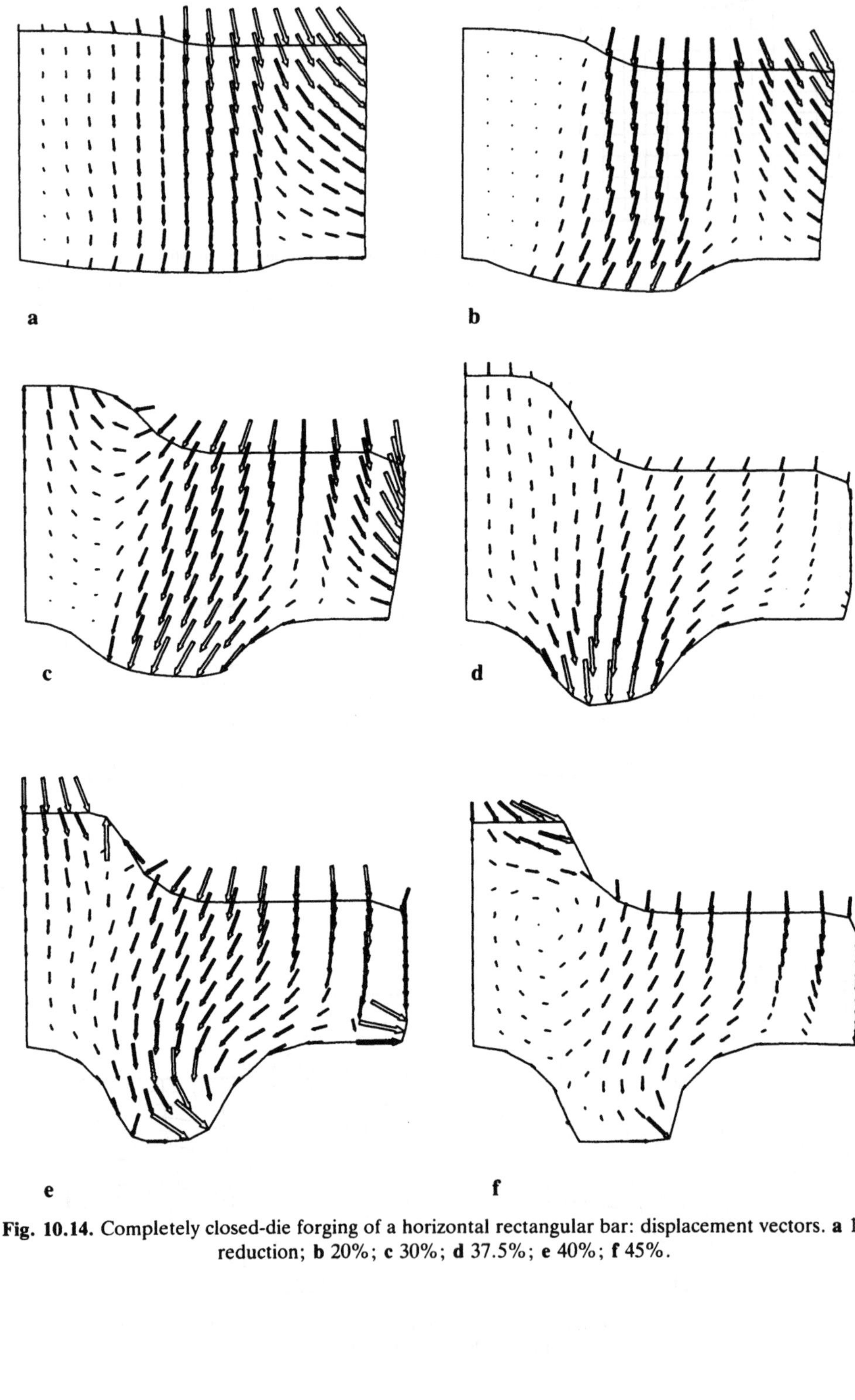

Fig. 10.14. Completely closed-die forging of a horizontal rectangular bar: displacement vectors. **a** 10% reduction; **b** 20%; **c** 30%; **d** 37.5%; **e** 40%; **f** 45%.

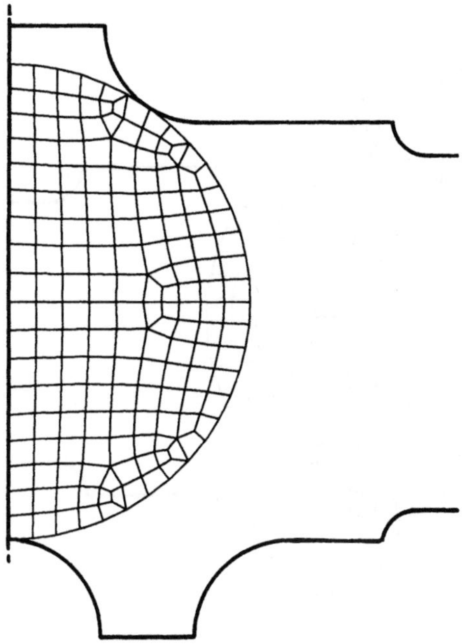

Fig. 10.15. Initial finite-element mesh and die profiles for closed-die forging of a circular bar (dimensions in mm). The die profiles are as in Fig. 10.5.

10.3.4 Forging Simulation with Remeshing

In this example we return to open-die forging using the die profiles already seen. The workpiece is again a 200 mm diameter bar as shown in Fig. 10.15.

The sequence of distorted grids for the process is illustrated in Fig. 10.16 which shows how successive remeshing has allowed the complete process to be modelled. In this example four remeshing stages are required, summarized in Table 10.1. The new meshes in this example are not based on any error assessments or self-adaptive schemes but have been chosen to illustrate various features of remeshing forging problems.

The initial mesh of 162 elements was selected to have a fairly uniform element size. This mesh performed reasonably well until 40% reduction (Fig. 10.16d), where elements in the region "a" became too extended with material apparently crossing into the dies. The new mesh used in Fig. 10.16e used extra elements to improve the modelling of the deformation around the curved die profiles but did not need as many elements within the main body of the workpiece model. This mesh, however, did not deform very far until similar problems appear around

Table 10.1. Details of remeshing stages and new meshes

Mesh	Elements	Reduction at Remeshing (%)	Illustrated in Fig.
1	162	0	10.16 a–d
2	96	41	10.16 e–f
3	100	47	10.16 g–h
4	124	55	10.16 i–j
5	205	60	10.16 k

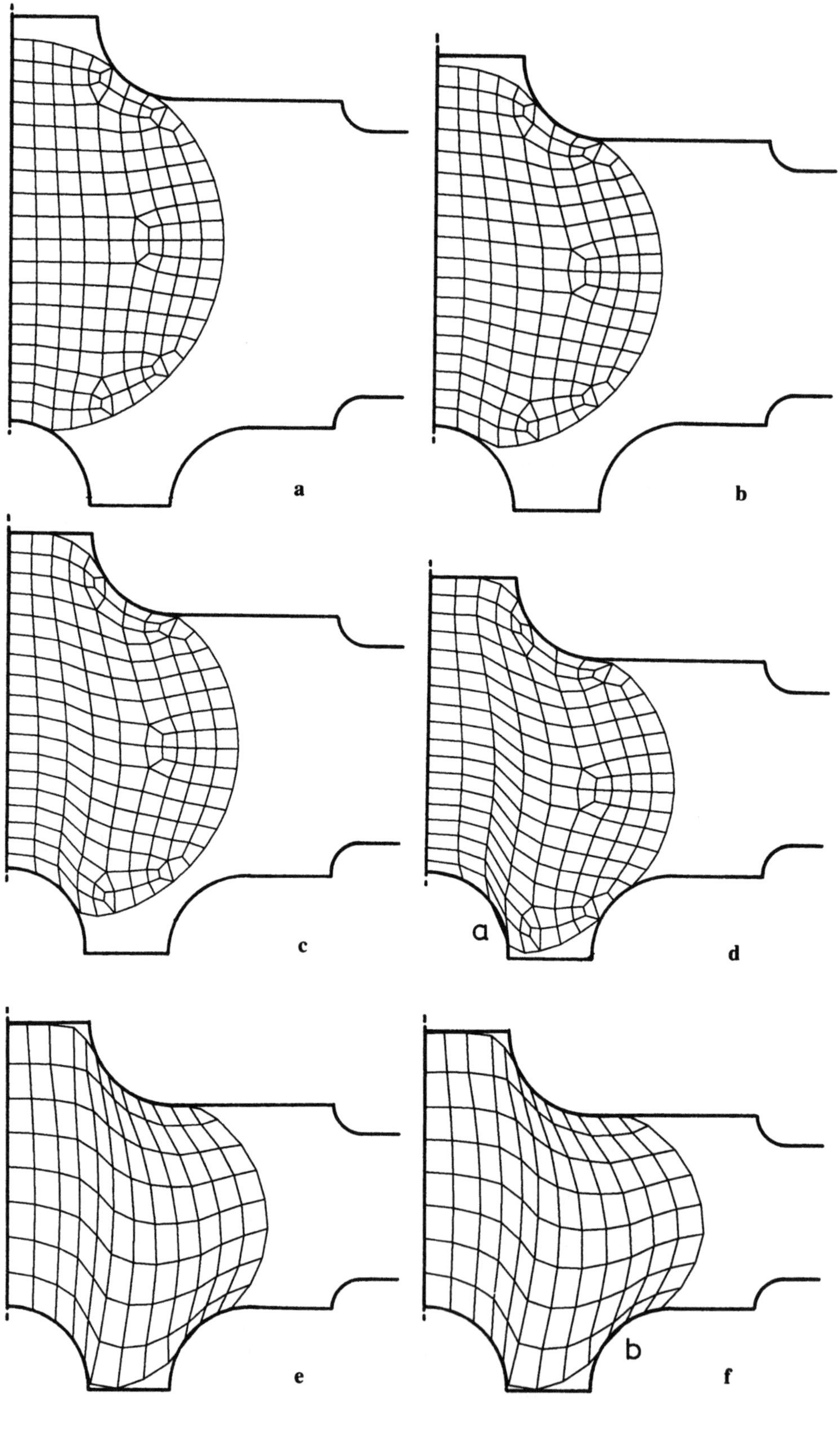
a
b
c
a
d
e
b
f

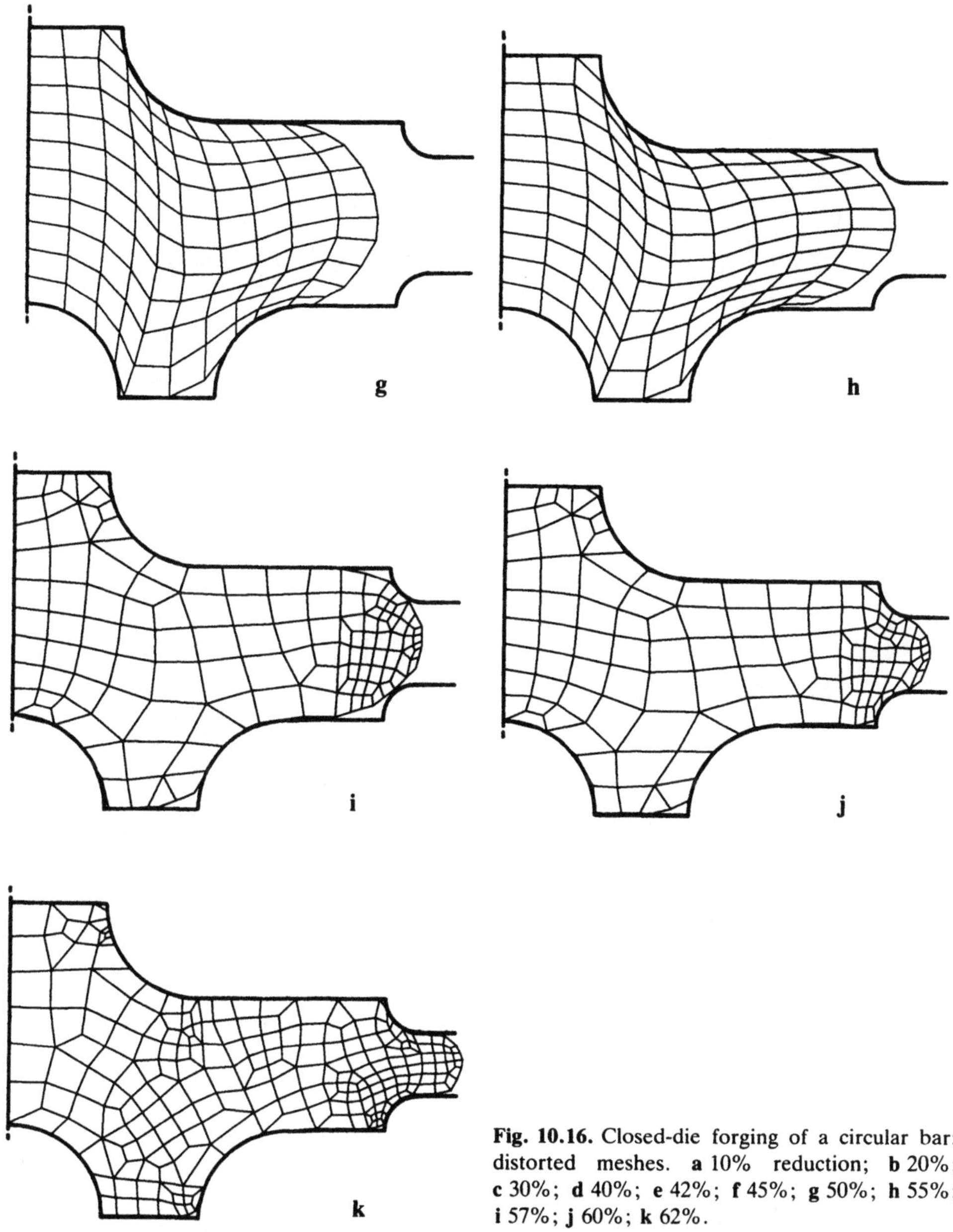

Fig. 10.16. Closed-die forging of a circular bar: distorted meshes. **a** 10% reduction; **b** 20%; **c** 30%; **d** 40%; **e** 42%; **f** 45%; **g** 50%; **h** 55%; **i** 57%; **j** 60%; **k** 62%.

another part of the curved profile (region "b" in Fig. 10.16f). A minor rearrangment of elements is made for the third mesh used in Fig. 10.16g–h, but again only a small amount of deformation is possible. To continue further more significant changes in the mesh are required particularly as the material approaches the small radii either side of the flash region (see Fig. 10.16i–j). The final remeshing allows the process simulation to be completed as shown in Fig. 10.16k.

Selected flow patterns for the simulation of this process are shown in Fig. 10.17. As the circular cross-section allows the upper die cavity to be filled very quickly there are not the severe changes in flow patterns observed in the earlier analyses.

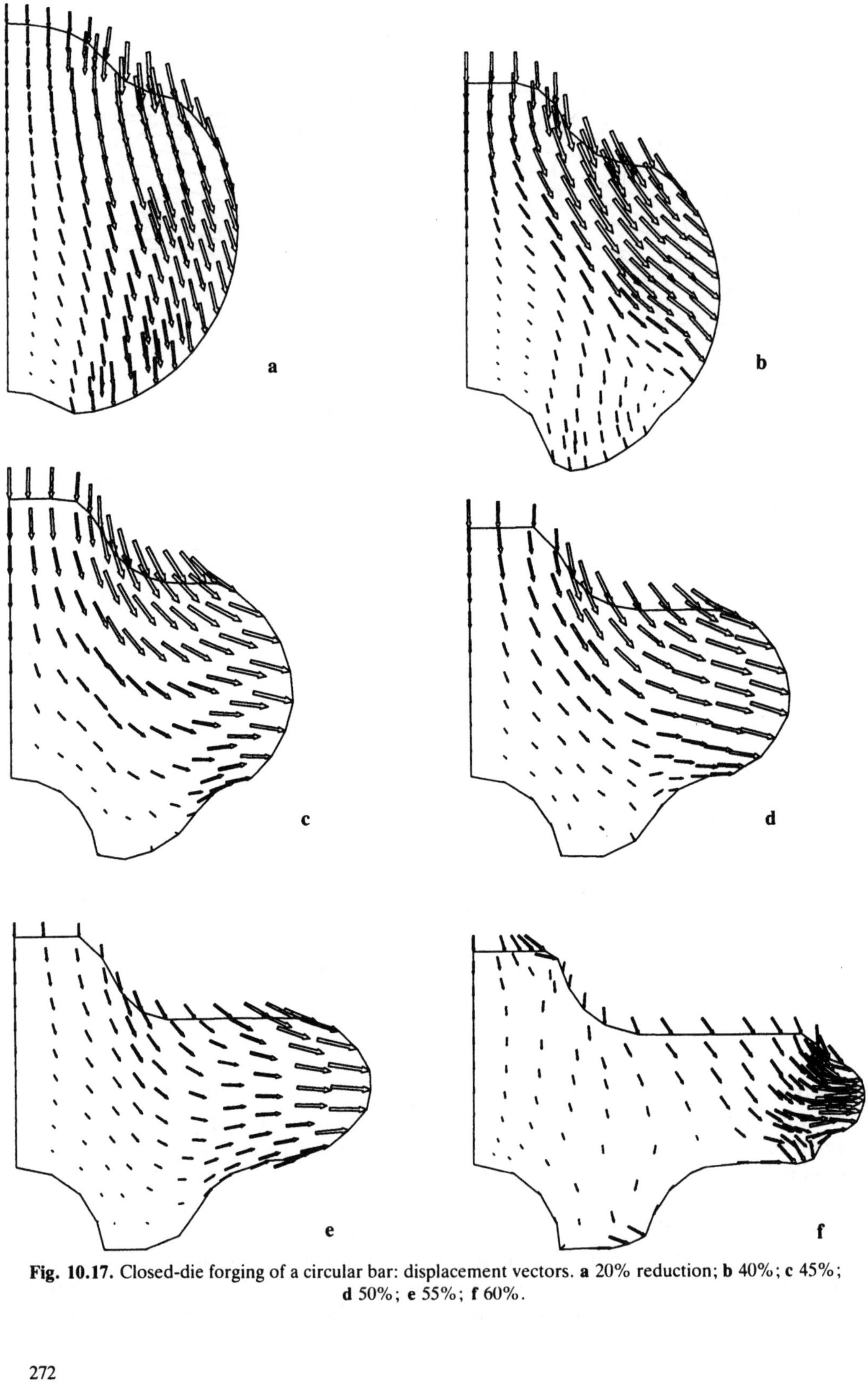

Fig. 10.17. Closed-die forging of a circular bar: displacement vectors. **a** 20% reduction; **b** 40%; **c** 45%; **d** 50%; **e** 55%; **f** 60%.

The remeshing example here clearly shows how successive remeshing allows the completion of a full simulation. It also demonstrates that any self-adaptive scheme that is used should be based on appropriate error estimates to ensure that suitable meshes are produced. This is a difficult process in two-dimensional modelling, and becomes a real challenge in the modelling of complex three-dimensional forging shapes.

10.4 Conclusions

A number of examples of modelling forging processes using an elastic-plastic finite-element program have been presented. These have served to illustrate the potential of the technique for obtaining detailed information, on flow and plastic strain for example, that is almost impossible to obtain experimentally. The examples have been chosen specifically to illustrate some of the difficulties that may arise in trying to model forging processes. Remeshing has been used as a means of overcoming some of these problems.

Acknowledgements. Dr P. Hartley is currently seconded part-time to the Interdisciplinary Research Centre in Materials for High Performance Applications, at the University of Birmingham. Mr S.M. Roberts is an SERC-sponsored research student.

References

Cottrell AM (1967) An introduction to metallurgy. Edward Arnold

De Garmo EP, Temple Black J, Kohser RA (1984) Materials and processes in manufacturing. Macmillan

Feldmann HD (1961) Cold forging of steel. Hutchinson

Kalpakjian S (1989) Manufacturing engineering and technology. Addison-Wesley

Pillinger I, Hartley P, Sturgess CEN, Rowe GW (1987) Thermo-mechanical modelling of metalforming using the finite-element method. In: Owen DRJ, Hinton E, Oñate E (eds) International conference on computational plasticity. Pineridge Press, pp 1073–1086

Rowe GW (1977) Principles of industrial metalworking processes. Edward Arnold

Rowe GW, Sturgess CEN, Hartley P, Pillinger I (1991) Finite-element plasticity and metalforming analysis. Cambridge University Press

Van Houtte P, Van Bael A, Aernoudt E, Pillinger I, Hartley P, Sturgess CEN (1989) The introduction of anisotropic yield loci derived from texture measurements in elasto-plastic finite-element calculations. In: Khan AS, Tokuda M, Advances in plasticity '89. Proceedings of the 2nd international symposium on plasticity. Pergamon, pp 111–114

11 **Rolling Process Modelling**

J.G. Lenard and M. Pietrzyk

11.1 Introduction

Flat rolling is a process in which the thickness of a flat sheet is reduced by passing it between two counter-rotating cylinders which are separated by a fixed distance. During forming, a schematic diagram of which is shown in Fig. 11.1, the metal elongates but does not spread in significant measure. The process may be performed at room temperature on a cold mill or at high temperature on a hot strip mill, both of which may consist of one or more mill stands. Following hot rolling, the cleaned strip may be cold rolled again, after which heat treatment and further forming may follow.

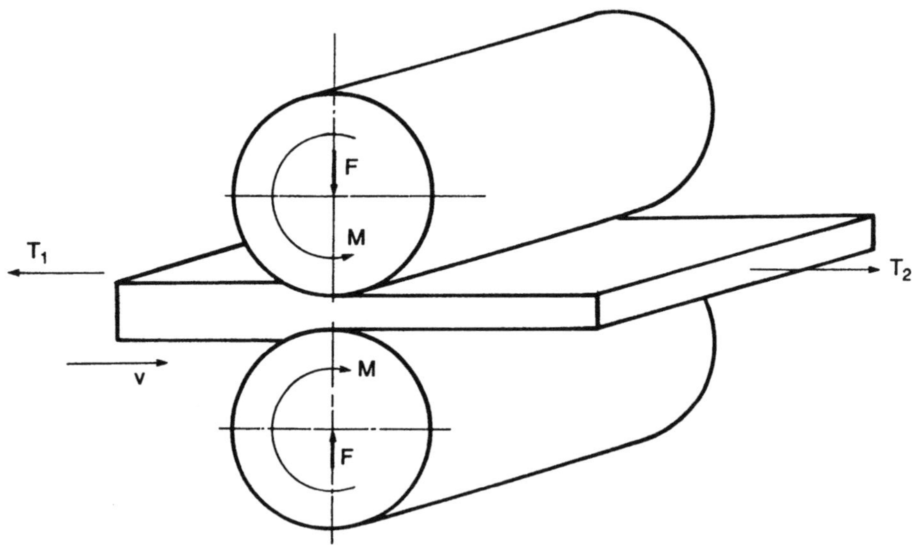

Fig. 11.1. Schematic diagram of the flat rolling process.

The hot and cold flat rolling processes are among the most often used, most economical and most efficient of all techniques available to produce sheets, plates and strips. Leon (1987) points out that Canada and the USA produced a total of about 43 million tons of sheet and strip in 1985. While the consumption of steel per capita appears to be levelling off, the contribution of the steel industry to the gross national products of developed countries remains very significant. In that context, research and development aimed at improving the quality of rolled products, reducing costs and increasing cost effectiveness are of prime importance. The abundance of mathematical models describing the process underlines the above comments. Further, understanding the limitations of the available models, possible improvements, their comparative evaluation and the experimental substantiation of their predictive capabilities may also aid the work of the steel-mill engineer.

Mathematical models of the flat rolling process may be conveniently divided into two groups. The first one includes the traditional technique of Orowan (1943) and its simplified versions (Bland and Ford 1948, Sims 1954). In each of these a one-dimensional equation of equilibrium is combined with the constitutive relations of the rolled metal, an equation defining the roll–strip contact surface and a description of the nature of friction there. The temperature of the workpiece appears only through its effect on the properties of the material.

A different formulation of the governing equations of flat rolling was developed by Tselikov (1939), presented by Tselikov and Grishkov (1970). Tselikov's assumptions include the substitution of the arc of contact by its chord, Coulomb's model of friction and a constant value of the yield strength.

The second group encompasses the finite-element models, some of which include analyses of the thermal as well as mechanical events during processing. Recent developments concerning research on flat rolling have been presented by Hartley et al. (1989).

In what follows, these aspects of modelling will be discussed in some detail. After a description of what needs to be included in a model the traditional, one-dimensional formulations and a discussion of their predictive ability are presented. The criteria under which more rigorous analysis is necessary are given, followed by an exposition of a thermal–mechanical finite-element model and its ability to simulate hot, warm and cold flat rolling. The chapter closes with an analysis of the heat transfer coefficient at the roll/strip contact surface and its effect on the computations.

11.2 Mathematical Models

As the strip enters the roll gap it is first deformed elastically. It speeds up; the relative velocity between the roll and the strip is such that frictional forces draw the metal in. The criterion of plastic flow governs the manner in which the transformation from elastic to plastic states happens. As Fig. 11.2 shows, the two regions are separated by a surface called the elastic–plastic interface. The strip proceeds through the roll gap; more plastic flow occurs until finally at exit – if rolling without tensions is considered – the roll pressure is removed. The strip is then unloaded elastically. It is observed that the relative velocities of the roll and the strip change during the process. As the strip is accelerating forward, it reaches the surface velocity of the roll at the "no-slip" or neutral point. From then on, as further compression occurs, the strip speeds up and the direction of friction changes in such a way that it now retards motion.

11.2.1 Traditional Models

The traditionally accepted model of hot and cold flat rolling is due to Orowan (1943). Orowan's model is based on a one-dimensional equation of equilibrium which is combined with a yield criterion, frictional conditions and Hitchcock's formula (1935) to evaluate the radius of the flattened but still circular roll contour and an appropriate description of the material's resistance to deformation. Numerical integration then allows computation of the roll pressures and hence the roll separating forces and roll torques (Alexander 1972).

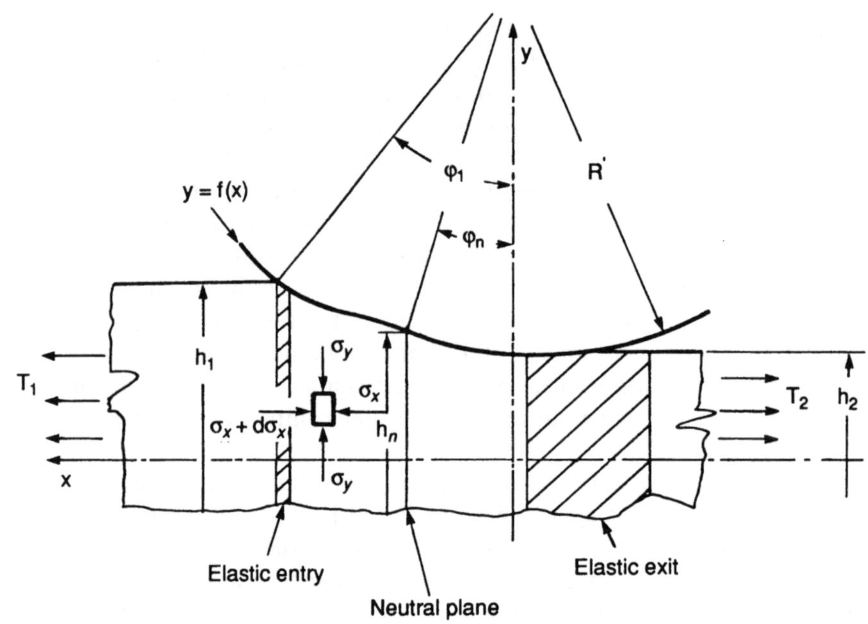

Fig. 11.2. Elastic and plastic states in the deformation zone.

Corrections for the effects of elastic entry and exit are also included in Alexander's treatment of Orowan's analysis.

Use of Orowan's model is time consuming and before the advent of the computer age it was preferable to employ simplifications which would allow the derivation of closed-form solutions.

For cold rolling Bland and Ford (1948) used the "small angle" approximation to speed up computations. For hot rolling Sims (1954) and Ford and Alexander (1963) gave useful formulae. A thorough discussion of these along with some comparative calculations has been given by Alexander (1972). Tselikov's model (1939), as given by Tselikov and Grishkov (1970) also leads to easy-to-use, closed-form expressions. The shortcomings of the Orowan model and its variants are concerned with the description of the boundary conditions of the process. Use of the Hitchcock formula to account for roll deformation has been shown to lead to erroneous results (Jortner et al. 1958, Golten 1969); since it assumes that the roll remains circular during the pass its use leads to incorrect appreciation of the size of the roll gap contact length. As well, the frictional conditions used by Orowan have been shown to be unsatisfactory (van Rooyen and Backofen 1957, Al-Salehi et al. 1973, Jeswiet and Rice 1982, Lim and Lenard 1984, Karagiozis and Lenard 1985). As mentioned above, the entry and exit conditions are accounted for in the model. However, those corrections do not determine the growth of the contact length caused by the elastic conditions at those locations.

11.2.2 Refinement of the Orowan Model (Roychoudhuri and Lenard 1984)

The underlying philosophy guiding the development of this method of computation is to achieve maximum accuracy and consistency while using a minimum of arbitrary assumptions. The model consists of several parts. These include:

1. *Plastic flow of the rolled strip in the roll gap.* Homogeneous compression and plane-strain conditions are assumed to exist in the plastically flowing strip. The independent variable is now chosen to be the direction of rolling instead of the angular variable around the roll. This reduces the number of divisions required for the numerical integration of the differential equations. In Atreya and Lenard (1979), the use of 50 slabs yielded sufficient accuracy; in Roychoudhuri and Lenard (1984), 200 divisions were required. This compares well with the number of divisions needed by Alexander's (1972) solution of Orowan's equations.
2. *Elastic compression and unloading of the strip at entry and exit.* The theory of elasticity is used to determine the one-dimensional stress distribution at the entry to and exit from the roll gap. Through the strain field the actual roll gap contact length is then determined. The Huber–Mises criterion is used to locate the elastic–plastic boundaries.
3. *Flattening of the work roll.* Michell's (1900) solution of the biharmonic equation is used to analyse the stress and strain distributions in the work roll.
4. *Frictional conditions at the roll–material interface.* The model allows easy use of the data on the coefficient of friction, developed by Lim and Lenard (1984) and Karagiozis and Lenard (1985).
5. *Material's resistance to deformation.* The material's constitutive behaviour may be modelled as the researcher sees fit. This may include the development of empirical equations or a multidimensional matrix, from which flow strength values are retrieved when required (Lenard et al. 1987). It is pointed out that Orowan's model also allows the use of any of these approaches.

11.2.2.1 Roll Deformation

The schematic loading diagram of a work roll is shown in Fig. 11.3. The loading consists of the roll pressure p and the interfacial friction stress q. These loads must be balanced by a statically equivalent force system, the location of which would depend on the type of mill considered. For four- or six-high mills the backup rolls would supply the necessary balance, while for two-high mills the reaction forces would be applied through the bearings of the roll journals. As the present analysis is concerned with a two-dimensional treatment of the work rolls of a two-high mill and roll bending is neglected, the balancing loads are taken to be distributed over the rolls with $\xi = \pi/2$. This condition ensures that the midplanes of the rolls remain undeformed and stationary.

Following Michell (1900), the stress and strain distributions in the cylinder in a state of plane strain may be calculated from a stress function, given by

$$\Phi = c_0 r^2 + d_1 r^3 \sin\phi + d_2 r^3 \cos\phi + \sum_{n=2}^{\infty} [(a_{1n} r^n + b_{1n} r^{n+2}) \sin(n\phi) + (a_{2n} r^n + b_{2n} r^{n+2}) \cos(n\phi)] \qquad (11.1)$$

where the coefficients of terms singular at the origin were taken to equal zero. In Eq. (11.1) r and ϕ are the coordinates in a cylindrical system. The radial and shear stresses are then obtained from

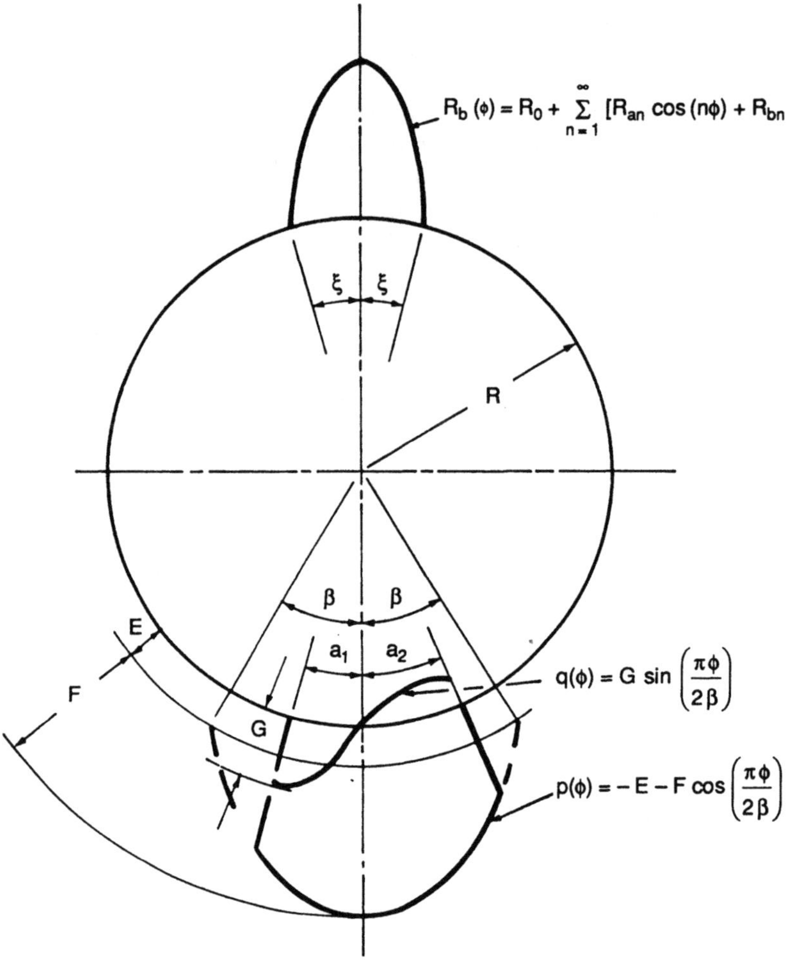

Fig. 11.3. Schematic loading diagram of the work roll (Roychoudhuri and Lenard 1984).

$$\sigma_r = \frac{1}{r}\frac{\partial \Phi}{\partial r} + \frac{1}{r^2}\frac{\partial^2 \Phi}{\partial \phi^2} \tag{11.2}$$

and

$$\tau_{r\phi} = -\frac{\partial}{\partial r}\left(\frac{1}{r}\frac{\partial \Phi}{\partial \phi}\right) \tag{11.3}$$

respectively, subject to boundary conditions, given by

$$\sigma_r = p(\phi) \quad \text{and} \quad \tau_{r\phi} = q(\phi) \quad \text{at} \quad r = R \tag{11.4}$$

for the contact region, where R is the roll radius. Expressions similar to these are used at the locations of the forces providing the reactions required for balance.

The coefficients in Eq. (11.1) are determined next by representing the normal and shear loading on the roll – $p(\phi)$ and $q(\phi)$ – in terms of Fourier series

$$p(\phi) = p_{a0} + \sum_{n=1}^{\infty} [p_{an} \cos(n\phi) + p_{bn} \sin(n\phi)] \tag{11.5}$$

$$q(\phi) = q_{a0} + \sum_{n=1}^{\infty} [q_{an} \cos(n\phi) + q_{bn} \sin(n\phi)] \tag{11.6}$$

where the coefficients are obtained from the Euler formulae. For the normal pressure distribution then

$$p_{a0} = \frac{1}{2\pi} \int_{\alpha_1}^{\alpha_2} p(\phi) d\phi + \int_{-\pi/2}^{\pi/2} R_b(\phi) d\phi \tag{11.7}$$

$$p_{an} = \frac{1}{\pi} \int_{\alpha_1}^{\alpha_2} p(\phi) \cos(n\phi) d\phi + \int_{-\pi/2}^{\pi/2} R_b(\phi) \cos(n\phi) d\phi \tag{11.8}$$

and

$$p_{bn} = \frac{1}{\pi} \int_{\alpha_1}^{\alpha_2} p(\phi) \sin(n\phi) d\phi + \int_{-\pi/2}^{\pi/2} R_b(\phi) \sin(n\phi) d\phi \tag{11.9}$$

and for the shear stress distribution

$$q_{a0} = \frac{1}{2\pi} \int_{\alpha_1}^{\alpha_2} q(\phi) d\phi \tag{11.10}$$

$$q_{an} = \frac{1}{\pi} \int_{\alpha_1}^{\alpha_2} q(\phi) \cos(n\phi) d\phi \tag{11.11}$$

and

$$q_{bn} = \frac{1}{\pi} \int_{\alpha_1}^{\alpha_2} q(\phi) \sin(n\phi) d\phi \tag{11.12}$$

The roll pressure distribution in the roll gap may be expressed as a cosine function in the form

$$p(\phi) = -E - F \cos\frac{\pi\phi}{2\beta} \tag{11.13}$$

shown in Fig. 11.3, or alternatively, it may be given as a set of calculated points. In Eq. (11.13) E and F are constants. The function R_b in equations (11.7), (11.8) and (11.9) represents the reactions required to keep the work roll in equilibrium. This can be calculated by expressing the reactions in terms of Fourier series:

$$R_b(\phi) = R_0 + \sum_{n=1}^{\infty} [R_{an} \cos(n\phi) + R_{bn} \sin(n\phi)] \tag{11.14}$$

where again, the coefficients are obtained by the Euler formulae.

In Eqs (11.10)–(11.12) the shear stress distribution is represented by

$$q(\phi) = G \sin\frac{\pi\phi}{2\beta} \tag{11.15}$$

The values E, F and G are indicated in Fig. 11.3. As shown, the maximum value of the roll pressure is given by $E + F$; the period is 2β; entry and exit locations are designated by α_1 and α_2 respectively. G stands for the maximum value of the friction stress. Finally, the coefficients of the stress function expansion – and hence, those for the stresses and strains – are obtained as

$$
\begin{aligned}
c_0 &= \frac{p_{a0}}{2} \\
d_1 &= \frac{p_{b1}}{2r_b} = -\frac{q_{a1}}{2r_b} \\
d_2 &= \frac{p_{a1}}{2r_b} = \frac{q_{b1}}{2r_b} \\
a_{1n} &= \frac{np_{bn} - (n-2)q_{an}}{2n(n-1)r_b^{n-2}} \\
b_{1n} &= \frac{p_{bn} - q_{an}}{2(n+1)r_b^n} \\
a_{2n} &= -\frac{np_{an} - (n-2)q_{bn}}{2n(n-1)r_b^{n-2}} \\
b_{2n} &= \frac{p_{an} + q_{an}}{2(n+1)r_b^n}
\end{aligned}
\tag{11.16}
$$

and the plane strain form of Hooke's law is used to determine the strain components.

A comparison of the ratio of the projected contact length, l'_d, to the one based on rigid rolls and rigid-plastic strips, l_d, as a function of reduction, obtained by Hitchcock's (1935) formula, by the complex variable method of Pietrzyk and Zakrzewski (1978) and by the integration of the biharmonic equation (Roychoudhuri and Lenard 1984) is given for a low carbon and an alloy steel in Fig. 11.4.

The ratio l'_d/l_d is plotted on the ordinate and the reduction is given along the abscissa. It is observed that the assumptions of elliptical pressure distribution and circular, cylindrical rolls underestimate roll flattening significantly for both steels and that the method of complex variables and the integration of the biharmonic equation give results that are in close agreement. The last method will be used in connection with the finite-element approach and in all further calculations in the present chapter.

11.2.2.2 Roll Pressure

In order to supply the required loads for the calculation of the deformed roll shape, the stresses in the plastically flowing strip need to be determined. This involves the derivation and the solution of the equations of equilibrium in both plastic and elastic regions and matching the resulting distributions of stresses at the elastic–plastic interfaces.

The shape of the roll–strip contact surface is not known at this point. The computation of the roll pressure distribution must therefore be combined with that of the deformed roll contour and the accuracy of the iterative technique is controlled by the definition of a suitable error-norm.

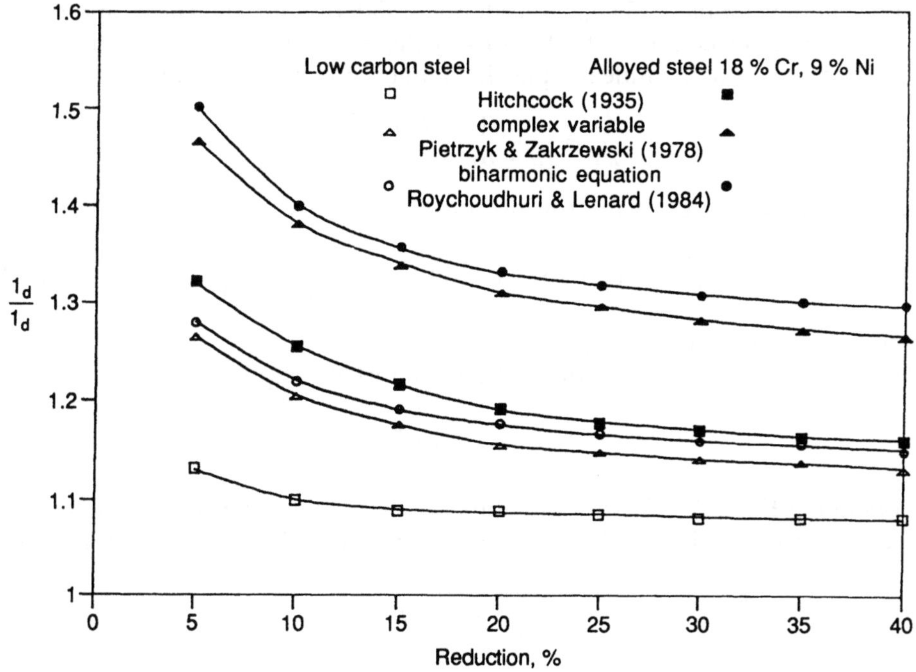

Fig. 11.4. Length of contact during flat rolling (Pietrzyk and Lenard 1989a).

Plastic Region. The equation of equilibrium in the longitudinal direction is obtained as

$$\frac{d}{dx}\left[h\left(p - 2k \mp \tau\frac{dy}{dx}\right)\right] = 2\left(p\frac{dy}{dx} \pm \tau\right) \tag{11.17}$$

where h is strip thickness and k is shear yield stress; the upper sign holds for the exit and the lower sign for the entry region. This convention will be adhered to in all subsequent developments.

The deformed roll shape is given by a function $y = f(x)$. The strip thickness at an arbitrary location is then expressed by

$$h = 2(y + C) \tag{11.18}$$

where C is half of the centre-to-centre distance of the rolls.

The frictional shear stresses at the roll–strip interface may be expressed by the traditional Coulomb–Amonton relationship, according to which $\tau = \mu p$, where μ is a coefficient of friction. In practice this coefficient is usually taken to be a constant. If the adhesion theory of friction is taken to be valid – and as Rabinowicz (1965) points out, this is almost universally accepted – then the frictional coefficient should be related to material and process parameters, (Lim and Lenard 1984, Karagiozis and Lenard 1985) and a functional relationship

$$\mu = \mu\,(\text{material and process parameters, independent variables}) \tag{11.19}$$

should be used. This relation may be an empirical equation or values of μ may be supplied directly in a digital form. Since not nearly enough data are available to allow Eq. (11.19) to be used with confidence, μ = constant is employed in the present development. Then, considering a slab of finite thickness, approximating the shape of its contact with the rolls by the relation $y = a + bx$ and assuming that it behaves in a rigid-ideally plastic manner, Eq. (11.17) reduces to

$$\frac{dp}{dx} = \frac{\pm p\mu(1 + b^2) + 2kb}{(a + bx + C)(1 \mp \mu b)} \tag{11.20}$$

integration of which gives the roll pressure as

$$p = c_1 (a + bx + C)^{\pm \frac{\mu(1 + b^2)}{b(1 \mp \mu b)}} \mp \frac{2kb}{\mu(1 + b^2)} \tag{11.21}$$

where c_1 is a constant of integration.

The boundary condition at the roll–strip interface requires that

$$\sigma_y = p \mp \mu p \frac{dy}{dx} \tag{11.22}$$

Combining Eqs (11.22), with the Huber–Mises yield criterion and the assumed shape of the roll–strip interface then gives another relationship for p and σ_x in the form

$$p = \frac{\sigma_x + 2k}{1 \mp \mu b} \tag{11.23}$$

Eliminating the roll pressure between Eqs (11.21) and (11.23) gives a relationship for the stress in the direction of rolling as

$$\sigma_x = 2k\left[-1 \mp \frac{b(1 \mp \mu b)}{\mu(1 + b^2)}\right] + c_1(1 \mp \mu b)(a + bx + C)^{\pm \frac{\mu(1 + b^2)}{b(1 \mp \mu b)}} \tag{11.24}$$

Knowing σ_{xi} at the boundary of the ith slab, the constant c_1 may be found. Thus σ_{xi+1} is calculated for the next slab by ensuring horizontal equilibrium between the slabs

$$\sigma_{xi+1} = 2k_i\left[-1 \mp \frac{b_i(1 \mp \mu b_i)}{\mu(1 + b_i^2)}\right] + \left[\sigma_{xi} + 2k_i \pm \frac{2k_i b_i(1 \mp \mu b_i)}{\mu(1 + b_i^2)}\right] \times \left(\frac{a_i + b_i x_{i+1} + C}{a_i + b_i x_i + C}\right)^{\pm \frac{\mu(1 + b_i^2)}{b_i(1 \mp \mu b_i)}} \tag{11.25}$$

and the roll pressure on slab i is calculated from Eq. (11.23) as

$$p_i = \frac{\sigma_{xi} + 2k_i}{1 \mp \mu b_i} \tag{11.26}$$

When a straight line segment approximating the roll shape becomes horizontal, i.e. $b = 0$, Eq. (11.21) is no longer valid. This condition exists near the exit region and may also occur if the roll becomes indented. Thus, Eq. (11.20) for this case reduces to

$$\frac{dp}{dx} = \pm \frac{\mu p}{a + C} \tag{11.27}$$

and a derivation similar to the one above gives the two equations for the roll pressure and the stress in the x direction respectively, as

$$p_i = \sigma_{xi} + 2k_i \tag{11.28}$$

and

$$\sigma_{xi+1} = (\sigma_{xi} + 2k_i) \exp\left(\pm\mu \frac{x_{i+1} - x_i}{a_i + C}\right) - 2k_i \tag{11.29}$$

Elastic Regions. The following derivation is to determine the surface pressure distribution as a function of the gauge profile in the elastic zones. In Fig. 11.3 the elastic entry (compression) and elastic exit (recovery) zones are shown along with the plastic zone. It is assumed that if the vertical pressure loading in the elastic exit zone were removed, the strip would recover elastically to a uniform height h_2 from h_{2e}.

Here the elastic exit is analysed. The elastic entry is a mirror image, i.e. the terms multiplied by the coefficient of friction should be multiplied by -1. The derivation now makes use of the equation of equilibrium in the y direction

$$\sigma_y = p\left(\mu \frac{dy}{dx} + 1\right) \tag{11.30}$$

and in the x direction

$$\frac{d\sigma_x}{dx} = -\frac{\sigma_x \dfrac{dy}{dx}}{y + C} + \frac{p\left(\mu - \dfrac{dy}{dx}\right)}{y + C} \tag{11.31}$$

Hooke's law for the plane-strain case is now used to express the strain in the y direction, integration of which defines the change of thickness, $\delta = h - h_u$, as

$$\delta = \int_0^h \epsilon_y \, dy = \frac{1 - \nu^2}{E} \sigma_y h - \frac{\nu(1 + \nu)}{E} \sigma_x h \tag{11.32}$$

Here use was made of the fact that the only independent variable is x. The term h_u designates the undeformed strip thickness. It is equal to h_1, at entry and h_2 at exit. In Eq. (11.32) E is Young's modulus and ν represents the Poisson ratio of the strip. At the end of the region $h = h_2$ and since no more contact with the roll exists

$$\sigma_y = 0 \tag{11.33}$$

As well, the stress in the x direction must equal the back tension, so at that location

$$\sigma_x h_2 = - T_2 \tag{11.34}$$

where T_2 designates the back tension stress per unit width. Since the total change of thickness at the recovery region is

$$\delta = h_2 - h_u \tag{11.35}$$

where h_2 is the thickness after unloading, the exit thickness becomes

$$h_2 = h_u + \frac{\nu(1 + \nu)}{E} T_2 \tag{11.36}$$

Combining Eqs (11.18) and (11.33) with (11.36) yields

$$\sigma_y = \frac{E[2(y + C) - h_u] + 2\sigma_x\nu(1 + \nu)(y + C)}{2(1 - \nu^2)(y + C)} \tag{11.37}$$

Substituting (11.37) in (11.30) and eliminating the roll pressure between that result and Eq. (11.31) gives

$$\frac{d\sigma_x}{dx} = -\sigma_x\left[\frac{\dfrac{dy}{dx}}{y + C} - \frac{\nu\left(\mu - \dfrac{dy}{dx}\right)}{(1 - \nu)(y + C)\left(\mu\dfrac{dy}{dx} + 1\right)}\right]$$

$$+ \frac{1}{2}E\frac{[2(y + C) - h_u]\left(\mu - \dfrac{dy}{dx}\right)}{(1 - \nu^2)(y + C)^2\left(\mu\dfrac{dy}{dx} + 1\right)} \tag{11.38}$$

integration of which gives the stress in the elastic recovery zone in the x direction as

$$\sigma_x = -\frac{Eh_u(\mu - b)(a + bx + C)^{-1}}{2(1 - \nu^2)(\mu b + 1)b(\omega - 1)}$$

$$+ \frac{E(\mu - b)}{(1 - \nu^2)(\mu b + 1)b\omega} + c_1(a + bx + C)^{-\omega} \tag{11.39}$$

where

$$\omega = 1 - \frac{\nu(\mu - b)}{b(1 - \nu)(\mu b + 1)}$$

and c_1 is a constant of integration, determined by ensuring horizontal equilibrium of adjacent slabs. The stress in the x direction is then calculated and σ_y is found from Eq. (11.32). Finally Eq. (11.30) yields the roll pressure.

For the entry region the expressions for p and σ_x remain the same except for the following changes

$$\mu = -\mu, \quad h_2 = h_1, \quad T_2 = T_1$$

The interface between the elastic and plastic zones is determined by checking where and when the Huber–Mises yield criterion is satisfied; see also Eqs (11.40) and (11.41), which give the boundary conditions at entry to the plastic zone as

$$p_{p1} = 2k_1 - \sigma_{x1} + \mu p_1\frac{dy}{dx} \tag{11.40}$$

and at exit from the plastic zone

$$p_{p2} = 2k_2 + \sigma_{x2} \tag{11.41}$$

In Eqs (11.40) and (11.41) σ_{x1} and σ_{x2} designate the stresses, at entry to and exit from the plastic zone, respectively.

11.2.2.3 Roll Force and Roll Torque

The roll separating force is obtained by integrating the roll pressure distribution over the arc of contact which includes the elastic compression and recovery zones. The expression

$$F = p\left[\int_{x_1}^{x_n}\left(1+\mu\frac{dy}{dx}\right)dx + \int_{x_n}^{x_2}\left(1-\mu\frac{dy}{dx}\right)dx + \int_{x_1}^{x_n}\left(\mu-\frac{dy}{dx}\right)dx + \int_{x_n}^{x_2}\left(\mu+\frac{dy}{dx}\right)dx\right] \quad (11.42)$$

is used where x_1 and x_2 define the boundaries of the roll gap. The roll torque is computed by

$$M = p\int_{x_1}^{x_n}\left[x - y\frac{dy}{dx} + \mu\left(y + x\frac{dy}{dx}\right)\right]dx - p\int_{x_n}^{x_2}\left[x - y\frac{dy}{dx} - \mu\left(y + x\frac{dy}{dx}\right)\right]dx \quad (11.43)$$

11.2.3 Comparison of Predictive Capabilities

The predictive capabilities of the mathematical models, described above, are now compared to the experimental data of Shida and Awazuhara (1973) obtained during cold rolling of two types of steel strips with the stress–strain relationships given by

$$\text{steel 1:} \quad \sigma = 324(1 + 19.2\epsilon)^{0.295} \quad (11.44)$$

$$\text{steel 2:} \quad \sigma = 358(1 + 22.7\epsilon)^{0.3} \quad (11.45)$$

The roll radius was 65 mm. The frictional coefficient used in all calculations was 0.08. The entry thickness of the strip was 1 mm for steel 1 and 0.5 mm for steel 2. The results of measurements and calculations are given in Figs 11.5–11.8. In all of the figures roll forces and torques per unit width are plotted on the ordinates and the reduction in percent is given along the abscissa. A general conclusion is that, at least at low reductions, all five techniques of computation give results that are "in the ballpark". Orowan's (1943) technique as programmed by Alexander (1972) proved to be reasonable in predicting both forces and torques. The method of Roychoudhuri and Lenard (1984) did not distinguish itself, as its torque predictions for both steels were poor. The Bland and Ford (1948) analysis gave results very close to those of Orowan (1943).

The coefficient of friction was taken to equal 0.08 in all of these calculations. As is often the practice, this coefficient could be treated as a free parameter, and the predictions of the models could be improved by a judicious choice of μ. The frictional coefficient may also be measured however, and when the experimentally established values are used in the calculations, the predictive capabilities of the models increase significantly. This is shown in Table 11.1 where the data regarding μ, developed by Lim and Lenard (1984) and Karagiozis and

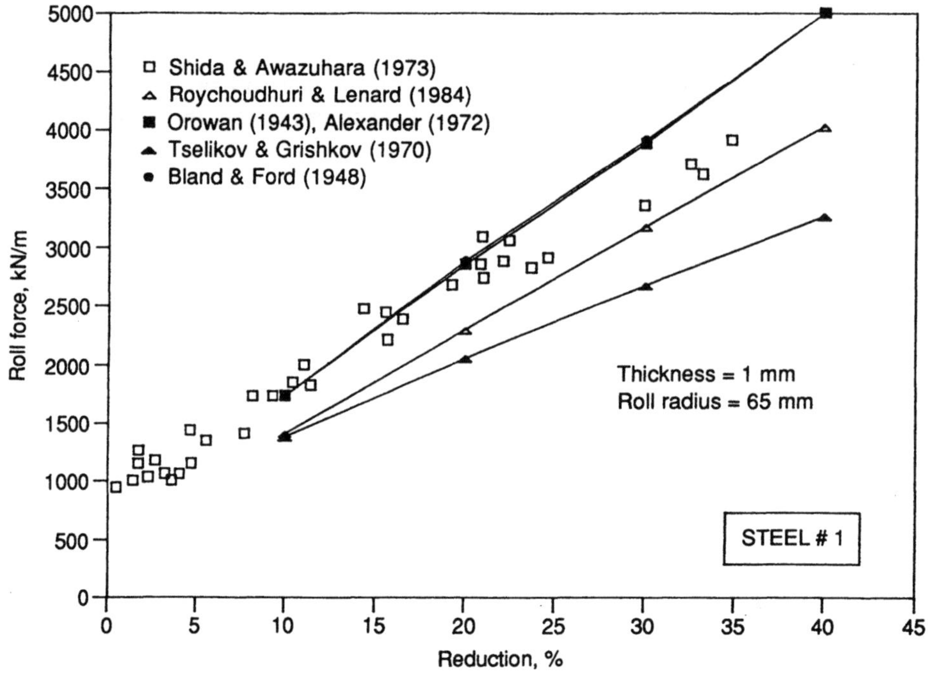

Fig. 11.5. Roll force vs reduction for steel 1 (Lenard 1987).

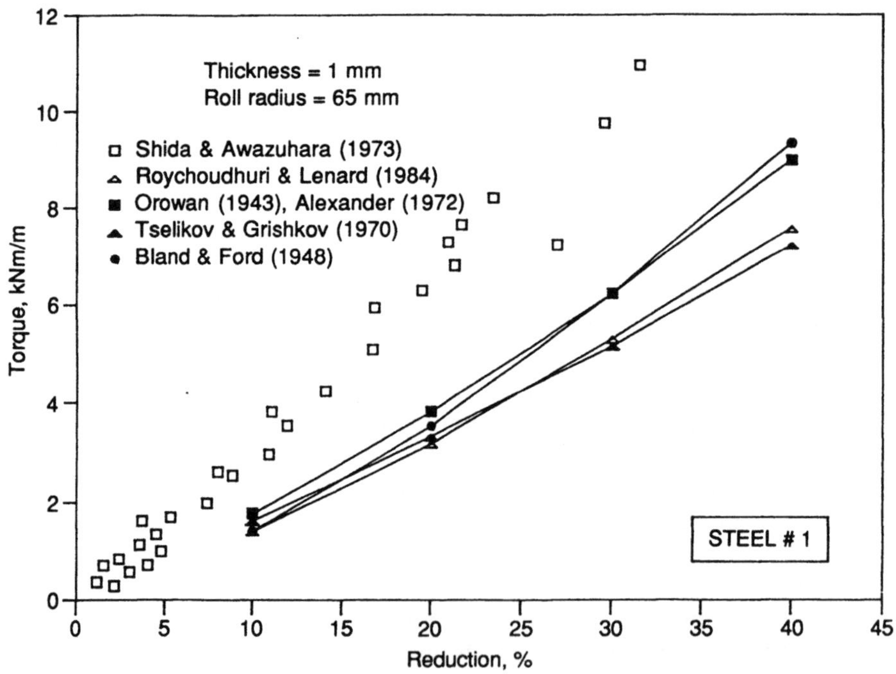

Fig. 11.6. Roll torque vs reduction for steel 1 (Lenard 1987).

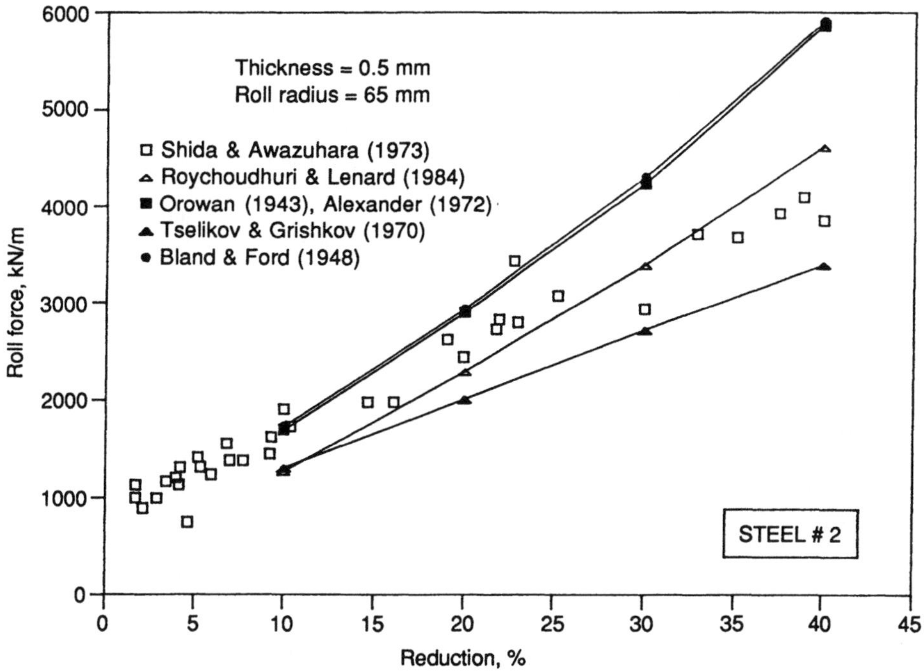

Fig. 11.7. Roll force vs reduction for steel 2 (Lenard 1987).

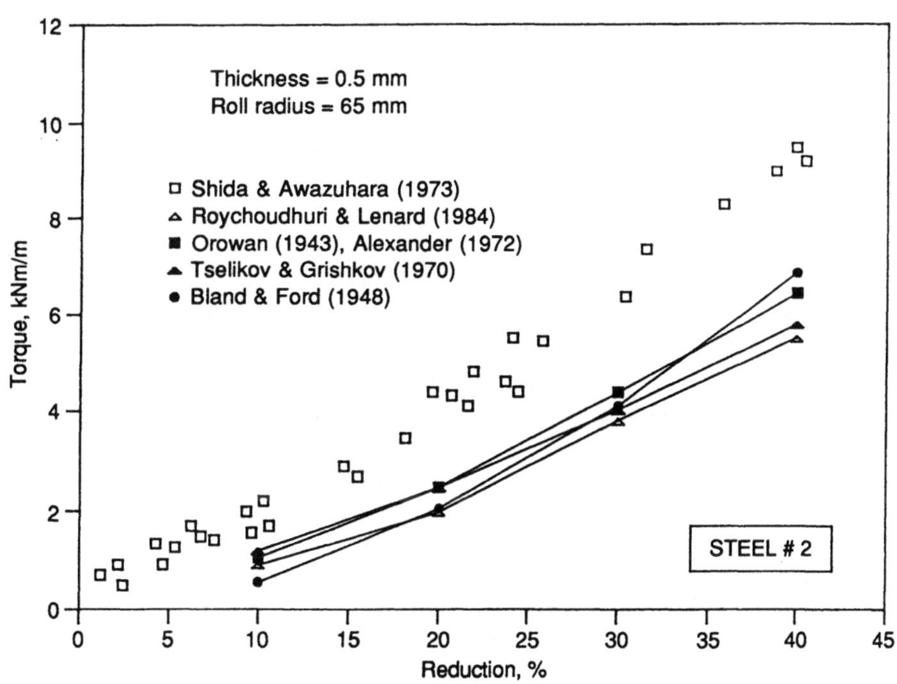

Fig. 11.8. Roll torque vs reduction for steel 2 (Lenard 1987).

Table 11.1. Comparison of measured and calculated roll forces (N mm^{-1}) (Pietrzyk and Lenard 1989b)

Material	Measured	Orowan (1943) μ = const.	Refined model	
			μ = const.	$\mu = \mu(\phi)$
Al 5052–H34	3651	4956	5098	4061
Al 6061–T6	3869	4752	4926	3845
Al 2024–T3	4460	4060	3364	4323

Lenard (1985), are used to demonstrate the validity of that claim. When the coefficient of friction values are digitized and included in the model, noticeable improvements in the predictive capabilities are observed.

It may be concluded at this stage that, as far as the models' ability to compute roll separating forces and roll torques is concerned, all perform in a satisfactory manner. Variations are observed, however, and none is considered to be significant. Inclusion of experimentally determined data concerning friction at the roll–strip interface enhances the ability of the model of Roychoudhuri and Lenard (1984) to calculate roll forces. It is probably premature to attempt to pick the best model. A more likely approach may be to choose the most appropriate model for a particular situation. For large roll radius-to-strip thickness ratios the slab models appear to be satisfactory. For large reductions, roll deformation, realistic friction conditions and the effects of the elastic entry/exit regions need to be considered.

11.2.4 Thermal–Mechanical Finite-Element Model

The assumption of homogeneous compression during a strip's passage through the roll gap is reasonable and leads to negligibly small errors when large roll/strip thickness ratios, typical of cold rolling or hot rolling in the finishing train, are considered. The errors may grow appreciably when slabs or shapes, such as bars, rods or rails, are processed where the nonhomogeneity of the deformation is likely to be very pronounced. Use of average temperatures defining temperature dependent properties may also lead to erroneous appreciation of the parameters of the process. It is then necessary to make use of the finite-element technique to analyse the problem.

The theory necessary to develop computer simulation of transient viscoplastic metal deformation problems coupled with heat transfer analysis has been presented by several authors.

Only a few publications deal with the rolling process however, which presents difficulties connected with the existence of the neutral point and with the cyclic thermal state of the work rolls. Almost all approaches are based on the Fourier equation and it is observed that the sophistication of various solutions was stimulated by the development of computer techniques. The scope of the published work ranges from a highly simplified approach yielding an approximate closed-form solution (Lee et al. 1963) to complex numerical methods giving more accurate predictions.

Wilmotte et al. (1973) calculated the temperature distribution in a strip using the finite-difference method, assuming a uniform increase of temperature in the roll gap and employing Carslaw and Jaeger's (1959) equation to calculate cooling during contact with the work rolls. Bryant and Heselton (1982) carried out similar calculations, neglecting the effect of friction and assuming uniform heat for plastic deformation and uniform temperature across the width of the strip. Bryant and Chiu (1982a,b) developed models for roll temperature calculations in two-high and four-high mills. A more sophisticated solution was presented by Lahoti et al. (1978) who also applied finite-different equations of heat transfer, taking into account nonuniform heat generation in the deformation zone. They determined the strain rate field from a simplified velocity field, assuming uniform strain distribution across the strip. Further, they included the temperature rise of the work roll in the computations. A similar solution with uniform strain-rate distribution across the thickness was presented by Sheppard and Wright (1980).

A still more complex model, describing the thermal behaviour of the strip and the roll simultaneously, was developed by Tseng (1984a,b). He presented numerical solutions for cold and hot rolling, using data obtained from direct measurements of power as input for the heat generated in the deformation zone.

It seems that more advanced approaches were developed for other forming processes which do not present major difficulties in the simulation of metal flow itself. Among a number of finite-element solutions, the one presented by Rebelo and Kobayashi (1980) needs to be pointed out. The authors simulated viscoplastic deformation and heat transfer, introducing separate functionals for the equilibrium and for the energy balance equations. The solution takes into account both positive and negative effects of dislocation density on the yield stress and on the heat generation which are due to strain hardening and to the annihilation occurring from the diffusion driven climb of dislocations. The method was successfully applied to the analysis of viscoplastic deformation and heat transfer in the solid cylinder compression and ring compression processes.

The first application of the finite-element method to analyse the temperature distribution in a strip during hot rolling appears to be due to Zienkiewicz et al. (1981). Several other models of no less significance have been given by Lahoti and Altan (1975), Sigalla (1957), Weber (1973), Pavlossoglou (1981a,b), Cerni et al. (1963), Sellars (1985) and Bertrand-Corsini et al. (1988).

In what follows, the thermal–mechanical model of flat rolling of Pietrzyk and Lenard (1988) will be described. As the title implies, the model consists of two components. All mechanical events, including the determination of deformations, velocities, strains, strain rates and stresses are handled by a rigid-plastic finite-element formulation. The temperature field, computed by the thermal portion, is then used to recalculate all temperature-dependent properties and parameters based on which the first set of dependent variables is recalculated. The iterations continue until the desired accuracy is reached.

11.2.4.1 Mechanical Component

When no body or inertial forces are present the equilibrium equations are

$$\nabla \boldsymbol{\sigma} = 0 \tag{11.46}$$

Writing Eq. (11.46) in a weak form gives

$$\int_V \nabla \boldsymbol{\sigma} \delta \mathbf{v} dV = 0 \tag{11.47}$$

where V is control volume. Applying the divergence theorem to Eq. (11.47) yields

$$\int_V \nabla \boldsymbol{\sigma} \delta \dot{\boldsymbol{\epsilon}} dV = \int_S \mathbf{F} \delta \mathbf{v} dS \tag{11.48}$$

where $\mathbf{F}$ is the vector of tractions, $\boldsymbol{\sigma}$ is the Cauchy stress vector, $\dot{\boldsymbol{\epsilon}}$ is the vector of strain rates and S is the boundary surface, and

$$\mathbf{F} = \boldsymbol{\sigma} \cdot \mathbf{n} \tag{11.49}$$

where $\mathbf{n}$ is the unit normal. Continuity is imposed by the condition

$$\int_V \delta \sigma_{\mathrm{m}} \dot{\epsilon}_{\mathrm{v}} dV = 0 \tag{11.50}$$

where $\dot{\epsilon}_{\mathrm{v}}$ is the volumetric strain rate and σ_{m} the average stress. Discretizing Eq. (11.48) with the condition (11.50), using the Levy–Mises flow rule

$$\dot{\boldsymbol{\epsilon}} = \frac{3}{2} \frac{\dot{\epsilon}_i}{\dot{\sigma}_i} \boldsymbol{\sigma}' \tag{11.51}$$

where $\boldsymbol{\sigma}'$ is the vector of deviatoric stress and $\dot{\sigma}_i$ and $\dot{\epsilon}_i$ are the effective stress and strain rate, and solving the resulting equations, a velocity field is obtained from which the strain rates and stresses are determined (Pietrzyk 1983).

11.2.4.2 The Thermal Component

The equation governing steady-state convective–conductive heat transfer in a flowing material is

$$\nabla^{\mathrm{T}} (k \nabla T) + Q - \rho c_{\mathrm{p}} \mathbf{v} \nabla T = 0 \tag{11.52}$$

where Q is the rate of heat production due to plastic work, T is temperature, ρ is density, $\mathbf{v}$ is the vector of velocity and c_{p} is specific heat. The rate of heat generation may be taken to equal the work of plastic deformation. If energy accumulated due to strain hardening is to be accounted for, Q may be expressed as given by Rebelo and Kobayashi (1980):

$$Q = \int_V \left[\boldsymbol{\sigma}' \delta \dot{\boldsymbol{\epsilon}} dV - \beta A \dot{\epsilon}_i + \beta B \kappa \exp\left(- \frac{D}{KT} \right) \right] dV \tag{11.53}$$

where A and B are material constants, κ is the dislocation density, D is the energy of self-diffusion and K is Boltzmann's constant. When the convective term in Eq. (11.52) is significant, application of the standard Galerkin–Bubnov discretization method leads to meaningless oscillations in the solution. Zienkiewicz (1977) gives a limiting value of the Peclet number as 2, above which the oscillatory solutions appear. In order to overcome this difficulty a special type of upwinded weighting functions should be introduced during discretization (Heinrich and

Zienkiewicz 1977, Heinrich et al. 1977, Zienkiewicz 1977). In the majority of strip rolling processes the velocities are high, the convective term in Eq. (11.52) is of significant magnitude, and use of the upwinded weighting functions is necessary.

Discretization of Eq. (11.52) in the finite-element manner with the upwinded weighting functions and introduction of the relevant boundary conditions yields (Zienkiewicz 1977)

$$\mathbf{HT} + \mathbf{P} = 0 \tag{11.54}$$

where $\mathbf{T}$ is the vector of nodal temperature and

$$H_{ij} = \int_V [(\nabla w_i)k(\nabla N_j) + \rho c_p w_i \mathbf{v} \nabla N_j] dV + \int_S \alpha w_i N_j dS \tag{11.55}$$

$$P_i = -\int_V w_i Q dV - \int_S w_i(\alpha T_0 + q) dS \tag{11.56}$$

where N_j are shape functions and w_i are weighting functions.

The temperature field in the roll is also simulated by Eq. (11.54). The rate of heat generation is then taken to equal zero.

Eq. (11.54) can be used to simulate thermal events in all rolling processes possessing steady-state temperature fields, such as hot and cold flat rolling. A three-dimensional version of this solution is necessary to model temperature variations during rolling of square, round, diamond and oval sections as well as during shape rolling, where steady-state temperature fields also exist. However, these solutions require large computer memory and computational time. There are some rolling processes in which plane-strain deformation can be assumed with reasonable accuracy, but still three-dimensional temperature fields need to be considered. These include all cases of flat, hot rolling of rectangular sections, when plane-strain deformation is a reasonable approximation but intensive cooling of the edges does not allow the application of two-dimensional temperature models. A simplified solution for these cases was proposed by Pietrzyk and Lenard (1988) which is based on the modelling of a two-dimensional nonsteady-state temperature field in a plane which is perpendicular to the strip axis and is allowed to move with the actual strip velocity. The general diffusion equation which is solved now is

$$\nabla^T (k \nabla T) + Q - \rho c_p \frac{\partial T}{\partial t} = 0 \tag{11.57}$$

Discretization of Eq. (11.57) yields

$$\mathbf{HT} + C^T \frac{\partial}{\partial t} \mathbf{T} + \mathbf{P} = 0 \tag{11.58}$$

where

$$H_{ij} = \int_V (\nabla N_i) k (\nabla N_j) dV + \int_S \alpha N_i N_i dS \tag{11.59}$$

$$P_i = \int_V N_i Q dV + \int_S N_i (\alpha T_0 + q) dS \tag{11.60}$$

$$C_{ij} = \int_V N_i c_p \rho N_j dV \tag{11.61}$$

where α is the heat transfer coefficient and T_0 is the roll temperature.

Assuming that the nodal temperatures are linear with respect to time over Δt and applying the weighted residual method gives the following set of linear equations (Pietrzyk and Lenard 1988)

$$\left(2\mathbf{H} + \frac{3}{\Delta t}\mathbf{C}\right)\mathbf{T}_{i+1} = \left(-\mathbf{H} + \frac{3}{\Delta t}\mathbf{C}\right)\mathbf{T}_i - 3\mathbf{P} \tag{11.62}$$

where $\mathbf{T}_i$ is a known vector of nodal temperatures at the beginning of the time interval Δt and $\mathbf{T}_{i+1}$ is an unknown vector of nodal temperatures after Δt. Solution of Eqs (11.62) in the plane perpendicular to the axis of the sample and moving with the actual sample velocity leads to a three-dimensional temperature field in the roll gap using a two-dimensional approach. This nonsteady state approach should also be used to simulate thermal events during rolling of short samples or blooms, where cooling of both ends is of some importance.

11.2.4.3 Boundary Conditions

Both thermal and mechanical boundary conditions adequate for the rolling process are presented in Fig. 11.9. These include $v_y = 0$ and $\sigma_x = 0$ along the axis of symmetry, $\sigma_x = \sigma_y = 0$ on both entry and exit planes if no tensions are applied and $v_y = -v_x \tan \gamma$ on the contact surface. If rolling is performed with tensions then the horizontal components of the stress on the entry and exit planes are taken equal to the tension stresses. The horizontal surface in the front of the roll gap is treated as a free surface and the condition of zero normal velocity has to be satisfied there, achieved by correcting the shape of the free surface during calculations such that it coincides with a stream line (Dawson and Thompson 1978).

The location of the neutral point is not known in advance. The interfacial frictional stresses are therefore expressed in terms of the relative velocities of the strip and the roll (Chen and Kobayashi 1978) in the form

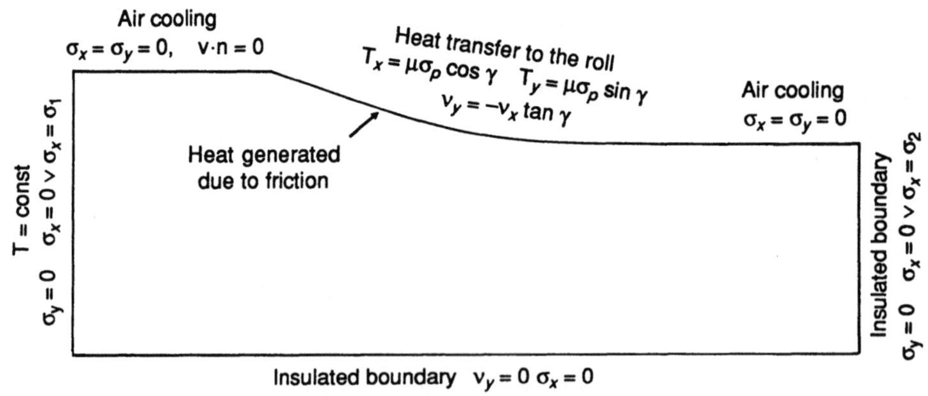

Fig. 11.9. Thermal and mechanical boundary conditions.

$$\tau = \frac{2}{\pi} \mu p \tan^{-1} \frac{v_s}{a} \tag{11.63}$$

where a is a constant, usually assumed to be three orders of magnitude less than the average slip velocity.

Thermal boundary conditions include heat conducted to the roll, heat generated due to friction forces and air cooling of the free surfaces. The exit plane and the surface of symmetry are treated as insulated boundaries and constant temperatures are assumed at the entry plane and at the entrance to the control volume of the roll. The temperature across the roll–workpiece interface is allowed to be discontinuous to account for the discontinuity created by the slip velocity between the surfaces and by the resistance to heat transfer. This latter effect is assumed to be constant.

Heat Transfer Coefficient. As Devadas and Samarasekera (1986) write, experimentally determined values of that coefficient have not been widely reported in the literature. Those that have been published – see, for example, Murata et al. (1984) who gave 7000–10 000 W m^{-2} K^{-1} for rolling without lubrication and with scale present on the surface and 23 000–81 000 W m^{-2} K^{-1} for no scale but with a lubricant, or Stevens et al. (1971) suggesting 38 700 W m^{-2} K^{-1} for hot rolling with water cooled rolls – show the dependence of α on process parameters. Preisendanz et al. (1967) have also provided values of α within the ranges mentioned above. It is also interesting to note that in a recent publication, Silvonen et al. (1987) used 70 000 W m^{-2} K^{-1} as the coefficient of heat transfer at the roll–strip contact surface during plate rolling. These studies have been conducted considering industrial conditions. Harding (1976) has determined heat transfer coefficients in a laboratory setting, arriving at α = 2055 W m^{-2} K^{-1} for 700°C and α = 5100 W m^{-2} K^{-1} for 1100°C, indicating further that the number of parameters that influence the magnitude of the heat flux at the contact surface is large and the choice of the interface heat transfer coefficient for use in modelling is still problematic.

Experimental investigation of the hot rolling process in laboratory conditions (Karagiozis and Lenard 1988) led to a heat transfer coefficient of 4800 W m^{-2} K^{-1} when a constant roll temperature of 40°C is assumed in the calculations (Pietrzyk and Lenard 1989b), close to those measured by Harding (1976). During hot rolling the temperature of the roll increases significantly, however. To investigate the effect of this rise on the predictions the finite-element model, accounting for the increase, was used to calculate the time–temperature profiles during hot rolling of low-carbon steel using various heat transfer coefficients. The results of calculations are compared with the measurements in Fig. 11.10 and the conclusion that α = 13 000 W m^{-2} K^{-1} gives the best simulation of heating and cooling rates is arrived at. This value of the heat transfer coefficient is used in all further calculations for hot rolling of steel in laboratory conditions. In industrial conditions significantly larger values need to be employed.

Slightly different values of the heat transfer coefficient have been reported for warm and hot rolling of aluminium. Semiatin et al. (1987) measured this coefficient using a two-die experiment and a ring compression test. They reported α to be between 4500 and 7500 W m^{-2} K^{-1} in the two-die experiment, depending on pressure, and α between 10 000 and 13 000 W m^{-2} K^{-1} in slow compression and

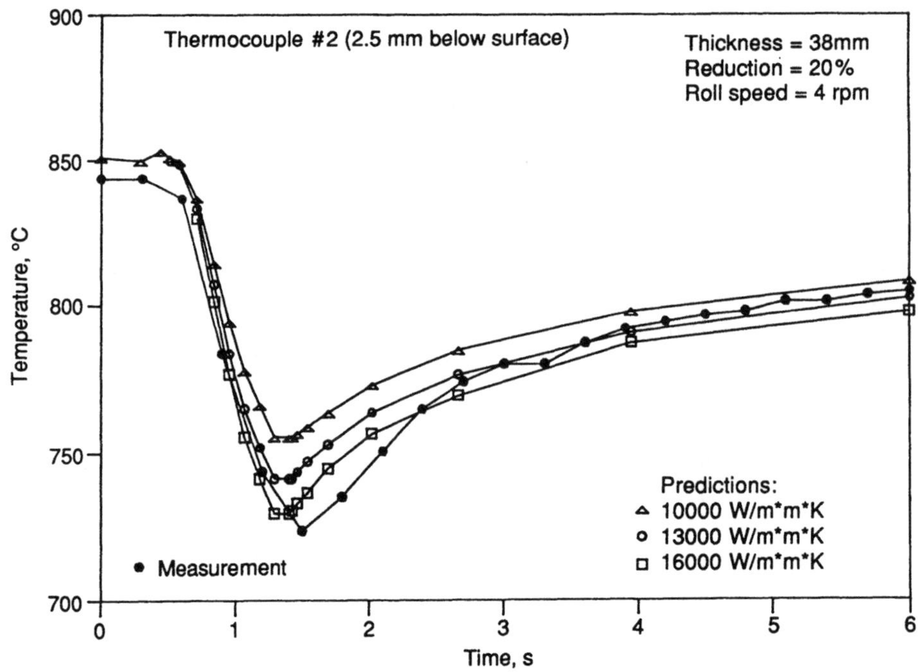

Fig. 11.10. The effect of the heat transfer coefficient on the model's predictions.

between 17 500 and 20 000 W m^{-2} K^{-1} in fast compression. Independently, Smelser and Thompson (1987) gave a value of 30 000 W m^{-2} K^{-1} for rolling of aluminium samples at 250°C. Calculations carried out with constant roll temperature led to α = 20 000 W m^{-2} K^{-1} (Pietrzyk and Lenard 1989b), which is close to the value obtained by Semiatin et al. (1987). However, when roll temperature calculations are introduced (Pietrzyk and Lenard 1990) the best agreement between measured and calculated time–temperature profiles is obtained for α = 30 000 W m^{-2} K^{-1} (Fig. 11.11), identical to that reported by Smelser and Thompson (1987).

Investigation of the thermal events in the cold rolling process shows that the heat transfer coefficient is difficult to determine by measurement of temperatures and that its influence on the predictions is not very significant (Pietrzyk and Lenard 1989b).

11.2.4.4 Substantiation of the Predictions

The model described above was used to simulate metal flow and heat transfer during hot rolling of low carbon steel and warm and cold rolling of aluminium. Rigid-plastic behaviour of the material was assumed in the velocity field calculations. The yield strength of the material was introduced as a function of temperature, strain and strain rate. In warm rolling processes plane strain compression tests were used to determine the flow strength. In hot rolling of low-carbon steel the equations of Shida (1974), which give the flow strength as a function of carbon content, strain, strain rate and temperature, were used. For cold rolling of aluminium the strain hardening curve obtained in plane-strain compression was employed.

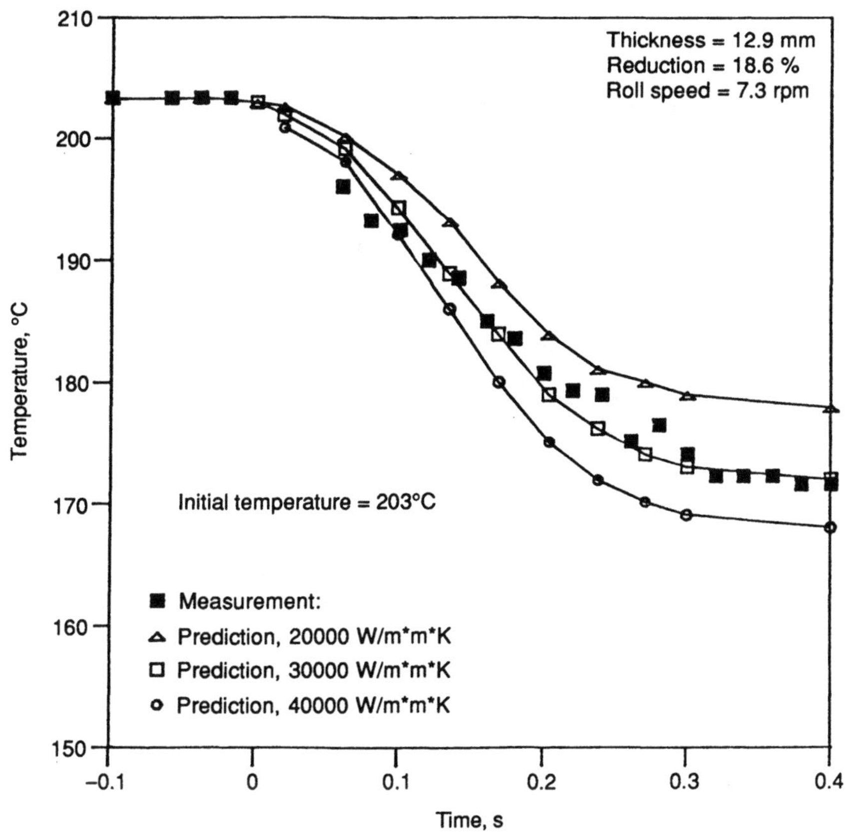

Fig. 11.11. The effect of the heat transfer coefficient on the model's predictions in warm rolling (Pietrzyk and Lenard 1990).

Predictions of the model were validated by comparing them with the results of measurements carried out in laboratory conditions. The details of the experimental programme, the equipment, materials and procedure have been given by Karagiozis and Lenard (1988). Briefly, low carbon steel slabs measuring 19 × 50.8 × 152.4 mm and 38 × 50.8 × 152.4 mm, were rolled on a two-high mill with work-rolls of 254 mm diameter at various temperatures and speeds to various reductions. The slabs had several thermocouples embedded in their tail end and by monitoring the thermocouples' output the time–temperature profile for each was determined. Aluminium samples of 13 mm thickness with two embedded thermocouples were used in the investigation of the warm rolling process at 150°C and 250°C. Cold rolling was performed on 19 and 13 mm thick aluminium samples with one centrally embedded thermocouple.

Typical comparisons of measured and calculated time–temperature profiles for hot rolling of low carbon steel are presented in Figs 11.12 and 11.13. In Fig. 11.12 the roll speed was 4 rpm and the entry temperature at the centre was 763°C. It is observed that using $\alpha = 13000$ W m^{-2} K^{-1} leads to cooling and heating rates and final temperatures that differ little from the measured ones. In Fig. 11.13 a thinner sample was used and again reasonable correspondence between the computed and recorded temperature values is noted. The differences between the predicted and measured cooling rates are somewhat larger than for the 38 mm

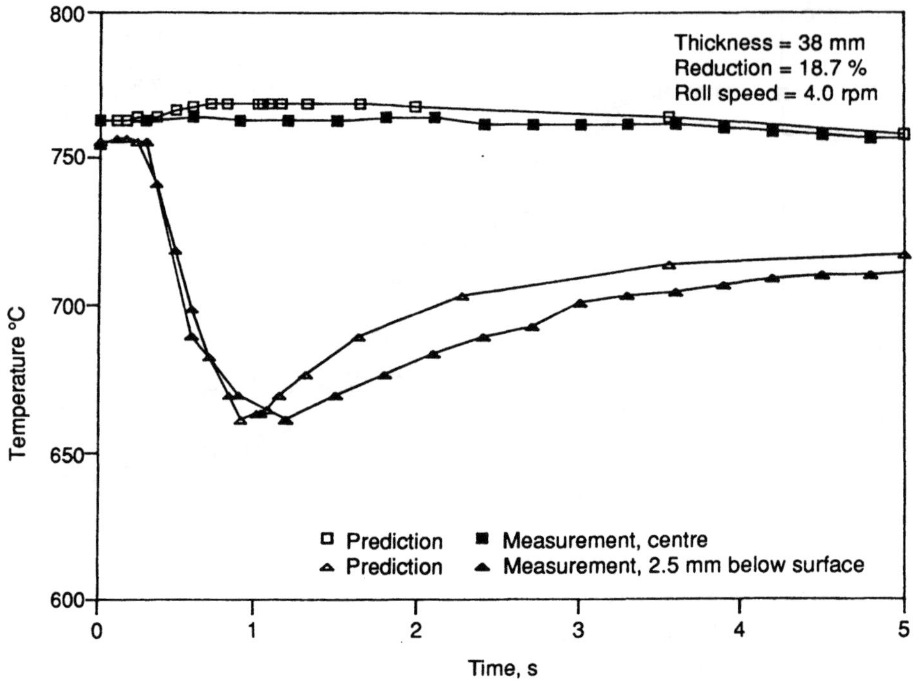

Fig. 11.12. Comparison of predictions and measurements during hot rolling (Pietrzyk and Lenard 1990).

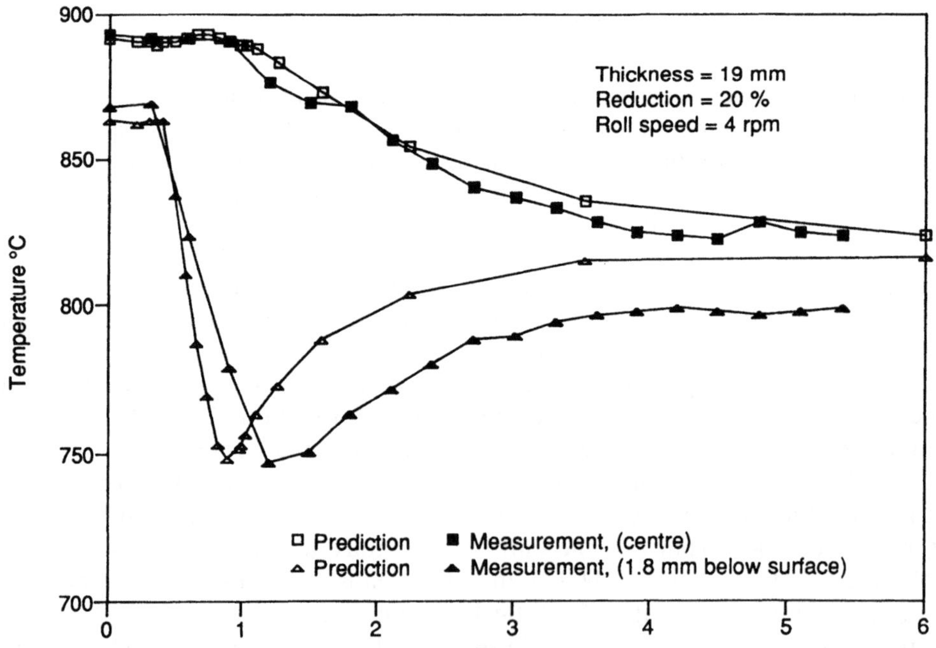

Fig. 11.13. Comparison of predictions and measurements during hot rolling (Pietrzyk and Lenard 1990).

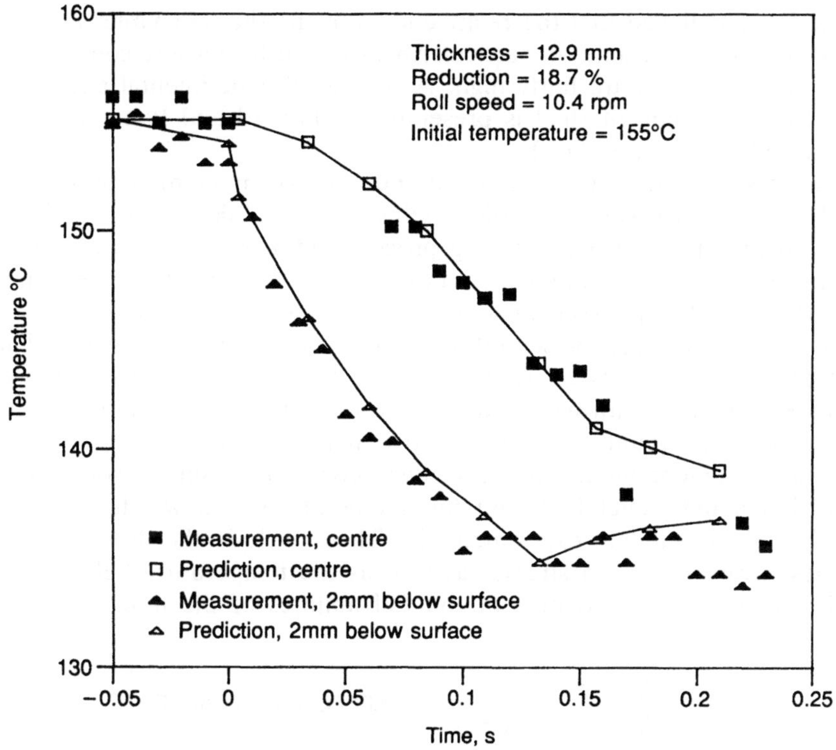

Fig. 11.14. Warm rolling of aluminium (Pietrzyk and Lenard 1990).

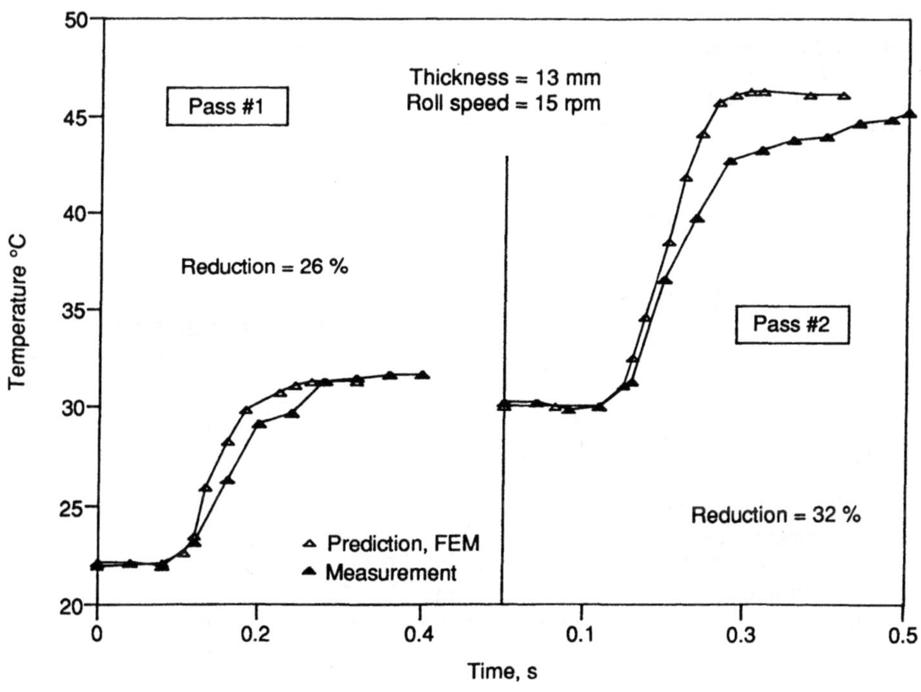

Fig. 11.15. Temperature rise during cold rolling of aluminium.

slabs and it is believed that the relative sizes of the thermocouples and the strip thicknesses are responsible, in particular near the surface where cooling rates and temperature gradients are fairly high. A full set of experimental and theoretical results for hot rolling of steel is presented by Lenard and Pietrzyk (1989) and Pietrzyk and Lenard (1989a,b).

Similar results obtained during warm and cold rolling aluminium are shown in Figs 11.14 and 11.15. A heat transfer coefficient $\alpha = 30\,000 \text{ W m}^{-2} \text{ K}^{-1}$ was used in the simulation of the warm rolling process and a reasonably good agreement between measured and predicted cooling rates and final temperatures was obtained. Some further comparisons of measurements and calculations carried out for this process are presented by Pietrzyk and Lenard (1989b). The cold rolling process was simulated using $\alpha = 7500 \text{ W m}^{-2} \text{ K}^{-1}$ and it is observed that the model predicts deformation heating very well. More results for the cold rolling process are found in Pietrzyk and Lenard (1989a).

The finite-element model of the rolling process may supply some additional information about metal flow and the temperature field which, however, are more difficult to validate experimentally. These include the distribution of the velocities, strain rates and stresses as well as the temperature field in both roll and the strip. In order to present the full abilities of the model some typical

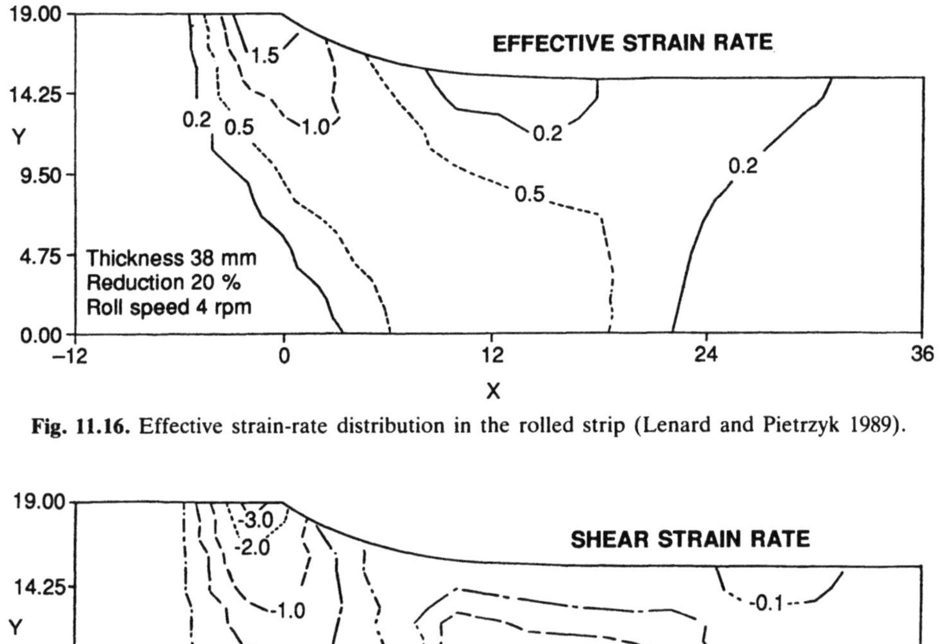

Fig. 11.16. Effective strain-rate distribution in the rolled strip (Lenard and Pietrzyk 1989).

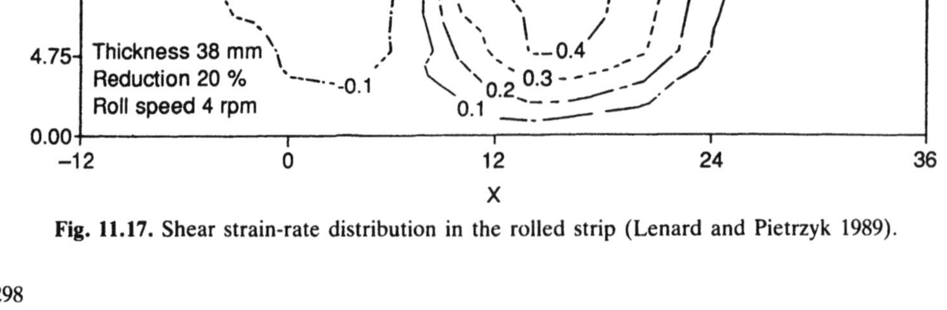

Fig. 11.17. Shear strain-rate distribution in the rolled strip (Lenard and Pietrzyk 1989).

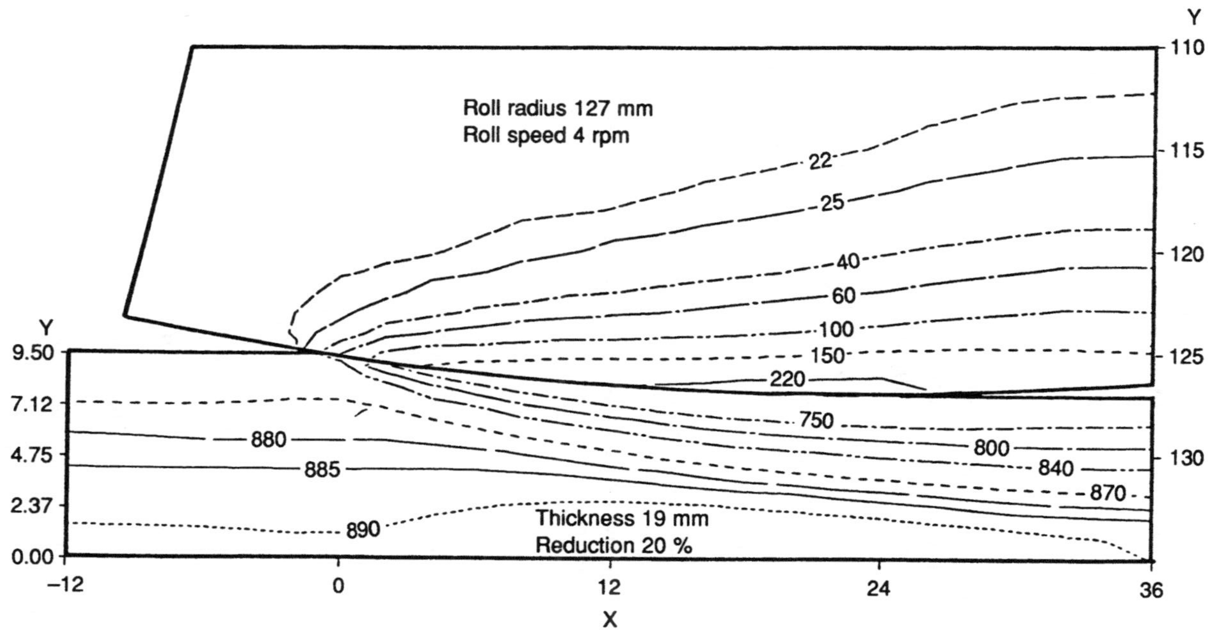

Fig. 11.18. Temperature distribution in the strip and the roll (Pietrzyk and Lenard 1990).

results of calculations are given in Figs 11.16–11.18. Distribution of the effective strain rate in the roll gap during rolling of a 38 mm thick sample is presented in Fig. 11.16. A significant variation of the strain rate across the sample is observed, indicating that use of average strain rates in calculating rate dependent properties may lead to significant errors. Fig. 11.17 shows the distribution of the shear strain rate in the roll gap and the temperature distribution in both the work roll and the strip is presented in Fig. 11.18. A very sharp temperature gradient close to the contact surface is observed in addition to some heating in the central part of the sample.

11.3 Conclusions

Concerning the question "Which model is to be used under a specific set of circumstances?", a preliminary answer may be given, of course subject to corrections as computational and experimental techniques improve and more data are generated.

If hot strip rolling is considered and only a general "feel" for roll forces and torques is needed, Sims' method, using an average temperature in the pass, should be adequate. If thermal–mechanical processing of microalloyed strips, including adaptive control of the hot strip mill and predictions of the evolution of the microstructure are required, use of at least a two-dimensional thermal–mechanical model is suggested. The length of computations in such a model, of course, precludes its use in an "on-line" situation. Research to overcome this problem – perhaps by precalculating, storing, and retrieving numbers as required – should be undertaken.

References

Alexander JM (1972) On the theory of rolling. Proc R Soc Lond A 326:535–563

Al-Salehi FAR, Firbank TC, Lancaster PR (1973) An experimental determination of the roll pressure distribution in cold rolling. Int J Mech Sci 15:693–710

Atreya A, Lenard JG (1979) A study of cold strip rolling. ASME J Engng Mat Technol 101:129–134

Bertrand-Corsini C, David C, Bern A et al. (1988) A three dimensional thermomechanical analysis of steady flows in hot forming processes. Application to hot flat rolling and hot shape rolling. In: Chenot JL, Oñate E (eds) Modelling of metal forming processes. Kluwer Academic Publishers, pp 271–279

Bland DR, Ford H (1948) The calculation of roll force and torque in cold strip rolling with tensions. Proc Inst Mech Engrs 159:144–153

Bryant GF, Chiu TSL (1982a) Simplified roll-temperature model: convective cooling. Met Technol 9:478–484

Bryant GF, Chiu TSL (1982b) Simplified roll-temperature model: spray-cooling and stress effects. Met Technol 9:485–492

Bryant GF, Heselton MO (1982) Roll-gap temperature models for hot mills. Met Technol 9:469–477

Carslaw HS, Jaeger JC (1959) Conduction of heat in solids. Oxford University Press, London

Cerni S, Weinstein AS, Zorowski C (1963) Temperatures and thermal stresses in the rolling of metal strip. In: Iron and Steel Engineering Yearbook, pp 165–171

Chen CC, Kobayashi S (1978) Rigid-plastic finite-element analysis of ring compression. Appl Num Meth Forming Proc 28:163–174

Dawson PR, Thompson EG (1978) Finite element analysis of steady state elasto-visco-plastic flow by the initial stress rate method. Int J Num Meth Engng 12:47–57

Devadas C, Samarasekera IV (1986) Heat transfer during hot rolling of steel strip. Ironmaking Steelmaking 13:311–321

Ford H, Alexander JM (1963) Simplified hot rolling calculations. J Inst Metals 92:397–403
Golten JW (1969) Analysis of cold rolling with particular reference to roll deformation. PhD dissertation, University of Wales, Swansea, UK
Harding RA (1976) PhD dissertation, University of Sheffield, Sheffield, UK
Hartley P, Sturgess CEN, Liu C, Rowe GW (1989) Experimental and theoretical studies of workpiece deformation, stress and strain during flat rolling. Int Mater Rev 34(1):19–34
Heinrich JC, Zienkiewicz OC (1977) Quadratic finite element schemes for two-dimensional convective transport problems. Int J Num Meth Engng 11:1831–1844
Heinrich JC, Huyacorn AR, Mitchell AR, Zienkiewicz OC (1977) An 'Upwind' finite element scheme for two-dimensional convective transport equation. Int J Num Meth Engng 11:131–143
Hitchcock JH (1935) Roll neck bearings. ASME research publication
Jeswiet J, Rice WB (1982) The design of a sensor for measuring normal pressure and friction stress in the roll gap during cold rolling. Proceedings NAMRC X, pp 130–134
Jortner D, Osterle JF, Zorowski CF (1958) An analysis of the mechanics of cold strip rolling. In: Iron and steel engineering yearbook, pp 403–414
Karagiozis AN, Lenard JG (1985) The effect of material properties on the coefficient of friction in cold rolling. In: Proceedings Eurotrib 85, Lyon, pp 2–13
Karagiozis AN, Lenard JG (1988) Temperature distribution in a slab during hot rolling. ASME J Engng Mater Technol 110:17–21
Lahoti GD, Altan T (1975) Prediction of temperature distribution in axisymmetric compression and torsion. ASME J Engng Mater Technol 97:113–120
Lahoti GD, Shah SN, Altan T (1978) Computer-aided analysis of the deformations and temperatures in strip rolling. ASME J Engng Ind. 100:159–166
Lee PW, Sims RB, Wright H (1963) A method for predicting temperatures in continuous hot strip mills. J Iron Steel Inst 201:270–279
Lenard JG (1987) Study of the predictive capabilities of mathematical models of flat rolling. In: Proc 4th Int steel rolling conference, Deauville, France, pp 3.4.1–4.10
Lenard JG, Pietrzyk M (1989) The predictive capabilities of a thermal model of flat rolling. Steel Res 60:403–406
Lenard JG, Wang F, Nadkarni G (1987) The role of constitutive formulation in the analysis of hot rolling. ASME, J Engng Mater Technol 109:343–349
Leon HW (1987) Steel strip-casting programme. Report to Industrial Materials Research Institute, Boucherville, Canada
Lim L-S, Lenard JG (1984) Study of friction in cold strip rolling. ASME J Engng Mater Technol 106:139–146
Michell JH (1900) Some elementary distributions of stress in three dimensions. Proc Lond Math Soc 32:23–35
Murata K, Morise H, Mitsutsuka M, Naito H, Komatsu T, Shida S (1984) Heat transfer between metals in contact and its application to protection of rolls, Trans Iron Steel Inst Jpn 24:B-309
Orowan E (1943) The calculation of roll pressure in hot and cold flat rolling. Proc Inst Mech Engrs 150:140–167
Pavlossoglou J (1981a) Mathematical model of the thermal field in continuous hot rolling of strip and simulation of the process. Arch Eisenhuttenwes 52:153–158
Pavlossoglou J (1981b) Mathematical model of the thermal field of the rolls and the strip in continuous hot rolling including heat losses under protective atmosphere and without water spray cooling. Arch Eisenhuttenwes 52:275–279
Pietrzyk M (1983) Walcowanie wyrobow plaskich – modele matematyczne. Metal Odlew no 97 (special issue)
Pietrzyk M, Lenard JG (1988) Experimental substantiation of modelling heat transfer in hot flat rolling. In: Jacobs HR (ed) Proceedings 25th National Heat Transfer Conference, Houston, HTD-96, vol 3, pp 47–53
Pietrzyk M, Lenard JG (1989a) A study of heat transfer during flat rolling. In: Thompson EG, Wood RD, Zienkiewicz OC, Samuelsson A (eds) Proceedings NUMIFORM'89, Fort Collins, pp 343–348
Pietrzyk M, Lenard JG (1989b) Boundary conditions in hot/cold flat rolling, Proceedings Conf. COMPLAS'89, Barcelona
Pietrzyk M, Lenard JG (1990) The effect of the temperature rise of the roll on the simulation of the flat rolling process. J Mater Proc Technol 22(2):177–190
Pietrzyk M, Zakrzewski J (1978) Methods to determine roll elastic flattening. Metal Odlew 4:107–124
Preisendanz H, Schuler P, Koschnitzke K (1967) Calculation of temperature variations at the surface

of hot ingots during contact with rolls. Arch Eisenhuttenwes 38:205–213
Rabinowicz E (1965) Friction and wear of materials. John Wiley, New York
Rebelo N, Kobayashi S (1980) A coupled analysis of viscoplastic deformation and heat transfer – II. Int J Mech Sci 22:707–718
Roychoudhuri R, Lenard JG (1984) A mathematical model of cold rolling – experimental substantiation. In: Proc 1st International Conference on Technology of Plasticity, Tokyo, pp 1138–1145
Sellars CM (1985) Computer modelling of hot working processes. Mat Sci Technol 1: 325–332
Semiatin SL, Collins EW, Wood VE, Altan T (1987) Determination of the interface heat transfer coefficient for non-isothermal bulk-forming processes. ASME J Engng Ind. 109:49–57
Sheppard T, Wright DS (1980) Structural and temperature variations during rolling of aluminium slabs. Metals Technol 7:274–281
Shida S (1974) Effect of carbon content, temperature and strain rate on compressive flow stress of carbon steels. Hitachi research laboratory report, pp 1–9
Shida S, Awazuhara H (1973) Rolling load and torque in cold rolling. Jpn Soc Technol Plast 14:267–278
Sigalla A (1957) The cooling of hot steel strip with water jets. J Iron Steel Inst 186:90–93
Silvonen A, Malinen A, Korhonen AS (1987) A finite element study of plane strain hot rolling. Scand J Metal 16:103–108
Sims RB (1954) The calculation of roll force and torque in hot rolling mills. Proc Inst Mech Engrs 168:191–200
Smelser RE, Thompson EG (1987) Validation of flow formulation for process modelling. In: Nakazawa S, Willam K, Rebelo N (eds) Proceedings symposium on advances in inelastic analysis, ASME Winter Annual Meeting, Boston, AMD-88, PED-28, pp 273–282
Stevens PG, Ivens KP, Harper PJ (1971) Increasing work-roll life by improved roll cooling practice. J Iron Steel Inst 209:1–11
Tselikov AI (1939) The influence of external friction and tension on roll face pressure. Metallurgiya, 6:61–76
Tselikov AI, Grishkov, AI (1970) Tieoriya Prokatki, Metallurgiya, Moscow
Tseng AA (1984a) A numerical heat transfer analysis of strip rolling. ASME J Heat Trans 106:512–517
Tseng AA (1984b) A generalized finite difference scheme for convection dominated metal forming problems. Int J Num Meth Engng 20:1885–1900
van Rooyen GT, Backofen WA (1957) Friction in cold rolling. J Iron Steel Inst 186:235–244
Weber KK (1973) Mathematical modelling of temperature variations during the hot rolling of plate and strips. Neue Hutte 18:285–295
Wilmotte S, Mignon J, Economopoulos M, Thomas G (1973) Model of the evolution of the temperature of the strip in the hot strip mill. CRM 36:35–44
Zienkiewicz OC (1977) The finite element method. McGraw Hill
Zienkiewicz OC, Oñate E, Heinrich JC (1981) A general formulation for coupled thermal flow of metals using finite elements. Int J Num Meth Engng 17:1497–1514

12 **Extrusion and Drawing**

K. Mori

12.1 Introduction

In extrusion and drawing, a cross-section of a material is reduced by passing through a die opening having a desired shape. Steady-state deformation is caused in these processes except for the start and the end of deformation. For the analysis of steady-state deformation in extrusion and drawing, the slab method, the slip line field method and the upper bound method have been applied.

Early applications of the finite-element method to extrusion and drawing processes were carried out with an elastic-plastic analysis in the early 1970s. Iwata et al. (1972) analysed axisymmetric and plane-strain deformation in hydrostatic extrusion by the elastic-plastic finite-element method. Nayak and Zienkiewicz (1972) and Gordon and Weinstein (1974) used the elastic-plastic finite-element method in analyses of plane-strain forward extrusion and drawing, respectively. Lee et al. (1973) calculated the distribution of residual stress in extruded billets. These analyses, however, were limited to small plastic deformation in the initial stages of deformation, because the elastic-plastic finite-element method was formulated on the basis of the infinitesimal deformation theory. To simulate more extensive plastic deformation, the elastic-plastic finite-element method was formulated on the basis of the finite deformation theory of plasticity. Lee and Mallett (1977) and Yamada and Hirakawa (1978) analysed plane-strain extrusion with curved dies by the method with finite deformation theory. The curved die can avoid the singularity of sharp metal flow around the die corners of the entry and the exit. The elastic-plastic finite-element method accumulates a large error in stress for elements passing around the die corners by the singularity. Bar extrusion (Argyris and Doltsinis 1979, Aravas 1986), backward extrusion (Hartley et al. 1980), hydrostatic extrusion (Xutang et al. 1983), residual stress in extruded bars (Hartley and Dehghani 1987), wire drawing (Gerhardt and Tekkaya 1987) and temperature distributions in wire drawing (Lugt and Huetink 1986) have been analysed by the elastic-plastic finite-element method.

The rigid-plastic finite-element method was applied to simulation of extrusion and drawing in the mid-1970s. Lung and Mahrenholtz (1973–74) simulated steady-state deformation in plane-strain drawing by the rigid-plastic finite-element method. Zienkiewicz and Godbole (1974) and Dawson (1978) treated strain-rate sensitivity of the material in extrusion in the viscoplastic approach. In the rigid-plastic analysis, the error of stress around the die corner does not accumulate, because the stress is determined at each state without referring to the stress at the previous stages. The rigid-plastic finite-element method has been employed for both the simulations of steady-state deformation and nonsteady-state deformation in extrusion and drawing. Bar extrusion (Kobayashi and Shah 1978 and Tayal and Natarajan 1981a), backward extrusion (Zienkiewicz and Godbole 1975 and Mori et al. 1983), three-dimensional extrusion (Zienkiewicz 1984), multipass extrusion (Chen and Kobayashi 1978), side extrusion (Tomita 1982), radial extrusion (Osen 1987), combined backward–forward extrusion (Tayal and Natarajan 1982), extrusion of composite materials (Tayal and Natarajan 1981b and Pacheco and Alexander 1982), temperature distribution in extrusion (Zienkiewicz et al. 1978), bar drawing (Chen et al. 1979), tube drawing (Tomita and Sowerby 1978) and temperature distribution in drawing (Asada et al. 1981) have been analysed by the rigid-plastic finite-element method. The occurrence of central bursting in extrusion has been predicted from the calculated stress and strain by the use of the criterion for ductile fracture (Oh et al. 1979 and Ayada et al. 1987).

In this chapter, examples for simulation of extrusion and drawing by the rigid-plastic finite-element method are given.

12.2 Method of Modelling

12.2.1 Rigid-Plastic Finite-Element Method

The present rigid-plastic finite-element method is formulated on the basis of the plasticity theory for a material with slight compressibility (Osakada et al. 1982). In the plasticity theory, the material is assumed to obey the yield criterion

$$\bar{\sigma} = \sqrt{\tfrac{1}{2}\{(\sigma_x - \sigma_y)^2 + (\sigma_y - \sigma_z)^2 + (\sigma_z - \sigma_x)^2 + 6(\tau_{xy}^2 + \tau_{yz}^2 + \tau_{zx}^2)\} + g\sigma_m^2} \quad (12.1)$$

where $\bar{\sigma}$ is the equivalent stress, σ_m is the hydrostatic stress and g is a small positive constant (0.01–0.0001). The volume of the material changes during plastic deformation because of the pressure sensitivity, and thus the stress can be calculated directly from the strain rates

$$\sigma_{ij} = \frac{\bar{\sigma}}{\dot{\bar{\epsilon}}}\left\{\tfrac{2}{3}\dot{\epsilon}_{ij} + \delta_{ij}\left(\frac{1}{g} - \frac{2}{9}\right)\dot{\epsilon}_v\right\} \quad (12.2)$$

where $\dot{\bar{\epsilon}}$ is the equivalent strain rate, $\dot{\epsilon}_v$ is the volumetric strain rate and δ_{ij} is the Kronecker delta. The equivalent strain rate for this material is expressed by

$$\dot{\bar{\epsilon}} = \sqrt{\tfrac{2}{9}\{(\dot{\epsilon}_x - \dot{\epsilon}_y)^2 + (\dot{\epsilon}_y - \dot{\epsilon}_z)^2 + (\dot{\epsilon}_z - \dot{\epsilon}_x)^2 + \tfrac{3}{2}(\dot{\gamma}_{xy}^2 + \dot{\gamma}_{yz}^2 + \dot{\gamma}_{zx}^2)\} + \frac{1}{g}\dot{\epsilon}_v^2} \quad (12.3)$$

To obtain the exact solution, the following functional Φ for the material is minimized

$$\Phi = \int_V \left[\int^{\dot{\bar{\epsilon}}} \bar{\sigma} d\dot{\bar{\epsilon}}\right] dV + \int_{S_f} \tau_f \Delta v dS \quad (12.4)$$

where τ_f is the frictional shear stress and Δv is the relative velocity between the workpiece and the tool. The functional takes account of not only the work-hardening characteristic of the flow stress but also the strain-rate sensitivity.

The equivalent strain rate is expressed as the function of the volumetric strain rate from the second term in the square root of the right-hand side of Eq. (12.3). Therefore, the condition of incompressibility is approximately satisfied by a small positive value of g in the minimization of the functional. In the present method, the constraint by the condition of incompressibility is incorporated into the functional. The present method is analogous to the penalty function method in the rigid-plastic finite-element scheme.

12.2.2 Frictional Boundary Conditions

Since the law of Coulomb friction has been often used as the frictional criterion in metal forming, it is convenient to employ the law for practical purposes. In this case, the frictional shear stress τ_f is expressed by using a coefficient of friction μ as

$$\tau_f = \mu p \tag{12.5}$$

where p is the contact pressure. The frictional shear stress τ_f is approximated by adopting the contact pressure at the previous stage of iteration in the minimizing procedure of the functional. The frictional shear stress is approximated to be $\bar{\sigma}/\sqrt{3}$ when μp is larger than $\bar{\sigma}/\sqrt{3}$.

The rigid-plastic finite-element method can also deal with a constant frictional shear factor m as

$$\tau_f = mk \tag{12.6}$$

where k is the shear yield stress.

12.2.3 Treatment of Rigid Zone

In the rigid-plastic finite-element approach, the whole region of the material is assumed to be deforming plastically. In continuous processes such as extrusion and drawing, however, nondeforming regions are always attached to the regions before entering the die opening and after passing through the opening. The calculated equivalent strain rates of the elements in those nondeforming regions are extremely small. In the simulation of forming processes with nondeforming regions, the convergence of the solution deteriorates and there is more scatter in the stress values, because the stress components have the term of the equivalent strain rate in the denominator. A linear constitutive equation was presumed for the regions having small equivalent strain rate (Chen et al. 1979). The use of the two constitutive equations for the fully plastic and the nearly rigid regions, however, complicates the calculation, because the locations of the rigid zones are not known beforehand.

Let us assume the material to be strain-rate sensitive or viscous in the low strain-rate region (Osakada et al. 1982), and the following constitutive equation is employed:

$$\bar{\sigma}' = \frac{\bar{\sigma}}{\sqrt{1 + (e/\dot{\bar{\varepsilon}})^2}} \tag{12.7}$$

where e is a small positive constant compared with the equivalent strain rate in the plastically deforming region.

From Eq. (12.4), the functional for this material in the rigid-plastic analysis is expressed by

$$\Phi' = \int_V \bar{\sigma}\sqrt{\dot{\bar{\varepsilon}}^2 + e^2}\, dV + \int_{S_f} \tau_f \Delta v dS \tag{12.8}$$

Since the simultaneous equations obtained by differentiating the functional with respect to the nodal velocities have the term of $1/\sqrt{\dot{\bar{\varepsilon}}^2+e^2}$, which never becomes infinite, high convergence of the functional is attained. It appears that the effect of the value e is negligible in the fully plastic region. This method utilizes the strain-rate sensitivity in the low strain-rate region.

12.2.4 Steady-State Scheme

In the simulation of steady-state deformation by the elastic-plastic finite-element method, the calculation is started from an initial state for the front end of the workpiece, and is repeated until a steady state is reached. In the nonsteady-state scheme, the element mesh is fixed to the material (Lagrangian formulation). The steady-state scheme can also be adopted in the rigid-plastic finite-element simulation. In this scheme, the stream lines and the flow stress distribution are iteratively renewed until convergence is attained. The computing time for the steady-state scheme is generally shorter than that for the nonsteady-state scheme. Since the elastic-plastic finite-element method requires the integration of stress-rate components along the stream lines, it is difficult to apply the steady-state scheme under these conditions.

The procedure of the steady-state scheme for work-hardening materials (Mori et al. 1982) is given as follows.

1. The distributions of velocity and equivalent strain rate for the material with the distribution of the assumed equivalent stress are calculated.
2. Stream lines are located by using the velocity distribution, and the equivalent strain distribution is calculated by integrating the equivalent strain rate along the stream lines starting from the entry boundary.
3. By using the equivalent strain distribution, the equivalent stress distribution is calculated from the flow curve, and the solution is again obtained.
4. Until the calculated result converges, 2 and 3 are repeated.

For the analysis of nonwork-hardening materials, the iterative calculation for converging the flow stress distribution is not needed, because the flow stress is known beforehand. In the multipass steady-state process, the equivalent strain for the material before entering the die opening is given by the result of the previous pass.

In the steady-state scheme, the element mesh may be fixed in space (Eulerian formulation). However, the integration of the equivalent strain is simplified, if the element mesh is modified to be parallel to the stream lines during the iterative calculations.

12.2.5 Temperature Distribution

In metal forming, the distribution of temperature in the material is changed due to generation of heat that arises from plastic deformation and friction, and by transfer of heat to the surrounding air and to the tools. Plastic deformation of the material is affected by the distribution of temperature, because the flow stress is a function of temperature. Thus, the temperature effect is important for the simulation of hot working and high speed forming. The finite-element method can be applied to the simulation of temperature distributions in metal forming

(Asada et al. 1981). When the finite-element method is used to determine the temperature distribution, a common element mesh with the deformation simulation is possible.

The differential equation governing heat conduction for a steady state is given by

$$\alpha\left(\frac{\partial^2 T}{\partial x^2}+\frac{\partial^2 T}{\partial y^2}+\frac{\partial^2 T}{\partial z^2}\right)-c\rho\left\{\frac{\partial}{\partial x}(v_x T)+\frac{\partial}{\partial y}(v_y T)+\frac{\partial}{\partial z}(v_z T)\right\}+\dot{q}=0 \quad (12.9)$$

where α is the thermal conductivity, T is the temperature, c is the specific heat, ρ is the density, v_x, v_y and v_z are velocity components and $\dot{q}$ is the heat generation rate. The first, second and third terms in Eq. (12.9) represent the heat conduction, the mass transfer and the heat generation, respectively. The heat generation due to deformation and friction is obtained from the deformation analysis. The boundary conditions for the heat transfer are provided by

$$\alpha\left(\frac{\partial T}{\partial x}l_x+\frac{\partial T}{\partial y}l_y+\frac{\partial T}{\partial z}l_z\right)-c\rho T(v_x l_x+v_y l_z+v_z l_z)+h(T-T_a)=0 \quad (12.10)$$

where l_x, l_y and l_z are the direction cosines of the tool surface and h and T_a are the heat transfer coefficient and the temperature of the air or the coolant, respectively.

The differential equation of (12.9) is formulated by using the Galerkin method as follows:

$$[A]\{T\}=\{B\} \quad (12.11)$$

$$[A]=\int_V \alpha\left(\frac{\partial\{N\}^{\mathrm{T}}}{\partial x}\frac{\partial\{N\}}{\partial x}+\frac{\partial\{N\}^{\mathrm{T}}}{\partial y}\frac{\partial\{N\}}{\partial y}+\frac{\partial\{N\}^{\mathrm{T}}}{\partial z}\frac{\partial\{N\}}{\partial z}\right)dV$$
$$-\int_V c\rho\left(\frac{\partial\{N\}^{\mathrm{T}}}{\partial x}\{N\}\{v_x\}\{N\}+\frac{\partial\{N\}^{\mathrm{T}}}{\partial y}\{N\}\{v_y\}\{N\}+\frac{\partial\{N\}^{\mathrm{T}}}{\partial z}\{N\}\{v_z\}\{N\}\right)dV$$
$$+\int_{S_h} h\{N\}^{\mathrm{T}}\{N\}\,dS \quad (12.12)$$

$$\{B\}=\int_V \{N\}^{\mathrm{T}}\dot{q}dV+\int_{S_h} h\{N\}^{\mathrm{T}}T_a\,dS \quad (12.13)$$

where N is the shape function of the element and $\{T\}$ is the vector of the nodal temperature. Eq. (12.11) is the linear simultaneous equation of the nodal temperatures. The boundary conditions of Eq. (12.10) are included in Eq. (12.11).

12.3 Extrusion

12.3.1 Steady-State Deformation in Forward Extrusion

Steady-state deformation of a bar in extrusion with a conical die is simulated by the steady-state scheme described in Sect. 12.2.4. The bar is divided into quadrilateral ring elements with four nodes. In the element mesh shown in Fig.

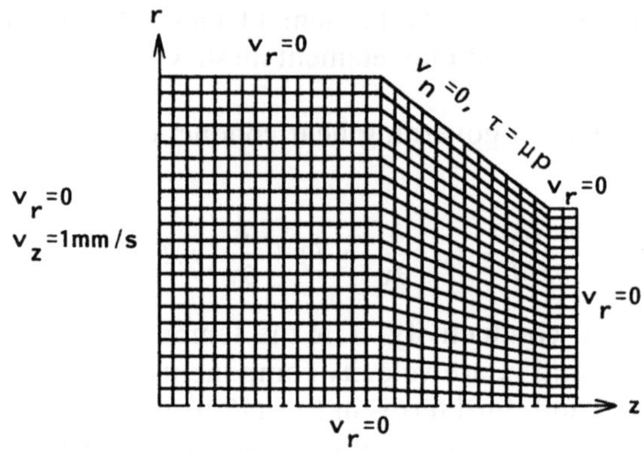

Fig. 12.1. Finite-element mesh and boundary conditions for simulation of steady-state extrusion of bar.

12.1, the boundary conditions used for the simulation of steady-state extrusion are given. In the rigid-plastic finite-element method for a slightly compressible material, the velocity component in the axial direction at the exit boundary is not prescribed as a boundary condition. Therefore, the stress components in steady-state deformation can be calculated by the slight compressibility. To simulate metal flow around the die corners, a finite-element mesh is employed.

The process geometry and the working conditions used for the simulation of steady-state extrusion are summarized as follows;

- Initial diameter of bar 20 mm
- Reduction in area $R=0.64$
- Die semicone angle $\beta=38.7°$
- Coefficient of friction $\mu=0.1$
- Flow curve (aluminium) $\bar{\sigma}=127\bar{\varepsilon}^{0.27}$ MPa

The flow curve was measured from the simple compression of an annealed aluminium cylinder. The value of g in Eq. (12.1) was 1/10 000 and the value of e in Eq. (12.7) was 5/1000, the maximum equivalent strain rate in the previous stage of iteration in the minimizing procedure of the functional.

The distribution of the calculated equivalent strain rate in the workpiece for steady-state extrusion is illustrated in Fig. 12.2. The equivalent strain rate represents the magnitude of local plastic deformation. Plastic deformation concentrates near the corners of the die entry and exit.

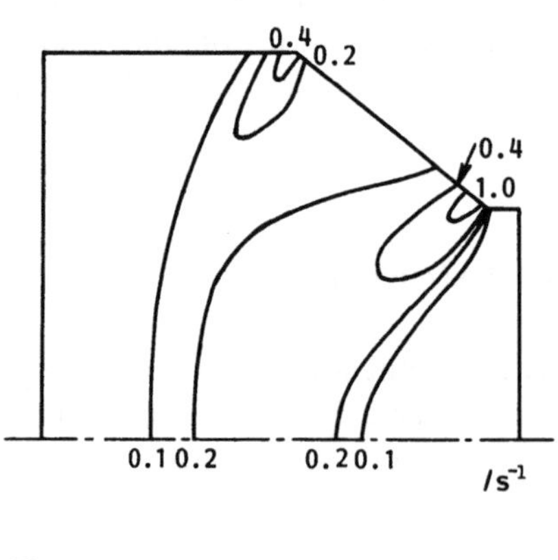

Fig. 12.2. Distribution of equivalent strain rate in workpiece for steady-state extrusion of aluminium bar ($R=0.64$, $\beta=38.7°$, $\mu=0.1$).

The distribution of the equivalent strain is obtained by integrating the equivalent strain rate along the stream lines (see Fig. 12.3). The equivalent strain for the extruded bar exhibits the highest value at the outer surface.

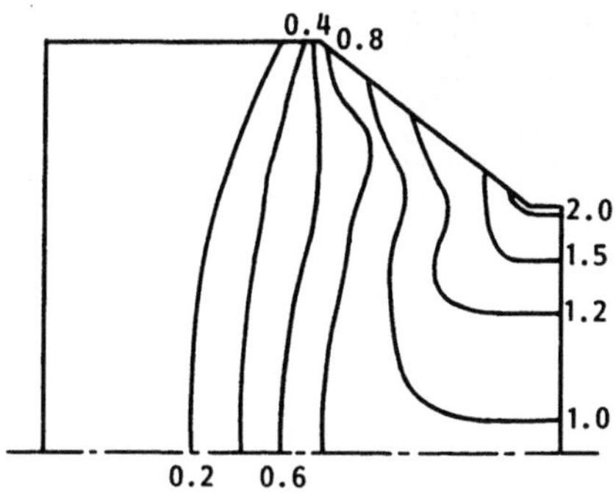

Fig. 12.3. Distribution of equivalent strain in workpiece for steady-state extrusion (R=0.64, β=38.7°, μ=0.1).

12.3.2 Nonsteady-State Deformation in Forward Extrusion

In extrusion and drawing, nonsteady-state deformation occurs in the vicinity of the front and the rear ends of the workpiece. It is important to predict the deformation behaviour in the nonsteady-state regions, because these are cut off as a crop loss. A stepwise calculation for incremental deformation is carried out to simulate nonsteady-state deformation. The working conditions for the simulation of nonsteady-state deformation are the same as those for the steady-state deformation in Sect. 12.3.1.

The distortions of the grid pattern for nonsteady-state deformation near the front end of the bar for two friction conditions at the die–workpiece interface are shown in Fig. 12.4. The centre of the front end advances as the ram stroke increases. The bulging of the front end is more marked for the larger value of friction.

12.3.3 Backward Extrusion

In forming processes, the direction of metal flow changes sharply at the corners of tools. The basic finite-element method is not suitable for dealing with the sharp change in flow direction, because an ordinary element allows only a smooth change of velocity. To model the sharp change in direction of metal flow with a small number of elements in the rigid-plastic finite-element simulation, the element with a singular point at the die corner was proposed by Mori et al. (1983). The element consists of a quadrilateral subelement *ijlm* and a triangular subelement *jkl* as shown in Fig. 12.5. At the punch corner in backward extrusion, node *m* moves downward with the bottom surface of the punch and node *l* moves upward along the side surface of the punch. If the position of node *m* coincides with that of *l* as shown in Fig. 12.5b, the element has a singular point of velocity at the corner. In this element, the velocity changes continuously from the plane *mi* to the plane *lj*. The elements without a singular point are used in the workpiece except for the punch corners. When two quadrilateral subelements for the element with a singular point are used by adding a nodal point at the punch corner, the

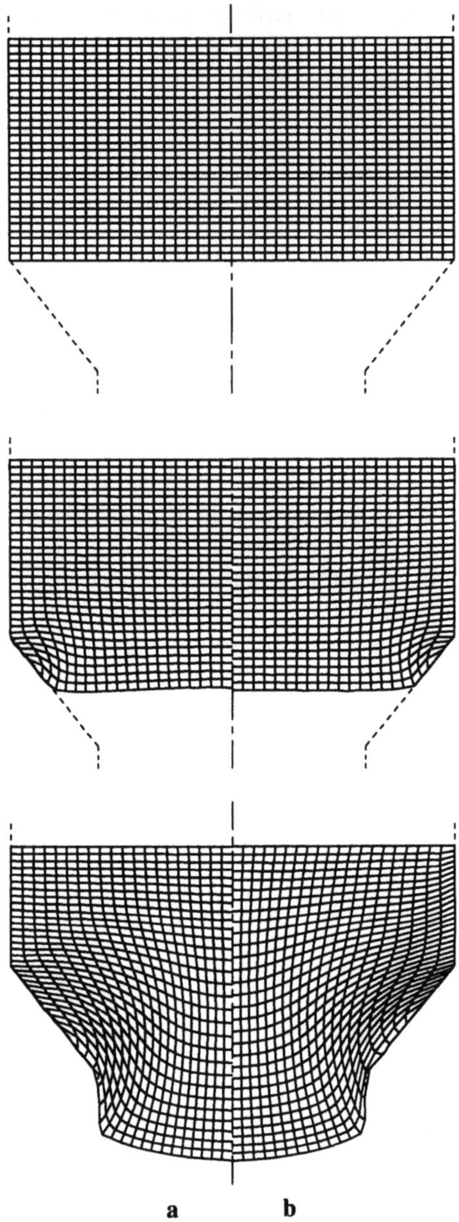

a **b**

Fig. 12.4. Distortions of grid pattern for **a** μ=0.1 and **b** μ=0.2 in nonsteady-state extrusion of bar (R=0.64, β=38.7°).

treatment becomes easier because of the same shape function of each of the elements.

In forward and backward extrusion processes, the workpiece is severely deformed in the vicinity of the tool corner. To simulate large and severe deformation by the finite-element method, spatially fixed elements (Fig. 12.6) were proposed by Mori et al. (1983). The spatially fixed elements are useful to avoid severely distorted shapes of elements by plastic deformation. When these elements are used, the equivalent strain and the distortion of the initial grid are

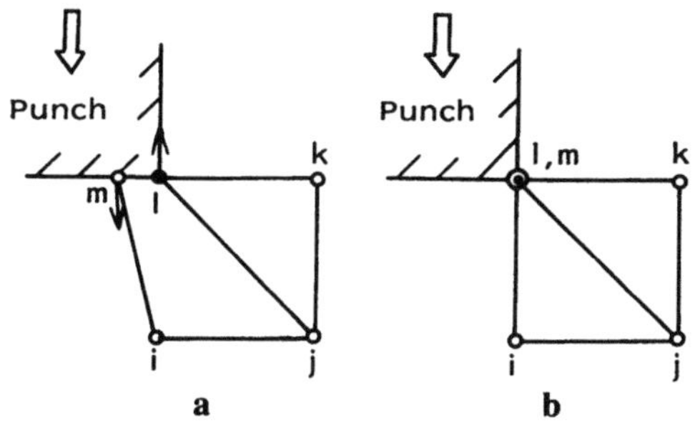

Fig. 12.5. Treatment of velocity boundary condition at corner of punch in backward extrusion of can. **a** Triangular sub-element and quadrilateral sub-element and **b** element with a singular point.

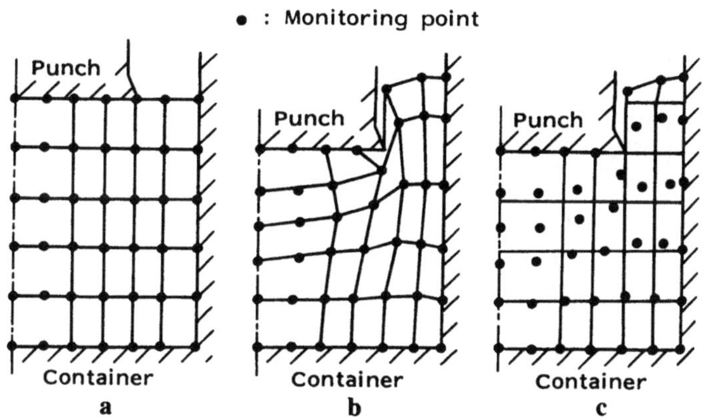

Fig. 12.6. Treatment of severe plastic deformation. **a** Initial state; **b** subsequent stage of deformation; **c** spatially fixed elements.

traced by the monitoring points embedded in the workpiece in the initial state.

Axisymmetric backward extrusion of an aluminium can with a flat punch is simulated by using the element with a singular point and the spatially fixed elements. The distorted grid patterns on the meridional plane obtained by calculation and by experiment are shown in Fig. 12.7. The grid patterns were obtained from the positions of the monitoring points, not from the element mesh. Extremely severe plastic deformation is successfully simulated.

Tools used in metal forming are usually rounded at the corners to allow smooth metal flow. Elements for simulating the metal flow around the round corners of the tools in metal forming have been presented by Mori et al. (1987). The element has a round boundary fitting to the surface of the tool corner and satisfies the exact boundary condition for metal flow along the round corner (see Fig. 12.8).

The distributions of the calculated equivalent strain-rate in the workpieces in backward extrusion of aluminium cans with round corner punches are illustrated in Fig. 12.9. The elements with a round boundary are located in the punch corner in each step of deformation. The concentration of plastic deformation is relaxed as the corner radius of the punch increases.

12.3.4 Benchmark

To examine the capability of a variety of computers, a benchmark was carried out in the simulation of steady-state forward extrusion shown in Fig. 12.1. The

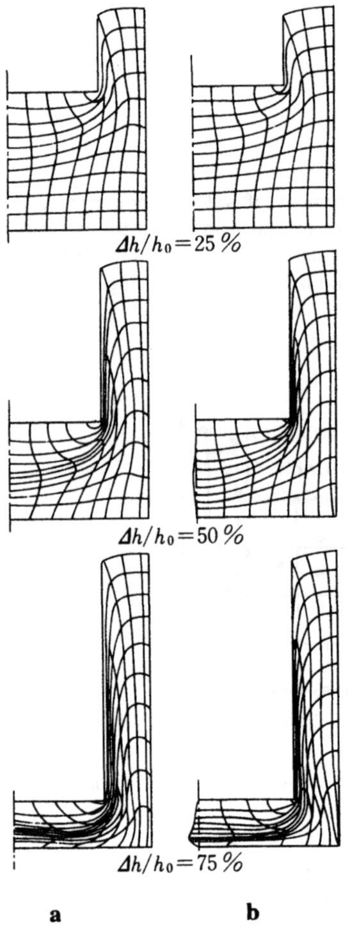

Fig. 12.7. Comparison of distorted grid patterns between **a** calculation and **b** experiment in backward extrusion of aluminium can (R=0.44, μ=0.13).

Table 12.1. Time for execution of benchmark problem of steady-state extrusion

Computer	Execution Time (s)	Ratio
CISC workstation Sony NEWS 711 (68020,16.7MHZ, 2MIPS)	51588	1
RISC workstation Sumitomo Electric Workstation S-P200 (R2000, 16.7MHz, 12MIPS, 1.8MFLOPS)	4179	1/12
Large scale computer NEC ACOS-930/10 (24MIPS)	997	1/52
Mini-Supercomputer Ardent Titan 1CPU (16MFLOPS)	3690	1/14
Supercomputer Fujitsu FACOM VP-400E (1.7GFLOPS)	461	1/112

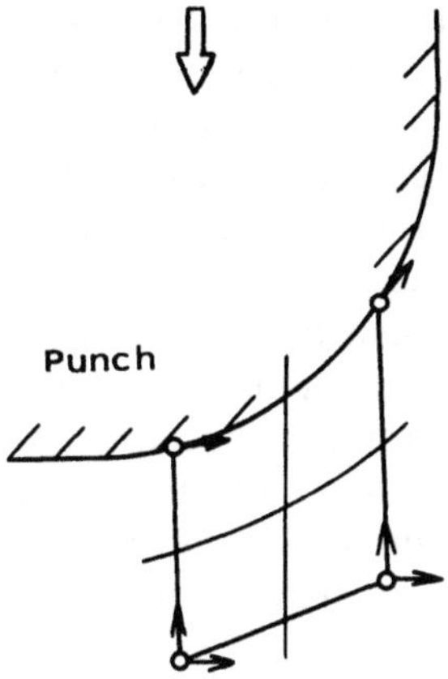

Fig. 12.8. Element with a round boundary in round corner of tool.

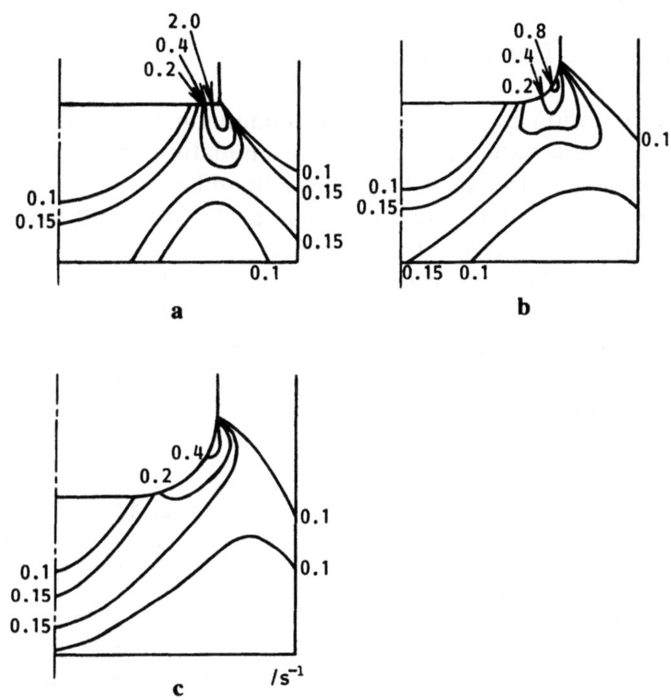

Fig. 12.9. Distributions of equivalent strain rate in workpiece for punch corner radii of **a** 0 mm, **b** 2.5 mm and **c** 5.5 mm in backward extrusion of aluminium can (R=0.44, μ=0.06).

number of elements for the problem is 600. The computer code is written in FORTRAN language and is of about 1200 lines.

The execution time for the benchmark problem is given in Table 12.1. The RISC (Reduced Instruction Set Computer) workstation has a RISC microprocessor in which the computational speed is heightened by decreasing the number of instructions. The microprocessor MC68020 used in the CISC (Complex Instruction Set Computer) workstation is popular in the field of workstations. The RISC workstation takes about four times longer than the large scale computer. The workstations can execute almost the same scale of the computer code as that for the large scale computer without modification of the code, because they have

4M–200M byte main memory and 10M–1G byte virtual memory in a hard disk unit. The editing, debugging and computer graphics of the workstations for the development of computer code are superior to those for the large scale computers. The computational speed of the workstations is rapidly increasing because of the development of new microprocessors. Thus, the workstation is expected to be of major importance as a hardware base for finite-element simulation.

High computational speed of the supercomputers is obtained by vectorization of the computer code. The supercomputer and the mini-supercomputer could not achieve the high computational speed in the benchmark problem, because the number of variables in this problem was too small to take advantage of vectorization facilities. The supercomputer and the mini-supercomputer are more suitable for large-scale computation such as in three-dimensional problems.

12.4 Drawing

12.4.1 Steady-State Deformation

Steady-state deformation in drawing of an aluminium bar with a conical die is simulated by the steady-state scheme. For the simulation of drawing, the velocity component in the axial direction at the entry boundary is a variable because of the slight compressibility. The reduction in area R is 0.19 and the coefficient of friction μ at the die–workpiece interface is 0.1.

The distributions of the hydrostatic stress in the workpiece in bar drawing for two semi die angles are illustrated in Fig. 12.10. As the die angle increases, the hydrostatic stress becomes tensile in the region near the centre of the bar. Large tensile stress in the central region can cause the occurrence of central bursting cracks in multipass extrusion and drawing.

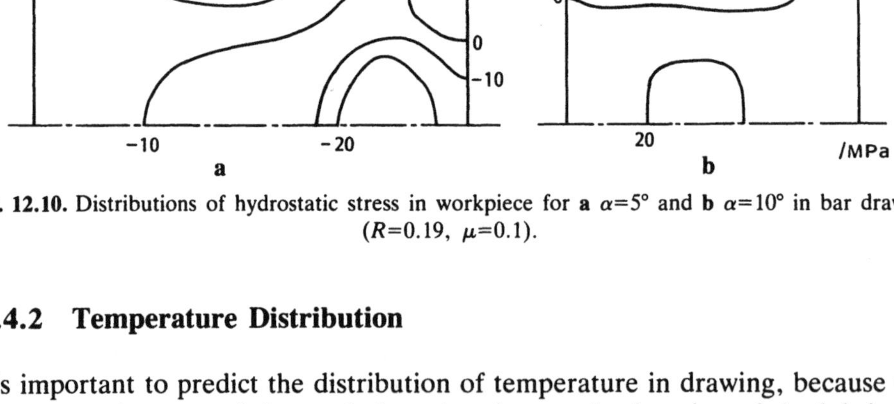

Fig. 12.10. Distributions of hydrostatic stress in workpiece for **a** $\alpha=5°$ and **b** $\alpha=10°$ in bar drawing ($R=0.19$, $\mu=0.1$).

12.4.2 Temperature Distribution

It is important to predict the distribution of temperature in drawing, because the temperature increase of the workpiece deteriorates the function of the lubricant. The distribution of temperature in steady-state drawing of a stainless steel tube with a plug is calculated by the finite-element method for heat conduction as described in Sect. 12.2.5. The heat generation is calculated by the rigid-plastic finite-element method. The reduction in area is 0.29, the die semicone angle $\beta=10°$ and the drawing speed is 0.5 m s^{-1}. The frictional shear factors for the die and the plug are 0.1 and 0.2, respectively. The thermal conductivities for the tools and the workpiece are 42 W m^{-1} K^{-1} and 16 W m^{-1} K^{-1}, respectively. The flow stress of the workpiece is assumed to be temperature insensitive.

The distribution of the calculated temperature in the workpiece and the tools in tube drawing is illustrated in Fig. 12.11. The temperature increase of the plug is extremely large. Large temperature increases can often bring about galling and seizure at the tool–workpiece interface. It is not easy to supply sufficient lubrication between the plug and the tube.

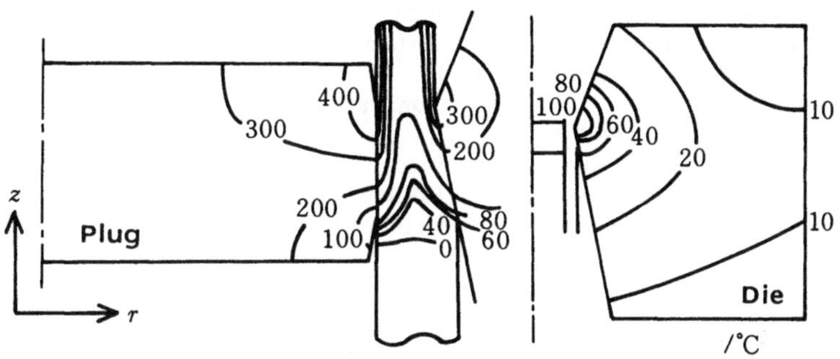

Fig. 12.11. Distribution of temperature in workpiece in tube drawing ($R=0.29$, $\alpha=10°$, $m=0.1$(die), 0.2(plug), drawing speed 0.5 m s^{-1}).

12.5 Conclusion

A discussion of the rigid-plastic finite-element method and its application in extrusion and drawing has been presented. The basic theory has been introduced followed by comments on how to deal with boundary conditions, rigid zones and temperature distributions. Steady-state and nonsteady-state schemes are compared. Examples of the simulation of forward extrusion, backward extrusion and drawing demonstrate the capabilities of the rigid-plastic approach. Attention has been given to the development of special elements to overcome severe restrictions to modelling of metal flow around sharp tool corners.

References

Aravas N (1986) The analysis of void growth that leads to central bursts during extrusion. J Mech Phys Solids 34:55–79

Argyris JH, St Doltsinis J (1979) On the large strain inelastic analysis in natural formulation, part 1: quasistatic problems. Comp Meth Appl Mech Engng 20:213–251

Asada Y, Mori K, Yoshikawa K, Osakada K (1981) Analysis of temperature distribution in wire and tube drawing processes by the finite element method. J Jpn Soc Tech Plast, 22:488–494 (in Japanese)

Ayada M, Higashino T, Mori K (1987) Central bursting in extrusion of inhomogeneous materials. In: Proceedings 2nd ICTP, Stuttgart, vol 1, pp 553–558

Chen CC, Kobayashi S (1978) Deformation analysis of multi-pass bar drawing and extrusion. Ann CIRP 27:151–155

Chen CC, Oh SI, Kobayashi S (1979) Ductile fracture in axisymmetric extrusion and drawing, Part 1: Deformation mechanics of extrusion and drawing. Trans ASME J Engng Ind 101:23–35
Dawson PR (1978) Viscoplastic finite element analysis of steady-state forming processes including strain history and stress flux dependence. Appl Num Meth Forming Proc 28:55–66
Gerhardt J, Tekkaya AE (1987) Simulation of drawing-processes by the finite element method. In: Proceedings 2nd ICTP, Stuttgart, vol 2, pp 841–848
Gordon JL, Weinstein AS (1974) A finite element analysis of the plane strain drawing problem. In: Proceedings 2nd NAMRC, pp 194–208
Hartley CS, Dehghani M (1987) Residual stresses in axisymmetrically formed products. Proceedings 2nd ICTP, Stuttgart, vol 1, pp 605–611
Hartley P, Sturgess CEN, Rowe GW (1980) Prediction of deformation and homogeneity in rim-disc forging. J Mech Work Tech 4:145–154
Iwata K, Osakada K, Fujino S (1972) Analysis of hydrostatic extrusion by the finite element method. Trans ASME J Engng Ind 94:697–703
Kobayashi S, Shah SN (1978) The matrix method for the analysis of metal-forming processes. In: Burke JJ, Weiss V (eds) Advances in deformation processing. Plenum, pp 51–98
Lee CH, Iwasaki H, Kobayashi S (1973) Calculation of residual stresses in plastic deformation processes. Trans ASME J Engng Ind 95:283–291
Lee EH, Mallett RL (1977) Stress and deformation analysis of the metal extrusion process. Comp Meth Appl Mech Engng 10:339–353
Lugt J, Huetink J (1986) Thermal mechanically coupled finite element analysis in metal-forming processes. Comp Meth Appl Mech Engng 54:145–160
Lung M, Mahrenholtz O (1973–74) A finite element procedure for analysis of metal forming processes. Trans CSME 2:31–36
Mori K, Osakada K, Oda T (1982) Simulation of plane-strain rolling by the rigid-plastic finite element method. Int J Mech Sci 24:519–527
Mori K, Osakada K, Fukuda M (1983) Simulation of severe plastic deformation by finite element method with spatially fixed elements. Int J Mech Sci 25:775–783
Mori K, Osakada K, Ieda T, Fukuda M (1987) Finite element simulation of metal flow around round corner of tool. In: Proceedings 2nd ICTP, Stuttgart, vol 2, pp 1105–1110
Nayak GC, Zienkiewicz OC (1972) Elasto-plastic stress analysis. A generalization for various constitutive relations including strain softening. Int J Num Meth Engng 5:113–135
Oh SI, Chen CC, Kobayashi S (1979) Ductile fracture in axisymmetric extrusion and drawing, part 2: workability in extrusion and drawing. Trans ASME J Engng Ind 101:36–44
Osakada K, Nakano J, Mori K (1982) Finite element method for rigid-plastic analysis of metal forming – formulation for finite deformation. Int J Mech Sci 24:459–468
Osen W (1987) Possibilities and limitations of cold extrusion processes combined with radial extrusion. In: Proceedings 2nd ICTP, Stuttgart, vol 1 pp 575–582
Pacheco LA, Alexander JM (1982) On the hydrostatic extrusion of copper-covered aluminium rods. In: Pittman JFT et al. (eds) Numerical methods in industrial forming processes. Pineridge Press, pp 205–216
Tayal AK, Natarajan R (1981a) Extrusion of rate-sensitive materials using a viscoplastic constitutive equation and the finite element method. Int J Mech Sci 23:89–98
Tayal AK, Natarajan R (1981b) A finite element analysis of axisymmetric extrusion of composite rods. Int J Mach Tool Des Res 21:227–235
Tayal AK, Natarajan R (1982) Analysis of combined backward-forward extrusion by the finite element method. In: Pittman JFT et al. (eds) Numerical methods in industrial forming processes. Pineridge Press, pp 247–255
Tomita Y (1982) A rigid plastic finite element perturbation method for the effective prediction of the influence of the change in parameters in the constitutive equation upon deformation behaviour. Int J Mech Sci 24:711–716
Tomita Y, Sowerby R (1978) An approximate analysis for studying the deformation mechanics of rate sensitive materials. Int J Mech Sci 20:361–371
Xutang W, Qijian J, Shueisheing X (1983) Elasto-plastic finite element analysis of hydrostatic extrusion with various mathematically contoured dies. In: Proceedings 24th MTDR, Manchester, pp 51–57
Yamada Y, Hirakawa H (1978) Large deformation and instability analysis in metal forming process. Appl Num Meth Forming Proc 28:27–38
Zienkiewicz OC (1984) Flow formulation for numerical solution of forming processes. In: Pittman JFT et al. (eds) Numerical analysis of forming processes. John Wiley, pp 1–44

Zienkiewicz OC, Godbole PN (1974) Flow of plastic and visco-plastic solids with special reference to extrusion and forming processes. Int J Num Meth Engng 8:3–16

Zienkiewicz OC, Godbole PN (1975) A penalty function approach to problems of plastic flow of metals with large surface deformations. J Strain Analysis 10:180–183

Zienkiewicz OC, Onate E, Heinrich JC (1978) Plastic flow in metal forming. Appl Num Meth Forming Proc 28:107–120

13 Numerical Modelling of Sheet-Metal-Forming Problems

E. Oñate and C. Agelet de Saracibar

13.1 Introduction

Stamping of sheet-metal parts by means of punches and dies is a standard manufacturing process. However, despite its broad application in industry, the design of forming processes is still largely based on experimental techniques, such as the use of circular grid systems (Keeler 1968) or forming limit diagrams (Hecker 1973, Keeler 1974).

The development of reliable analytical procedures to predict the behaviour of sheet-metal deformation processes has encountered many serious obstacles. Together with the nonlinearity of the material, other important effects like the unsteady nature of the process, the large magnitude of the strains involved and the importance of contact and frictional effects at the tool–sheet interface make the study of stamping processes so complex that its analysis justifies the use of sophisticated numerical algorithms and usually leads to large-scale computer requirements.

Numerical solutions were originally obtained for the simpler axisymmetric case using the finite-difference method (Woo 1965a,b, 1968). Soon finite-element-based solutions took the lead as it became evident that this procedure has a better ability to deal with complex three-dimensional shapes.

The finite-element formulations for sheet-metal-forming problems can be roughly classified in two categories: (1) Those based on incremental large strain elastic-plastic/viscoplastic theory (Wifi 1976, Kim et al. 1978, Wang and Budiansky 1978, Honnor 1985, Bellet 1988, Batoz et al. 1989) and (2) those using a rigid-plastic/viscoplastic approach (Zienkiewicz et al. 1978, Oñate and Zienkiewicz 1983, Bonet 1989, St Doltsinis 1989, Wood et al. 1989).

Option (1) termed hereafter *solid approach* is of considerable theoretical complexity. However, it can predict the evolution of both elastic and plastic strains during the sheet deformation and it also allows the direct computation of elastic spring-back effects. On the other hand, option (2), usually known as *flow approach*, is conceptually more simple, as the final equations simplify to those of standard elasticity. Also the flow approach predicts accurately the sheet deformation and the total (nonlinear) strains. This approach has the drawback that it neglects elastic effects and therefore spring-back effects cannot be directly obtained. However, they can be approximated by adequate a posteriori algorithms (Zienkiewicz et al. 1978, Zienkiewicz 1984).

Both solid and flow approaches have been successfully used in practice using *continuum* and *shell* type finite elements. So far the latter have been more popular and many membrane and shell formulations have been developed for sheet-metal-forming analysis (Wang and Budiansky 1978, Oñate and Zienkiewicz 1983, Bellet 1988, Batoz et al. 1989, Hambrecht et al. 1989).

The problem of contact between sheet and tool has been traditionally solved with algorithms based on penalty forms (Hallquist 1976, Simo et al. 1985, Wriggers and Simo 1985, Nour-Omid and Wriggers 1985, St Doltsinis 1989) and Lagrangian multipliers (Bathe and Chaudary 1985, Simo et al. 1985, Wriggers and Simo 1985, Nour-Omid and Wriggers 1985).Friction effects have been usually treated using the standard Coulomb law (Oden and Pires 1983) and also kinematic friction models (Bellet 1988).

In this chapter we present an overview of the solid and flow approaches for the analysis of sheet-metal-forming problems. Material degradation effects due to the development of microscopic voids are also taken into account in both cases. Particular emphasis is put in the presentation of a selective membrane-bending shell flow formulation developed by the authors. This allows the use of membrane or bending elements in different parts of the sheet according to the specific features of the deformation pattern.

Contact and friction effects are also treated and a generalized contact algorithm is presented.

The chapter concludes with examples of applications of some of the methods presented to axisymmetric and three-dimensional sheet-metal-forming problems.

13.2 Introductory Concepts

As previously mentioned, the analysis of sheet-metal-forming problems can be attempted from two conceptually different views:

1. *Solid approach using incremental theory for elastic-plastic/viscoplastic metals.* This approach uses a total or updated Lagrangian description of the motion. The basic variables are the displacements of the deforming body and these are related to the strains by standard nonlinear kinematic expressions (Zienkiewicz 1977, Kleiber 1989). On the other hand, the constitutive equation relating stresses and strains must be expressed in rate (incremental) form. Finally, the equilibrium equations can be written pointwise through the adequate differential equations to be satisfied in the volume and the boundary of the body, or in a global sense through the principle of virtual work (see Table 13.1).
2. *Flow approach based on rigid-plastic/viscoplastic theory.* In the deformation of a metal during forming operations the nonlinear (plastic/viscoplastic) strains are much larger than the elastic ones. This allows the possibility of neglecting the latter and to use a simpler rigid-plastic/viscoplastic model for the analysis. It can be shown (Zienkiewicz et al. 1978) that the resulting governing equations for this case are identical to those of a non-Newtonian fluid. This justifies the use of a *flow approach* in which the main variables are the velocities of the deforming body defined in an Eulerian frame typical of fluid flow problems. Also the strain rates are linearly dependent on the velocities in a standard manner. The constitutive equation now relates stress and strain rates and the equilibrium equations are identical to those of the solid approach if displacements and strains are now replaced by velocities and strain rates respectively.

It is interesting to note that overall governing equations of the flow approach are also analogous to those of standard infinitesimal elasticity (Zienkiewicz et al.

Table 13.1. Basic equations of solid and flow approaches

	Solid approach	Flow approach
Basic variables	Displacements $\mathbf{u}$	Velocities $\mathbf{v}$
Kinematic equations	$\boldsymbol{\epsilon} = (\mathbf{L} + \mathbf{L}(\mathbf{u}))\mathbf{u}$	$\dot{\boldsymbol{\epsilon}} = \mathbf{L}\mathbf{v}$
Equilibrium equations	$\mathbf{L}^{\mathrm{T}}\boldsymbol{\sigma} + \mathbf{b} = 0$ in volume V $\mathbf{M}^{\mathrm{T}}\ \boldsymbol{\sigma} = \mathbf{t}$ in boundary Γ_t $\mathbf{b}$: body forces, $\mathbf{t}$: surface loads $\mathbf{M}$: contains components of unit normal to Γ_t	
Virtual work	$\int_V \delta\boldsymbol{\epsilon}^{\mathrm{T}}\,\boldsymbol{\sigma} dV = \int_V \delta\mathbf{u}^{\mathrm{T}}\mathbf{b}\, dV + \int_{\Gamma_t} \delta\mathbf{u}^{\mathrm{T}}\mathbf{t} d\Gamma$	$\int_V \delta\dot{\boldsymbol{\epsilon}}^{\mathrm{T}}\,\boldsymbol{\sigma} dV = \int_V \delta\mathbf{v}^{\mathrm{T}}\mathbf{b} dV + \int_{\Gamma_t} \delta\mathbf{v}^{\mathrm{T}}\mathbf{t} d\Gamma$
Constitutive equations	$\overset{\triangle}{\boldsymbol{\sigma}} = \mathbf{D}_{\mathrm{T}}\dot{\boldsymbol{\epsilon}}$	$\boldsymbol{\sigma} = \mathbf{D}\dot{\boldsymbol{\epsilon}}$

1978). This analogy can be directly exploited to simplify further the computational formulation, as will be discussed later.

A summary of the basic equations of the solid and flow approaches is given in Table 13.1.

In the next sections we will comment on the basic differences between these two approaches for the analysis of sheet metal forming problems.

13.3 Solid Approach: Basic Equations for Nonlinear Deformation of Porous Metals

It is well established that in ductile metals, softening and ultimate failure by nucleation, growth and coalescence of microscopic voids are important fracture mechanisms. Voids nucleate mainly by particle fracture or by interfacial decohesion and subsequently the voids grow due to intensive plastic straining when the ligaments between neighbouring voids have thinned down sufficiently.

In a general nonlinear elastoplastic porous metal the total strains are obtained as the sum of the elastic and plastic parts in the standard manner, i.e.

$$\epsilon_{ij} = \epsilon_{ij}^{e} + \epsilon_{ij}^{p} \tag{13.1}$$

and the stresses satisfy the following general yield criterion written in the stress space as

$$F(\boldsymbol{\sigma}, \mathbf{q}, \sigma_{\mathrm{M}}) = F(I_1, J_2', \mathbf{q}, \sigma_{\mathrm{M}}) = 0 \tag{13.2}$$

where the Cauchy stresses $\boldsymbol{\sigma}$ are taken as *free* variables, $\mathbf{q} = [\boldsymbol{\epsilon}^{\mathrm{p}}, f]^{\mathrm{T}}$ is the vector of the *internal plastic* variables with f being the void volume fraction (porosity), σ_{M} is the yield stress of the undamaged magterial, taken as an *explicit* variable, and I_1 and J_2' are the usual stress invariants defined as

$$I_1 = \sigma_{kk}, \qquad J_2' = \sqrt{\tfrac{3}{2} s_{ij} s_{ij}} \tag{13.3}$$

where $s_{ij} = \sigma_{ij} - \frac{1}{3}\sigma_{kk}\delta_{ij}$ is the deviatoric Cauchy stress.

The *evolution laws* for the internal variables can be written as

$$\dot{\boldsymbol{\epsilon}}^{\mathrm{p}} = \dot{\lambda} \frac{\partial G}{\partial \boldsymbol{\sigma}} \tag{13.4a}$$

$$\dot{f} = \dot{\lambda} H_f(\boldsymbol{\sigma}, \mathbf{q}, \sigma_{\mathrm{M}}) + \boldsymbol{\beta} : \overset{\triangle}{\boldsymbol{\sigma}} \tag{13.4b}$$

where G is the plastic potential and $\overset{\triangle}{\boldsymbol{\sigma}}$ denotes an objective (frame invariant) corotational rate of $\boldsymbol{\sigma}$.

The uniaxial strain-hardening law for the undamaged material can be defined in rate form as

$$\dot{\sigma}_{\mathrm{M}} = \dot{\lambda} H_{\sigma_{\mathrm{M}}}(\boldsymbol{\sigma}, \mathbf{q}, \sigma_{\mathrm{M}}) \tag{13.5}$$

Functions H_f, $\boldsymbol{\beta}$ and $H_{\sigma_{\mathrm{M}}}$ in (13.4)–(13.5) must be adequately defined and this is done in the next section.

13.3.1 Evolution of Porosity

We will consider here the development of porosity due to nucleation ($\dot{f}_{\mathrm{n}}$) and growth ($\dot{f}_{\mathrm{g}}$) of microscopic voids. Coalescence effects can be dealt with in an ad hoc a posteriori numerical manner and this will be discussed in a later section. Thus

$$\dot{f} = \dot{f}_{\mathrm{n}} + \dot{f}_{\mathrm{g}} \tag{13.6}$$

It is accepted that the void growth is due to the change in the macroscopic volumetric plastic rate of deformation, i.e.

$$\dot{f}_{\mathrm{g}} = (1 - f)\dot{\epsilon}^{\mathrm{p}}_{kk} \tag{13.7}$$

On the other hand, it is assumed that void nucleation is stress controlled only and stated in general terms as

$$\dot{f}_{\mathrm{n}} = \frac{k}{\sigma_{\mathrm{M}}} (\dot{\sigma}_{\mathrm{M}} + \tfrac{1}{3}\dot{\sigma}_{kk}) \tag{13.8}$$

where k is a parameter describing the rate of material transformed to voids per unit change of stress. The nonzero value of k is used only when $(\sigma_{\mathrm{M}} + \sigma_{kk}/3)$ exceeds in the current time increment its previous maximum.

Expression (13.6) can be written in the form (13.4b) using (13.7) and (13.8) with

$$H_{\mathrm{f}} = (1 - f)\frac{\partial G}{\partial \boldsymbol{\sigma}} : \mathbf{1} + \frac{k}{\sigma_{\mathrm{M}}} H_{\sigma_{\mathrm{M}}}$$

$$\boldsymbol{\beta} = \frac{k}{3\sigma_{\mathrm{M}}} \mathbf{1} \tag{13.9}$$

where $\mathbf{1}$ is the second-order unit tensor.

13.3.2 Strain-Hardening Law

Hardening of the matrix material can be described by the relation

$$\dot{\bar{\sigma}}_{\mathrm{M}} = h_{\sigma_{\mathrm{M}}} \dot{\bar{\epsilon}}^{\mathrm{p}}_{\mathrm{M}} \tag{13.10}$$

where $h_{\sigma_{\mathrm{M}}}$ denotes the uniaxial hardening coefficient of the undamaged material and $\dot{\bar{\epsilon}}^{\mathrm{p}}_{\mathrm{M}}$ is the equivalent plastic strain rate. This can be obtained from the assumption that the rate of plastic work appearing macroscopically is exclusively a result of the plastic flow in the matrix occupying the volume fraction $(1 - f)$. This gives

$$\boldsymbol{\sigma} : \dot{\boldsymbol{\epsilon}}^{\mathrm{p}} = (1 - f)\, \bar{\sigma}_{\mathrm{M}} \dot{\bar{\epsilon}}^{\mathrm{p}}_{\mathrm{M}} \tag{13.11}$$

Using (13.4a) and (13.10) we obtain

$$\dot{\bar{\sigma}}_{\mathrm{M}} = \dot{\lambda} \frac{h_{\sigma_{\mathrm{M}}}}{(1 - f)\, \bar{\sigma}_{\mathrm{M}}} \boldsymbol{\sigma} : \frac{\partial G}{\partial \boldsymbol{\sigma}} = \dot{\lambda} H_{\sigma_{\mathrm{M}}} \tag{13.12}$$

from which $H_{\sigma_{\mathrm{M}}}$ can be deduced.

13.3.3 Elastoplastic Model

After yielding the stresses must remain on the yield surface. This allows us to obtain the parameter $\dot{\lambda}$ in this case from the consistency condition $\dot{F} = 0$ as

$$\dot{\lambda} = \frac{1}{H} \left(\frac{\partial F}{\partial \boldsymbol{\sigma}} + \frac{\partial F}{\partial f} \boldsymbol{\beta} \right) : \overset{\triangle}{\boldsymbol{\sigma}} \tag{13.13}$$

with

$$H = - \left(\frac{\partial F}{\partial_{\sigma_{\mathrm{M}}}} H_{\sigma_{\mathrm{M}}} + \frac{\partial F}{\partial f} H_{\mathrm{f}} \right) \tag{13.14}$$

The constitutive equation is initially written in the standard rate form relating stress rates with the corresponding *elastic* strain rates as

$$\overset{\triangle}{\boldsymbol{\sigma}} = \mathbf{D} : (\dot{\boldsymbol{\epsilon}} - \dot{\boldsymbol{\epsilon}}^{\mathrm{p}}) \tag{13.15}$$

where $\mathbf{D}$ is the elastic constitutive matrix. Combining now (13.4a), (13.13) and (13.15) we can rewrite this as

$$\overset{\triangle}{\boldsymbol{\sigma}} = \mathbf{D}_{\mathrm{T}} : \dot{\boldsymbol{\epsilon}} \tag{13.16}$$

where the tangent constitutive matrix is given by

$$\mathbf{D}_T = \mathbf{D} - \frac{1}{H^*} \left[\mathbf{D} : \left(\frac{\partial G}{\partial \boldsymbol{\sigma}} \right) \right] \otimes \left[\mathbf{D} : \left(\frac{\partial F}{\partial \boldsymbol{\sigma}} + \frac{\partial F}{\partial f} \boldsymbol{\beta} \right) \right] \tag{13.17}$$

with

$$H^* = H + \left(\frac{\partial F}{\partial \boldsymbol{\sigma}} + \frac{\partial F}{\partial f} \boldsymbol{\beta} \right) : \mathbf{D} : \frac{\partial G}{\partial \boldsymbol{\sigma}} \tag{13.18}$$

In metals, associated flow conditions ($F = G$) are usually assumed. Note that for this case matrix $\mathbf{D}_{\mathrm{T}}$ is *still nonsymmetric*, unless $\boldsymbol{\beta} = \mathbf{0}$ or $\partial F / \partial f = 0$ (i.e. $f = 0$).

13.3.4 Elasto-viscoplastic Model

In the viscoplastic case stresses are allowed to lie *outside* the yield surface. The flow rule is now identical to (13.4a) with the parameter $\dot{\lambda}$ defined explicitly in terms of the overstress as (Perzyna 1966, Zienkiewicz and Cormeau 1974)

$$\dot{\lambda} = \gamma\langle\phi(F)\rangle \tag{13.19}$$

where γ is a fluidity parameter and $\langle \cdot \rangle$ denotes the MacCauley bracket operator, i.e. $\langle\phi\rangle = \phi$ if $\phi > 0$, otherwise $\langle\phi\rangle = 0$. An exponential form $\phi(F) = F^n$ is usually chosen for metals (Zienkiewicz 1977).

The constitutive equation is obtained in this case by directly substituting the value of $\dot{\lambda}$ in (13.4a) and this one into (13.15).

13.3.5 Particularization for Gurson's Void Model

Gurson (1977) has proposed a yield condition for a randomly voided material with spherical (for three-dimensional (3D) problems) or circular cylindrical voids (for plane-stress problems) as

$$F = \bar{\sigma}_{\mathrm{M}} - \sigma_{\mathrm{M}} = 0 \tag{13.20}$$

$$\bar{\sigma}_{\mathrm{M}} = \left(\frac{3}{2}\frac{s_{ij}s_{ij}}{w}\right)^{\frac{1}{2}}; \quad w = 1 + f^2 - 2f\cosh\left(\frac{\sigma_{kk}}{2\sigma_{\mathrm{M}}}\right) \tag{13.21}$$

Assuming associated flow ($F = G$) the flow rule (13.4a) can be written as

$$\dot{\epsilon}^{\mathrm{p}}_{ij} = \underbrace{\dot{\lambda}\frac{\partial G}{\partial s_{ij}}}_{\dot{e}^{\mathrm{p}}_{ij}} + \underbrace{\dot{\lambda}\frac{\partial G}{\partial \sigma_{kk}}\delta_{ij}}_{\frac{1}{3}\dot{\epsilon}^{\mathrm{p}}_{kk}\,\delta_{ij}} = \dot{\lambda}\frac{3s_{ij}}{2\bar{\sigma}_{\mathrm{M}}\,w} + \dot{\lambda}\frac{\bar{\sigma}_{\mathrm{M}}}{\sigma_{\mathrm{M}}}\frac{fS}{2w}\delta_{ij} \tag{13.22}$$

where $\dot{e}^{\mathrm{p}}_{ij} = \dot{\epsilon}^{\mathrm{p}}_{ij} - \frac{1}{3}\dot{\epsilon}^{\mathrm{p}}_{kk}\,\delta_{ij}$ is the deviatoric strain rate and $S = \sinh A$ with $A = \sigma_{kk}/2\sigma_{\mathrm{M}}$.

Note that in (13.22) $\sigma_{\mathrm{M}} \neq \sigma_{\mathrm{M}}$ has been assumed, thus allowing for a viscoplastic deformation.

The flow rule (13.22) can be inverted in the deviatoric strain-rate space to give

$$s_{ij} = \frac{2\bar{\sigma}_{\mathrm{M}}w}{3\dot{\lambda}}\dot{e}^{\mathrm{p}}_{ij} \tag{13.23}$$

Substituting now the value of s_{ij} in the yield function (13.20) we obtain

$$\dot{\lambda} = \left(\frac{2}{3}\dot{e}^{\mathrm{p}}_{ij}\dot{e}^{\mathrm{p}}_{ij}\,w\right)^{\frac{1}{2}} \equiv \dot{e}^{\mathrm{p}}_{\mathrm{M}} \tag{13.24}$$

Eq. (13.24) provides the definition of the *equivalent deviatoric strain rate* in the porous metal (Agelet de Saracibar and Oñate 1989, Oñate et al. 1989).

Using (13.22) we can finally obtain from (13.9), after some algebra, the expression for the equivalent plastic strain rate as

$$\dot{\epsilon}^{\mathrm{p}}_{\mathrm{M}} = \dot{e}^{\mathrm{p}}_{\mathrm{M}}\frac{w + fAS}{w(1-f)} \tag{13.25}$$

For the zero porosity case ($f = 0$, $w = 1$) we have

$$\dot{\epsilon}^{\mathrm{p}}_{\mathrm{M}} = \dot{e}^{\mathrm{p}}_{\mathrm{M}} = \dot{\epsilon}^{\mathrm{p}} = \dot{e}^{\mathrm{p}} = \left(\frac{2}{3}\,\dot{\epsilon}^{\mathrm{p}}_{ij}\,\dot{\epsilon}^{\mathrm{p}}_{ij}\right)^{\frac{1}{2}} \tag{13.26}$$

which coincides with the equivalent strain rate of classical von Mises plasticity theory (Zienkiewicz et al. 1978). Eqs (13.22)–(13.25) are essential for the derivation of the stress–strain-rate equations for the rigid-plastic/viscoplastic cases presented in the next section.

13.4 Rigid-Plastic/Viscoplastic Model: Flow Approach

The constitutive equations for porous metals can be extremely simplified both from the practical and conceptual point of view if elastic deformations are neglected. This assumption is very reasonable for the analysis of metal-forming problems and it has been successfully used to obtain finite-element solutions of practical extrusion, rolling and sheet-metal-forming processes (Zienkiewicz et al. 1978, Oñate and Zienkiewicz 1983).

If elastic deformations are neglected the total strains coincide with the plastic/viscoplastic ones, i.e.

$$\dot{\epsilon}_{ij} = \dot{\epsilon}^{\mathrm{p}}_{ij}, \qquad \dot{e}_{ij} = \dot{e}^{\mathrm{p}}_{ij}, \qquad \dot{\epsilon}_{kk} = \dot{\epsilon}^{\mathrm{p}}_{kk} \tag{13.27}$$

Therefore we can suppress the superscript p from all strain measures.

By directly inverting the flow rule (13.22) the constitutive equations can be written for both rigid-plastic and viscoplastic models in the form

$$\sigma_{ij} = 2\bar{G}\left(\dot{\epsilon}_{ij} + \frac{\bar{\nu}}{1 - 2\bar{\nu}}\,\dot{\epsilon}_{kk}\,\delta_{ij}\right) \tag{13.28}$$

Eq. 13.28 is analogous to the stress–strain constitutive relationship of standard elasticity with the nonlinear strain rates playing now the role of the elastic strains and the parameters $\bar{G}$ and $\bar{\nu}$ that of the elastic shear modulus and the Poisson's ratio, respectively. The form of these parameters for plastic and viscoplastic metals is given in Table 13.2.

The flow-elasticity analogy extends also to the strain rate–velocity relationships and to the equilibrium equations (see Table 13.2). Consequently the problems of rigid-plastic/viscoplastic deformation of metals can be solved by directly using standard finite-element codes written for the elasticity case in which displacements and strains are replaced by velocities and strain rates respectively, and the shear modulus and the Poisson's ratio by parameters $\bar{G}$ and $\bar{\nu}$, respectively. Note in Table 13.2 the stress dependence of $\bar{G}$ and $\bar{\nu}$ which implies that the numerical solution must be found iteratively.

The term *flow approach* arises by noting that (13.28) is also identical to a constitutive equation of a viscous fluid with the parameter $\bar{G}$ playing the role of the non-Newtonian flow viscosity and $\bar{\nu}/(1-2\bar{\nu})$ being the compressibility of the fluid. Note that for the nonporous case $\bar{\nu} = 1/2$ and $\bar{\nu}/(1-2\bar{\nu}) = \infty$, thus recovering the classical expressions of incompressible flow for a von Mises material (Zienkiewicz et al. 1978).

Table 13.2. Equivalent elastic parameters $\bar{G}$ and $\bar{\nu}$ for a porous and nonporous rigid-plastic/viscoplastic metal ($B = fS/2A$, $K = (1/\gamma)^m$, $m = 1/n$, $S = \sinh A$, $A = \sigma_{kk}/2\sigma_M$)

	Rigid-viscoplastic	Rigid-plastic ($\gamma = \infty$)
Constitutive equation	$\sigma_{ij} = 2\bar{G}\left(\dot{\epsilon}_{ij} + \frac{\bar{\nu}}{1-2\bar{\nu}}\dot{\epsilon}_{kk}\delta_{ij}\right)$	
Porous metal	$\bar{G} = \frac{\sigma_M w}{3\dot{\bar{e}}_M} + \frac{1}{3}Kw(\dot{\bar{e}}_M)^{m-1}$	$\bar{G} = \frac{\sigma_M w}{3\dot{\bar{e}}_M}$
	$\bar{\nu} = \frac{1 - B[1 + \frac{K}{\sigma_.}(\dot{\bar{e}}_M)^m]^2}{2 + B[1 + \frac{K}{\sigma_M}(\dot{\bar{e}}_M)^m]^2}$	
Nonporous metal ($f = 0$, $w = 1$)	$\bar{G} = \frac{\sigma_M + \left(\frac{\dot{\bar{\epsilon}}}{\gamma}\right)^m}{3\dot{\bar{\epsilon}}}$	$\bar{G} = \frac{\sigma_M}{3\dot{\bar{\epsilon}}}$
	$\bar{\nu} = \frac{1}{2}$	$\bar{\nu} = \frac{1}{2}$

13.5 Finite-Element Discretization: Computational Aspects

For both solid and flow approaches the resulting nonlinear equilibrium equation can be written after finite-element discretization in the form (see Table 13.3)

$$\mathbf{r}(\mathbf{a},\mathbf{x},t) = \mathbf{p}(\mathbf{a},\mathbf{x}) - \mathbf{f}(t,\mathbf{x}) = 0 \tag{13.29}$$

where $\mathbf{r}$, $\mathbf{p}$ and $\mathbf{f}$ stand for the vectors of residual forces, internal forces and external forces, respectively, $\mathbf{a}$ are displacements and velocities in the solid and flow approaches, respectively, $\mathbf{x}$ is the Cartesian coordinate vector and t represents the time in the flow approach and the load increment in the solid approach.

Eq. (13.29) can be iteratively solved for the values of vector $\mathbf{a}$. For the kth iteration we have

$$\Delta\mathbf{a}^k = -\left[{}^{t+\Delta t}\mathbf{H}^k\right]^{-1}\,{}^{t+\Delta t}\mathbf{r}^k \tag{13.30}$$

where $\mathbf{H}$ is an adequate iteration matrix. Vector $\mathbf{a}$ is subsequently updated as

$${}^{t+\Delta t}\mathbf{a}^{k+1} = {}^{t+\Delta t}\mathbf{a}^k + \Delta\mathbf{a}^k \tag{13.31}$$

The next step is to compute the new stress field. In the flow approach the stresses are *directly* obtained from the updated velocity field (see Table 13.3). In the solid approach the computation of the stresses implies the integration of the rate constitutive equation in a consistent manner (Ortiz and Popov 1985).

The final step is to update the sheet geometry. This can be simply done from the displacement and velocity fields as shown in Table 13.3. At this stage the void volume fraction and the sheet thickness are also updated and the sheet–tool contact and friction conditions are checked.

Table 13.3. Finite-element solution algorithm for solid and flow approaches

	Solid approach	Flow approach
Basic discretization	$\mathbf{u} = \mathbf{N}\mathbf{a}$	$\mathbf{v} = \mathbf{N}\mathbf{a}$
	$\dot{\boldsymbol{\epsilon}} = \mathbf{B}_s\dot{\mathbf{a}}$ $\overset{\triangle}{\boldsymbol{\sigma}} = \mathbf{D}_T\dot{\boldsymbol{\epsilon}} = \mathbf{D}_T\mathbf{B}_s\dot{\mathbf{a}}$	$\dot{\boldsymbol{\epsilon}} = \mathbf{B}_f\mathbf{a}$ $\boldsymbol{\sigma} = \mathbf{D}\dot{\boldsymbol{\epsilon}} = \mathbf{D}\mathbf{B}_f\mathbf{a}$
Equilibrium equation	$\mathbf{r} = \mathbf{p} - \mathbf{f} = 0$	
	$\mathbf{p} = \int_v \mathbf{B}_s^T \boldsymbol{\sigma} dV$	$\mathbf{p} = \left(\int_v \mathbf{B}_f^T \mathbf{D}\mathbf{B}_f dV\right)\mathbf{a} = \mathbf{K}\mathbf{a}$
	$\mathbf{f} = \int_v \mathbf{N}^T\mathbf{b} dV + \int_s \mathbf{N}^T\mathbf{t}\, ds$	
Solution procedure	$\Delta\mathbf{a}^k = -[{}^{t+\Delta t}\mathbf{H}^k]^{-1}({}^{t+\Delta t}\mathbf{r}^k)$ ${}^{t+\Delta t}\mathbf{a}^{k+1} = {}^{t+\Delta t}\mathbf{a}^k + \Delta\mathbf{a}^k$	
	${}^{t+\Delta t}\boldsymbol{\sigma}^{k+1}$ are computed by integrating the constitutive equations	${}^{t+\Delta t}\boldsymbol{\sigma}^{k+1} = \mathbf{D}\mathbf{B}_f\, {}^{t+\Delta t}\mathbf{a}^{k+1}$
Geometry updating	${}^{t+\Delta t}\mathbf{x}^{k+1} = {}^{t+\Delta t}\mathbf{x}^k + \Delta\mathbf{a}^k$	${}^{t+\Delta t}\mathbf{x}^{k+1} = {}^t\mathbf{x} + {}^t\mathbf{a}(1-\theta)\,\Delta t + {}^{t+\Delta t}\mathbf{a}^{k+1}\theta\Delta t$ $\theta = 0$ explicit solution $\theta \neq 0$ implicit solution
	Compute error norm $E = \dfrac{\lVert {}^{t+\Delta t}\mathbf{r}^{k+1} \rVert}{\lVert {}^{t+\Delta t}\mathbf{f}^{k+1} \rVert}$ $E.GT.TOL. \xrightarrow{\text{yes}}$ continue iterations $k = k + 1$ $\downarrow$ no UPDATE GEOMETRY CHECK CONTACT INITIATE NEW LOAD INCREMENT	

The updated geometrical and mechanical properties of the sheet are used to compute the new increments $\Delta\mathbf{a}^k$ of the next iteration from (13.30). The process is continued until convergence is achieved. This is usually measured by the satisfaction of (13.29) using a mean quadratic norm for the residual forces. A summary of the solution process for both the solid and flow approaches is given in Table 13.3.

The detailed discussion of the computational aspects of the finite-element solution falls outside the scope of this chapter. Details of the different matrices and vectors appearing in Table 13.3 can be found in Zienkiewicz (1977). Also in the next section the particular form of the axisymmetric shell formulation for the flow case will be given.

Nevertheless the following remarks should be noted at this point.

1. The iteration matrix $\mathbf{H}$ in the solid approach is usually taken as the tangent stiffness matrix computed as $\mathbf{H} = \partial\mathbf{p}/\partial\mathbf{a}$. The solution algorithm coincides in this case with the standard Newton–Raphson iteration scheme.
2. For the flow approach the selection of the iteration matrix $\mathbf{H}$ can be made on the basis of *secant* or *tangent* approaches. The expression of $\mathbf{H}$ for the secant case coincides with that of the finite-element stiffness matrix $\mathbf{K}$ for

standard infinitesimal elasticity (Oñate et al. 1989). The form of the exact tangent matrix is complex and generally nonsymmetric and approximate simpler forms are used (Oñate et al. 1989).

3. The effects of void coalescence can be numerically accounted for in the following way. According to Tvergaard (1982) coalescence takes place at $f \simeq 0.20$. Thus an extremely simple yet effective numerical scheme can be used to reproduce this phenomenon if, at points for which $f = 0.2$ a proportional increase in f is assumed for a number of fixed incremental steps ($\simeq 10$) up to $f = 1.0$ for which the material carrying capacity in that point is effectively zero (Kleiber 1984a,b,1986).

13.6 Viscous Shell Approach

The viscous shell approach is based on the flow assumptions, i.e. the elastic strains are neglected and the total strains in the deforming sheet are plastic (or viscoplastic). As already mentioned this assumption yields a complete analogy between the resulting *viscous flow* equations with those of standard infinitesimal (nonlinear) elasticity. Consequently the finite-element equations for the viscous shell are readily obtained by replacing in any standard elastic shell finite-element formulation the displacement and strains by velocities and strain rates, respectively and the shear modulus and Poisson's ratio by parameters $\bar{G}$ and $\bar{\nu}$ of the previous section, respectively.

The basis of the success of the viscous shell approach therefore lies in the efficiency of the analogous elastic shell formulation. In the next section we will present a formulation for axisymmetric problems using simple two troncoconical shell elements and a selective membrane–bending approach.

13.6.1 Axisymmetric Viscous Shell Formulation

The following formulation is based on Reissner–Mindlin axisymmetric shell theory (Zienkiewicz et al. 1978). Fig. 13.1 shows a typical geometry of an axisymmetric sheet discretized in two-node troncoconical elements.

The velocity field, after discretization an be expressed as

$$\mathbf{u} = \begin{Bmatrix} u \\ w \\ \theta \end{Bmatrix} = \sum_{i=1}^{2} \mathbf{N}_i \mathbf{a}_i \tag{13.32}$$

with

$$\mathbf{N}_i = N_i \mathbf{I}_3 \quad \text{and} \quad \mathbf{a}_i = \begin{Bmatrix} u_i \\ w_i \\ \theta_i \end{Bmatrix} \tag{13.33}$$

with u_i, w_i and θ_i being the two global velocities and the angular velocity of node i, respectively and N_i the linear shape function of node i (Fig. 13.1).

The generalized strain rate and resultant stress vectors in each element can be expressed as

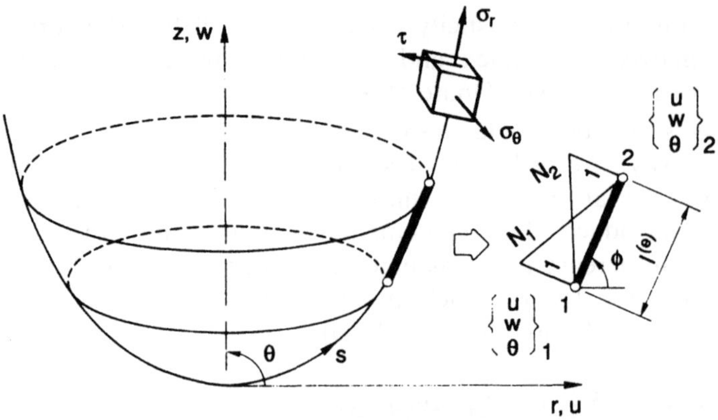

Fig. 13.1. Axisymmetric shell. Discretization in axisymmetric linear elements.

$$\dot{\boldsymbol{\epsilon}} = \begin{Bmatrix} \dot{\boldsymbol{\epsilon}}_m \\ \dot{\boldsymbol{\epsilon}}_b \\ \dot{\boldsymbol{\epsilon}}_s \end{Bmatrix} = \sum_{i=1}^{2} \mathbf{B}_i \mathbf{a}_i \tag{13.34}$$

$$\boldsymbol{\sigma} = \begin{Bmatrix} \boldsymbol{\sigma}_m \\ \boldsymbol{\sigma}_b \\ \boldsymbol{\sigma}_s \end{Bmatrix} = \mathbf{D}\dot{\boldsymbol{\epsilon}} = \mathbf{D} \sum_{i=1}^{2} \mathbf{B}_i \mathbf{a}_i \tag{13.35}$$

where subscripts m, b and s denote membrane, bending and shear terms respectively. The expression of the above vectors and matrices is shown in Table 13.4.

After adequate substitution of (13.32)–(13.35) in the rate of virtual work the following system of equations can be obtained:

$$\mathbf{r} = \mathbf{K}(\mathbf{a})\mathbf{a} - \mathbf{f} = 0 \tag{13.36}$$

The element contributions to the stiffness matrix $\mathbf{K}$ and the nodal forces vector $\mathbf{f}$ are

$$\mathbf{K}_{ij}^{(e)} = 2\pi \int_{l^{(e)}} \mathbf{B}_i^T \mathbf{D} \mathbf{B}_j r ds \tag{13.37}$$

$$\mathbf{f}_i^{(e)} = 2\pi \int_{l^{(e)}} \mathbf{N}_i \mathbf{t} r ds + 2\pi r_i \mathbf{p}_i \tag{13.38}$$

where $l^{(e)}$ is the element length, r the radial distance and $\mathbf{t}$ and $\mathbf{p}_i$ surface and point load vectors, respectively.

Note the nonlinearity of (13.36) due to the stress and strain dependence of parameters $\bar{G}$ and $\bar{\nu}$ (see Table 13.2). Eq. (13.36) can be solved using a direct iteration scheme or an incremental-iterative predictor–corrector algorithm (Oñate et al. 1989, Agelet de Saracibar and Oñate 1990).

It has been shown by Oñate et al. (1983, 1988) that for a successful use of this formulation for thin sheet analysis the element stiffness matrix of (13.37) must be numerically computed using a single Gaussian integration point. In practice this simply implies obtaining $\mathbf{K}_{ij}^{(e)}$ explictly as

Table 13.4. Axisymmetric viscous shell: generalized strain rate–velocity and constitutive relationships

$$\dot{\boldsymbol{\epsilon}} = \begin{Bmatrix} \dot{\boldsymbol{\epsilon}}_{\mathrm{m}} \\ \dot{\boldsymbol{\epsilon}}_{\mathrm{b}} \\ \dot{\boldsymbol{\epsilon}}_{\mathrm{s}} \end{Bmatrix} = \begin{Bmatrix} \dot{\epsilon}_r \\ \dot{\epsilon}_\theta \\ --- \\ \dot{\chi}_r \\ \dot{\chi}_\theta \\ --- \\ \dot{\gamma} \end{Bmatrix} = \begin{Bmatrix} C\dfrac{\partial u}{\partial s} + S\dfrac{\partial w}{\partial s} \\ \dfrac{u_0}{r} \\ --- \\ -\dfrac{\partial \theta}{\partial s} \\ -\dfrac{\theta \cos\phi}{r} \\ --- \\ -S\dfrac{\partial u}{\partial s} + C\dfrac{\partial w}{\partial s} - \theta \end{Bmatrix} = \sum_{i=1}^{2} \mathbf{B}_i \mathbf{a}_i$$

$$\mathbf{B}_i = \begin{Bmatrix} \mathbf{B}_{\mathrm{m}_i} \\ \mathbf{B}_{\mathrm{b}_i} \\ \mathbf{B}_{\mathrm{s}_i} \end{Bmatrix} = \begin{bmatrix} C\dfrac{\partial N_i}{\partial s} & S\dfrac{\partial N_i}{\partial s} & 0 \\ \dfrac{N_i}{r} & 0 & 0 \\ - & - & - \\ 0 & 0 & -\dfrac{\partial N_i}{\partial s} \\ 0 & 0 & -\dfrac{N_i C}{r} \\ - & - & - \\ -S\dfrac{\partial N_i}{\partial s} & C\dfrac{\partial N_i}{\partial s} & -N_i \end{bmatrix}$$

$$S = \sin\phi, \quad C = \cos\phi$$

$$\boldsymbol{\sigma} = [N_r, N_\theta, M_r, M_\theta, Q]^{\mathrm{T}} = \mathbf{D}\dot{\boldsymbol{\epsilon}} = \mathbf{D}\sum_{i=1}^{2} \mathbf{B}_i \mathbf{a}_i$$

$$\mathbf{D} = \begin{bmatrix} \mathbf{D}_{\mathrm{m}} & \mathbf{D}_{\mathrm{mb}} & 0 \\ \mathbf{D}_{\mathrm{mb}} & \mathbf{D}_{\mathrm{b}} & 0 \\ 0 & 0 & D_{\mathrm{s}} \end{bmatrix} = \int_{-t/2}^{t/2} \begin{bmatrix} \hat{\mathbf{D}}_{\mathrm{m}} & z'\hat{\mathbf{D}}_{\mathrm{m}} & \mathbf{0} \\ z'\hat{\mathbf{D}}_{\mathrm{m}} & (z')^2\hat{\mathbf{D}}_{\mathrm{m}} & \mathbf{0} \\ \mathbf{0} & \mathbf{0} & D_{\mathrm{c}} \end{bmatrix} dz'$$

$$\hat{\mathbf{D}}_{\mathrm{m}} = \frac{2\bar{G}}{1-\bar{\nu}} \begin{bmatrix} 1 & \bar{\nu} \\ \bar{\nu} & 1 \end{bmatrix}, \quad D_{\mathrm{c}} = \bar{G}$$

$$\mathbf{K}_{ij}^{(e)} = 2\pi \bar{\mathbf{B}}_i^{\mathrm{T}} \overline{\mathbf{D}} \bar{\mathbf{B}}_j \bar{r} l^{(e)} \tag{13.39}$$

where the bar denotes values at the element midpoint. The expression of $\bar{\mathbf{B}}_i$ is readily obtained by substituting the terms N_i and $\partial N_i/\partial s$ in Table 13.4 by $\frac{1}{2}$ and $(-1)^i/l^{(e)}$, respectively.

13.6.2 Selective Bending–Membrane Formulation

More recently, Oñate and Agelet de Saracibar (1988) and Agelet de Saracibar and Oñate (1990) have shown the possibilities of the viscous shell formulation for deriving a selective bending–membrane approach. Here, bending and membrane elements are used in a selective manner in accordance with the specific features of the sheet deformation pattern at each deforming step, i.e. membrane elements are used in zones where the shell is basically under tension conditions, whereas in zones where bending effects are predominant (i.e. sharp corners) a bending formulation is chosen. To explain this procedure we will consider a sheet discretized in flat viscous shell elements. An example of this are the troncoconical elements for axisymmetric shell analysis studied in the previous section or any of the flat triangular elements traditionally used for 3D shell analysis (Zienkiewicz 1977). For any of these elements the stiffness matrix **K** of (13.37) can be written in the form

$$\mathbf{K}^{(e)} = \mathbf{K}_{m}^{(e)} + \mathbf{K}_{b}^{(e)} + \mathbf{K}_{mb}^{(c)} + \mathbf{K}_{s}^{(e)} \tag{13.40}$$

where subscripts m, b, mb and s denote membrane, bending, membrane–bending coupling and shear contributions to the total stiffness matrix of the element. An example of the form of the above matrices for the simple two node axisymmetric viscous element is

$$\mathbf{K}_{m_{ij}}^{(e)} = 2\pi\bar{r}l^{(e)}\,\bar{\mathbf{B}}_{m_i}^{T}\,\bar{\mathbf{D}}_{m}\bar{\mathbf{B}}_{m_j} \tag{13.41a}$$

$$\mathbf{K}_{b_{ij}}^{(e)} = 2\pi\bar{r}l^{(e)}\,\bar{\mathbf{B}}_{b_i}^{(T)}\,\bar{\mathbf{D}}_{b}\bar{\mathbf{B}}_{b_j} \tag{13.41b}$$

$$\mathbf{K}_{mb_{ij}}^{(e)} = 2\pi\bar{r}l^{(e)}\,[\bar{\mathbf{B}}_{m_i}^{T}\,\bar{\mathbf{D}}_{mb}\bar{\mathbf{B}}_{b_j} + \bar{\mathbf{B}}_{b_i}^{T}\,\bar{\mathbf{D}}_{mb}\bar{\mathbf{B}}_{m_j}] \tag{13.41c}$$

$$\mathbf{K}_{s_{ij}}^{(e)} = 2\pi\bar{r}l^{(e)}\,\bar{\mathbf{B}}_{s_i}^{T}\,\bar{\mathbf{D}}_{s}\bar{\mathbf{B}}_{s_j} \tag{13.41d}$$

where the different **B** and **D** matrices can be obtained from Table 13.4.

Eq. (13.40) allows us to compute the energy rates for the element at each deformation stage once the velocity field has been found from (13.36). Thus we can define for each element

$$\dot{U}^{(e)} = \tfrac{1}{2}\,[\mathbf{a}^{(e)}]^{T}\,\mathbf{K}^{(e)}\mathbf{a}^{(e)} \tag{13.42a}$$

$$\dot{U}_{m}^{(e)} = \tfrac{1}{2}\,[\mathbf{a}^{(e)}]^{T}\,\mathbf{K}_{m}^{(e)}\,\mathbf{a}^{(e)} \tag{13.42b}$$

where $\dot{U}^{(e)}$ and $\dot{U}_{m}^{(e)}$ stand for the total energy rate and the energy rate due to membrane effects only.

Clearly, under membrane conditions the value of $\dot{U}_{m}^{(e)}$ will be approximately equal to that of $\dot{U}^{(e)}$. This simple fact allows us to define the following scheme for choosing a membrane or bending formulation for the element:

1. At the onset of the deformation process the bending formulation is used for all elements in the sheet (Fig. 13.2a).
2. Once the velocity solution has been obtained for each new deforming configuration, the following energy ratio parameter is computed for each element:

$$\zeta^{(e)} = \frac{\dot{U}_{m}^{(e)}}{\dot{U}^{(e)}} \tag{13.43}$$

3. If $\zeta^{(e)} < \zeta_0$ where ζ_0 is a prescribed value close to unity the effect of bending is assumed to be significant and the full element stiffness matrix of (13.40) is retained.

4. If $\zeta^{(e)} > \zeta_0$ the element is assumed to work under membrane conditions. This implies the following operations:
   (a) A new stiffness matrix is computed for the element as
   $$\mathbf{K}^{(e)} = \mathbf{K}_m^{(e)} \tag{13.44}$$
   where $\mathbf{K}_m$ contains the contribution of the membrane terms to the stiffness matrix.
   (b) The "transition" nodes shared by a patch of different elements working under bending and membrane conditions *retain the rotational degrees of freedom*. However, these contribute to the stiffness matrix of the bending elements only (Fig. 13.2b). Examples of this combination of membrane and bending behaviour are very common in framed structures with internal pin-joints (Fig. 13.3).

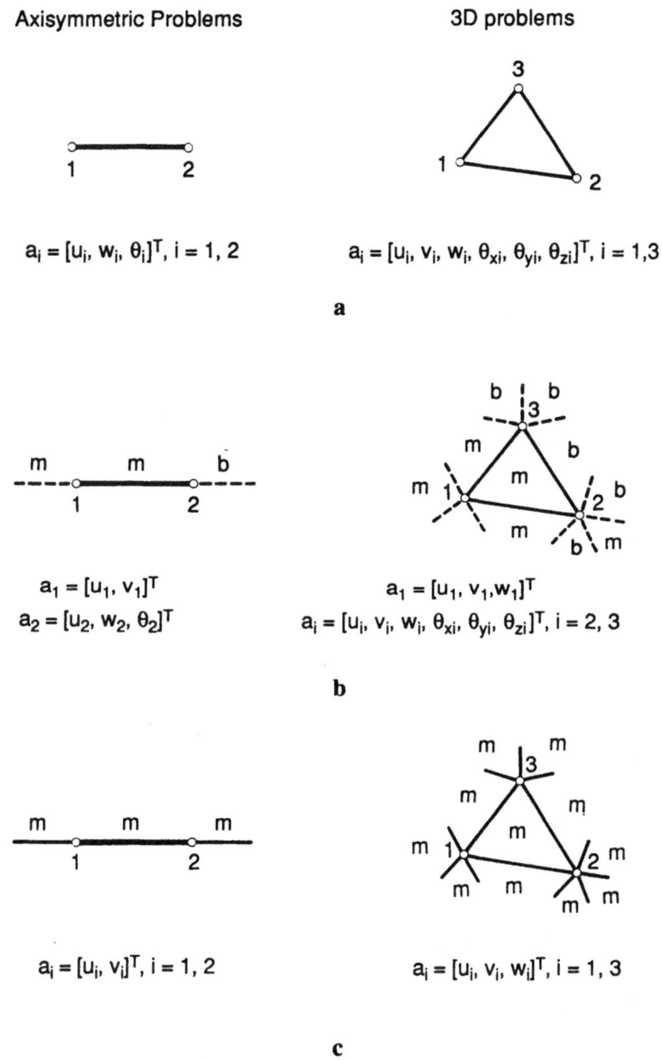

Fig. 13.2. Nodal variables for **a** bending (b), **b** membrane (m) and **c** transition bending–membrane elements for axisymmetric and 3D sheet-forming analysis.

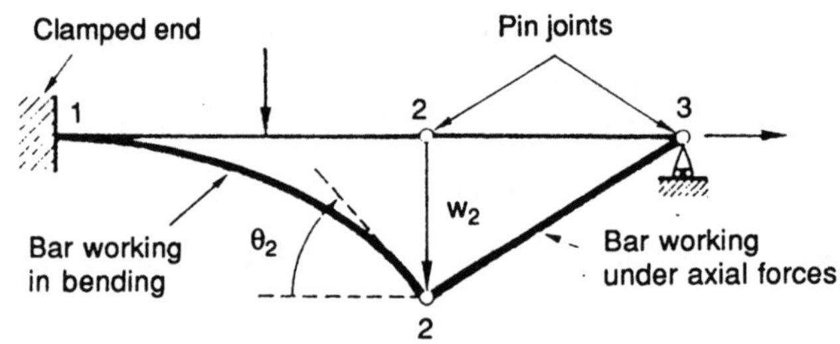

Fig. 13.3. Beam combining bending and axial elements.

(c) If *all* elements surrounding a node work under membrane conditions, the rotational degrees of freedom of the node can be eliminated (Fig. 13.2c). This can be simply accomplished in practice by deleting the rows and columns corresponding to the nodal rotations in the element stiffness matrix.

Figs 13.4 and 13.5 show schematic examples of the different situations described above for 2D axisymmetric and 3D sheet-forming processes.

13.6.3 Change from Membrane to Bending Formulation

The contact of some parts of the sheet with sharp boundaries can lead to a drastic change from membrane to a bending situation. To detect these changes the value of the energy ratio ζ must be continuously checked throughout the whole deformation of the sheet. This implies keeping a record of the nodal rotation rates at each deforming stage so that the bending energy rate can be adequately computed. The problem arises in membrane (or transition bending–membrane elements) in which the nodal rotation variables have been eliminated from all (or some) nodes. This difficulty can be overcome by computing first an *estimated* value of the element rotation from the element geometric changes between two consecutive deforming configurations. The corresponding nodal rotations are subsequently computed by a simple averaging of the rotations of all elements adjacent to the node.

This procedure is extremely simple for axisymmetric problems as shown in Fig. 13.6a. For 3D problems the situation is somehow more complicated, particularly if large rotations are involved. However, assuming small rotations between two consecutive configurations the problem is much simpler, since the components of the element rotations can be obtained from tensor transformation relating the known components of the normal vectors to the element at two consecutive configurations (Fig. 13.6b).

13.7 Contact and Friction

Contact and friction play an important role in many sheet-metal-forming processes. The analysis of such effects involves two main steps. In the first one a contact-searching algorithm must be used in order to detect the kinematic incompatibilities produced between the interacting bodies. In the second step these kinematic

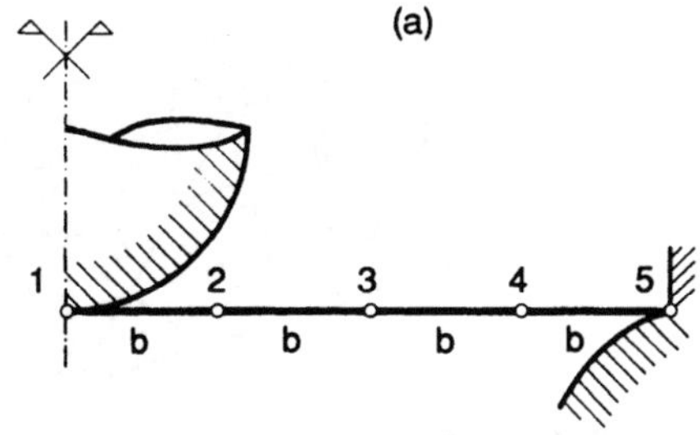

Degrees of freedom:

$a_i = [u_i, w_i, \theta_i]^T, i = 1, \cdots, 5$

Full $K^{(e)}$ for all elements

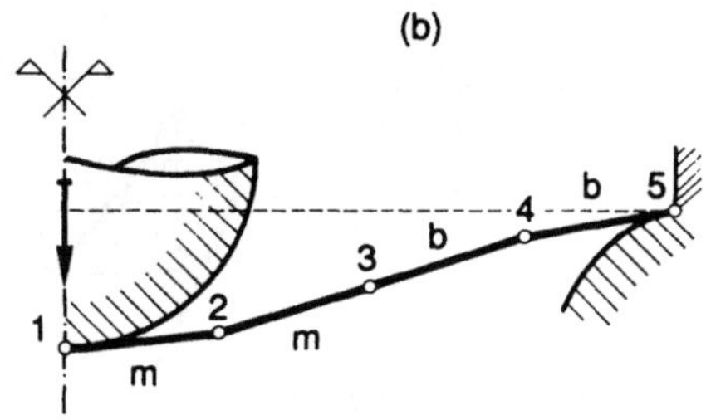

Degrees of freedom:

$a_i = [u_i, w_i]^T, i = 1, 2$

$a_i = [u_i, w_i, \theta_i]^T, i = 3, 4, 5$

Element	*Stiffness matrix*
1 – 2	$K_m^{(e)}$ eliminating rows and columns of θ_1 and θ_2
2 – 3	$K_m^{(e)}$ eliminating the rows and columns of θ_2
3 – 4 and 4 – 5	Full $K^{(e)}$

Fig. 13.4. Axisymmetric sheet-forming problem using selectively bending (b) and membrane (m) elements.

incompatibilities are eliminated by introducing the contact kinematic restrictions in the finite-element formulation.

Contact-searching algorithms consume much CPU time. Use of contact-searching procedures and efficient algorithms are highly recommended in order to optimize CPU time, specially for large-scale computations.

Within the framework of the finite-element analysis there are two well-established formulations which can be used to introduce the kinematic restrictions imposed by the interacting bodies. These are the Lagrange multiplier and the penalty methods. It is interesting to note that both methods can be considered as a particular case of a perturbed Lagrangian formulation (Simo et al. 1985, Wriggers and Simo 1985). Using standard variational procedures the *exact linearization* of the first variation of the perturbed (total) functional leads to the

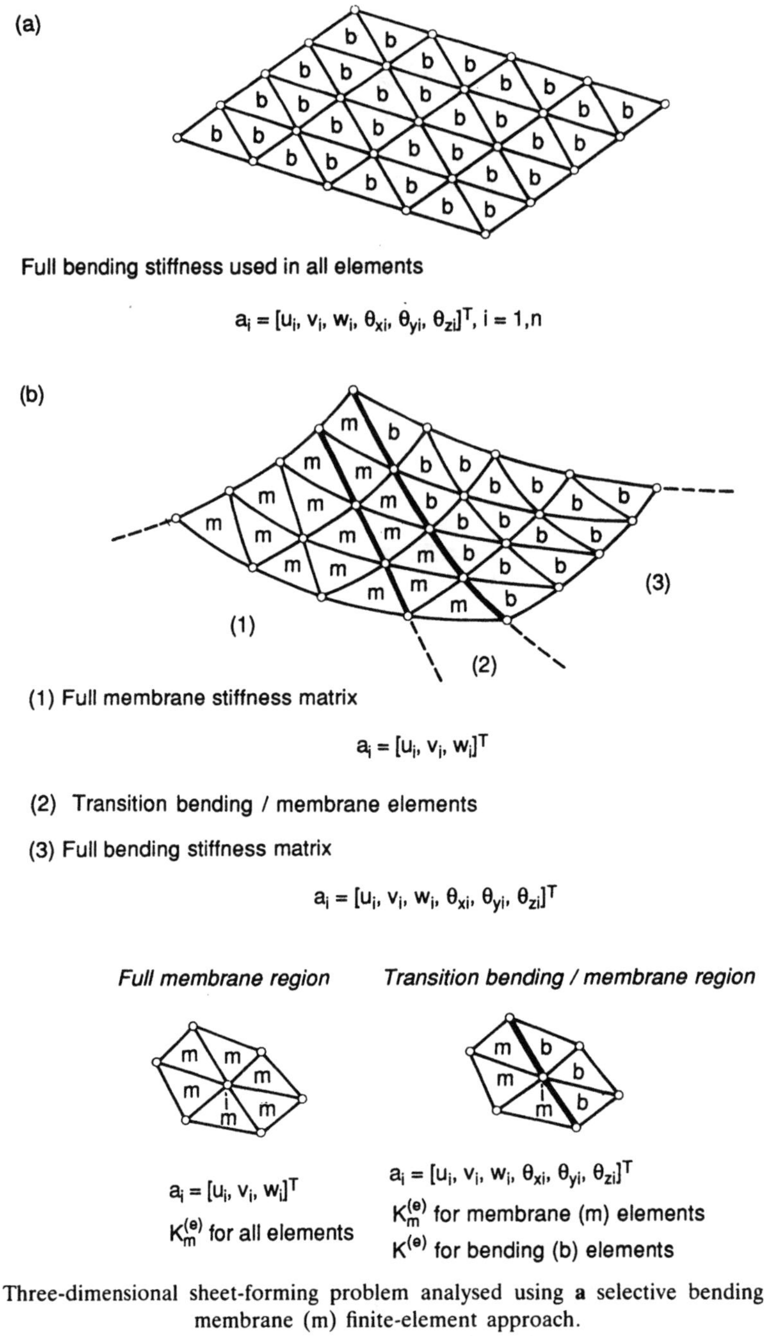

Fig. 13.5. Three-dimensional sheet-forming problem analysed using **a** selective bending (b) and **b** membrane (m) finite-element approach.

contact *consistent residual* and *consistent tangent operators*, for a fully nonlinear kinematic.

Frictional effects can be treated on the basis of Coulomb or kinematic friction models. In the Coulomb model, a variational formulation can be obtained from the expression of the first variation of an *unknown* friction functional (Ju et al. 1987). In the kinematic friction model the formulation is obtained from the

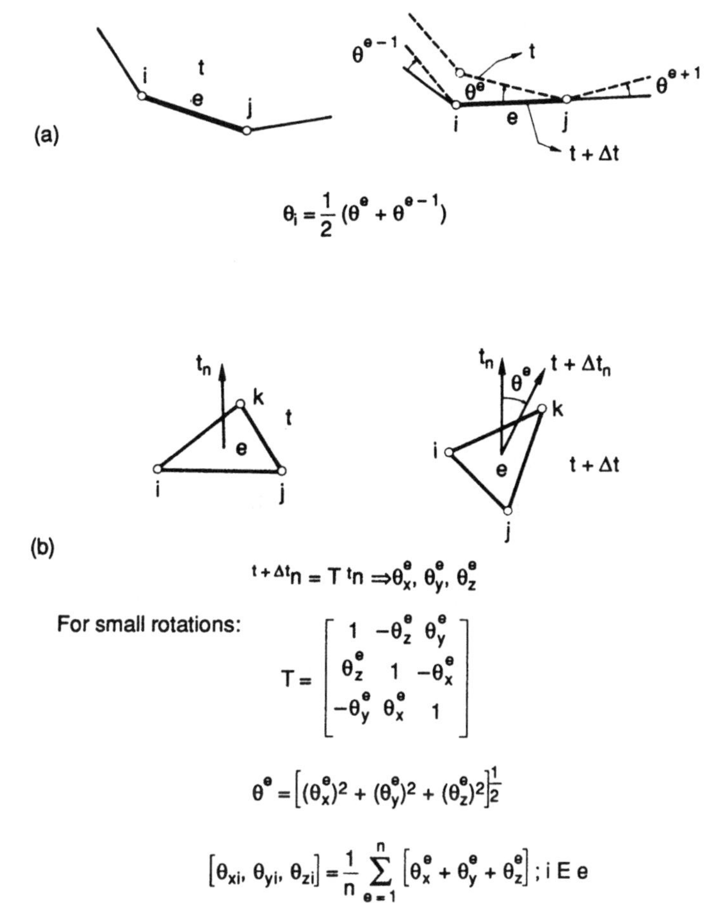

Fig. 13.6. Computation of element and nodal rotations from geometrical changes. **a** Axisymmetrical problems; **b** 3D problems.

expression of a *known* friction functional. In both cases the exact linearization of the first variation of the functional leads to the frictional consistent residual and tangent operators (Ju et al. 1987).

Use of consistent tangent operators is essential in order to preserve the *asymptotic quadratic rate of convergence* of Newton's method (Simo et al. 1985, Wriggers and Simo 1985).

13.7.1 Contact Formulation

For the fully nonlinear contact problem, the kinematic restriction between two interacting bodies takes the form

$$g := (\mathbf{\Phi}^2 - \mathbf{\Phi}^1) \cdot \mathbf{n} \geq 0 \text{ on } \partial B_c \tag{13.45}$$

where $\mathbf{\Phi}^2$, $\mathbf{\Phi}^1$ refers to the configurations of the bodies $\mathbf{B}^1$ and $\mathbf{B}^2$ respectively, $\mathbf{n}$ is the normal to the contact surface $\partial B_c = \partial\mathbf{\Phi}^1 (B^1) \cap \partial\mathbf{\Phi}^2 (B^2)$, and g is the gap function.

The restriction (13.45) can be introduced on the basis of a contact-perturbed Lagrangian functional of the form

$$\pi_c(\Phi^1, \Phi^2) = \int_{\partial B_c} \lambda \left(g - \frac{\lambda}{2\epsilon}\right) ds \tag{13.46}$$

where λ refers to the contact tractions on the surface ∂B_c, and ϵ is a perturbation parameter.

Within the framework of the finite-element method, the perturbed Lagrangian functional can be obtained by adding the discretized version of (13.46) to the usual total potential energy of the bodies (Wriggers and Simo 1985).

$$\pi_\epsilon(\mathbf{u}, \boldsymbol{\lambda}) = \bar{\pi}(\mathbf{u}) + \boldsymbol{\lambda} \cdot \mathbf{g} - \frac{1}{2}\boldsymbol{\lambda} \cdot \boldsymbol{\lambda} \tag{13.47}$$

where $\mathbf{u} \in \mathbb{R}^n$ are the nodal displacement vector, $\boldsymbol{\lambda} \in \mathbb{R}^s$ the nodal contact forces, $\mathbf{g} \in \mathbb{R}^s$ the nodal gaps vector and $\bar{\pi}(\mathbf{u})$ the total potential energy of the bodies.

Restricting our attention to 2D problems using a linear surface discretization, let us consider a slave node "s" that has penetrated a master segment defined by the nodes (m_1, m_2).

The current gap and contact position for the slave node s are (Fig. 13.7)

$$g_s = (\mathbf{x}_s - \mathbf{x}_{m_1}) \cdot \mathbf{n} \tag{13.48}$$

$$\alpha = \frac{1}{\|\Delta \mathbf{x}_m\|}(\mathbf{x}_s - \mathbf{x}_{m_1}) \cdot \mathbf{t} \tag{13.49}$$

where $\mathbf{t}$ and $\mathbf{n}$ are the unit local tangent and normal to the master segment and $\Delta \mathbf{x}_m = \mathbf{x}_{m_2} - \mathbf{x}_{m_1}$.

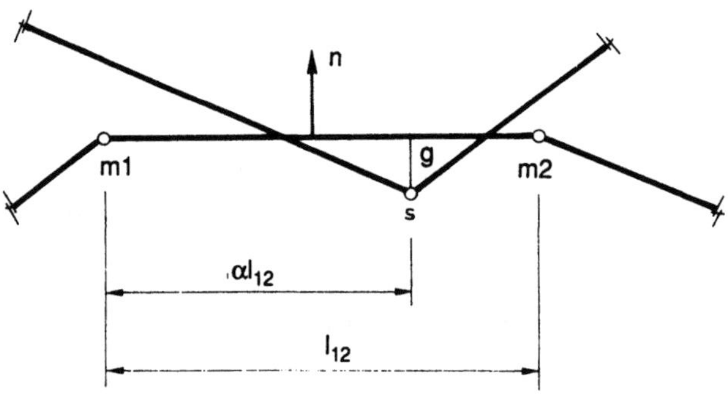

Fig. 13.7. Geometry of a slave node s with a master segment m_1–m_2.

13.7.2 First Variation of the Functional

The first variation of the functional (13.47) leads to the following set of equations:

$$D\pi_\epsilon \cdot \boldsymbol{\eta} = D\bar{\pi} \cdot \boldsymbol{\eta} + \boldsymbol{\lambda} \cdot (D\mathbf{g} \cdot \boldsymbol{\eta}) = 0 \tag{13.50a}$$

$$D\pi_\epsilon \cdot \delta\boldsymbol{\lambda} = \left(\mathbf{g} - \frac{1}{\epsilon}\boldsymbol{\lambda}\right) \cdot \delta\boldsymbol{\lambda} = 0 \tag{13.50b}$$

where the use of directional derivatives is exploited.

Using the directional derivatives of the tangent and normal vectors

$$D\mathbf{t}\cdot\boldsymbol{\eta} = \frac{1}{\|\Delta\mathbf{x}_m\|} - (\mathbf{n}\otimes\mathbf{n})\cdot\Delta\boldsymbol{\eta}_m \tag{13.51}$$

$$D\mathbf{n}\cdot\boldsymbol{\eta} = -\frac{1}{\|\Delta\mathbf{x}_m\|}(\mathbf{t}\otimes\mathbf{n})\cdot\Delta\boldsymbol{\eta}_m \tag{13.52}$$

where $\Delta\boldsymbol{\eta}_m = \boldsymbol{\eta}_{m_2} - \boldsymbol{\eta}_{m_1}$, the variational equations (13.50) take the form

$$D\pi_\epsilon(\mathbf{u},\boldsymbol{\lambda})\cdot\boldsymbol{\eta} = \left(D\bar{\pi}(\mathbf{u}) + \bigcup_{s=1}^{S}\lambda_s\mathbf{r}_s\right)\cdot\boldsymbol{\eta} = 0 \tag{13.53a}$$

$$D\pi_\epsilon(\mathbf{u},\boldsymbol{\lambda})\cdot\delta\boldsymbol{\lambda} = \left(\mathbf{g} - \frac{1}{\epsilon}\boldsymbol{\lambda}\right)\cdot\delta\boldsymbol{\lambda} = 0 \tag{13.53b}$$

where S is the number of slave nodes and $\mathbf{r}_s$ is defined according to the expression

$$\mathbf{r}_s\cdot\boldsymbol{\eta} = \mathbf{n}\cdot(\boldsymbol{\eta}_s - \boldsymbol{\eta}_m) \tag{13.54}$$

where $\boldsymbol{\eta}_m = (1-\alpha)\boldsymbol{\eta}_{m_1} + \alpha\boldsymbol{\eta}_{m_2}$ with α given by (13.49).

13.7.3 Second Variation of the Functional

The second variation of the functional (13.47) leads to the following expressions:

$$D(D\pi_\epsilon(\mathbf{u},\boldsymbol{\lambda})\cdot\boldsymbol{\eta})\cdot\Delta\mathbf{u} = \boldsymbol{\eta}\cdot\left(\mathbf{K}_T + \bigcup_{s=1}^{S}\mathbf{K}_s\right)\cdot\Delta\mathbf{u} \tag{13.55a}$$

$$D(D\pi_\epsilon(\mathbf{u},\boldsymbol{\lambda})\cdot\boldsymbol{\eta})\cdot\Delta\boldsymbol{\lambda} = \boldsymbol{\eta}\cdot\left(\bigcup_{s=1}^{S}\mathbf{r}_s\right)\cdot\Delta\boldsymbol{\lambda} \tag{13.55b}$$

$$D(D\pi_\epsilon(\mathbf{u},\boldsymbol{\lambda})\cdot\delta\boldsymbol{\lambda})\cdot\Delta\mathbf{u} = \delta\boldsymbol{\lambda}\cdot\left(\bigcup_{s=1}^{S}\mathbf{r}_s\right)\cdot\Delta\mathbf{u} \tag{13.55c}$$

$$D(D\pi_\epsilon(\mathbf{u},\boldsymbol{\lambda})\cdot\delta\boldsymbol{\lambda})\cdot\Delta\boldsymbol{\lambda} = \delta\boldsymbol{\lambda}\cdot\left(\frac{1}{\epsilon}\mathbf{I}\right)\cdot\Delta\boldsymbol{\lambda} \tag{13.55d}$$

where $\mathbf{K}_s$ is defined by the expression

$$\begin{aligned}\boldsymbol{\eta}\cdot\mathbf{K}_s\cdot\mathbf{u} = -\frac{\lambda_s}{\|\Delta\mathbf{x}_m\|}[&(\boldsymbol{\eta}_s - \boldsymbol{\eta}_m)\cdot(\mathbf{t}\otimes\mathbf{n})\,\Delta\mathbf{u}_m + \\ &\Delta\boldsymbol{\eta}_m\cdot(\mathbf{n}\otimes\mathbf{t})\,(\mathbf{u}_s - \mathbf{u}_m) + \\ &\frac{g_s}{\|\Delta\mathbf{x}_m\|}\Delta\boldsymbol{\eta}_m\cdot(\mathbf{n}\otimes\mathbf{n})\Delta\mathbf{u}_m]\end{aligned} \tag{13.56}$$

with $\Delta\mathbf{u}_m = \mathbf{u}_{m_2} - \mathbf{u}_{m_1}$.

13.7.4 Consistent Operators

The linearized expression of the first variation of the functional leads to the following consistent residual and tangent operators for the contact element (s, m_1, m_2)

$$\mathbf{R}_{c_s} = \left[\lambda_s \mathbf{N}_s, -\frac{1}{\epsilon}\lambda_s + g_s \right]^{\mathrm{T}} \tag{13.57a}$$

$$\mathbf{K}_{c_s} = \begin{bmatrix} -\dfrac{\lambda_s}{\|\Delta \mathbf{x}_m\|}\left(\mathbf{T}_s \otimes \mathbf{N}_m + \mathbf{N}_m \otimes \mathbf{T}_s + \dfrac{g_s}{\|\Delta \mathbf{x}_m\|}\mathbf{N}_m \otimes \mathbf{N}_m\right) & \mathbf{N}_s \\ \mathbf{N}_s^{\mathrm{T}} & -\dfrac{1}{\epsilon} \end{bmatrix} \tag{13.57b}$$

where the following operators have been introduced:

$$\mathbf{N}_s = [\mathbf{n}, -(1-\alpha)\mathbf{n}, -\alpha\mathbf{n}]^{\mathrm{T}} \tag{13.58a}$$

$$\mathbf{T}_s = [\mathbf{t}, -(1-\alpha)\mathbf{t}, -\alpha\mathbf{t}]^{\mathrm{T}} \tag{13.58b}$$

$$\mathbf{N}_m = [0, -\mathbf{n}, \mathbf{n}]^{\mathrm{T}} \tag{13.58c}$$

The residual force vector $\mathbf{R}_{c_s}$ and tangent matrix $\mathbf{K}_{c_s}$ for the contact element are added to the expressions of $\mathbf{r}$ and $\mathbf{H}$ respectively, in the global nonlinear solution scheme described in Section 13.5.

13.7.5 Lagrange Multiplier Method

The consistent residual and tangent operator for the Lagrange multiplier method can be obtained by setting the penalty parameter $\epsilon \rightarrow \infty$ in (13.57):

$$\mathbf{R}_{c_s} = [\lambda_s \mathbf{N}_s, g_s]^{\mathrm{T}} \tag{13.59a}$$

$$\mathbf{K}_{c_s} = \begin{bmatrix} -\dfrac{\lambda_s}{\|\Delta \mathbf{x}_m\|}\left(\mathbf{T}_s \otimes \mathbf{N}_m + \mathbf{N}_m \otimes \mathbf{T}_s + \dfrac{g_s}{\|\Delta \mathbf{x}_m\|}\mathbf{N}_m \otimes \mathbf{N}_m\right) & \mathbf{N}_s \\ \mathbf{N}_s^{\mathrm{T}} & 0 \end{bmatrix} \tag{13.59b}$$

13.7.6 Penalty Method

The consistent residual and tangent operator for the penalty method can be obtained by static condensation of $\Delta\lambda_s$ at the element level and setting $\lambda_s = \epsilon g_s$. This gives

$$\mathbf{R}_{c_s} = \epsilon g_s \mathbf{N}_s \tag{13.60a}$$

$$\mathbf{K}_{c_s} = \epsilon \left[\mathbf{N}_s \otimes \mathbf{N}_s - \frac{g_s}{\|\Delta \mathbf{x}_m\|}\left(\mathbf{T}_s \otimes \mathbf{N}_m + \mathbf{N}_m \otimes \mathbf{T}_s + \frac{g_s}{\|\Delta \mathbf{x}_m\|}\mathbf{N}_m \otimes \mathbf{N}_m\right)\right] \tag{13.60b}$$

13.7.7 Velocity Approach

In the flow formulation, the main variables are the velocities. In this case the discretized version of the perturbed Lagrangian functional, takes the form

$$\pi_\epsilon(\mathbf{v},\boldsymbol{\lambda}) = \bar{\pi}(\mathbf{v}) + \boldsymbol{\lambda} \cdot \mathbf{g} - \frac{1}{2\epsilon}\boldsymbol{\lambda} \cdot \boldsymbol{\lambda} \tag{13.61}$$

where $\mathbf{v} \in \mathbb{R}^n$ is the nodal velocities vector and $\boldsymbol{\lambda}$, $\mathbf{g}$, ϵ and $\bar{\pi}$ have the same meaning as in (13.47).

In order to express the gap in terms of the velocities a time integration scheme of the form

$$^{t+\Delta t}\mathbf{x}^{i+1} = {}^{t+\Delta t}\mathbf{x}^{\text{trial}} + \theta\Delta t^{t+\Delta t}\mathbf{v}^{i+1} \tag{13.62}$$

where

$$^{t+\Delta t}\mathbf{x}^{\text{trial}} = {}^{t}\mathbf{x} + (1-\theta)\Delta t^{t}\mathbf{v} \tag{13.63}$$

is considered.

As in the solid approach, the *exact* linearization of the first variation of (13.61) leads to the *consistent* residual and tangent operators.

It is interesting to note that these operators take exactly the same expressions as in the solid approach, the only difference being that the $\mathbf{T}_s$, $\mathbf{N}_s$ and $\mathbf{N}_m$ operators are defined now as

$$\mathbf{T}_s = \theta\Delta t[\mathbf{t}, -(1-\alpha)\mathbf{t}, -\alpha\mathbf{t}]^{\mathrm{T}} \tag{13.64a}$$

$$\mathbf{N}_s = \theta\Delta t[\mathbf{n}, -(1-\alpha)\mathbf{n}, -\alpha\mathbf{n}]^{\mathrm{T}} \tag{13.64b}$$

$$\mathbf{N}_m = \theta\Delta t[0, -\mathbf{n}, \mathbf{n}]^{\mathrm{T}} \tag{13.64c}$$

13.7.8 Treatment of Friction

We will consider here a local linear friction formulation based on the well known Coulomb model. A description of more complex nonlocal and nonlinear models can be found in Oden and Pires (1983).

For the three-node contact element (s, m_1, m_2) the incremental slip takes the form

$$gt_s = (\mathbf{x}_s - \mathbf{x}_{m_1})\cdot\mathbf{t} - \alpha_0\|\Delta\mathbf{x}_m\| \tag{13.65}$$

where α_0 is the contact coordinate at the last iteration step.

The first variation of the *unknown* frictional functional using a penalty contact formulation, takes the form

$$D\pi_{\mathrm{f}}\cdot\boldsymbol{\eta} = -\bigcup_{s=1}^{S}\mu\epsilon g_s\,(Dgt_s\cdot\boldsymbol{\eta}) \tag{13.66}$$

where μ is the Coulomb friction coefficient. The directional derivative of the incremental slip is

$$Dgt_s\cdot\boldsymbol{\eta} = (\boldsymbol{\eta}_s - \boldsymbol{\eta}_m^0)\cdot\mathbf{t} + \frac{g_s}{\|\Delta\mathbf{x}_m\|}\Delta\boldsymbol{\eta}_m\cdot\mathbf{n} \tag{13.67}$$

where $\boldsymbol{\eta}_m^0 = (1-\alpha_0)\,\boldsymbol{\eta}_{m_1} + \alpha_0\boldsymbol{\eta}_{m_2}$.

Substituting (13.67) into (13.66) the first variation of the frictional functional takes the form

$$D\pi_{\mathrm{f}}\cdot\boldsymbol{\eta} = -\bigcup_{s=1}^{S}\mu\epsilon g_s\left[(\boldsymbol{\eta}_s - \boldsymbol{\eta}_m^0)\cdot\mathbf{t} + \frac{g_s}{\|\Delta\mathbf{x}_m\|}\Delta\boldsymbol{\eta}_m\cdot\mathbf{n}\right] \tag{13.68}$$

and the second variation is

$$D(D\pi_f \cdot \boldsymbol{\eta}) \cdot \mathbf{u} = \bigcup_{s=1}^{S} \left\{ -\mu\epsilon \left[(\boldsymbol{\eta}_s - \boldsymbol{\eta}_m^0) \cdot \mathbf{t} + \frac{g_s}{\|\Delta \mathbf{x}_m\|} \Delta\boldsymbol{\eta}_m \cdot \mathbf{n} \right] \otimes [\mathbf{n} \cdot (\mathbf{u}_s - \mathbf{u}_m)] - \right.$$

$$\mu\epsilon \frac{g_s}{\|\Delta \mathbf{x}_m\|} \left[(\boldsymbol{\eta}_s - \boldsymbol{\eta}_m^0) \cdot \mathbf{n} - \frac{g_s}{\|\Delta \mathbf{x}_m\|} \Delta\boldsymbol{\eta}_m \cdot \mathbf{t} \right] \otimes [\mathbf{n} \cdot \Delta \mathbf{u}_m)] -$$

$$\left. \mu\epsilon \frac{g_s}{\|\Delta \mathbf{x}_m\|} [\Delta\boldsymbol{\eta}_m \cdot \mathbf{n}] \otimes \left[\mathbf{n} \cdot (\mathbf{u}_s - \mathbf{u}_m) - \frac{g_s}{\|\Delta \mathbf{x}_m\|} \mathbf{t} \cdot \Delta \mathbf{u}_m \right] \right\} \tag{13.69}$$

Using (13.68) and (13.69) the *exact* linearization of (13.66) leads to the following *consistent* residual and tangent operators

$$\mathbf{R}_{f_s} = -\mu\epsilon g_s \left(\mathbf{T}_s + \frac{g_s}{\|\Delta \mathbf{x}_m\|} \mathbf{N}_m \right) \tag{13.70a}$$

$$\mathbf{K}_{f_s} = -\mu\epsilon \left(\mathbf{T}_s + \frac{g_s}{\|\Delta \mathbf{x}_m\|} \mathbf{N}_m \right) \otimes \mathbf{N}_s - \tag{13.70b}$$

$$\mu\epsilon \frac{g_s}{\|\Delta \mathbf{x}_m\|} \left[\mathbf{N}_m \otimes \mathbf{N}_s + \mathbf{N}_s \otimes \mathbf{N}_m - \frac{g_s}{\|\Delta \mathbf{x}_m\|} (\mathbf{T}_m \otimes \mathbf{N}_m + \mathbf{N}_m \otimes \mathbf{T}_m) \right]$$

where $\mathbf{T}_m = [0, -\mathbf{t}, \mathbf{t}]^T$ and $\mathbf{T}_s$, $\mathbf{N}_s$, $\mathbf{N}_m$ have already been defined in (13.5).

The expressions of the residual and tangent operators for the flow approach are identical to that of (13.70) with $\mathbf{T}_m = \theta \Delta t \mathbf{T}_m$ and the expression of $\mathbf{T}_s$, $\mathbf{N}_s$ and $\mathbf{N}_m$ given now by (13.64).

13.8 Use of Constant Spatial Velocity Field

In most sheet problems the "spatial" velocity field does not change much between two consecutive solutions once the forming process reaches a well-developed stage. Consider, for example, the case shown by Baynham and Zienkiewicz (1982) of the deep drawing of a circular sheet with a flat bottom punch. Fig. 13.8 shows the shape of the blank at two positions of the punch (δ apart) for a well-developed process (that is one where punch displacement is greater than the sum of the punch and die profile radii). If the radial and axial components of the instantaneous velocity for the two blank geometries are plotted against the radial coordinate (see Fig. 13.8) it is found that the shape of the curve is very similar for the two positions.

If, however, the velocity of a particular point of the blank is compared at two punch travel positions (for example, by plotting the velocity against the radius of the blank in the first position) it is found that some parts of the blank undergo a severe change of velocity. Such a comparison is made in Fig. 13.8 and, clearly, the points which undergo the greatest change of velocity are those near the die profile radius.

Thus, if a Lagrangian approach is employed in which the velocity of material points are used to update the blank geometry, the time steps Δt must be small. On the other hand, if an Eulerian approach is used in which the constant spatial velocity field is used to update the geometry, then the same velocity field can be

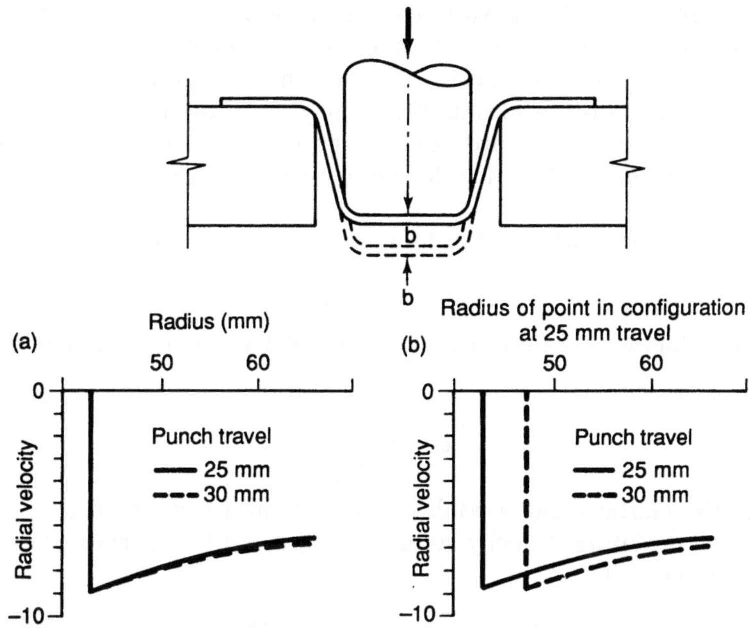

Fig. 13.8. Radial velocity: **a** spatial velocity field; **b** material velocity field for punch positions δ apart (Baynham and Zienkiewicz 1982).

used for a greater length of time and so the number of resolutions can be reduced significantly.

This latter method can be combined with a constant time-step algorithm, as follows:

1. Once the velocity field has been obtained for a blank position, a record is kept of the spatial velocity field in that particular position, which is taken as "initial" in the following updating of the blank.
2. The blank geometry is updated using the spatial field with a constant time step. After each updating the geometry of the blank is checked and modified so that it follows the tooling and punch profiles.
3. Step 2 is repeated a few times using the same spatial velocity field. However, after a number of time steps a new converged velocity field must be computed for a more precise evaluation of punch force and sheet strains.

13.9 Inclusion of Elastic Effects in the Viscous Voided Shell Formulation

The viscous shell formulation presented in previous sections exclude elastic effects by the nature of the constitutive relations assumed in the viscous flow approach. This implies that the removal of loads is not accompanied by an elastic spring-back. Therefore the resulting residual stresses cannot be calculated.

One simple procedure for recovery of residual stresses is to assume that the removal of the forces from the final viscous configuration is purely elastic and is resolved by a small deformation elastic (or elastoplastic) program. This procedure, originally presented by Zienkiewicz et al. (1978) introduces some approximations since the influence of the elastic deformation in the final sheet shape is ignored. This, however, is of little significance in most sheet metal problems.

Nevertheless, inclusion of elastic effects may be of interest in some situations and below we generalize a procedure suggested by Zienkiewicz (1984) and Thompson and Berman (1984) for dealing with elastic effects in the viscous shell formulation.

If elastic effects are included the total strain rate is given by

$$\dot{\epsilon}_{ij} = \dot{\epsilon}_{ij}^{(e)} + \dot{\epsilon}_{ij}^{(p)} \tag{13.71}$$

On the other hand, if we assume that the elastic deformations are incompressible we can write

$$\dot{\epsilon}_{ij}^{(e)} = \frac{1}{2G} \overset{\triangle}{s}_{ij} \tag{13.72}$$

where G is the elastic shear modulus and $\overset{\triangle}{s}_{ij}$ is the corotational (i.e. Jaumann) derivate of the deviatoric Cauchy stress. For updated Lagrangian problems such derivatives are given by

$$\overset{\triangle}{s}_{ij} = w_{ik} s_{kj} + w_{jk} s_{ki} + \frac{\partial s_{ij}}{\partial t} \tag{13.73}$$

where

$$w_{ij} = \tfrac{1}{2}(u_{j,i} - u_{i,j}) \tag{13.74}$$

is the rigid-body rotation rate caused by the motion. Eq. (13.73) simply takes into account stress changes associated with rotation of the axes of the material. The assumption of compressible elastic effects is also possible and this simply implies a modification of (13.72) to account for the effect of Poisson's ratio.

From (13.28) and (13.71) we obtain

$$\boldsymbol{\sigma} = \mathbf{D}(\dot{\boldsymbol{\epsilon}} - \dot{\boldsymbol{\epsilon}}^{(e)}) \tag{13.75}$$

where $\mathbf{D}$ is the equivalent elasticity matrix of the viscous voided material expressed in terms of $\bar{G}$ and $\bar{\nu}$ parameters, as explained in previous sections.

Eq. (13.75) shows that if the elastic strain rate is assumed to be known from the previous iteration the problem is similar to the classical initial stress method in which the additional stress terms $-\mathbf{D}\dot{\boldsymbol{\epsilon}}^{(e)}$ are added to the equilibrium flow equations as an "unbalanced force" and connected.

Thus, the discretized equilibrium equations have the form

$$\mathbf{K}\mathbf{a} = \mathbf{f} + \mathbf{f}_0 \tag{13.76}$$

where $\mathbf{K}$ and $\mathbf{f}$ are the secant stiffness matrix and the nodal force vector for the flow approach respectively and the extra force vector $\mathbf{f}_0$ is given for the axisymmetric viscous shell case as

$$\mathbf{f}_0 = 2\pi \int_{l^{(e)}} \mathbf{B}^{\mathrm{T}} \mathbf{D} \dot{\boldsymbol{\epsilon}}^{(e)} \, r ds \tag{13.77}$$

where $\dot{\boldsymbol{\epsilon}}^{(e)}$ is obtained from Eqs (13.72)–(13.74).

13.10 Examples

13.10.1 Hemispherical Stretching of a Voided Circular Sheet

The first example is the hemispherical stretching of a circular isotropic sheet for which experimental results are available (Woo 1965b).

The geometrical configuration of the problem is shown in Fig. 13.9. Fifty linear elements have been used for the analysis. The uniaxial stress–effective strain curve of the matrix material is given by

$$\bar{\sigma} = 5.4 + 27.8\bar{\epsilon}^{0.504} \text{ tons in}^{-2}, \quad \bar{\epsilon} < 0.36 \tag{13.78}$$

$$\bar{\sigma} = 5.4 + 24.4\bar{\epsilon}^{0.504} \text{ tons in}^{-2}, \quad 0.36 < \bar{\epsilon} \tag{13.79}$$

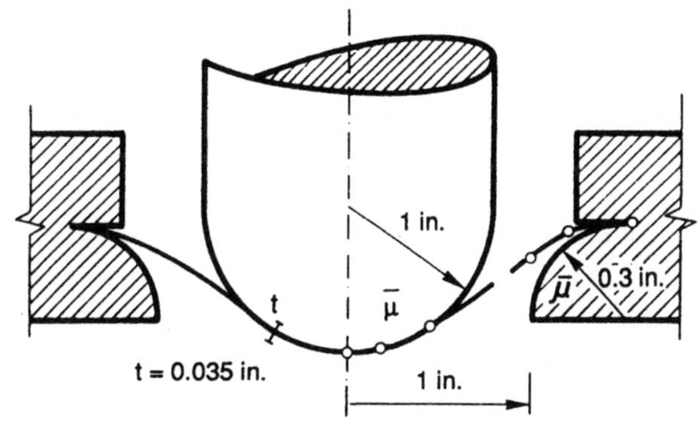

Fig. 13.9. Hemispherical punch stretching. Geometry of punch and blank.

Fig. 13.10 shows results for the punch load–displacement curves obtained for zero porosity and two friction coefficients of 0.04 and 0.2. Numerical results obtained for $\mu = 0.04$ agree well with the experimental and numerical results reported by Woo (1965b) for this problem. Good coincidence is obtained with numerical results obtained by Honnor (1985) and Belingardi (1989, personal communication) for the same problem using an elastoplastic formulation.

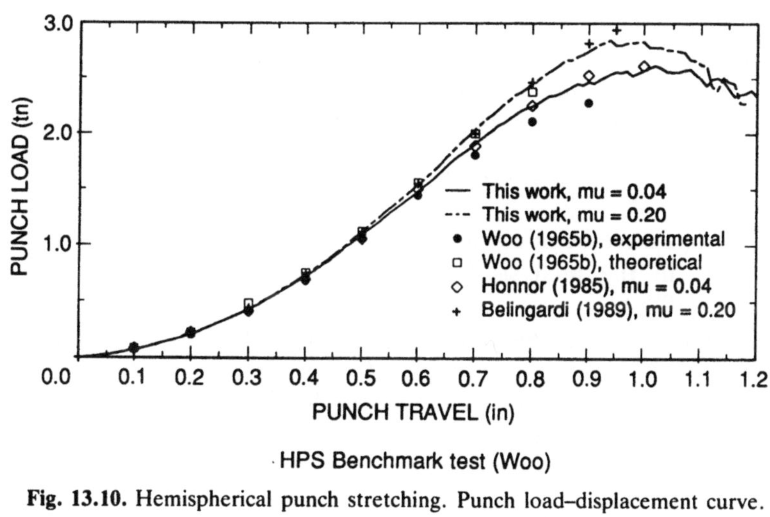

Fig. 13.10. Hemispherical punch stretching. Punch load–displacement curve.

The effect of taking into account porosity changes is shown in Fig. 13.11, where results for the punch load–displacement curve for $f_0 = 0.01$ and 0.1 and $k = 0$ are presented. It can be seen that an increase of initial porosity leads to a reduction of the load carrying capacity of the sheet.

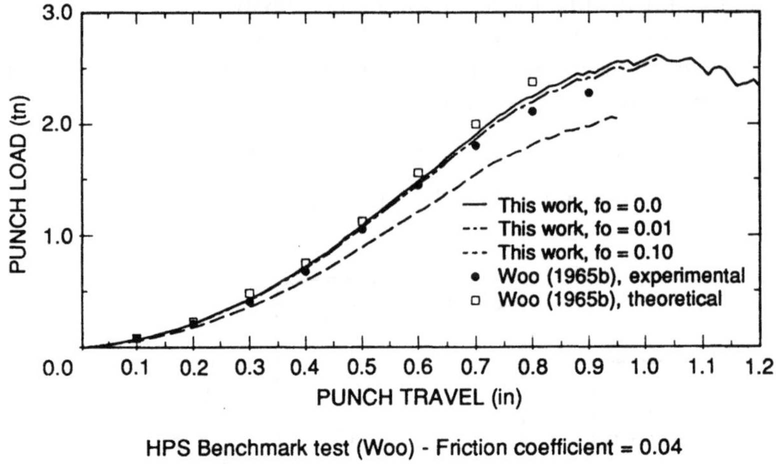

Fig. 13.11. Hemispherical punch stretching. Effect of porosity in the punch load–displacement curve.

Figs 13.12 and 13.13 show the distribution of the radial and hoop strains and that of the sheet thickness and porosity for different punch travels and two friction coefficients of $\mu = 0.04$ and 0.2. A value of the initial porosity $f_0 = 0.01$ and $k = 0$ has been taken in both cases. It can be clearly seen in those figures the well known localization effect of radial and thickness strains induced by high friction values. We have to note that numerical results for the strain distribution of the low friction case ($\mu = 0.04$) compare very well with the experimental values reported by Woo (1965b).

13.10.2 Benchmark Test for Hemispherical Punch Stretching

This example corresponds to a benchmark test for numerical simulation of sheet-metal-forming processes, recently proposed by Wagoner et al. (1988). The geometry of the sheet, punch and die is shown in Fig. 13.14. The sheet material has been assumed to be rigid plastic with a strain-hardening law given by

$$\bar{\sigma} = 589(10^{-4} + \bar{\epsilon})^{0.216}\,\text{MPa} \tag{13.80}$$

No porosity effects have been considered. Fourteen axisymmetric linear elements have been used in the analysis following the mesh spacing as suggested by Wagoner et al. (1988).

Fig. 13.15 shows the radial- and hoop-strain distributions obtained using a membrane and a full bending formulation for two friction coefficients of 0.0 and 0.3 respectively. It can be seen that the strain distributions for both formulations are almost identical with the exception of some small discrepancies near the edge and the pole. This shows that a membrane formulation could be successfully used throughout the analysis.

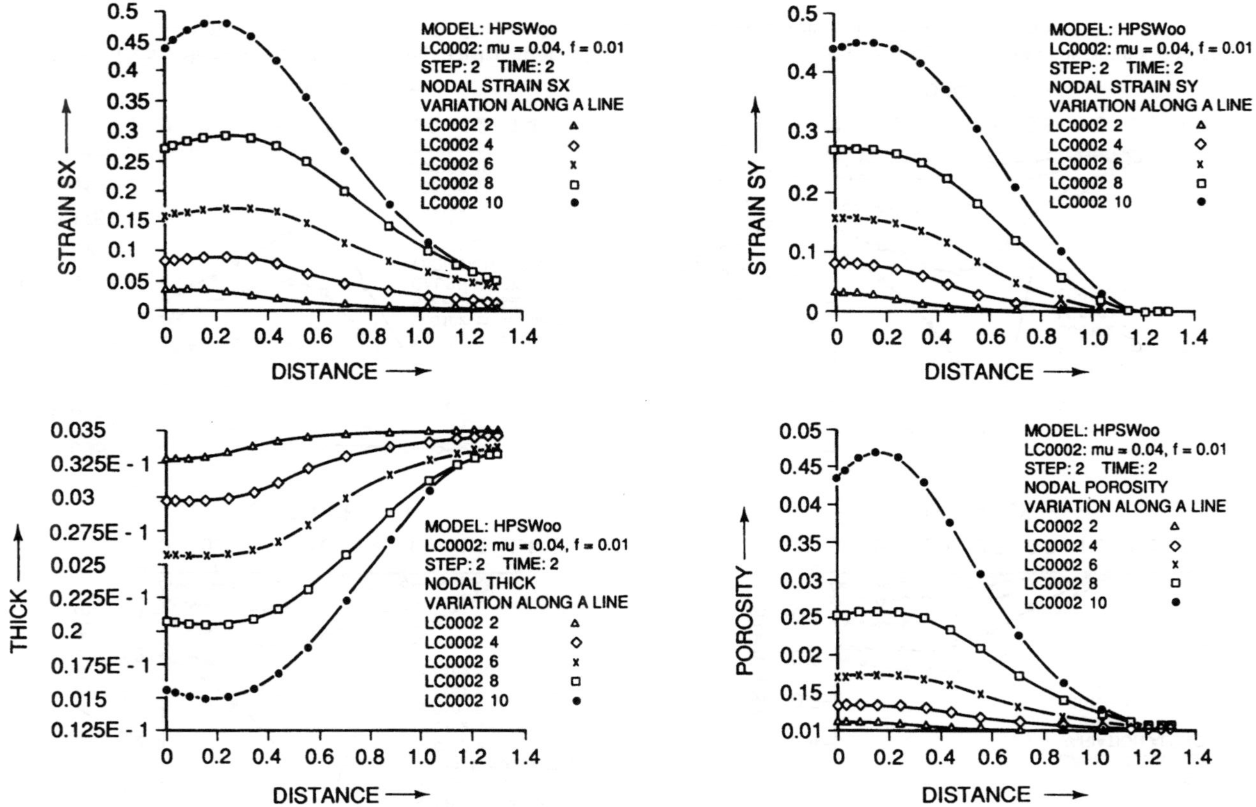

Fig. 13.12. Hemispherical punch stretching. Distribution of radial (S_x) and hoop strains (S_y), sheet thickness and porosity for $\mu = 0.04$, $f_0 = 0.01$ and $k = 0$. Curves show values for punch travels of 0.2, 0.4, 0.6, 0.8 and 1.0 in.

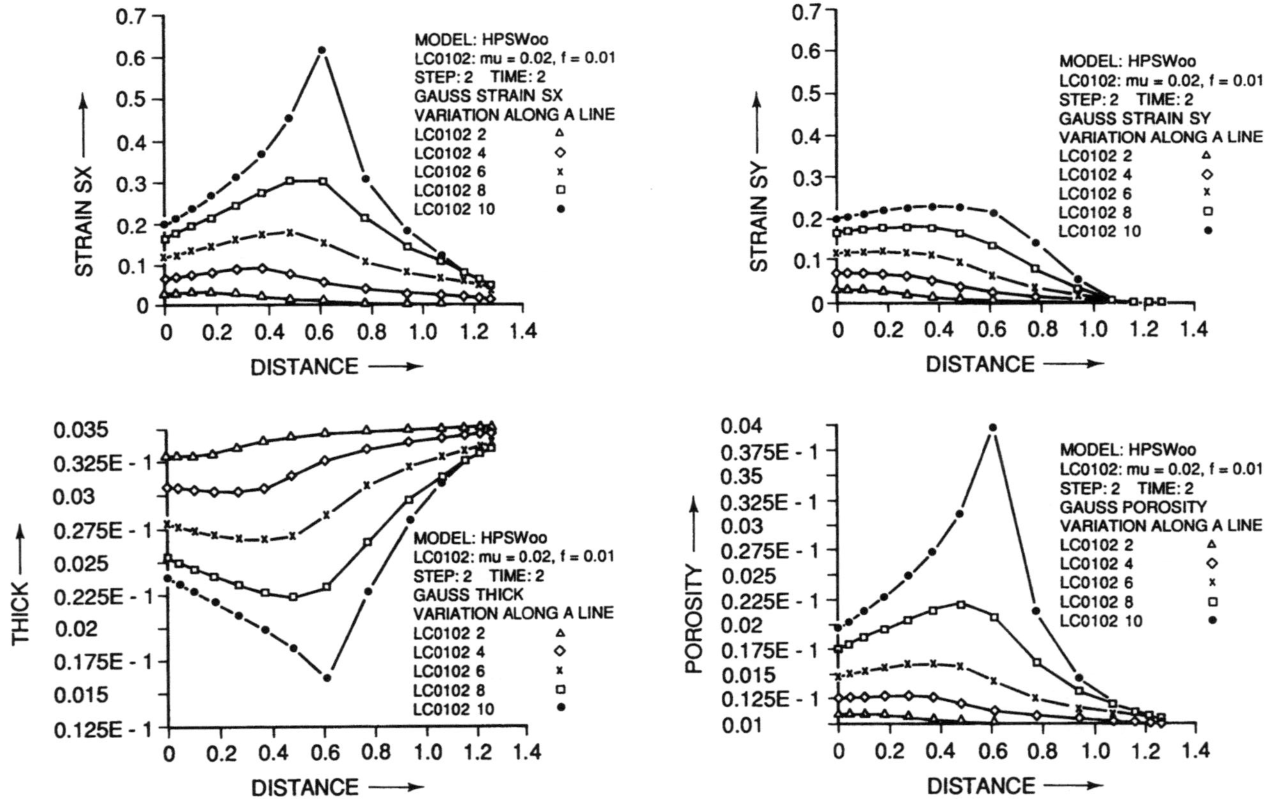

Fig. 13.13. Hemispherical punch stretching. Distribution of radial (S_x) and hoop strains (S_y), sheet thickness and porosity for $\mu = 0.02$, $f_0 = 0.01$ and $k = 0$. Curves show values for punch travels of 0.2, 0.4, 0.6, 0.8 and 1.0 in.

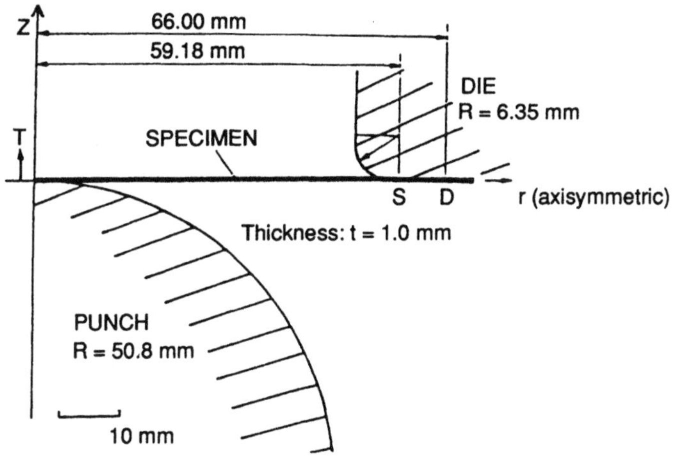

Fig. 13.14. Geometry of the sheet and the tools for the Hemispherical Punch Stretching (HPS) benchmark test (Wagoner et al. 1988).

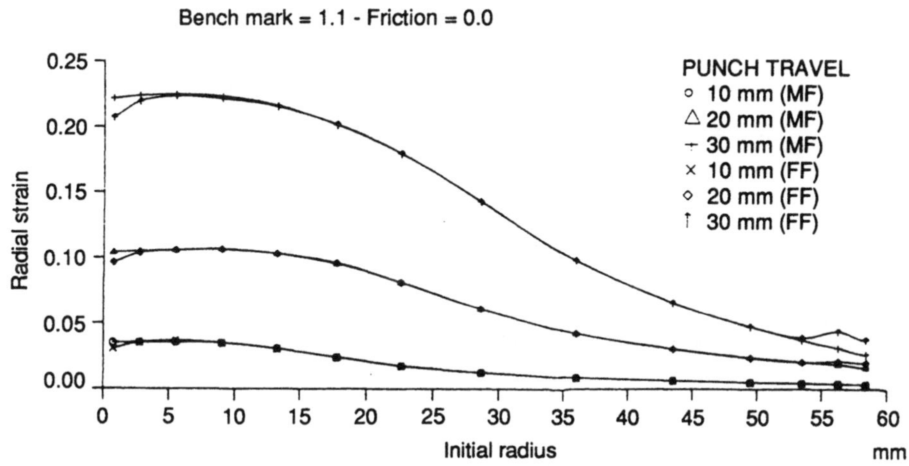

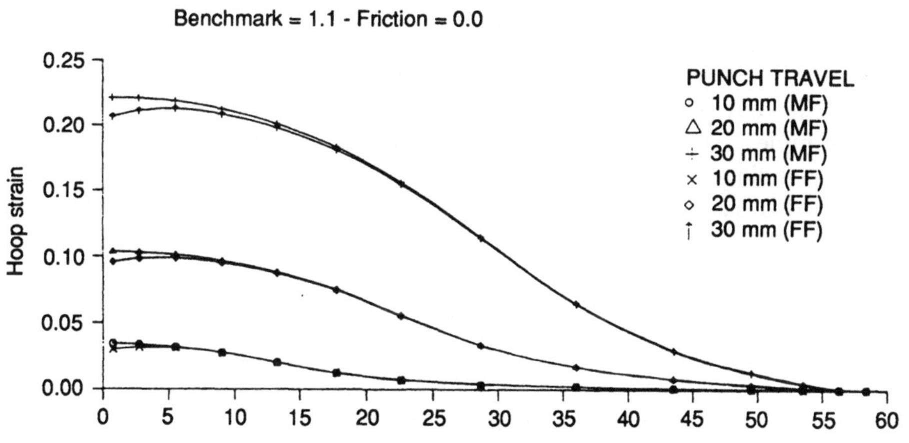

Fig. 13.15a. HPS benchmark test. Comparison between membrane (MF) and bending (FF) formulation. Radial and hoop strains for a friction coefficient of 0.0.

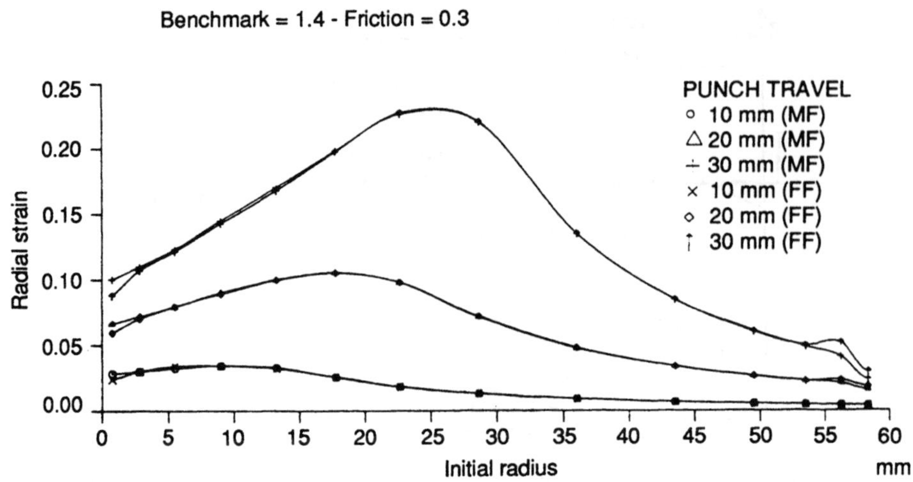

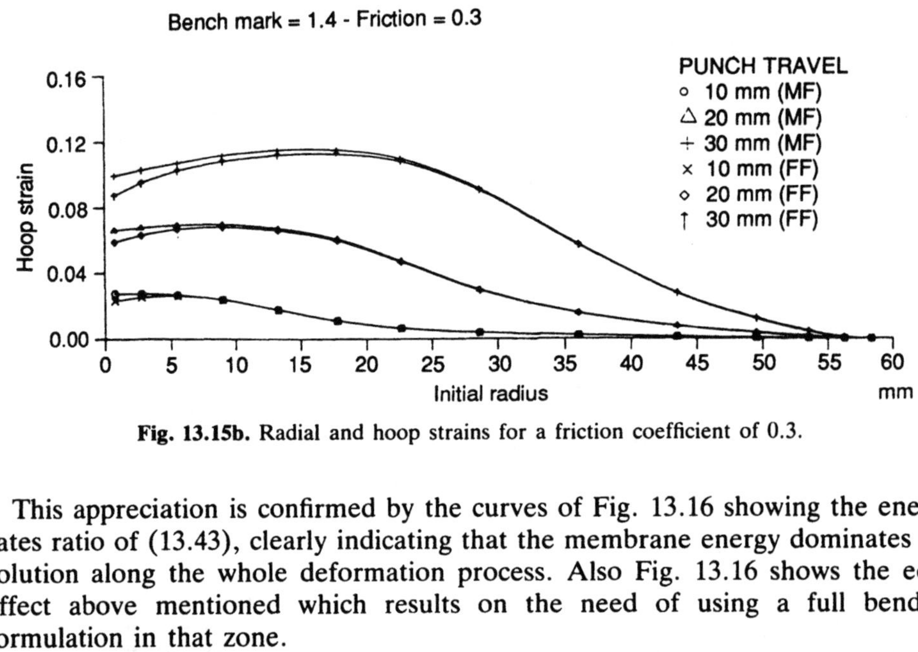

Fig. 13.15b. Radial and hoop strains for a friction coefficient of 0.3.

This appreciation is confirmed by the curves of Fig. 13.16 showing the energy rates ratio of (13.43), clearly indicating that the membrane energy dominates the solution along the whole deformation process. Also Fig. 13.16 shows the edge effect above mentioned which results on the need of using a full bending formulation in that zone.

This example has also been successfully analysed using the selective membrane–bending approach presented on Section 13.6. Numerical results obtained in this case coincide exactly with those shown on Fig. 13.15 for the bending formulation as expected.

13.10.3 Superplastic Forming of a Thin Sheet Under Plane Strain Conditions

Figure 13.17 shows the geometry of the tool. Plane-strain conditions have been assumed in the analysis. This simply implies neglecting circumferential effects in the axisymmetric shell formulation. Forty-nine two-node 1D elements have been used in the analysis. The material constitutive parameters are $K = 1200$ MPa s^m

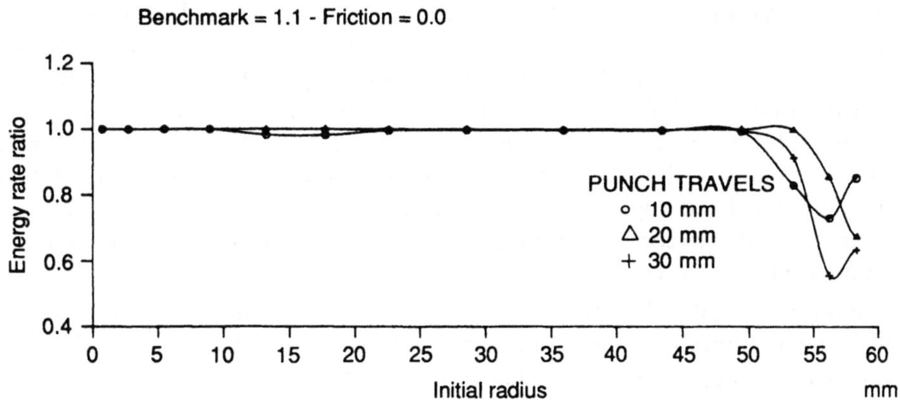

Fig. 13.16. HPS benchmark test. Energy rate ratio at different punch travels.

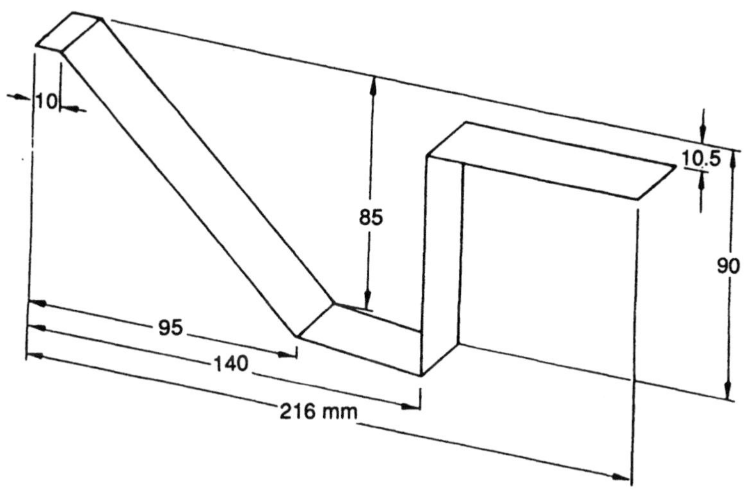

Fig. 13.17. Superplastic forming (SPF) of a plane strip. Geometry of the forming tool.

and $m = 0.6$. Contact between sheet and tool has been treated with a standard penalty formulation and full slipping condition (zero friction) at the contact regions has been assumed. A limit value for the energy ratio of $\zeta_0 = 0.99$ to define the change from bending to membrane condition has been chosen.

Figs 13.18–13.21 show the thickness distribution at different deformation stages using membrane, bending and selective bending–membrane formulations. Also the distribution of the energy rate ratio ζ at each deforming stage is shown in each figure.

The following points emerge from these figures:

1. The values of the energy rate ratio ζ over the sheet at the onset of deformation are in many points below the limit value $\zeta_0 = 0.99$, thus indicating a global influence of bending in the early deformation stage (Fig. 13.18). Bending elements are used in this case throughout most of the sheet as clearly shown in Fig. 13.18d. (Bending and membrane elements are represented by black and white zones respectively.)

2. As deformation progresses the condition $\zeta > \zeta_0$ is satisfied in many parts of the sheet and membrane elements can be used in these zones (white areas in Figs 13.19d–13.21d). Note that for well-developed stages membrane theory can be used in $\simeq$ 80% of the mesh, thus introducing considerably economies in the solution process. Exception to this rule are the shape corners where the low values of the energy rate ratio indicates the need of bending elements in those localized regions. It is also interesting to note that some elements near the bottom corners change from membrane to bending behaviour as contact with the tool progresses in that area.
3. Numerical results obtained for the thickness distribution with the selective bending–membrane formulation agree very well with those obtained with the full bending approach for all deforming stages.

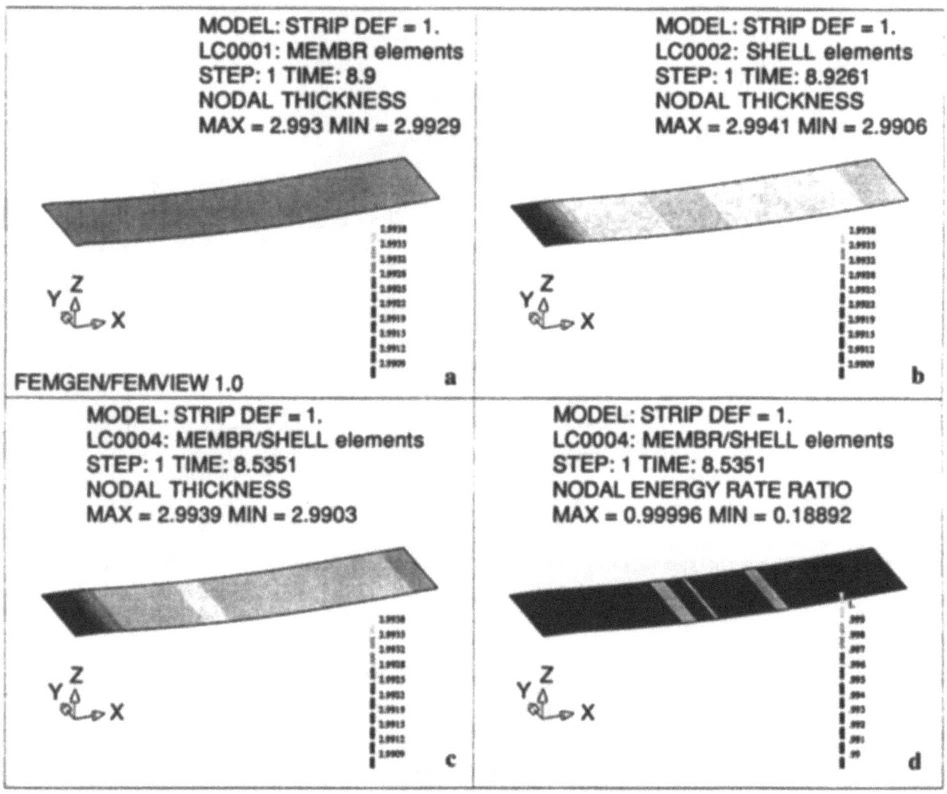

Fig. 13.18. Superplastic forming of a strip. Step 1. Time = 8.5 s. Thickness distributions: **a** membrane approach; **b** bending approach; **c** selective bending–membrane approach; **d** Energy rate ratio distribution.

Fig. 13.19. Superplastic forming of a strip. Step 3. Time = 503 s. Thickness distributions: **a** membrane approach; **b** bending approach; **c** selective bending–membrane approach; **d** Energy rate ratio distribution.

Fig. 13.20. Superplastic forming of a strip. Step 5. Time = 1500 s. Thickness distributions: **a** membrane approach; **b** bending approach; **c** selective bending–membrane approach; **d** Energy rate ratio distribution.

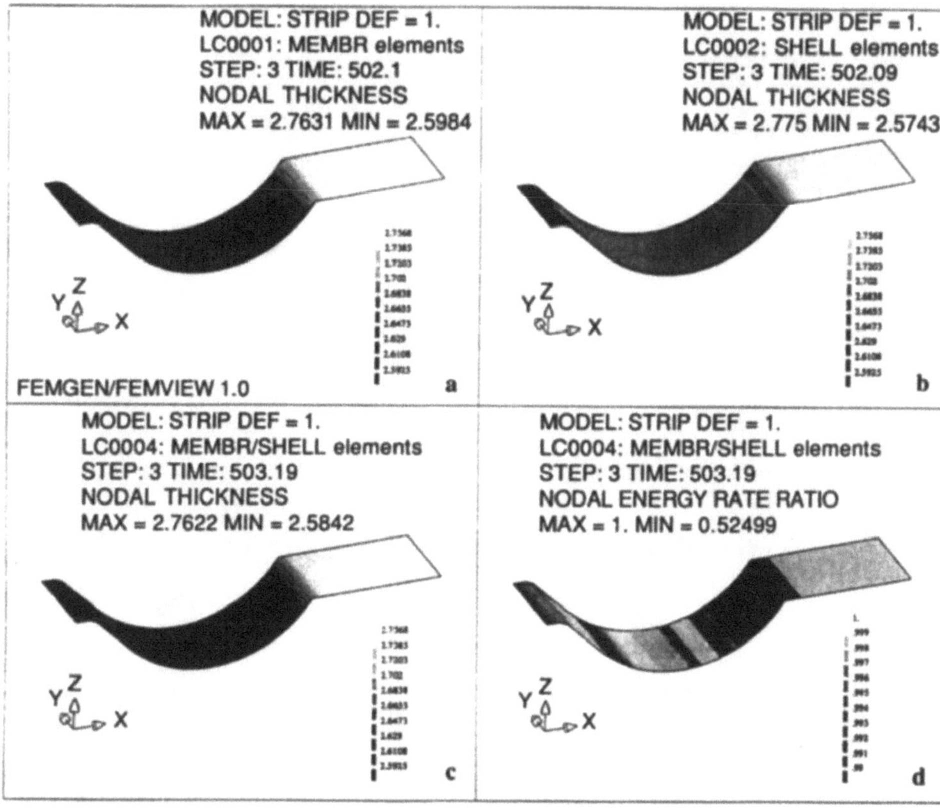
MODEL: STRIP DEF = 1.
LC0001: MEMBR elements
STEP: 3 TIME: 502.1
NODAL THICKNESS
MAX = 2.7631 MIN = 2.5984
FEMGEN/FEMVIEW 1.0
a
MODEL: STRIP DEF = 1.
LC0002: SHELL elements
STEP: 3 TIME: 502.09
NODAL THICKNESS
MAX = 2.775 MIN = 2.5743
b
MODEL: STRIP DEF = 1.
LC0004: MEMBR/SHELL elements
STEP: 3 TIME: 503.19
NODAL THICKNESS
MAX = 2.7622 MIN = 2.5842
c
MODEL: STRIP DEF = 1.
LC0004: MEMBR/SHELL elements
STEP: 3 TIME: 503.19
NODAL ENERGY RATE RATIO
MAX = 1. MIN = 0.52499
d

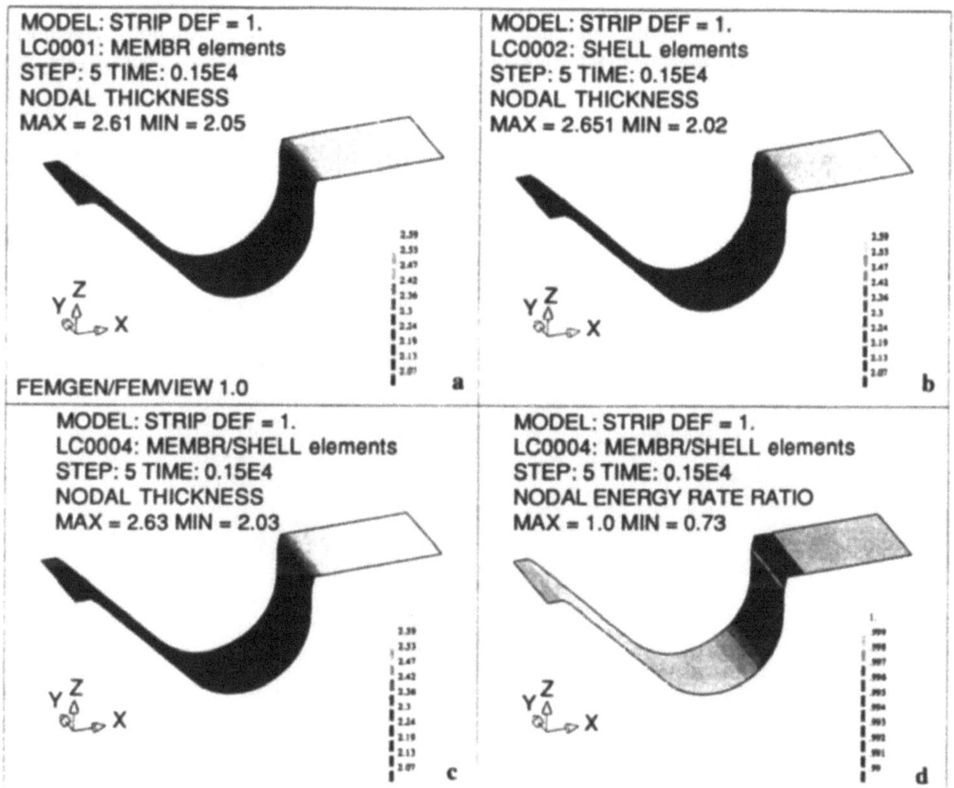
MODEL: STRIP DEF = 1.
LC0001: MEMBR elements
STEP: 5 TIME: 0.15E4
NODAL THICKNESS
MAX = 2.61 MIN = 2.05
FEMGEN/FEMVIEW 1.0
a
MODEL: STRIP DEF = 1.
LC0002: SHELL elements
STEP: 5 TIME: 0.15E4
NODAL THICKNESS
MAX = 2.651 MIN = 2.02
b
MODEL: STRIP DEF = 1.
LC0004: MEMBR/SHELL elements
STEP: 5 TIME: 0.15E4
NODAL THICKNESS
MAX = 2.63 MIN = 2.03
c
MODEL: STRIP DEF = 1.
LC0004: MEMBR/SHELL elements
STEP: 5 TIME: 0.15E4
NODAL ENERGY RATE RATIO
MAX = 1.0 MIN = 0.73
d

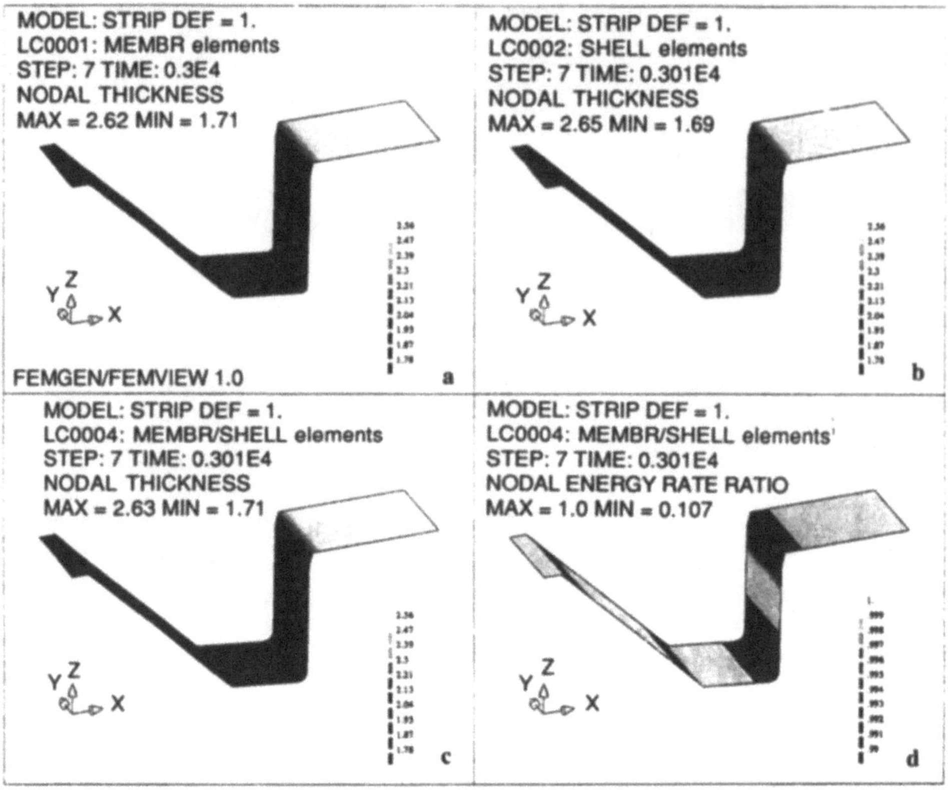

Fig. 13.21. Superplastic forming of a strip. Step 7. Time = 3000 s. Thickness distributions: **a** membrane approach; **b** bending approach; **c** selective bending–membrane approach; **d** Energy rate ratio distribution.

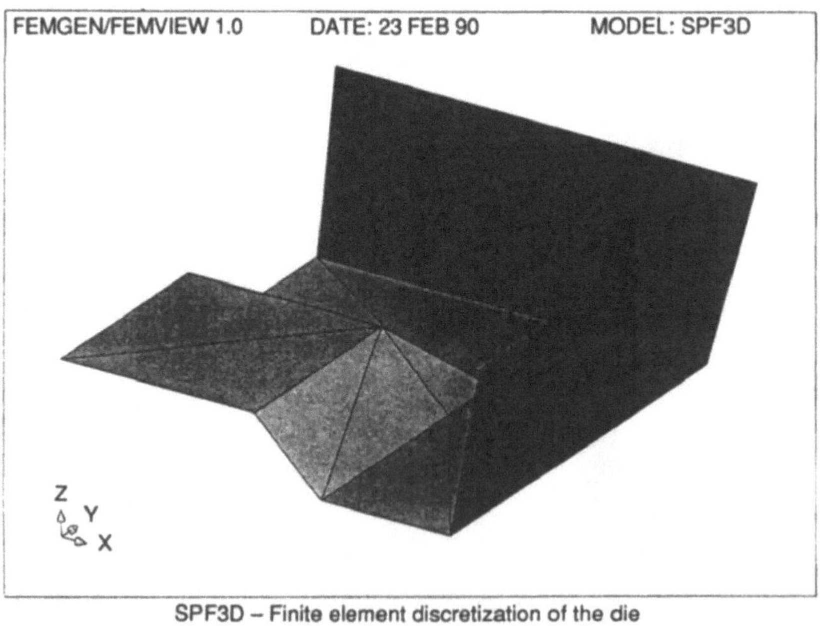

Fig. 13.22. Superplastic forming of a square pan. Discretization of the die.

13.10.4 Superplastic Forming of a 3D Pan

The last example shows the possibilities of the general flow formulation to deal with 3D superplastic forming processes. In this a square sheet is formed onto a three-dimensional die by gas pressure. Due to the symmetry only one quarter of the problem is considered. Fig. 13.22 shows the discretization of the die.

The viscoplastic constitutive behaviour of the material is characterized by the parameters $K = 125.893$ MPa s^m and $m = 0.3$. A full implicit time integration scheme is considered. Upon contact the sheet is assumed to slide along the die without friction.

In order to reduce the propensity of the material to undergo localized necking the pressure cycle is controlled by constraining the maximum equivalent plastic strain rate to the prescribed value $\dot{\bar{\epsilon}}_0 = 3 \times 10^{-4}\ s^{-1}$ by a trial-and-error approach (Bellet 1988, Bonet 1989, St Doltsinis 1989). The computed pressure cycle and the evolution of the maximum equivalent plastic strain rate along the process are shown in Figs 13.23 and 13.24 respectively. The requested part may be considered formed at $t = 4370$ s. At this stage the thickness of the sheet is reduced from initially 0.035 mm to a minimum of 0.0164 mm locally. Fig. 13.25 shows the equivalent plastic strain contours at different stages of the forming process.

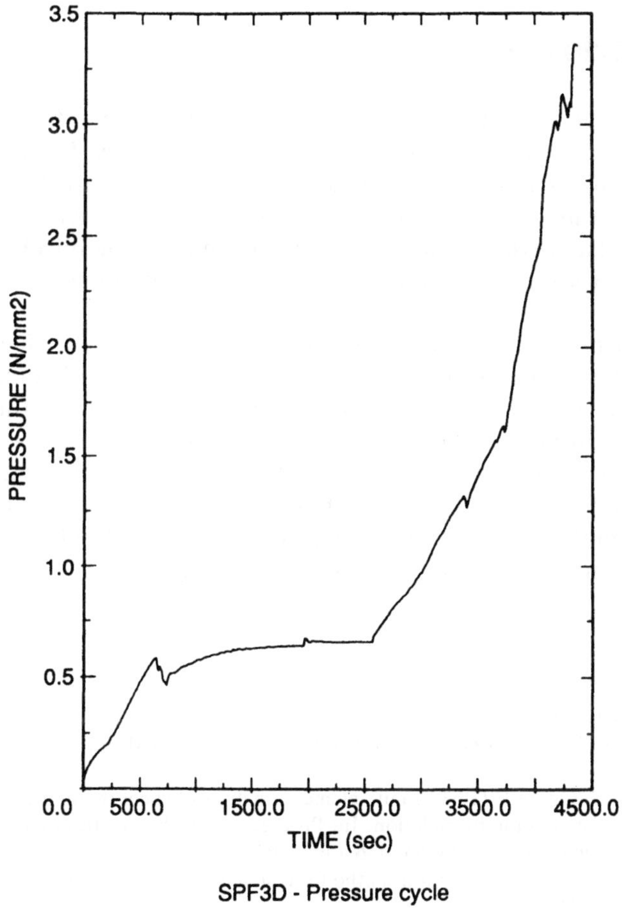

Fig. 13.23. Superplastic forming of a square pan. Pressure cycle.

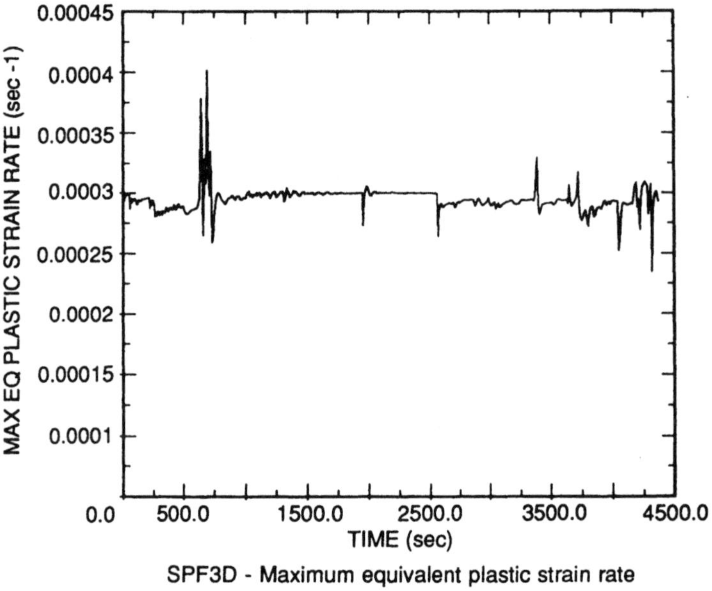

Fig. 13.24. Superplastic forming of a square pan. Maximum equivalent plastic strain rate.

13.11 Conclusion

This chapter has presented an overview of both the solid and flow approaches for the finite-element analysis of sheet-metal-forming processes. Material degradation effects have been discussed and particular emphasis given to the presentation of a selective membrane–bending shell flow formulation. Contact and friction effects have also been described. Applications of the technique to hemispherical stretching and superplastic forming in plane-strain and 3D conditions have been presented.

Acknowledgements. The authors would like to thank the financial support provided for this research by the European Economic Commission through BRITE contract No. RI-1B-0240-C (EL). Also the support of the Comisión de Investigación Científica y Técnica (CICYT) of the Ministerio de Educación y Ciencia of Spain through contract No. PB87-0603 is fully acknowledged. C. Agelet de Saracibar was also supported by a Fellowship from the Comissió Interdepartamental de Recerca i Innovació Tecnològica (CIRIT) of the Generalitat de Catalunya. This support is gratefully acknowledged.

References

Agelet de Saracibar C, Oñate E (1989) Plasticity models for porous metals. In: Owen DRJ, Hinton E, Oñate E (eds) Computational plasticity: models, software and applications. Pineridge Press, Swansea

Agelet de Saracibar C, Oñate E (1990) Finite element analysis of sheet metal forming problems using a selective bending/membrane formulation. In: Proceedings of the 3rd International Conference on Technology of Plasticity, Kyoto, Japan, 1–6 July 1990

Bathe KJ, Chaudary A (1985) A solution method for planar and axisymmetric contact problems. Int J Num Meth Engng 21:65–88

Batoz JL, Duroux P, Roelandt JM, Pol P (1989) A membrane bending finite element model for

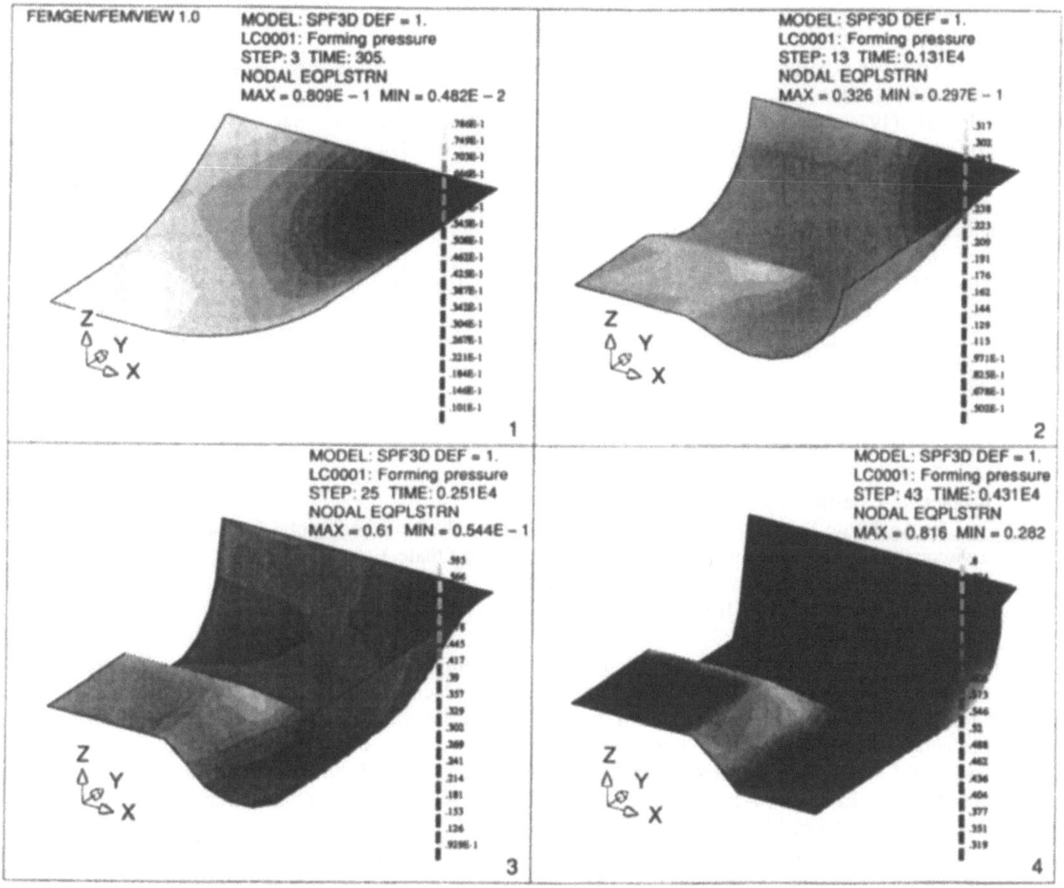

Fig. 13.25. Superplastic forming of a square pan. Equivalent plastic strain contours at different stages of the process.

sheet forming. In: Thompson EG et al. (eds) Numerical methods in industrial forming processes NUMIFORM 89. AA Balkema, pp 389–395

Baynham JMV, Zienkiewicz OC (1982) Developments in the finite element analysis of thin sheet drawing and direct redrawing processes using the rigid plastic approach. In Pittman JFT et al. (eds) Numerical methods in industrial forming processes. Pineridge Press, Swansea, pp 697–708

Bellet M (1988) Modélisation numérique du formage superplastique de tôles. PhD Thesis, Ecole des Mines de Paris, Sophia-Antipolis, France

Bonet J (1989) Finite element analysis of thin sheet superplastic forming. PhD Thesis, University College of Swansea, UK

Gurson AL (1977) Continuum theory of ductile rupture by void nucleation and growth, I. Yield criteria and flow rules for porous ductile media. J Engng Mater Tech 99:2–15

Hallquist JO (1976) A procedure for the solution of finite deformation contact-impact problems by the finite element method. Report UCRL-52066, Lawrence Livermore Laboratory

Hambrecht J, Choudhry S, Lee JK, Wagoner RH (1989) Numerical study of two dimensional sheet forming processes using bending, membrane and solid finite elements. In: Thompson EG et al. (eds) Numerical methods in industrial forming processes NUMIFORM 89. AA Balkema, pp 451–457

Hecker SS (1973) Formability of high-strength low-alloy steel sheets. Met Engng Quart 13:42

Honnor ME (1985) Nonlinear finite element analysis of axisymmetric shells applied to sheet metal forming. PhD Thesis, University College of Swansea, UK

Ju JW, Taylor RL, Cheng LY (1987) A consistent finite element formulation of nonlinear frictional

contact problems. In: Middleton et al. (eds) Numerical methods in engineering and transient analysis NUMETA'87. AA Balkema, D5, pp 1–13
Keeler SP (1968) Circular grid system – a valuable aid for evaluating sheet metal formability. SAE Paper No. 680092
Keeler SP (1974) Forming limit criteria-sheets. In: Proceedings 21st Sagamore Army Materials Research Conference, Lake Roquette, NY, 13–16 August 1974
Kim JH, Oh SL, Kobayashi S (1978) Analysis of stretching of sheet metals with hemispherical punches. Int J Mach Tool Des 18:219–226
Kleiber M (1984a) Numerical study on necking type bifurcations in void containing elastic plastic material. Int J Solid Struct 20:191–210
Kleiber M (1984b) Formulation and computational strategies for novel problems of plasticity and creep. Nucl Engng Des 79:321–341
Kleiber M (1986) On plastic localization and failure in plane strain void containing tensile bars. Int J Plast 2:205–221
Kleiber M (1989) Incremental finite element modelling in nonlinear solid mechanics. Ellis Horwood
Nour-Omid B, Wriggers P (1985) A two-level iteration method for solution of contact problems. Comput Meth Appl Mech Engng 54:131–144
Oden JT, Pires EB (1983) Nonlocal and nonlinear friction laws and variational principles for contact problems in elasticity. J Appl Mech 50:67–76
Oñate E, Agelet de Saracibar C (1988) Finite element analysis of sheet metal forming problems using a viscous voided shell formulation. In: Chenot JL and Oñate E (eds) Modelling of metal forming processes, Kluwer Academic Publishers, pp 163–178
Oñate E, Zienkiewicz OC (1983) A viscous shell formulation for the analysis of thin sheet metal forming. Int J Mech Sci 25:305–335
Oñate E, Kleiber M, Agelet de Saracibar C (1988) Plastic and viscoplastic flow of void containing metals: Applications to axisymmetric sheet forming problems. Int J Num Meth Engng 25:225–251
Oñate E, Agelet de Saracibar C, Dalin JB (1989) Finite element analysis of sheet metal forming problems using a selective viscous shell membrane formulation. In: Thompson EG et al. (eds) Numerical methods in industrial forming processes NUMIFORM 89. AA Balkema, pp 23–30
Ortiz M, Popov EP (1985) Accuracy and stability of integration algorithms for elastoplastic constitutive relations. Int J Num Meth Engng 21:1561–1576
Perzyna P (1966) Fundamental Problems in viscoplasticity. In: Recent advances in applied mechanics. Academic Press, New York, pp 243–377
Simo JC, Wriggers P, Taylor RL (1985) A perturbed Lagrangian formulation for the finite element solution of contact problems. Comput Meth Appl Mech Engng 50:163–180
St Doltsinis J (1989) Nonlinear concepts in the analysis of solids and structures. In: St Doltsinis J (ed) Advances in computational nonlinear mechanics. Springer-Verlag, Berlin Heidelberg New York, pp 1–80
Thompson EG, Berman HM (1984) Steady state analysis of elasto viscoplastic flow during rolling. In: Pittman JFT et al. (eds) Numerical methods in industrial forming processes. John Wiley, New York, pp 269–285
Tvergaard V (1982) On localization in ductile materials containing spherical voids. Int J Fract 18:237–252
Wagoner RH, Nakamachi E, Lee JK (1988) A benchmark test for sheet metal forming analysis. Technical Report, Ohio State University
Wang NM, Budiansky B (1978) Analysis of sheet metal stamping by a finite element method. J Appl Mech Trans ASME 45:73–82
Wifi AS (1976) An incremental complete solution of the stretch forming and deep drawing of a circular blank using a hemispherical punch. Int J Mech Sci 18:23–31
Woo DM (1965a) Analysis of the cup drawing process, J Mech Engng Sci 6:116–124
Woo DM (1965b) The stretch forming test. Engineer 220:876–880
Woo DM (1968) On the complete solution of the deep drawing problem. Int J Mech Sci 10:83–94
Wood RD, Bonet J, Wargadipura AHS (1989) Simulation of superplastic forming of thin sheet components by the finite element method. In: Thompson EG et al. (eds) Numerical methods in industrial forming processes NUMIFORM 89. AA Balkema, pp 85–95
Wriggers P, Simo JC (1985) A note on tangent stiffness for fully nonlinear contact problems. Commun Appl Num Meth 1:199–203
Zienkiewicz OC (1977) The finite element method. 3rd edn, McGraw-Hill

Zienkiewicz OC (1984) Flow formulation for numerical solution of forming processes. In: Pittman JFT et al. (eds) Numerical analysis of forming processes. John Wiley, New York, pp 1–45
Zienkiewicz OC, Cormeau IC (1974) Visco-plasticity and creep in elastic solids: A unified numerical solution approach. Int J Num Meth Engng 8:821–845
Zienkiewicz OC, Jain PC, Oñate E (1978) Flow of solids during forming and extrusion. Some aspects of numerical solutions. Int J Solids Struct 14:14–38

14 **Forming of Metal Matrix Composites**

C.K.H. Dharan and S. Kobayashi

14.1 Introduction

Beginning with the introduction of boron fibre-reinforced aluminium, metal matrix composites have been used extensively, and research in the area has increased rapidly in recent years. Metal matrix composites exhibit high specific strength (strength-to-weight ratio) and high specific modulus, in addition to a service temperature capability much higher than that of polymer matrix composites. In addition, they are excellent thermal conductors. Potentially, the ductility and good environmental resistance of metallic matrices can result in superior composite material products.

The major disadvantage of metal matrix composites is their very high cost. It is interesting to note, however, that the constituents of a metal matrix composite, for example, graphite fibre-reinforced aluminium, are much less expensive than the composite itself. This observation leads one to conclude that the high cost of manufacturing is the primary cause for the unusually high cost of the composite. This is not true for polymer matrix composites whose cost, due to the relative maturity of the associated manufacturing processes, is reasonably close to the costs of the fibre and matrix.

Generally, metal matrix composites consist of a ductile metal matrix containing strong, brittle fibres. Due to the ductility of the matrix, metal matrix composites exhibit a limited range of plastic deformation, unlike other types of composites which are typically brittle and which fail while still in the elastic state. Classical metal-forming processes like forging and rolling can, in theory, be applied to metal matrix composites as forming or fabrication techniques.

14.2 Consolidation and Deformation of Metal Matrix Composites

Continuous fibre metal matrix composites are usually consolidated from precursors, which consist of a bundle of continuous fibres coated by the matrix metal. A typical process for consolidation consists of laying up precursor metal matrix composite wires (for example, carbon fibres in an aluminium matrix) and diffusion bonding them with interlayed metallic foils in a hot press. The process is slow and is one of the primary reasons for the high cost of the finished product.

For rapid consolidation, rolling is a possible candidate process; however, the few attempts at rolling metal matrix composites have consisted of rolling sheets along the fibre direction. Due to their high stiffness, the fibres can be considered rigid compared to the matrix. This implies that the deforming process is primarily in-plane strain with zero strain along the fibre direction. Rolling in the fibre direction cannot, therefore, be a viable process since segments of fibre before and after the neutral point will move at different velocities. This will cause bending and buckling of the fibres, and will eventually lead to debonding and fracture (Salkind et al. 1969). It has been observed that plastic deformation of metal matrix composites, in general, causes debonding and fibre breakage prior to matrix cracking (Erturk et al. 1974, Erturk and Kuhn 1979). Despite these drawbacks, mild rolling of a consolidated metal matrix composite billet has been found to decrease the porosity of preforms, relieve residual stresses and work-harden the matrix, resulting in an increase in the tensile strength and the elastic range of the material (Getten and Ebert 1969).

If the precursor wires are rolled transverse to the fibre direction, however, the resulting plane-strain deformation is in the transverse plane which is macroscopically isotropic. Such a process has the potential for the rapid consolidation of metal matrix composite precursor wires into sheets. Large strains are not imposed along the brittle and high modulus fibre direction; the issue now becomes one of determining the conditions under which fibre-matrix debonding can occur during the rolling process.

This chapter describes the results of numerical modelling of metal matrix composites starting from an analysis of simple one-dimensional transverse compression of a metal matrix composite billet in which individual fibres are modelled (Hung 1990). The results show a significant influence of fibre arrangement on the flow stress, in addition to the effects of fibre volume fraction and die face friction. This discussion is followed by an analysis of the local stresses around an individual fibre during a deformation process such as rolling transverse to the fibres (Im 1990, Im and Dharan, 1990). The main goal here is to determine the stress state at the fibre–matrix interface; tensile stresses are known to be responsible for damage, primarily by bond failure and void growth. The results of this analysis suggest a method for suppressing fibre–matrix debonding damage.

14.3 Application of the Finite-Element Method to Metal Forming

The application of the finite-element method to metal-forming processes began with the work of Yamada et al. (1968) who conducted an elastic-plastic analysis, followed by Lee and Kobayashi (1973), who suggested a rigid-plastic model. The latter approach, also called flow formulation, has been applied to a variety of problems important to metal forming. Since these early efforts, Shah and Kobayashi (1974) broadened their initial approach by providing a method for determining the boundary of nearly rigid zones. This was followed by a velocity-dependent friction model which used the inverse tangent function to approximate the friction distribution near the neutral point (Chen and Kobayashi 1978), an advance that allowed one to eliminate the discontinuities between the die and workpiece at neutral points.

A user-friendly code, called Analysis of Large Plastic Incremental Deformation (ALPID), which was based on a rigid-viscoplastic formulation and applied to axisymmetric or plane-strain analysis of forging processes with arbitrarily shaped dies, was subsequently developed (Oh 1982). Preform design using a backward tracing scheme was proposed by Park et al., and applied to preform design for several metal-forming problems (Park et al. 1983, Hwang and Kobayashi 1984, 1986). Three-dimensional deformation modelling of open die forging (Park and Kobayashi 1984, Shiau 1987), and ring and shape rolling (Mori and Osakada 1984, Kim 1989), have also been studied. In recent years, research has moved from development of suitable numerical techniques used in finite-element analysis to application to advanced materials such as composites, and other application-oriented research such as optimal sequence design.

14.4 Equivalent Material with Orthotropic Properties

Metal matrix composites reinforced by continuous fibres may exhibit an appreciable amount of elastic-plastic deformation depending on the state of stress and temperature. Several publications on the theory of plasticity for anisotropic materials are available. Bahei-El-Din (1979) proposed a yielding criterion for a hardening rule assuming that plastic deformation in a metal matrix composite is dominated by the matrix. This theory was applied to composite laminates by Bahei-El-Din and Dvorak (1979) and Dvorak and Bahei-El-Din (1979). Also, finite-element calculations to obtain the yield surface were conducted by Dvorak et al. (1974). These techniques, however, deal mainly with laminate problems in plane stress, and are not appropriate for calculating stresses under large deformation conditions. Hill's (1950) yield criterion for generally anisotropic materials neglects the Bauschinger effect and implies that the material is homogeneous. Hill's criterion is adopted in the present approach because its associated flow rule is consistent with stress analysis in the plane-strain rolling process, and the results can be easily obtained and compared with other theories.

From a macroscopic point of view, one may assume that metal matrix composites are homogeneous and anisotropic materials. Hill's criterion is

$$2f(\sigma_{ij}) = F(\sigma_y - \sigma_z)^2 + G(\sigma_z - \sigma_x)^2 + H(\sigma_x - \sigma_y)^2 + 2L\tau_{yz}^2 + 2M\tau_{zx}^2 + 2N\tau_{xy}^2 = 1 \tag{14.1}$$

where *F, G, H, L, M, N* are parameters representing the current state of anisotropy. The yield criterion has this form when the principal axes of anisotropy are the axes of reference. If *X, Y, Z* are the tensile yield stresses in the principal directions of anistropy, it is easily shown that

$$\left.\begin{aligned}\frac{1}{X^2} &= G + H, \qquad 2F = \frac{1}{Y^2} + \frac{1}{Z^2} - \frac{1}{X^2}\\ \frac{1}{Y^2} &= H + F, \qquad 2G = \frac{1}{Z^2} + \frac{1}{X^2} - \frac{1}{Y^2}\\ \frac{1}{Z^2} &= F + G, \qquad 2H = \frac{1}{X^2} + \frac{1}{Y^2} - \frac{1}{Z^2}\end{aligned}\right\} \tag{14.2}$$

Also, if *P, Q, R* are the yield stresses in shear with respect to the principal axes of anisotropy, then

$$2L = \frac{1}{P^2}, \qquad 2M = \frac{1}{Q^2}, \qquad 2N = \frac{1}{R^2} \tag{14.3}$$

The equivalent stress σ for an anisotropic material may be defined as

$$\bar{\sigma} = \sqrt{\frac{3}{2}}\left[\frac{F(\sigma_y - \sigma_z)^2 + G(\sigma_z - \sigma_x)^2 + H(\sigma_x - \sigma_y)^2 + 2L\tau_{yz}^2 + 2M\tau_{zx}^2 + 2N\tau_{xy}^2}{F + G + H}\right]^{\frac{1}{2}} \tag{14.4}$$

For rotational symmetry of the anistropy about the z-axis, it can be shown that

$$N = F + 2H = G + 2H, \quad F = G, \quad L = M \tag{14.5}$$

In analogy with the Levy–Mises equations for isotropic materials, $f(\sigma_{ij})$ in Eq. (14.1) is the plastic potential, neglecting elastic strains. Then, the stress strain-increment relations, referred to the principal axes of anisotropy, are

$$\left.\begin{aligned} d\epsilon_x &= d\lambda\,[H(\sigma_x - \sigma_y) + G(\sigma_x - \sigma_z)] \\ d\epsilon_y &= d\lambda\,[F(\sigma_y - \sigma_z) + H(\sigma_y - \sigma_x)] \\ d\epsilon_z &= d\lambda\,[G(\sigma_z - \sigma_x) + F(\sigma_z - \sigma_y)] \\ d\gamma_{yz} &= d\lambda L \tau_{yz} \\ d\gamma_{zx} &= d\lambda M \tau_{zx} \\ d\gamma_{xy} &= d\lambda N \tau_{xy} \end{aligned}\right\} \tag{14.6}$$

It should be noted that $d\epsilon_x + d\epsilon_y + d\epsilon_z = 0$. If the stress is reversed, so also is the strain increment.

Hill's general yield criterion for anisotropic materials reduces to the von Mises yield criterion when the anisotropy is vanishingly small. This orthotropic formulation was implemented in the ALPID finite-element code (see the Appendix for details).

14.5 Effect of Fibre Arrangement on Transverse Compression Deformation

In transverse compression, the high strength of the fibres and the assumption of a perfect bond between the fibre and the matrix result in a restriction of plastic flow along the fibre direction. It is reasonable, therefore, to assume plane strain deformation. To determine the effect of the fibre arrangement, square and hexagonal fibre arrangements were modelled, in which both matrix and fibre materials are treated as rigid-plastic, with the yield strength of the fibre set at 100 times the yield strength of the matrix. Thus, the fibres are virtually nondeformable.

The finite-element mesh is shown in Fig. 14.1 with typical boundary conditions. Due to symmetry, only one-quarter of the domain is modelled. The loading is controlled by the die face velocity. A modified version of ALPID was run on a Cray XMP-14 supercomputer, and an IBM 3081 mainframe computer was used for post processing and plotting (Hung 1990).

Several models were created; each contained numbers of fibres ranging from 25 to 361 fibres. The fibre volume fraction in all cases was maintained at 40%. Loading was conducted at 0° and 45° to the fibre pattern for the square array, and 0° and 30° for the hexagonal array. Figs 14.2 and 14.3 show the fibre arrangements and finite-element meshes for the square and hexagonal patterns, respectively.

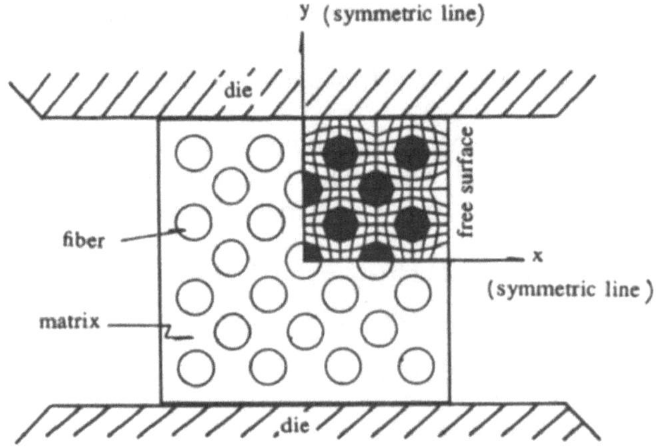

Fig. 14.1. Finite-element model and boundary conditions for simulation of a two-phase composite material.

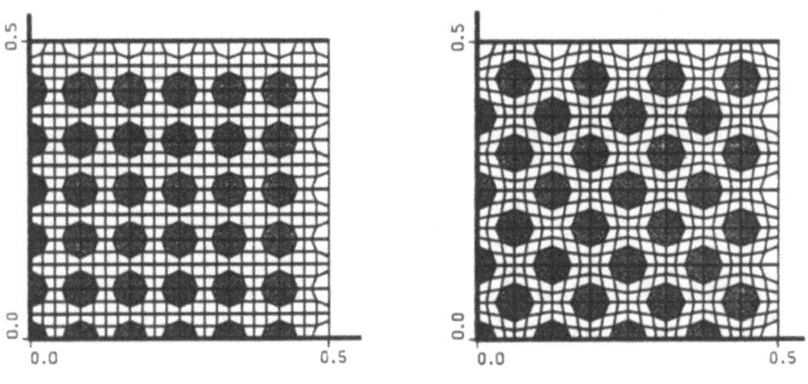

(b) 0° loading specimen mesh.

(c) 45° loading specimen mesh.

Fig. 14.2. Square pattern fibre arrangement.

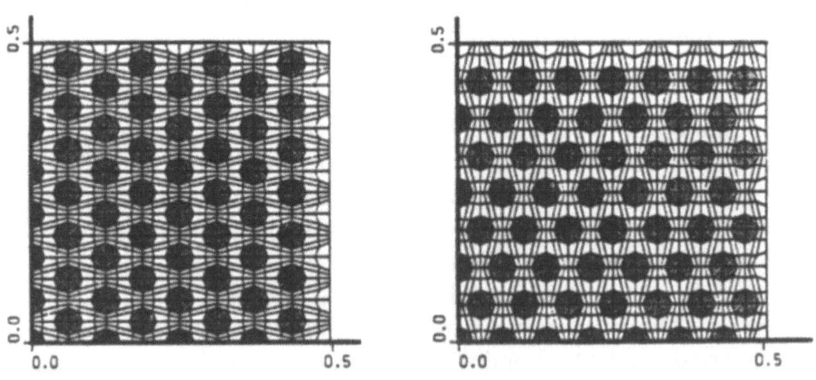

0° loading specimen mesh.

30° loading specimen mesh.

Fig. 14.3. Hexagonal pattern fibre arrangement.

14.5.1 Deformation Response

14.5.1.1 Square Array

Fig. 14.4a shows the deformation response of the square array arrangement loaded in the 0° direction as a function of the number of fibres; the flow stress approaches a constant value as the number of fibres in the model increases. The final deformation patterns for 20% reduction are shown in Fig. 14.4b. In the 0° loading direction, fibre–fibre interaction is large, resulting in greater constraint to the deformation, and there is a "work-hardening" effect, resulting in the flow stress reaching a value that is significantly higher than that of the matrix material.

Figs 14.5a and 14.5b show the same plots for the square array loaded in the 45° direction. Here fibre–fibre interaction is low, resulting in the flow stress having a value close to that of the unreinforced matrix. The effect of loading direction for the square array is compared further in Fig. 14.6 which shows the large difference between the 0° and 45° loading cases, especially at large strains, and the increase in the "work-hardening" effect with deformation.

Using the 0° and 45° loading axes responses in the equivalent orthotropic formulation, simulation of uniaxial compression under different friction conditions was conducted. The results are shown in Fig. 14.7 for 20% deformation. Comparing these results with the two-phase results given in Fig. 14.8, one notes that the equivalent material approach predicts the behaviour closely.

The effect of a matrix material boundary layer was also determined; such a layer often exists in metal matrix composite billets. For this case, there are differences between the equivalent material model response shown in Fig. 14.9 and the two-phase response given in Fig. 14.10, particularly at low friction. The spread between the layer in contact with the die faces and the centre of the billet is greater in the two-phase model compared to the equivalent material model. This is because the two-phase model more accurately models the low yield strength matrix layer which is less constrained than when containing fibres. One conclusion is that in simulating a metal matrix composite with equivalent properties, one may need to use several equivalent material properties at specific locations in the deforming body for more accurate representation.

14.5.1.2 Hexagonal Array

Response of the hexagonal array is similar to that of the square array deformed in the 45° direction; the effect of fibres is small, and the resulting flow stress of the composite is close to that of the matrix. There is a small difference between loading in the 0° direction and the 30° direction for the hexagonal array; in either case, fibre–fibre interaction is small. Fig. 14.11 shows the flow stress as a function of strain for the two cases; the 30° loading case shows a small increase in the flow stress due to slightly larger fibre–fibre interaction.

14.5.2 Effective Strain Distribution

Calculation of the effective strain fields produced during the deformation of the square array loaded along the geometry axis was conducted to reveal the existence

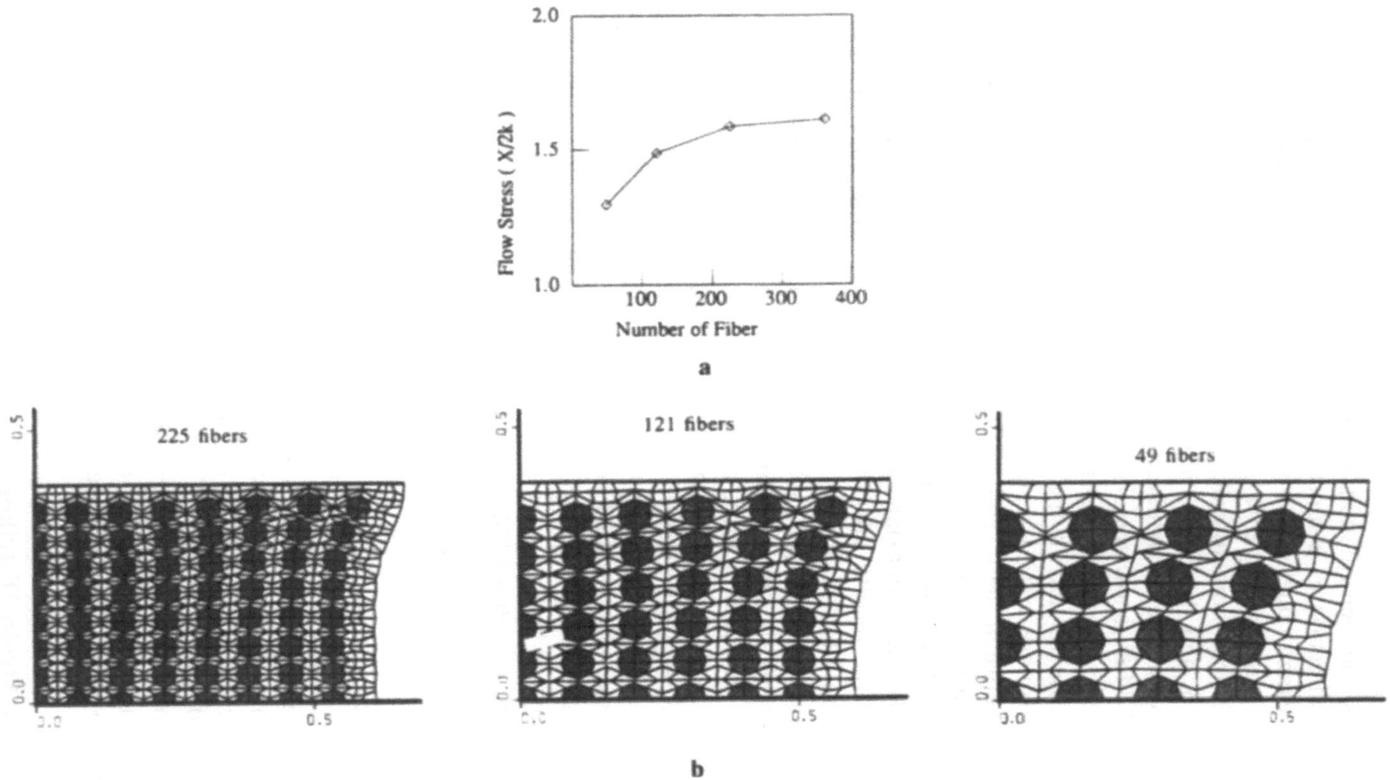

Fig. 14.4. Deformation of the square pattern fibre arrangement, composite loaded along the 0° direction. **a** Normalized apparent flow stress vs number of fibres. **b** Final deformation pattern for 20% reduction.

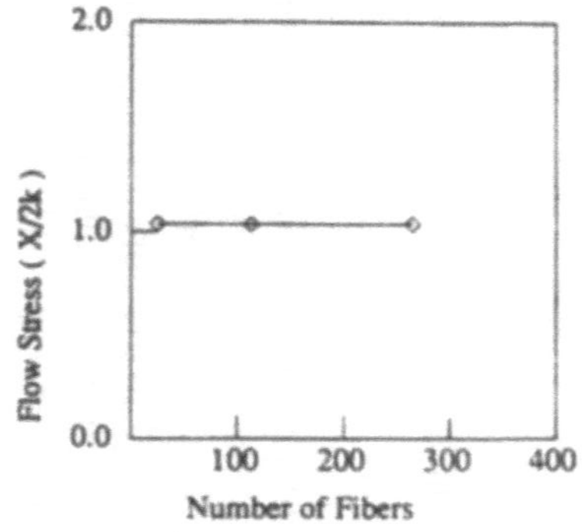

a

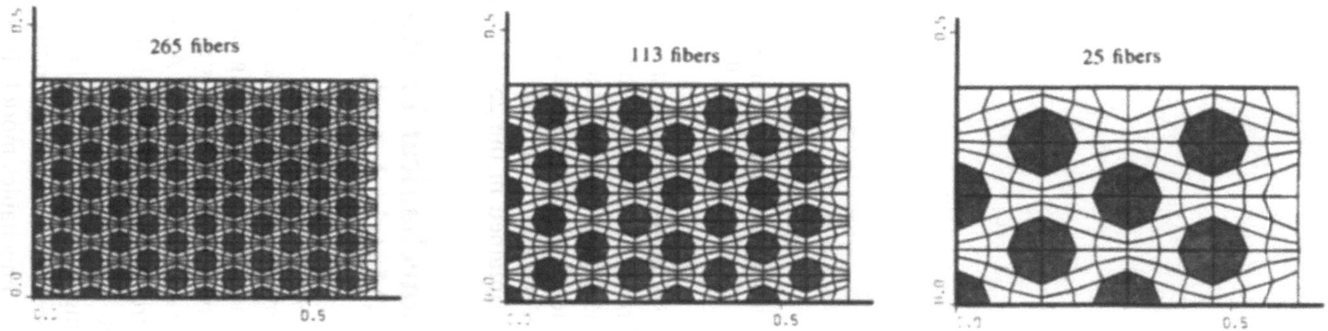

b

Fig. 14.5. Deformation of the square pattern fibre arrangement, composite loaded along the 45° direction. **a** Normalized apparent flow stress vs number of fibres. **b** Final deformation pattern for 20% reduction.

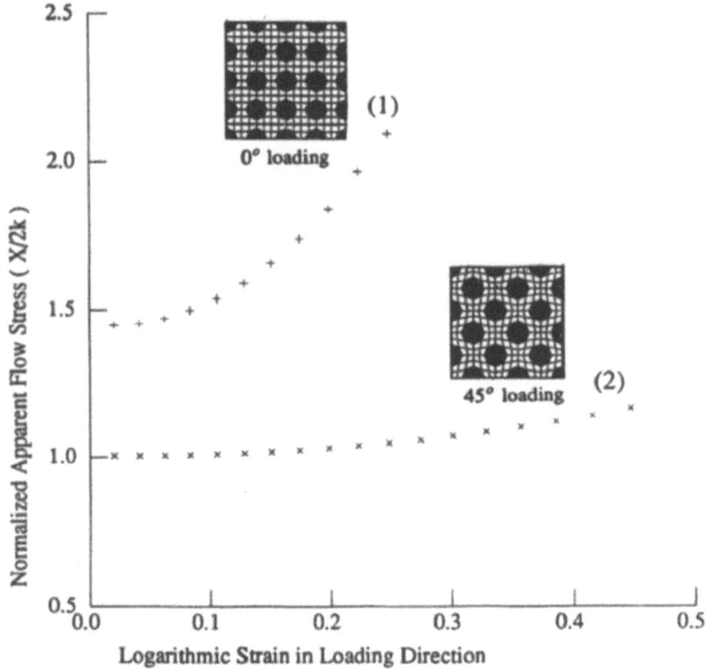

Fig. 14.6. Apparent flow stress vs strain of the square pattern, fibre arrangement composite loaded at 0° and 45°.

of discrete heterogeneous bands as shown in Fig. 14.12 (Hung 1990). The bands run along 45° directions and skirt the fibre–matrix interfaces of the fibres. The distribution of the bands becomes increasingly uniform as the friction coefficient increases. When loading of the square array is done at 45° to the fibre arrangement, however, the deformation bands become much less uniformly distributed; two major bands are obtained in the 25-fibre model shown in Fig. 14.13.

14.6 Micromechanical Finite-Element Analysis of Fibre–Matrix Interfacial Stresses

There is no theory at present that can predict the onset of debonding at the fibre–matrix interface. Debonding and void formation (Fig. 14.14) and fibre failure are known to limit the forming range. Erturk et al. (1974) have suggested the ratio of the two local principal stresses as a parameter that can be used to predict the onset of void formation. Argon et al. (1975) showed that the tensile normal radial stress on a cylindrical inclusion is the primary factor governing cavity formation.

In the following analysis, elastic-plastic finite-element analysis is performed to calculate the stresses on the fibre–matrix interface for several volume fraction cases using a micromechanics model. It is assumed that the principal stress ratios are known from analytical solutions, using an equivalent orthotropic material model, for example. The addition of beneficial stresses on the preform to reduce the tendency toward fibre–matrix debonding can also be estimated using this model.

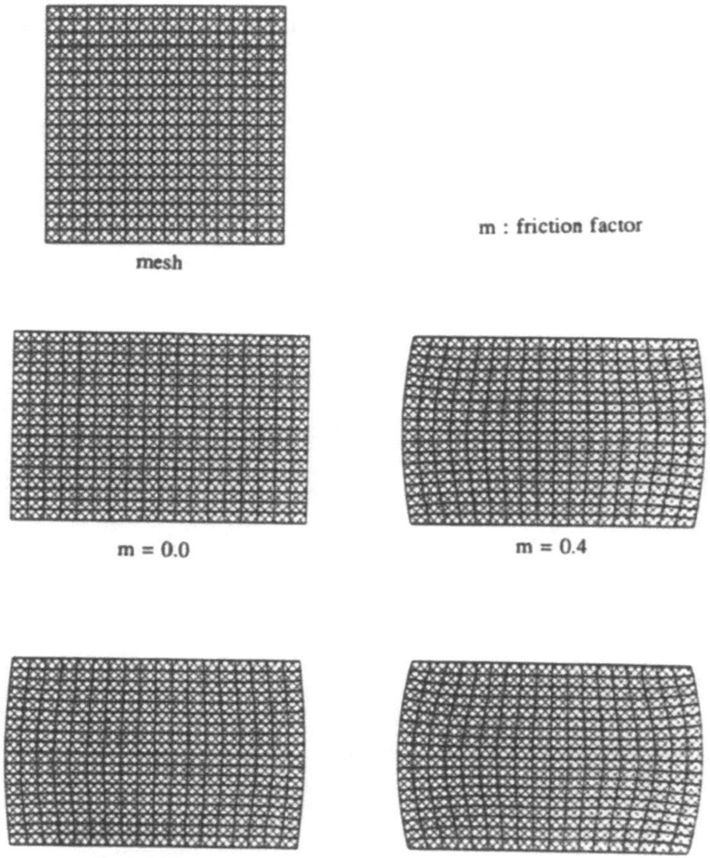

Fig. 14.7. Deformed shape of the equivalent orthotropic model as a function of friction coefficient (at 20% strain).

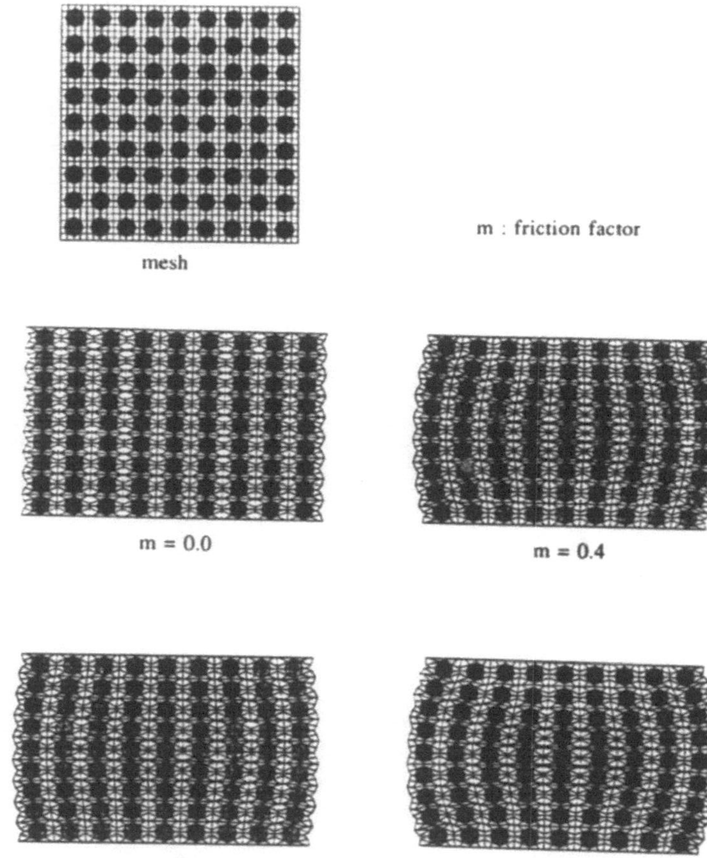

Fig. 14.8. Deformed shape of the two-phase model as a function of friction coefficient (at 20% strain).

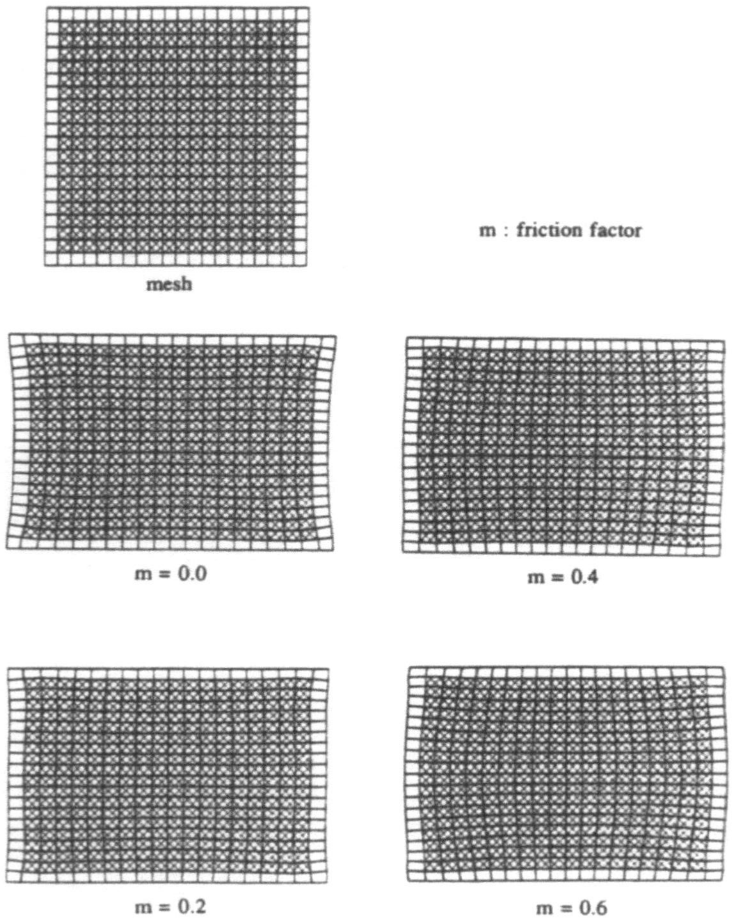

Fig. 14.9. Deformed shape of the equivalent orthotropic model with a matrix boundary layer as a function of friction coefficient (at 20% strain).

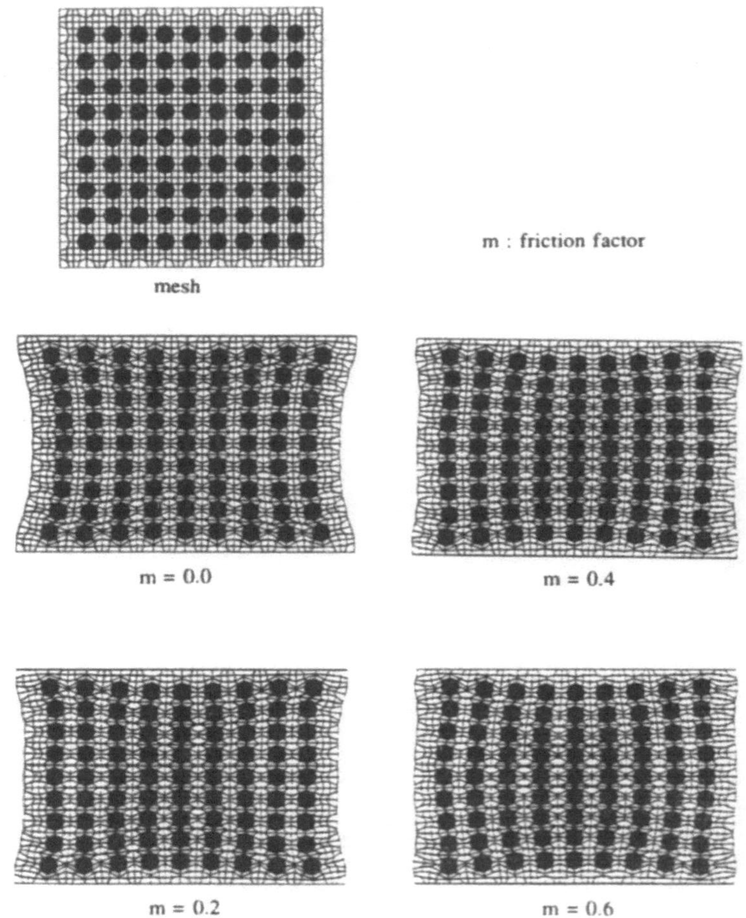

Fig. 14.10. Deformed shape of the two-phase model with an unreinforced matrix as a boundary layer (at 20% strain).

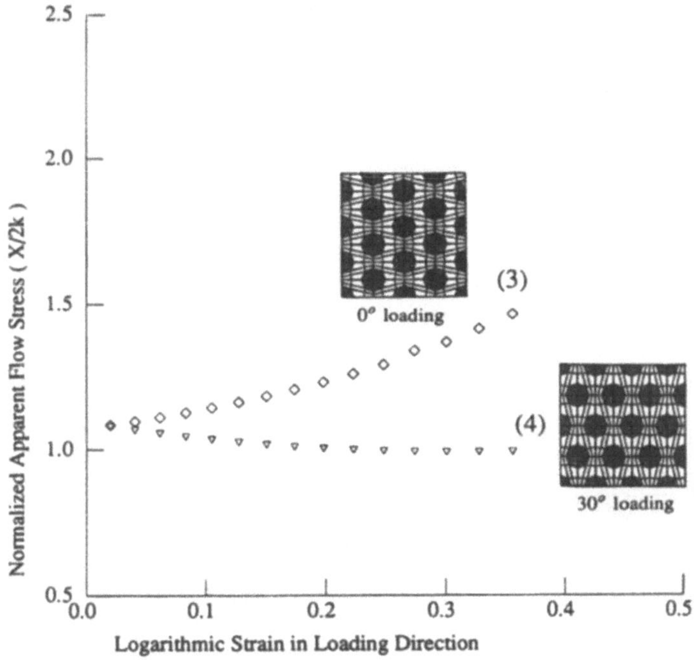

Fig. 14.11. Apparent flow stress vs strain of the hexagonal pattern fibre arrangement composite loaded at 0° and 30°.

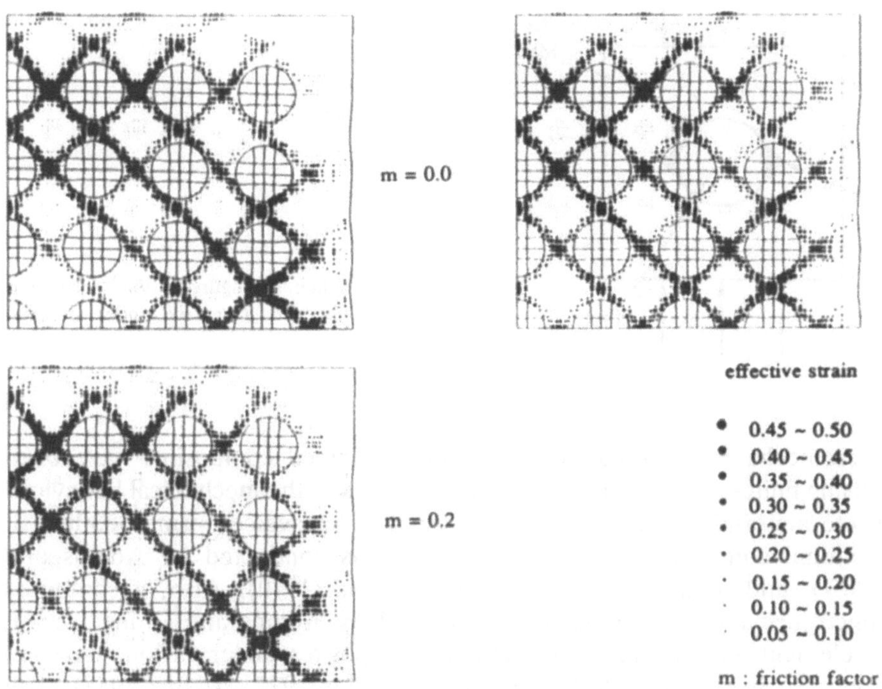

Fig. 14.12. Effective strain distribution (square pattern loaded along the axis) as a function of friction coefficient (5% strain). Note the heterogeneous nature of the deformation (Hung 1990).

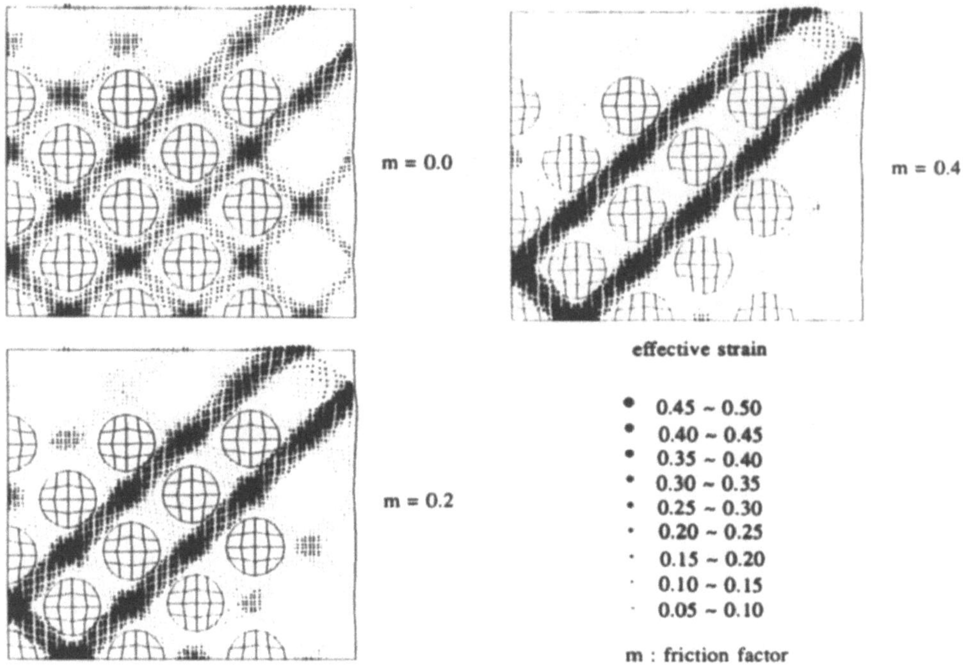

Fig. 14.13. Effective strain distribution (square pattern loaded at 45° to the square fibre pattern axis). Strain localization is increased relative to Fig. 14.12.

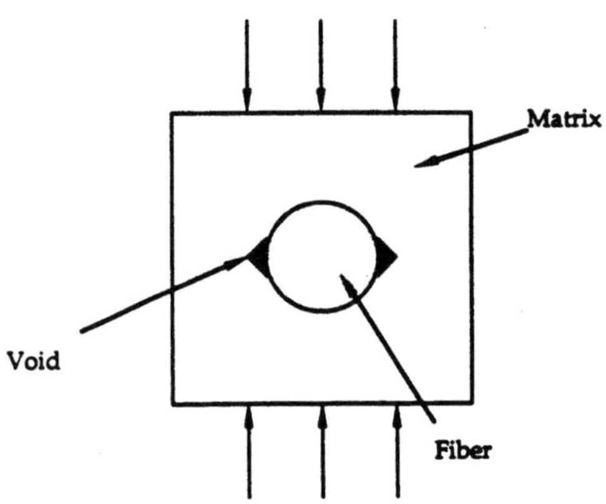

Fig. 14.14. Schematic figure of void formation around a fibre in a matrix under uniaxial loading.

The micromechanics of unidirectional fibre-reinforced composites have been analysed using finite-element analysis in investigations of the mechanical behaviour of composite materials. Finite-element analysis for transverse normal loading of a unidirectional composite in the elastic state was conducted by Adams and Doner (1967), and in the inelastic state by Adams (1970). Also, a combined loading micromechanical analysis was performed by Adams and Crane (1984), and finite-element analyses for unidirectional composites under various environments have also been conducted (Adams and Miller 1977, Adams 1987). The present problem was addressed by performing elastic-plastic finite-element analysis to determine the stress state around the interface. The finite-element program

FEAP, developed at the University of California, Berkeley, was used (Taylor 1989).

The unidirectionally reinforced metal matrix composite is assumed to consist of a square array of circular fibres in an infinite matrix as shown in Fig. 14.15. Perfect bonding between each fibre and the matrix is assumed. The overall metal matrix composite is assumed to be loaded by average stresses p and q. Every square cell may be assumed to be under the same loading conditions; a typical cell (ABCD) is chosen such that after deformation, the sides of each square cell remain straight and parallel to the original sides. The loading conditions on the cell ABCD are compressive average stresses p along AB and CD, and q along BC and AD. Due to the symmetry of loading and the geometry of the cell, only the first quadrant is modelled as shown in Fig. 14.16; the applied boundary conditions are also shown.

14.6.1 Results of the Micromechanics Analysis

The computations were performed for varying fibre volume fractions, ranging from 30% to 50% and for $q/p=0, 0.1, 0.2, 0.3, 0.4$. The radius of arc ED (Fig. 14.16) was adjusted to obtain the corresponding fibre volume fraction. The iteration in each step was conducted until the energy convergence factor was less than 1×10^{-11}.

Some of the computational results are shown in Figs 14.17 and 14.18 for fibre volume fractions of 0.3 and 0.5, respectively, and for various q/p ratios (the loading q is horizontal and the loading p is vertical in these figures). Plastic zones initiate under the load p in the upper left of the modelled quadrant and propagate

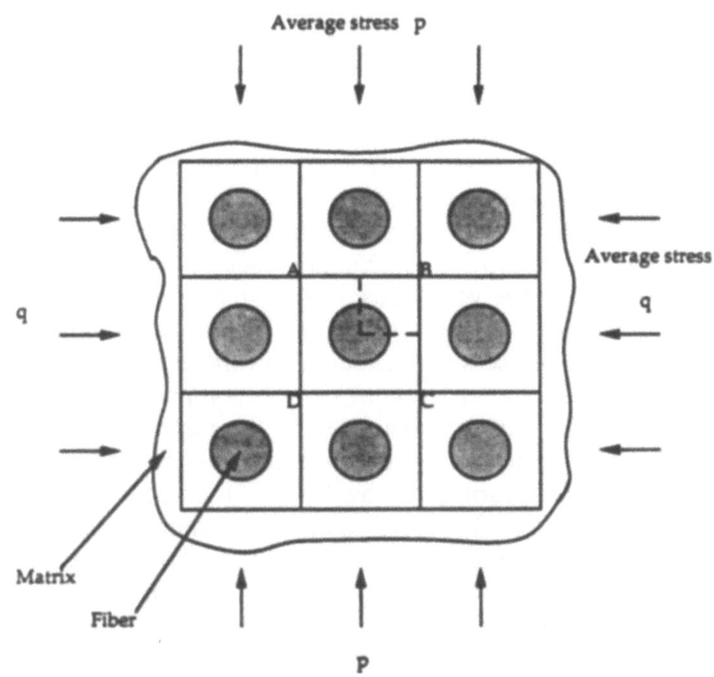

Fig. 14.15. Square array model with applied average stresses p and q.

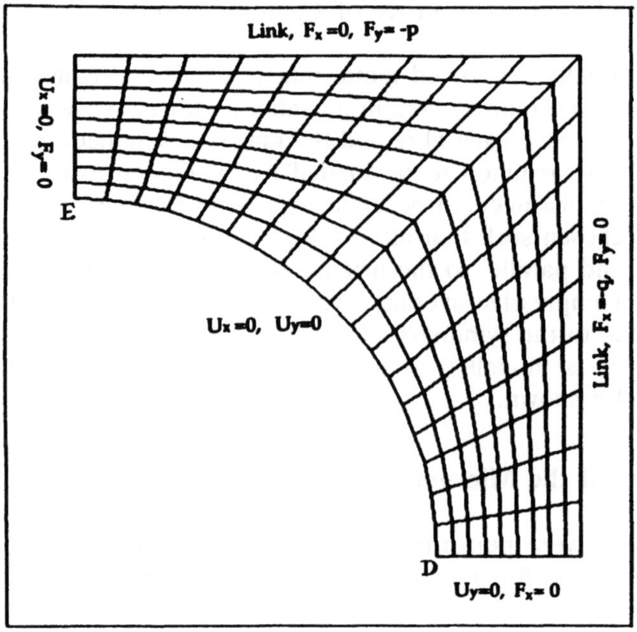

Fig. 14.16. Finite-element model and boundary conditions.

along the interface and to the lower right of the quadrant. The effect of increasing fibre volume fraction is to decrease the extent of the plastic zones for a given loading p (compare Fig. 14.17b with Fig. 14.18b); this effect follows from the fact that the lower fibre volume fraction material is more compliant.

The stresses at the interface are of interest since they can be used to define the conditions for interfacial failure during deformation. Shear stresses, $\tau_{r\theta}$, around the interface are plotted in Fig. 14.19 for a fibre volume fraction of 0.5. The stress is plotted as the normalized value of $\tau_{r\theta}/Y$, where Y represents the yield stress of matrix material. Here the angle around the fibre is taken from 0°, corresponding to the q loading axis (horizontal in Figs. 14.17 and 14.18), to 90°, which is aligned with the p loading axis. The shear stresses show little change with the applied stress ratio q/p. The maximum value of the shear stress is $0.577Y$, the shear yield strength in the von Mises' yield criterion, and is reached at about $\theta = 22.5°$. For other fibre volume fractions, the results are similar, reaching the same maximum value. To assess the tendency toward debonding, a critical shear stress for the fibre–matrix bond strength can be compared with these stresses. Since the maximum shear stress is bounded by the matrix yield strength, a low yield strength matrix is desirable to minimize fibre–matrix debonding.

The normal radial stress normalized by Y (σ_{rr}/Y) is shown in Fig. 14.20 for a fibre volume fraction of 0.5. One notes that the maximum tensile stress on the fibre–matrix interface is along the q loading axis (0°) and decreases with increasing q/p ratio. This behaviour is similar with other volume fractions, the maximum tensile stress decreasing with fibre volume fraction. The maximum tensile stress is thought to be responsible for the initiation of the types of voids shown schematically in Fig. 14.14. For $q/p = 0$, the fibre–matrix bond strength must be at least about 0.4 Y to preclude bond failure.

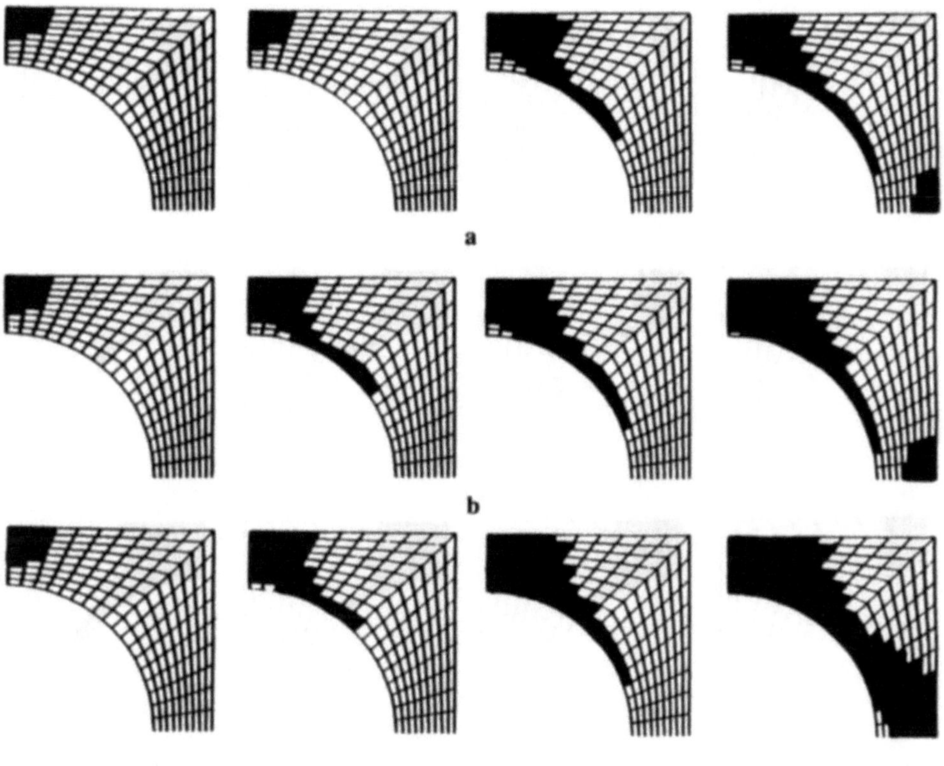

Fig. 14.17. Propagation of the plastic zone for fibre volume fraction = 0.3 and q/p ratios of **a** 0; **b** 0.1; **c** 0.4. The increase in loading is from left to right. The compressive stress p acts vertically downward on the top surface of the model.

Note that when the stress ratio q/p is around 0.3 or higher, the radial normal stress is always compressive, which would indicate that no void formation is possible. The circumferential stress distribution ($\sigma_{\theta\theta}/Y$) is plotted in Fig. 14.21, showing similar behaviour. Once again, the maximum tensile stress is at 0°, i.e., along the q loading axis, and is a maximum for low q/p.

The stress ratio in plane-strain deformation is thus shown to be an important parameter for developing forming criteria for metal matrix composites. Uniaxial deformation produces high tensile stresses. In rolling, $q/p = 0$ at the entry and the exit of the workpiece relative to the rolls. In the deformation zone, q/p can rise to about 0.6 which is desirable; however, since the workpiece was exposed to a stress state of $q/p = 0$ entry, debonding damage may already be present. To reduce the possibility of such damage from occurring, friction rolls can be added to the exit side of the rolls in order to maintain $q/p>0$ by applying in-plane compression to the workpiece (Im and Dharan 1990). Just the opposite is done in the rolling of metals, where tension stresses are applied via tensioning rolls to reduce roll forces and power.

14.7 Conclusions

Numerical modelling of the forming of metal matrix composites reveals the important influence of fibre arrangement on the strain fields in the material.

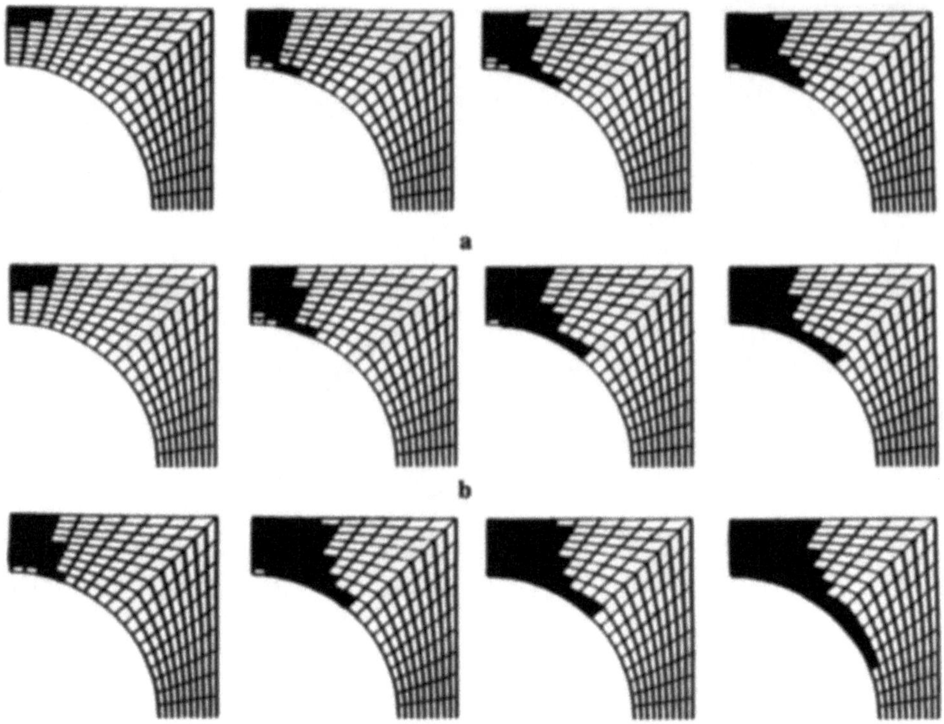

Fig. 14.18. Propagation of the plastic zone for fibre volume fraction = 0.5 and q/p ratios of **a** 0; **b** 0.1; **c** 0.4. The increase in loading is from left to right. The compressive stress p acts vertically downward on the top surface of the model.

Uniaxial deformation of a metal matrix composite with the fibres arranged in a square pattern exhibits a "work-hardening" effect in which the stress rises rapidly with strain. This effect is much less pronounced in the hexagonal fibre arrangement, the other fibre arrangement geometry modelled here. This effect is due to fibre–fibre interaction; a higher degree of plastic zone constraint occurs in the square fibre arrangement resulting in higher loads. The rigid-plastic analysis also shows that heterogeneous deformation bands (effective strain fields) are produced in a heterogeneous material containing rigid inclusions in a plastic matrix. The formation and distribution of these bands also depend on the fibre arrangement; a square fibre pattern shows a more uniform distribution of the deformation bands, while the same pattern deformed at 45° shows the bands to be much more localized.

An elastic-plastic numerical analysis of the stress distribution around an individual fibre in a metal matrix composite undergoing biaxial deformation shows that high radial tensile stresses are produced at the fibre–matrix interface perpendicular to the primary loading axis (applied compressive stress p). The magnitude of the radial stress at the fibre–matrix interface increases with fibre volume fraction; thus high volume fraction composites are more likely to develop debond damage. Void growth has been attributed to such tensile stresses. The analysis also shows that this stress reduces and becomes compressive when a lateral compressive stress q is applied; thus, the q/p ratio is an important parameter in the forming of metal matrix composites. When q/p is maintained greater than

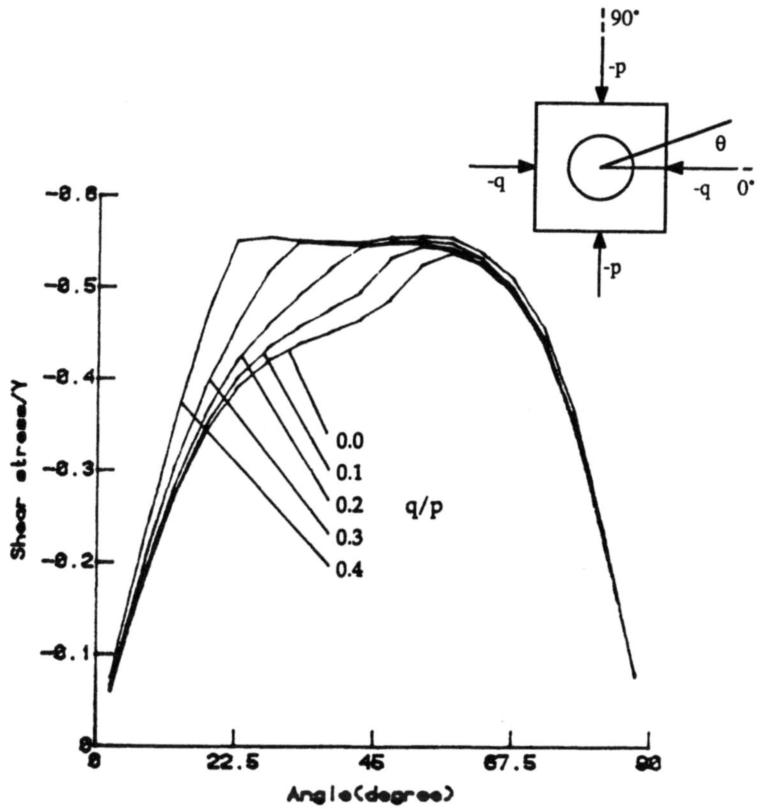

Fig. 14.19. The shear stress distribution at the fibre–matrix interface as a function of angular position around the fibre. The fibre volume fraction is 0.5. $\theta = 90°$ coincides with the p loading axis.

0.3 during the deformation process, the radial stress is compressive and therefore unlikely to produce debonding.

When applied to transverse rolling of a metal matrix composite, radial tensile stresses are developed at the fibre–matrix interface in the longitudinal (rolling) direction at workpiece entry to the rolls. Thus, weakly bonded fibres are likely to suffer debond damage at the start of rolling. The same condition exists at the exit plane. Keeping the workpiece in compression via friction rolls is suggested as a way to maintain a q/p ratio greater than zero (preferably at around 0.3); this approach is opposite to that used in the rolling of metals where strip tensioning is often employed to reduce roll forces and power.

A second criterion must also be met to minimize debonding damage; the shear stress at the fibre–matrix interface cannot exceed the shear strength of the interface. The micromechanical analysis shows that the shear stress at the interface is only mildly dependent on the q/p ratio. Its maximum value, however, is limited by the yield strength of the matrix. Thus, low yield strength materials are desirable, or, alternatively, the matrix material must be heat treated (for example, solution treated and quenched) to a soft condition prior to rolling. Another approach would be to raise the temperature during rolling in order to lower the yield strength of the matrix. This may require the use of heated rolls and preheating the workpiece; due attention must be paid, however, to the mechanisms

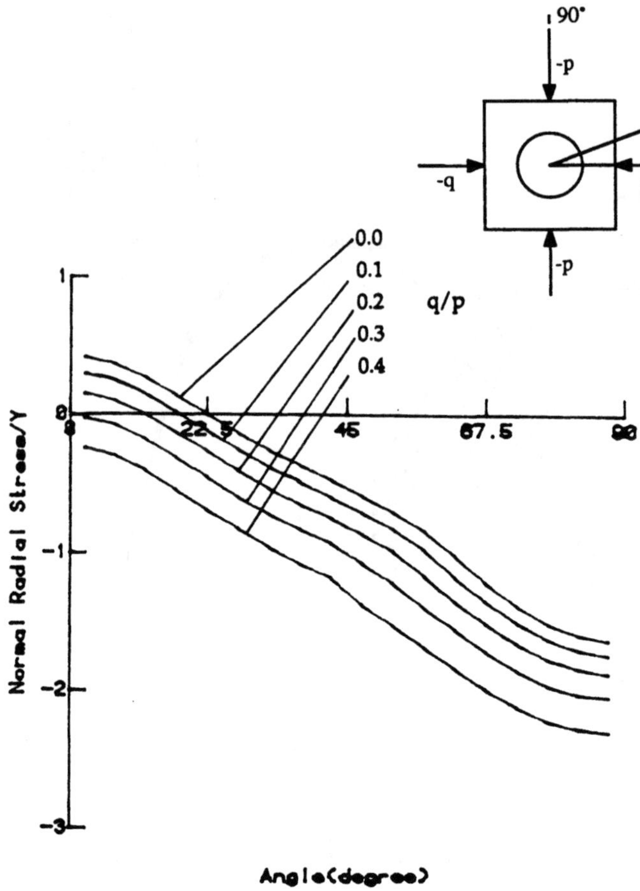

Fig. 14.20. The radial stress distribution at the fibre–matrix interface as a function of angular position around the fibre. The fibre volume fraction = 0.5. $\theta = 90°$, coincides with the p loading axis. Note that for $q/p = 0$, the radial stress is tensile between 0° and 22.5° (i.e. in the vicinity of the q loading axis).

of thermal degradation that can occur at fibre–matrix interfaces when metal matrix composites are exposed to elevated temperatures.

Acknowledgements. This research was supported in part by a grant from the National Science Foundation (MSS-8805427). The computations were conducted by K.H. Im and C. Hung.

Appendix. Plane-Strain Orthotropic Formulation

14.A.1 Yield Function and Effective Stress

Rewrite the anisotropic yield function Eq. (14.1) and define the effective stress as

$$2\hat{f}(\sigma_{ij}) = f(\sigma_y - \sigma_2)^2 + g(\sigma_2 - \sigma_x)^2 + h(\sigma_x - \sigma_y)^2 + 2l\tau_{yz}^2 + 2m\tau_{zx}^2 + 2n\tau_{xy}^2 = 2\bar{\sigma}^2 \tag{14.A1}$$

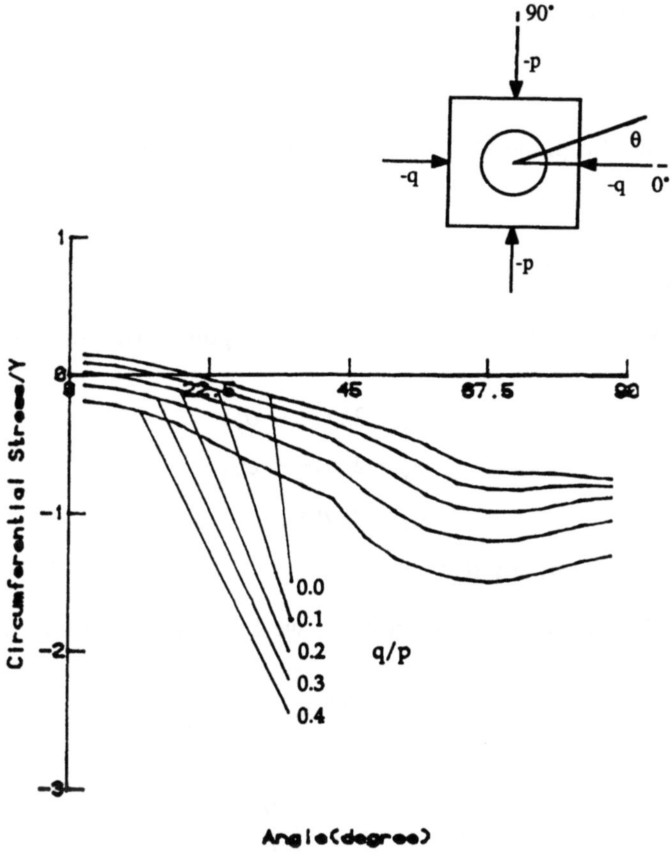

Fig. 14.21. The circumferential stress distribution at the fibre–matrix interface as a function of angular position around the fibre. The fibre volume fraction is 0.5. $\theta = 90°$ coincides with the p loading axis. The circumferential stress, like the radial stress, is tensile at low values for θ and q/p.

where f, g, h, l, m, n are dimensionless parameters characteristic of the current state of anisotropy, and x, y, z are the principal directions of anisotropy, respectively. Assume $\dot{\epsilon}_z = 0$ in the plane-strain condition, then

$$\sigma_z = \frac{g\sigma_x + f\sigma_y}{g + h} \tag{14.A2}$$

Also, in plane strain, $\tau_{yz} = \tau_{zx} = 0$ and $\sigma_z = \sigma_z(\sigma_x, \sigma_y)$, therefore, the yield function becomes

$$2\hat{f}(\sigma_{ij}) = \left(\frac{fg+gh+hf}{f+g}\right)(\sigma_x - \sigma_y)^2 + 2n\tau_{xy}^2 = 2\bar{\sigma}^2 \tag{14.A3}$$

Choose the effective stress as the plane-strain yield stress in the x-direction, i.e. $\bar{\sigma} = X_{p0}$, then $(fg + gh + hf)/(f + g) = 2$ and the yield function becomes

$$\hat{f}(\sigma_{ij}) = (\sigma_x - \sigma_y)^2 + n\tau_{xy}^2 = X_{p0}^2 \tag{14.A4}$$

The parameter n can be determined by using stress transformation in the 45° direction.

The yield function after transformation is then

$$\hat{f}(\sigma'_{ij}) = \frac{n}{4}(\sigma'_x - \sigma'_y)^2 + 4\tau'^2_{xy} = X^2_{p0} \tag{14.A5}$$

Denote X_{p45} as the plane-strain yield stress in the 45° direction, then

$$n = 4\frac{X^2_{p0}}{X^2_{p45}} \tag{14.A6}$$

14.A.2 Flow rule

Assume the yield function $\hat{f}(\sigma_{ij})$ is the plastic potential; the flow rules can be expressed as

$$\begin{aligned} \dot{\epsilon}_x &= 2\lambda(\sigma_x - \sigma_y) \\ \dot{\epsilon}_y &= 2\lambda(\sigma_y - \sigma_x) \\ \dot{\gamma}_{xy} &= 2\lambda\, n\, \tau_{xy} \end{aligned} \tag{14.A7}$$

14.A.3 Effective Strain Rate

Using the definition of plastic work rate, we have the following relations:

$$\dot{W}^p = \sigma_{ij}\dot{\epsilon}_{ij} = \sigma_{ij}\dot{\lambda}\frac{\partial f}{\partial \sigma_{ij}} = 2f\dot{\lambda} = \bar{\sigma}\dot{\bar{\epsilon}} \tag{14.A8}$$

Since

$$\begin{aligned} &\left[\left(\frac{\dot{\epsilon}_x - \dot{\epsilon}_y}{4}\right)^2 + \dot{\gamma}^2_{xy}\right]^{\frac{1}{2}} \\ &= [\dot{\lambda}^2(\sigma_x - \sigma_y)^2 + \dot{\lambda}^2 n\tau^2_{xy}]^{\frac{1}{2}} \\ &= \dot{\lambda}(f)^{\frac{1}{2}} \end{aligned} \tag{14.A9}$$

the effective strain rate becomes

$$\begin{aligned} \dot{\bar{\epsilon}} &= 2(f)^{\frac{1}{2}}\dot{\lambda} \\ &= \left(\tfrac{1}{4}(\dot{\epsilon}_x - \dot{\epsilon}_y)^2 + \frac{1}{n}\dot{\gamma}^2_{xy}\right)^{\frac{1}{2}} \\ &= (\dot{\epsilon}^2_x + \frac{1}{n}\dot{\gamma}^2_{xy})^{\frac{1}{2}} \end{aligned} \tag{14.A10}$$

and the proportionality constant $\dot{\lambda}$ becomes

$$\dot{\lambda} = \frac{1}{2}\frac{\dot{\bar{\epsilon}}}{\bar{\sigma}} \tag{14.A11}$$

References

Adams DF (1970) Inelastic analysis of a unidirectional composite subject to transverse normal loading. J Comp Mater 4:310–328

Adams DF (1987) A micromechanics analysis of the influence of the interface on the performance of polymer-matrix composites. J Reinf Plast Comp 6:66–88

Adams DF, Crane DA (1984) Combined loading micromechanical analysis of a unidirectional composite. Composites 15 (3):181–191

Adams DF, Doner DR (1967) Transverse normal loading of a unidirectional composite. J Comp Mater 1:152–164

Adams DF, Miller AK (1977) Hygrothermal microstresses in a unidirectional composite exhibiting inelastic material behavior. J Comp Mater 11:285–299

Argon AS, Im J, Safoglu R (1975) Cavity formation from inclusions in ductile fracture. Met Trans A 6:825–837

Bahei-El-Din YA (1979) Plastic analysis of metal–matrix composite laminates. PhD dissertation, Duke University, South Carolina

Bahei-El-Din YA, Dvorak GJ (1979) Plastic yielding at a circular hole in a laminated FP-Al plate. In: Modern developments in composite materials and structures. American Society of Mechanical Engineers, New York, pp 123–147

Chen CC, Kobayashi S (1978) Rigid-plastic finite element analysis of ring compression. In: Application of numerical methods to forming processes, vol AMD-28, American Society of Mechanical Engineers, New York

Dvorak GJ, Bahei-El-Din YA (1979) Elastic-plastic behavior of fibrous composites. J Mech Phys Solids 27:51–72

Dvorak GJ, Rao MSM, Tarn JQ (1974) Generalized initial yield surfaces for unidirectional composites. Trans ASME: J Appl Mech 41:249–253

Erturk T, Kuhn HA (1979) Use of forming criteria in forging complex shapes from metal–matrix composites. Trans ASME: J Eng Mat Tech 101:3–11

Erturk T, Kuhn HA, Lawley A (1974) Forging of metal–matrix composites – forming criteria. Met Trans A 5:2295–2303

Getten JR, Ebert LJ (1969) The cold rolling characteristics of aluminum–boron fiber composites. Trans ASM 62:869–878

Hill R (1950) Mathematical theory of plasticity. Oxford University Press, London

Hung C (1990) Process design of three-dimensional open-die forging and the deformation analysis of metal matrix composites. PhD dissertation, University of California, Berkeley

Hwang SM, Kobayashi S (1984) Preform design in plane strain rolling by the finite-element method. Int J Mach Tool Des Res 24 (4):253–266

Hwang SM, Kobayashi S (1986) Preform design in disk forging. Int J Mach Tool Des Res 26 (3):231–243

Im KH (1990) Forming criteria for metal matrix composites. PhD dissertation, University of California, Berkeley

Im KH, Dharan CKH (1990) Forming criteria based on fiber-matrix interfacial strength for metal matrix composites. Submitted for publication

Kim NS (1989) Computer-aided preform design in metal forming by the finite element method. PhD dissertation, University of California, Berkeley

Lee CH, Kobayashi S (1973) New solutions to rigid-plastic deformation problems using a matrix method. Trans ASME: J Eng Ind 95 (3):865–873

Mori K, Osakada K (1984) Simulation of three-dimensional deformation in rolling by the finite-element method. Int J Mech Sci 26 (9/10):515–525

Oh SI (1982) Finite element analysis of a metal forming process with arbitrary shaped dies. Int J Mech Sci 24 (8):479–493

Park JJ, Kobayashi S (1984) Three-dimensional finite-element analysis of block compression. Int J Mech Sci 26 (3):165–176

Park JJ, Rebelo N, Kobayashi S (1983) A new approach to preform design in metal forming with the finite element method. Int J Mach Tool Des Res 23 (1):71–79

Salkind M, George F, Tice W (1969) Some effects of cold rolling on the microstructure and properties of Al_3Ni whisker reinforced aluminum. Trans Met Soc AIME 245:2339–2345

Shah SN, Kobayashi S (1974) Rigid-plastic analysis of cold heading by the matrix method. In: Tobias SA, Koenigsberger K (eds) Proceedings 15th International Machine Tool Design Research Conference pp. 561–569

Shiau YC (1987) Three-dimensional finite element analyses of open-die forging and plate rolling. PhD Dissertation, University of California, Berkeley

Taylor RL (1989) FEAP – finite element analysis program manual. University of California, Berkeley

Yamada Y, Yoshimura N, Sakurai T (1968) Plastic stress-strain matrix and its application for the solution of elastic-plastic problems by finite-element method. Int J Mech Sci 10:343–354

15 Finite-Element Analysis of the Superplastic Forming of Thin Sheet Components

R.D. Wood and J. Bonet

15.1 Introduction

15.1.1 Superplasticity and Superplastic Forming

Certain materials, when deformed in tension at particular temperatures and strain rates exhibit very high elongations. Such a phenomenon is termed *superplasticity*. Elongations for superplastic materials are typically between 200% and 1000% and on occasions can be as much as 5000% (AGARD 1987). As a consequence of this extreme ductility, superplasticity can be utilized to devise forming processes that achieve near net-shape manufacturing. Whilst the superplastic properties of some alloys have been known for some 60 years it was not until the 1960s that the commercial potential of the phenomenon began to be realized. Superplastic forming (SPF) can be used to forge solid components such as turbine discs and die moulds or alternatively to blow form thick and thin sheet components. This chapter will be concerned with thin sheet forming.

Aluminium, titanium and nickel-based alloys are the most commonly used superplastic materials but many other alloys exhibit superplastic properties. In particular, microduplex stainless steel and ultrahigh carbon steel have been successfully superplastically formed albeit not at strain rates that would be of interest to, say, the automotive industry. Recently, superplastic ceramics and laminated composites have been developed and for ceramics it has been suggested (Hamilton and Paton 1988) that SPF may prove to be the only way to form such hard materials accurately.

Superplastic behaviour is characterized by the dependence of the flow stress upon the strain rate which is usually depicted by the logarithmic relationship shown in Fig. 15.1a. To produce maximum ductility, the SPF process must control the strain rate $\dot{\epsilon}$ so that the *strain-rate sensitivity index m* is a maximum (see Fig. 15.1a–c). Elongation is also dependent upon temperature, and forming must take place within a narrow temperature range (see Fig. 15.1d) to be successful. If these conditions can be met then, unlike conventional ductile materials, superplastic materials are much less susceptible to localized necking and can, importantly, be formed at flow stresses substantially lower than the mechanical yield stress.

The characteristics of superplastic behaviour are such that elastic deformation can be ignored and the material can be treated in terms of a viscoplastic constitutive relationship. Furthermore, since the pressure applied to the material is normally controlled so that the strain rate remains in the region of maximum m, for most applications a linear relationship between $\log \sigma$ and $\log \dot{\epsilon}$ proves to be accurate enough. Hence a simple two-parameter constitutive equation emerges as

$$\log \sigma = \log K + m \log \dot{\epsilon} \tag{15.1}$$

or equivalently

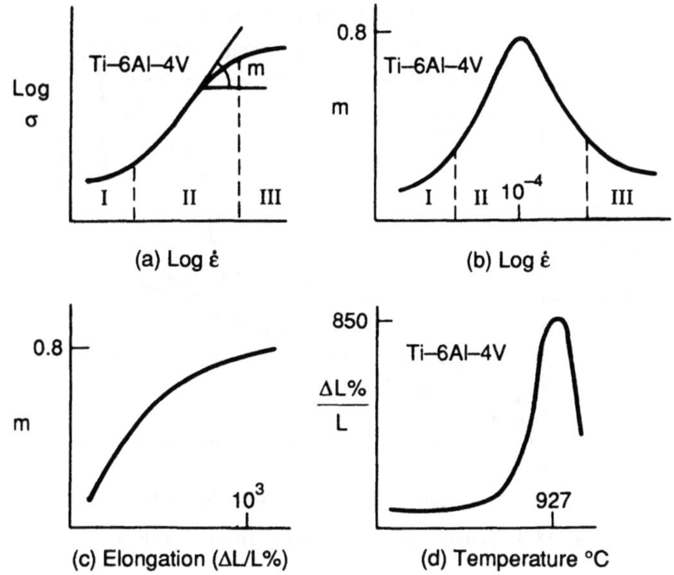

Fig. 15.1. Typical superplastic material characteristics.

$$\sigma = K\dot{\varepsilon}^m \tag{15.2}$$

In order to illustrate the influence of m upon the necking resistance of superplastic alloy, consider a tensile specimen with a cross-sectional inhomogeneity of 1% as shown in Fig. 15.2. Using the simplified equilibrium equation shown in Fig. 15.2 and assuming a linearized log σ–log $\dot{\varepsilon}$ relationship of the form given by Eq. (15.2), it is possible to study analytically the way in which the specimen would deform during a constant strain-rate test. The results obtained are summarized in Fig. 15.3 where the total strain in the neck region has been plotted against the total strain in the bar for different values of the strain-rate sensitivity index m. It transpires from this figure that high m values result in a slower growth of the inhomogeneity and consequently allow larger elongations to take place before fracture.

To be superplastic, metal alloys require a very fine grain size, typically less than 10 μm. During forming the grain size changes and modifies the superplastic properties of the material. Both static and deformation enhanced grain growth occur, the former being a function of time at a given temperature whilst the latter

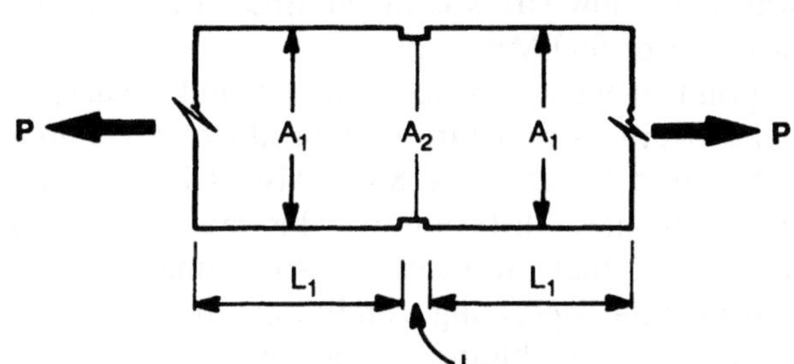

$f = (A_1^\circ - A_2^\circ)/A_1^\circ \qquad L_2 << L_1$

$\sigma_2(1-f)A_1^\circ c^{-\varepsilon_2} = \sigma_1 A_1^\circ c^{-\varepsilon_1}$

Fig. 15.2. Simplified tensile test model.

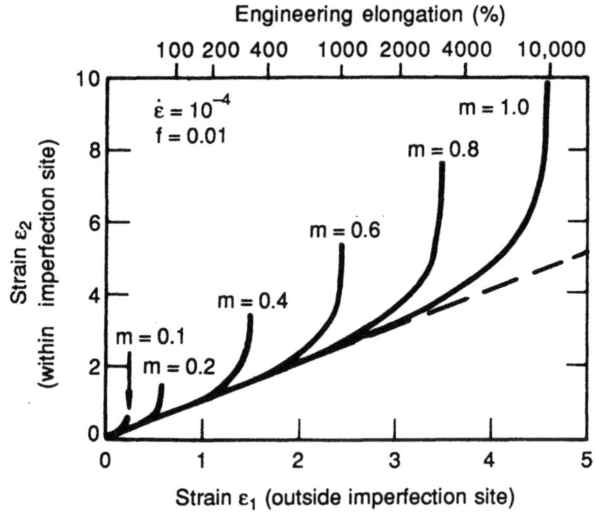

Fig. 15.3. Tensile test results for various m values.

depends primarily on the strain rate. Any meaningful constitutive equation must account for the evolution of the grain size. In addition it may be necessary to include the effects of cavitation (void growth), particularly in the case of aluminium alloys (Hamilton and Paton 1988). Although much has been written on the topic, the actual microstructural mechanism(s) which cause superplasticity are not the subject of any firm consensus. Discussions on superplasticity are contained in Hori et al. (1991), Hamilton and Paton (1988) and Padmanabham and Davies (1980).

This chapter will consider the numerical simulation of the superplastic blow forming of thin sheet components with examples being presented for the titanium alloy Ti-6Al-4V. The process involves slowly blowing a sheet, without fracturing, into a die to produce a very complex, light and strong component. Judicious use of diffusion bonding (DB) between adjacently forming sheets (Fig. 15.4), enables assemblages to be formed which are structurally more integral than the same part produced by traditional techniques such as welding, adhesive bonding or riveting.

For Ti-6Al-4V, for example, the forming temperature is about half the melting temperature, i.e. 927°C and the optimum strain rate is between 10^{-3} and 10^{-5} s^{-1}. The argon gas pressure sufficient for forming and diffusion bonding is of the order of 2 MPa, whilst the flow stress is of the order of 10 MPa, compared to a mechanical yield strength of 900 MPa.

From a practical point of view it is necessary to find a varying pressure cycle that will maintain the overall strain rate that produces the optimum strain rate sensitivity index m required for effective SPF. Also it is important to be able to predict the final sheet thickness distribution. This may well involve devising an initial thickness distribution that produces a desired final thickness that satisfies structural and environmental operating conditions. It is with respect to these problems that the predictive capabilities of the finite-element method can make an important contribution to the design of an SPF manufacturing process.

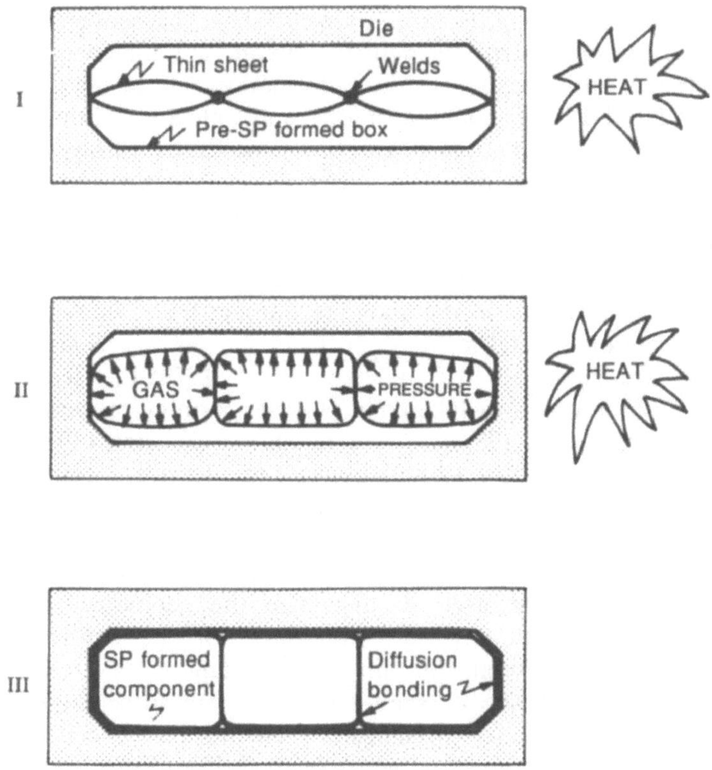

Fig. 15.4. Typical SPF/DB process.

15.1.2 Finite-Element Simulation of SPF Processes

15.1.2.1 Overview

Superplastic forming of a thin sheet is a process involving large deformations, large strain, nonlinear material behaviour and, usually, deformation-dependent boundary conditions. Consequently, the numerical analysis or simulation of such a highly nonlinear process is a formidable computational problem.

Essentially, the motion of the sheet is governed by two physical laws, namely the continuum equilibrium equation and the material or constitutive equation. Since superplastic forming is typically a very slow process, inertia effects can be ignored. The resulting equilibrium equation describes the balance at each point of the sheet between the internal stresses and the applied pressure. The constitutive equation, on the other hand, provides a relationship between the stresses, the deformation and deformation rate of the sheet. Unfortunately, due to the complex nature of forming processes, finding the motion of the sheet, or the shape as function of time, that satisfies both governing laws is practically impossible analytically.

In order to overcome these difficulties the finite-element method can be used to approximate the motion of the sheet by simple combinations of pre-established functions defined within a given element. Under such conditions a variational formulation will enable the equilibrium equation to be discretized as the balance at each node between the external forces, representing the applied pressure, and the internal forces, describing the stresses in the sheet. Moreover, recognition of the similarities between superplastic deformation and a non-Newtonian fluid flow, enables the stresses to be related to the deformation rate in the sheet, which in turn can be related to the nodal positions and velocities. Consequently, the complexities associated with stress integration procedures arising in elastoplastic or elastoviscoplastic analysis of other materials, can be avoided. Additionally, incorporating the constitutive equation into the discrete equilibrium equations yields a system of differential equations for the nodal positions, as functions of time, which can be integrated by means of a discrete time-stepping scheme.

Entitled the *flow formulation*, the above techniques have been used in the past to simulate conventional metal-forming processes such as extrusion, bulk forming, deep drawing, stretching, etc. (Zienkiewicz et al. 1978a,b, Zienkiewicz 1984, Wood et al. 1986), as well as as glass and polymer-forming processes (Nakazawa et al. 1982, Cormeau et al. 1984). In Argyris and St Doltsinis (1984) and Zhang et al. (1986) the general flow formulation techniques were employed for the numerical analysis of the superplastic forming of axisymmetric components, whilst in Chandrasekaran et al. (1985) and Rebelo and Wertheimer (1988) plane-strain problems were considered. In Argyris and St Doltsinis (1984) and Zhang et al. (1986) solid elements were used to discretize the geometry and motion of the material, together with a penalty approach for the enforcement of the incompressibility of the material. For thin-sheet forming, however, such generality is unnecessary and membrane or shell kinematics are sufficient to provide an adequate finite-element model. Membrane elements have been considered in Bellet and Chenot (1986), Bellet et al. (1987), Bellet (1988) and Bonet et al. (1988a–c), whereas in St Doltsinis et al. (1987) Mindlin shell elements were used and in Bonet and Wood (1987) and Bonet et al. (1989) Kirchhoff shell kinematics were assumed.

The flow formulation has been employed for the simulation of conventional thin-sheet-forming processes in Oñate and Zienkiewicz (1983) and Zienkiewicz et al. (1978b), while in Kobayashi and Kim (1978) and Wang (1984) rigid-plastic or rigid-viscoplastic models were used, and in Wang (1978), Wennerström et al. (1984) and Massoni et al. (1987, 1988) elastic effects were included. Different time-stepping schemes have been used in the past to simulate forming: Zienkiewicz et al. (1978a,b), Nakazawa et al. (1982), Oñate and Zienkiewicz (1983), Argyris and St Doltsinis (1984), Cormeau et al. (1984), Zienkiewicz (1984), Zhang et al. (1986), Bonet and Wood (1987) and St Doltsinis et al. (1987) employ explicit integration, whilst in Bellet and Chenot (1986), Bellet et al. (1987), Bellet (1988) and Bonet et al. (1988a–c) implicit schemes were prefered.

Finally, the numerical prediction of a pressure cycle which maintains an adequate optimum strain rate has been discussed in Chandrasekaran et al. (1985), St Doltsinis et al. (1987), Bellet (1988) and Bonet et al. (1988b,c). Usually, a simple a posteriori adjustment is made to the pressure after each time step in the solution procedure (St Doltsinis et al. 1987, Bonet 1989). In Chandrasekaran et al. (1985) and Rebelo and Wertheimer (1988), a similar technique is used, apparently in conjunction with the iterative scheme employed to solve the nonlinear equilibrium equations. Formulation details are only given in St Doltsinis et al. (1987), Rebelo and Wertheimer (1988) and Bonet et al. (1989), and in Chandrasekaran et al. (1985), St Doltsinis et al. (1987) and Bonet (1989) only an average constant optimum strain rate is maintained, albeit usually close to the target value. In Bonet et al. (1989) the targetted optimum strain-rate value is exactly maintained by introducing the magnitude of the pressure as an additional variable.

15.1.2.2 Solution Strategy for Thin Sheet SPF

The kinematic description of the sheet is influenced by a number of factors. For example, the forming sheet is undoubtably a shell subject to both membrane and bending load-carrying actions, nevertheless it is reasonable to assert that membrane effects dominate the behaviour. A practical consideration is that the component shape is likely to be sufficiently complex to demand the use of a large number of elements. In addition, complex finite-element meshes should be easy to generate and possibly refine as the computation progresses. The simple three-node triangular constant stress membrane element meets these requirements and also enables the stiffness matrices and nodal forces to be calculated explicitly.

The equations which derive from the discretization by the finite-element method are nonlinear functions of the nodal velocities and positions. Consequently a time integration (time-stepping) scheme must be introduced in order to calculate the change of geometry with respect to time and additionally some form of iteration, within the time integration scheme, must be employed to solve the nonlinear equilibrium equations. Previous experience (Oñate and Zienkiewicz 1983) obtained using the non-Newtonian fluid flow formulation for simulating thin-sheet axisymmetric punch forming indicated that an explicit forward Euler time-stepping scheme may be successful. Such a scheme renders the nonlinear equilibrium equations functions of velocity *only* with no implied dependence upon geometry. These equations may be solved using the method of successive substitutions (Picard iteration) (Oñate and Zienkiewicz 1983) or more efficiently using a Newton–Raphson scheme or one of its variants (Bonet and Wood 1987). Unfortunately in the present case of blow-formed nonaxisymmetric components prescribed velocity and axisymmetric constraints are absent and the forward Euler scheme becomes unstable. This suggests the use of unconditionally stable implicit schemes (Bonet et al. 1988c). Now the nonlinear equilibrium equations are functions of velocity with an implied (via the velocity) dependence upon geometry. At each time step the solution is normally achieved in two stages, known as the *predictor* and *corrector* phases. The predictor provides an initial guess for the velocity (and thence the geometry) which is subsequently corrected using a Newton–Raphson procedure until equilibrium is satisfied. Fig. 15.5 shows a schematic representation of the solution procedure where x refers to the geometry, v to the velocity, $\dot{\epsilon}_0$ the prescribed optimum strain rate and t the time. If the strain rate is not optimized then $\dot{\epsilon}_0$ would represent the prescribed pressure $p(t)$.

At each corrector stage a set of linear equations have to be solved, a choice having to be made between a direct or iterative solver. For large meshes iterative solvers are efficient but have the disadvantage that multiple solutions may not be processed as cheaply when compared to using a direct solver. In the present context, to calculate the pressure cycle for an SPF simulation requires two solutions per nonlinear iteration as opposed to one when the pressure cycle is prescribed. The solvers employed in this work were the direct LDU decomposition (Taylor 1985) and a preconditioned conjugate gradient (PCG) scheme (Bonet

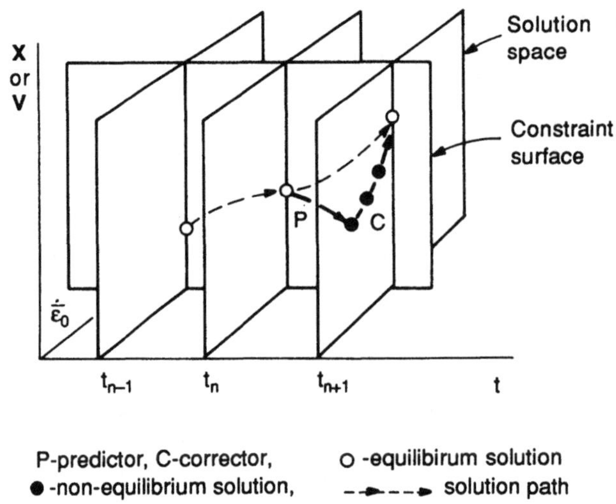

Fig. 15.5. Schematic representation of the numerical solution.

1989). The former requires substantial storage whilst the latter requires little storage but has to be used twice for pressure cycle calculation. Using a three-noded triangular membrane element for a typical free-blowing problem the PCG iteration is economical after about 1500 elements with prescribed pressure and at about 3000 elements when the pressure cycle is calculated.

As previously mentioned, SPF of thin sheets is a process involving progressive contact between the sheet and the die or, in the case of diffusion bonding, between adjacent sheets. From a practical perspective it is often sufficient to assume sticking contact between sheet and die and between adjacent sheets. Computationally, sticking die contact is relatively simple but intersheet contact involves the use of a penalty method of constraint (Wood et al. 1989).

15.1.3 Conclusion

The remainder of this chapter will consider the detailed formulation of the numerical simulation of the SPF of a thin sheet, beginning with the governing kinematic, constitutive and equilibrium equations. Using the three-noded triangular membrane element these equations will then be recast in discrete form. Recognizing these equations to be nonlinear differential equations in time leads to a discussion of various time integration schemes and to a derivation of the tangent matrix required for the nonlinear solution iterations.

Finally, a number of examples are presented to demonstrate the capabilities of the techniques.

15.2 Governing Equations

The governing kinematic, constitutive and equilibrium equations will be formulated using, where possible, a vector–matrix notation and a Cartesian coordinate system. Whilst it is possible to develop these equations using an elegant, compact and computationally more efficient tensorial description in terms of a convected coordinate system (Bonet 1989) the Cartesian-based derivation may be more accessible.

A key item of notation is the matrix derivative which is the derivative of one vector with respect to another. Typically let $\mathbf{a} = [a_1, a_2, a_3]^T$ and $\mathbf{b} = [b_1, b_2]^T$ then

$$\frac{\partial \mathbf{a}}{\partial \mathbf{b}} = \begin{bmatrix} \frac{\partial a_1}{\partial b_1} & \frac{\partial a_1}{\partial b_2} \\ \frac{\partial a_2}{\partial b_1} & \frac{\partial a_2}{\partial b_2} \\ \frac{\partial a_3}{\partial b_1} & \frac{\partial a_3}{\partial b_2} \end{bmatrix} \tag{15.3}$$

This notation will still apply even if $\mathbf{a}$ or $\mathbf{b}$ are scalars.

15.2.1 Kinematics

The kinematic description will be developed assuming that the sheet will be represented by flat three-node membrane elements with each element having a local Cartesian axes system $\mathbf{x} = [x_1, x_2, x_3]^T$, see Fig. 15.6 where x_2 lies in the plane of the element.

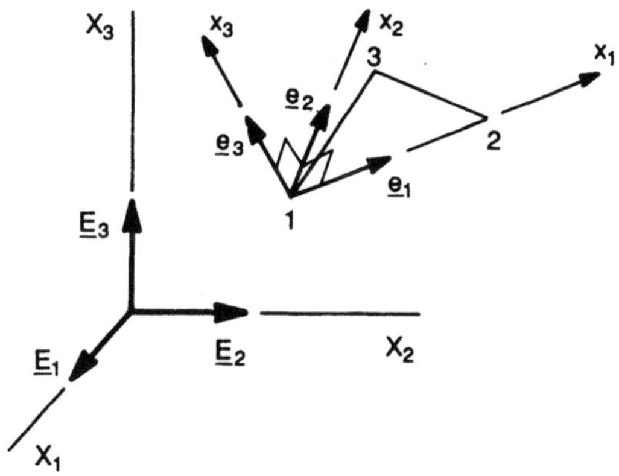

Fig. 15.6. Coordinate definitions.

In terms of the unit base vectors $\mathbf{E}_i$ and $\mathbf{e}_i$, in the global and local coordinate systems respectively, a typical local vector $\boldsymbol{v}$ is given in terms of its global equivalent $\mathbf{v}$ as

$$\boldsymbol{v} = \mathbf{R} \cdot \mathbf{v}, \quad \mathbf{R}_{ij} = \mathbf{e}_i \cdot \mathbf{E}_j, \quad i = 1,2 \;\; j = 1,3 \tag{15.4}$$

For a membrane the local velocity $\boldsymbol{v}$ is constant in the thickness x_3 direction.

In terms of the local velocity components the membrane strain rate $\dot{\boldsymbol{\epsilon}}$ is

$$\dot{\boldsymbol{\epsilon}} = \begin{Bmatrix} \dot{\epsilon}_{11} \\ \dot{\epsilon}_{22} \\ 2\dot{\epsilon}_{12} \end{Bmatrix} = \begin{Bmatrix} \frac{\partial v_1}{\partial x_1} \\ \frac{\partial v_2}{\partial x_2} \\ \frac{\partial v_1}{\partial x_2} + \frac{\partial v_2}{\partial x_1} \end{Bmatrix} \tag{15.5}$$

Later, in the development of the tangent matrix used in the solution, the derivatives of the global velocity with respect to the local coordinate system will be required; these are

$$\boldsymbol{\theta} = \begin{Bmatrix} \boldsymbol{\theta}_1 \\ \boldsymbol{\theta}_2 \end{Bmatrix}, \quad \boldsymbol{\theta}_i = \begin{bmatrix} \frac{\partial v_1}{\partial x_i} & \frac{\partial v_2}{\partial x_i} & \frac{\partial v_3}{\partial x_i} \end{bmatrix}^{\mathrm{T}} \tag{15.6}$$

Anticipating a constitutive equation involving incompressibility the out of plane local strain rate is

$$\dot{\epsilon}_{33} = -\,(\dot{\epsilon}_{11} + \dot{\epsilon}_{22}) \tag{15.7}$$

15.2.2 Constitutive Equations

The constitutive equations for superplastic behaviour are usually given as a relationship between the von Mises equivalent stress $\bar{\sigma}$ and equivalent strain rate $\dot{\bar{\epsilon}}$, equivalent strain $\bar{\epsilon}$ and grain size g:

$$\bar{\sigma} = \Phi(\dot{\bar{\epsilon}}, \bar{\epsilon}, g) \quad \text{or} \quad \dot{\bar{\epsilon}} = \Psi(\bar{\sigma}, \bar{\epsilon}, g) \tag{15.8a,b}$$

where, in terms of the deviatoric stress σ'_{ij} and strain rate $\dot{\epsilon}_{ij}$ and assuming incompressibility ($\dot{\epsilon}_{kk} = 0$)

$$\bar{\sigma} = \left(\frac{3}{2}\,\sigma'_{ij}\sigma'_{ij}\right)^{\frac{1}{2}}, \quad \dot{\bar{\epsilon}} = \left(\frac{2}{3}\,\dot{\epsilon}_{ij}\dot{\epsilon}_{ij}\right)^{\frac{1}{2}} \tag{15.9a,b}$$

Recognizing that superplasticity is essentially a viscoplastic phenomenon, a general relation between strain rate $\dot{\epsilon}_{ij}$ and Cauchy stress σ_{ij} can be written as

$$\dot{\epsilon}_{ij} = \gamma \langle \phi(F) \rangle \frac{\partial Q}{\partial \sigma_{ij}} \tag{15.10}$$

where F is a yield function, Q is the plastic potential and γ and the function ϕ derive from experimental data. Introducing J_2, flow theory gives

$$F = Q = (3J_2)^{\frac{1}{2}} - \sigma_y \tag{15.11}$$

where $J_2 = \frac{1}{2}\sigma'_{ij}\sigma'_{ij} = \bar{\sigma}^2/3$ and for superplasticity the yield (or threshold) stress σ_y is taken as zero. Eq. (15.10) may now be rewritten as

$$\dot{\epsilon}_{ij} = \gamma\phi(\bar{\sigma})\,\frac{3}{2\bar{\sigma}}\,\sigma'_{ij} \tag{15.12}$$

Substituting (15.12) into (15.9) enables the $\gamma\phi(\bar{\sigma})$ term to be evaluated as $\dot{\bar{\epsilon}}$ giving

$$\sigma'_{ij} = 2\mu\dot{\epsilon}_{ij}, \quad \mu = \frac{\bar{\sigma}}{3\dot{\bar{\epsilon}}} \tag{15.13a,b}$$

This is immediately recognized as the constitutive equation for non-Newtonian incompressible viscous flow, where μ is the nonlinear viscosity. Equation (15.13) is a general constitutive equation suitable for finite-element analysis provided incompressibility is maintained. If experimental data are obtained suitable for (15.8a) then the viscosity μ can be calculated directly, knowing the equivalent strain rate. Alternatively, if (15.8b) is available (Hamilton 1991), then $\bar{\sigma}$ may have to be found iteratively.

For the titanium alloy Ti-6Al-4V, uniaxial experimental results have been obtained as a series of log σ – log $\dot{\epsilon}$ curves for various grain sizes (Ghosh and Hamilton 1979). To accommodate these relations in the constitutive equation (15.13) requires the assumption that the stress and strain rate can be replaced by von Mises equivalent quantities. Additionally it is convenient, but not necessary, to employ a restricted range for which the strain-rate sensitivity index m is approaching a maximum value and for which the logarithmic relation between $\bar{\sigma}$ and $\dot{\bar{\epsilon}}$ is approximately linear (see Fig. 15.7). Consequently (15.8a) becomes

$$\bar{\sigma} = K(g)\dot{\bar{\epsilon}}^{m(g)} \qquad (15.14)$$

where $m = \partial \log \bar{\sigma} / \partial \log \dot{\bar{\epsilon}}$.

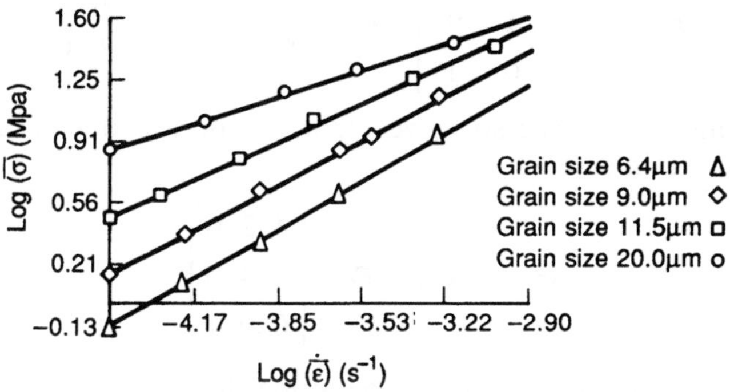

Fig. 15.7. Ti-6Al-4V linearized log σ–log $\dot{\epsilon}$ relations for different grain sizes.

The evolution of the grain size is based on the experimental evidence given in Ghosh and Hamilton (1979) as

$$\frac{g}{g_0} = \left(\frac{t}{600}\right)^N, \quad t \geqslant 600 \text{ s} \qquad (15.15)$$

where g_0 is the initial grain size for $t < 600$ s and $N = 1.8(\dot{\bar{\epsilon}} + 5 \times 10^{-5})^{0.237}$. Intermediate values of K and m can be interpolated from the information given in Fig. 15.7. From (15.13) and (15.14) the nonlinear viscosity can be found explicitly as

$$\mu = \tfrac{1}{3} K(g)\, \dot{\bar{\epsilon}}^{m(g)-1} \qquad (15.16)$$

For the particular case of an incompressible plane-stress membrane the constitutive equations are

$$\boldsymbol{\sigma}' = \mu \begin{bmatrix} 2 & 0 & 0 \\ 0 & 2 & 0 \\ 0 & 0 & 1 \end{bmatrix} \dot{\boldsymbol{\epsilon}} \qquad (15.17)$$

where the deviatoric stress vector $\boldsymbol{\sigma}'$ is given in terms of the Cauchy stress vector $\boldsymbol{\sigma}$ (see Fig. 15.8), as

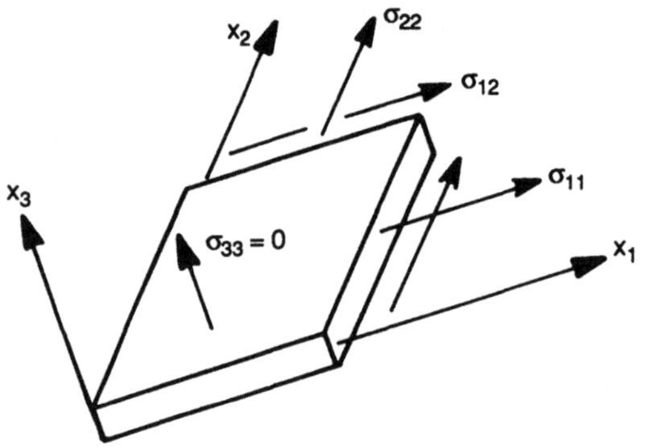

Fig. 15.8. Local Cauchy stresses.

$$\boldsymbol{\sigma}' = \begin{bmatrix} \frac{2}{3} & -\frac{1}{3} & 0 \\ -\frac{1}{3} & \frac{2}{3} & 0 \\ 0 & 0 & 1 \end{bmatrix} \boldsymbol{\sigma}, \quad \begin{aligned} \boldsymbol{\sigma}' &= [\sigma'_{11} \ \sigma'_{22} \ \sigma'_{12}]^{\mathrm{T}} \\ \boldsymbol{\sigma} &= [\sigma_{11} \ \sigma_{22} \ \sigma_{12}]^{\mathrm{T}} \end{aligned} \qquad (15.18\text{a,b,c})$$

From (15.17) and (15.18) the stress vector $\boldsymbol{\sigma}$ can be found as a function of the viscosity μ and strain rate $\dot{\boldsymbol{\epsilon}}$:

$$\boldsymbol{\sigma} = \mu \mathbf{D} \dot{\boldsymbol{\epsilon}}, \quad \mathbf{D} = \begin{bmatrix} 4 & 2 & 0 \\ 2 & 4 & 0 \\ 0 & 0 & 1 \end{bmatrix} \qquad (15.19)$$

Incompressibility is simply imposed by requiring that

$$hA^e = V_0^e \qquad (15.20)$$

where h and A^e are current element thickness and area respectively and V_0^e is the initial element volume.

15.2.3 Governing Virtual Work Equation

Although the sheet is deforming in real time the accelerations are negligible and inertia effects can be ignored in the equations of motion. The sheet is then in a state of quasistatic equilibrium under the action of the external pressure $\mathbf{p}$ and the internal stresses $\boldsymbol{\sigma}$. Equilibrium can be expressed in terms of the principle of virtual work as

$$\int_V \boldsymbol{\sigma}^{\mathrm{T}} \, \delta\dot{\boldsymbol{\epsilon}} \, dV - \int_A \mathbf{p}^{\mathrm{T}} \delta \mathbf{v} dA = 0 \qquad (15.21)$$

The virtual velocity field $\delta\mathbf{v}$ must satisfy compatibility and the kinematic boundary conditions. Anticipating that these integrals will eventually apply to a faceted membrane the virtual work equation can be recast in terms of the local virtual velocities $\delta\mathbf{v} = [\delta v_1, \ \delta v_2, \ \delta v_3]^{\mathrm{T}}$, which in accordance with membrane theory are constant through the sheet thickness. The virtual work equation is now rewritten in the standard manner as

$$\int_A \boldsymbol{\sigma}_m^{\mathrm{T}} \, \delta\dot{\boldsymbol{\epsilon}} dA - \int_A p \delta v_3 dA = 0 \qquad (15.22)$$

where the virtual strain-rate vector $\delta\dot{\boldsymbol{\epsilon}}$ is obtained from (15.5) by replacing $\boldsymbol{v}$ with $\delta\boldsymbol{v}$ and the membrane stress resultant vector, over the thickness h, is

$$\boldsymbol{\sigma}_{\mathrm{m}} = \begin{Bmatrix} N_{xx} \\ N_{yy} \\ N_{xy} \end{Bmatrix} = \int_{-h/2}^{h/2} \boldsymbol{\sigma}\, dx_3 \qquad (15.23)$$

15.3 Finite-Element Discretization

15.3.1 Equilibrium Equations

The sheet is discretized using the simple constant membrane stress triangle for which the local geometry $\boldsymbol{x}$ and velocity $\boldsymbol{v}$ are interpolated using shape functions N_a as

$$\boldsymbol{x} = \sum_{a=1}^{3} N_a \boldsymbol{x}_a, \qquad \boldsymbol{v} = \sum_{a=1}^{3} N_a \boldsymbol{v}_a \qquad (15.24\text{a,b})$$

where $\boldsymbol{x}_a$ and $\boldsymbol{v}_a$ are the local nodal positions and velocities respectively and N_a are functions of the local coordinates $\boldsymbol{x}$. The strain rate vector $\dot{\boldsymbol{\epsilon}}$ can be established in the usual manner as

$$\dot{\boldsymbol{\epsilon}} = \mathbf{B}\mathbf{v}^e \qquad (15.25\text{a})$$

where the strain-rate matrix is

$$\mathbf{B} = [\mathbf{B}_1, \mathbf{B}_2, \mathbf{B}_3], \qquad \mathbf{B}_a = \begin{bmatrix} \dfrac{\partial N_a}{\partial x_1} & 0 \\ 0 & \dfrac{\partial N_a}{\partial x_2} \\ \dfrac{\partial N_a}{\partial x_2} & \dfrac{\partial N_a}{\partial x_1} \end{bmatrix} \mathbf{R} \qquad (15.25\text{b,c})$$

and $\mathbf{v}^e$ is the *global* nodal velocity vector. The local gradient of the global velocities, $\boldsymbol{\theta}$, is interpolated in terms of the global nodal velocities $\mathbf{v}^e$ as

$$\boldsymbol{\theta} = \begin{Bmatrix} \boldsymbol{\theta}_1 \\ \boldsymbol{\theta}_2 \end{Bmatrix} = \mathbf{G}\mathbf{v}^e \qquad (15.26\text{a})$$

where the gradient matrix is

$$\mathbf{G} = [\mathbf{G}_1, \mathbf{G}_2, \mathbf{G}_3], \qquad \mathbf{G}_a = \begin{bmatrix} \dfrac{\partial N_a}{\partial x_1} & \mathbf{I}_{3\times 3} \\ \dfrac{\partial N_a}{\partial x_2} & \mathbf{I}_{3\times 3} \end{bmatrix} \qquad (15.26\text{b,c})$$

The *element* contribution to the virtual work equation can now be written in the global coordinate system as

$$(\delta\mathbf{v}^e)^{\mathrm{T}}\,[\mathbf{T}^e - \mathbf{F}^e] = 0 \qquad (15.27)$$

where the element internal nodal force vector $\mathbf{T}_a^e$ and the nodal equivalent external force vector $\mathbf{F}_a^e$ for node a are

$$\mathbf{T}_a^e = \int_{A^e} \mathbf{B}_a^{\mathrm{T}} \boldsymbol{\sigma}_m \, dA, \quad \mathbf{F}_a^e = \tfrac{1}{3} p A^e \mathbf{n}^e \tag{15.28a,b}$$

and the element normal vector $\mathbf{n}^e = \mathbf{e}_3/|\mathbf{e}_3|$.

The virtual work equation is the summation of (15.27) over all n_{el} elements which yields

$$\delta\mathbf{v}^{\mathrm{T}} \sum_{e=1}^{n_{el}} (\mathbf{T}^e - \mathbf{F}^e) = 0 \tag{15.29}$$

where the summation represents the standard assembly procedure and $\delta\mathbf{v}$ is the vector of all global nodal velocities. The arbitrary nature of these virtual velocities allows the discretized equilibrium equations finally to be written as

$$\mathbf{T}(\mathbf{x},\mathbf{v}) - p(t)\,\mathbf{F}(\mathbf{x}) = 0 \tag{15.30}$$

where $\mathbf{F}$ is now the assembled global external force vector *per unit pressure*; $\mathbf{T}$ is the assembled vector of internal nodal forces and $\mathbf{x}$ and $\mathbf{v}$ are the vectors of nodal positions and velocities respectively.

The internal force term $\mathbf{T}$ is a function of the membrane stress resultants $\boldsymbol{\sigma}_m$; these in turn, via the constitutive equations, are nonlinear functions of the strain-rate vector $\dot{\boldsymbol{\epsilon}}$. Furthermore, the strain rates are functions of the nodal velocities $\mathbf{v}$ and the shape function derivatives which are themselves functions of the geometry $\mathbf{x}$. In addition, $\mathbf{F}$ is obviously a function of the changing geometry. In short: the equilibrium equations are highly nonlinear with respect to both velocity and geometry.

15.3.2 Strain-Rate Constraint Equations

In the assembled equilibrium equation (15.30) the pressure parameter $p(t)$ may be prescribed or calculated. If the latter is the case then the equilibrium equations must be augmented by a (nonlinear) scalar equation, to be designated the *strain-rate constraint equation*, which enables $p(t)$ to be calculated. This additional equation is provided by requiring some function of the maximum equivalent strain rate to equal a target value $\dot{\bar{\epsilon}}_0$ judged to be optimum for SPF. Here the maximum strain rate will simply be taken as that occurring in a single element which is a function of $\mathbf{x}^e$ and $\mathbf{v}^e$:

$$L(\mathbf{x}^e, \mathbf{v}^e) = \dot{\bar{\epsilon}}_0 \tag{15.31}$$

Allowing for incompressibility, it can be shown that, for a single element,

$$L(\mathbf{x}^e,\mathbf{v}^e) = \left(\tfrac{1}{3}\,\dot{\boldsymbol{\epsilon}}^{\mathrm{T}}\mathbf{D}\dot{\boldsymbol{\epsilon}}\right)^{1/2} \tag{15.32a}$$

or

$$L(\mathbf{x}^e,\mathbf{v}^e) = \left(\tfrac{1}{3}(\mathbf{v}^e)^{\mathrm{T}}\mathbf{B}^{\mathrm{T}}\,\mathbf{D}\mathbf{B}\mathbf{v}^e\right)^{1/2} \tag{15.32b}$$

where it should be noted that $\mathbf{D}$ does *not* contain any constitutive information, see (15.19). Also observe that through the local coordinates $\boldsymbol{x}$ the strain matrix $\mathbf{B}$ is a function of the geometry $\mathbf{x}$.

15.3.3 Variation of the Equilibrium and Strain-Rate Constraint Equations

Having observed the nonlinear nature of the equilibrium equations it is worthwhile to pre-empt some aspects of the solution procedure and consider the variation of the equilibrium equations and the strain-rate constraint equation with respect to geometry $\mathbf{x}$ and velocity $\mathbf{v}$. Such variations result from the linearization of the residual force vector $\mathbf{R} = p\mathbf{F} - \mathbf{T}$ as

$$\mathbf{R}(\mathbf{x} + \Delta\mathbf{x}, \mathbf{v} + \Delta\mathbf{v}, p + \Delta p) \approx \mathbf{R}(\mathbf{x}, \mathbf{v}, p) - \mathbf{K}\,\Delta\mathbf{x} - \mathbf{C}\Delta\mathbf{v} + \Delta p\mathbf{F} \tag{15.33}$$

and the linearization of the strain-rate constraint error $s = \dot{\epsilon}_0 - L$ as

$$s(\mathbf{x} + \Delta\mathbf{x}, \mathbf{v} + \Delta\mathbf{v}) \approx s(\mathbf{x}, \mathbf{v}) - \mathbf{L}_x^T\,\Delta\mathbf{x} - \mathbf{L}_v^T\,\Delta\mathbf{v} \tag{15.34}$$

The tangent stiffness matrix and tangent viscous matrix, $\mathbf{K}$ and $\mathbf{C}$ respectively, are given by the derivatives of $\mathbf{R}$ with respect to $\mathbf{x}$ and $\mathbf{v}$ as

$$\mathbf{K} = \frac{\partial \mathbf{T}}{\partial \mathbf{x}} - p\frac{\partial \mathbf{F}}{\partial \mathbf{x}}; \quad \mathbf{C} = \frac{\partial \mathbf{T}}{\partial \mathbf{v}} \tag{15.35a,b}$$

and the geometry and velocity-dependent gradient vectors $\mathbf{L}_x$ and $\mathbf{L}_v$ respectively are

$$\mathbf{L}_x = \frac{\partial L}{\partial \mathbf{x}}; \quad \mathbf{L}_v = \frac{\partial L}{\partial \mathbf{v}} \tag{15.36a,b}$$

In order to understand the complexity of the fully developed tangent matrix it is necessary to examine the equilibrium equation $\mathbf{R} = \mathbf{0}$ in a little more detail. From (15.30) it is clear that the internal forces $\mathbf{T}$ are a function of the geometry $\mathbf{x}$ and, via the stress $\boldsymbol{\sigma}_m$ a function of geometry $\mathbf{x}$ and velocity $\mathbf{v}$. The equilibrium equations can be rewritten showing these dependences as

$$\mathbf{T}(\boldsymbol{\sigma}[\mu\,(\mathbf{x}, \mathbf{v}), \mathbf{v}, \mathbf{x}], \mathbf{x}) - p(t)\,\mathbf{F}(\mathbf{x}) = \mathbf{0} \tag{15.37}$$

Consequently the tangent stiffness matrix is properly written from (15.30) as

$$\mathbf{K} = \mathbf{K}_{\mu x} + \mathbf{K}_x + \mathbf{K}_\sigma - p\mathbf{K}_f \tag{15.38}$$

These various components of the tangent stiffness matrix are rather complicated due, primarily, to the dependence of the shape function derivatives, e.g. $\mathbf{G}$ in (15.26), upon the geometry vector $\mathbf{x}$. In the following development $\mathbf{K}_x$, $\mathbf{K}_f$ and $\mathbf{K}_{\mu x}$ will be ignored and only $\mathbf{K}_\sigma$ will be retained; this is expressed symbolically as

$$\mathbf{K}_\sigma = \left.\frac{\partial \mathbf{T}}{\partial \mathbf{x}}\right|_{\boldsymbol{\sigma}, \mathbf{G} = \text{constant}} \tag{15.39}$$

Observe that $\mathbf{K}_\sigma$ is the *initial stress matrix* and $\mathbf{K}_f$ is the *load correction matrix* familiar in geometrically nonlinear analysis. The tangent viscous matrix is

$$\mathbf{C} = \mathbf{C}_s + \mathbf{C}_\mu \tag{15.40a}$$

where

$$\mathbf{C}_s = \left.\frac{\partial \mathbf{T}}{\partial \boldsymbol{\sigma}}\frac{\partial \boldsymbol{\sigma}}{\partial \mathbf{v}}\right|_{\mu = \text{constant}} \qquad \mathbf{C}_\mu = \frac{\partial \mathbf{T}}{\partial \boldsymbol{\sigma}}\frac{\partial \boldsymbol{\sigma}}{\partial \mu}\frac{\partial \mu}{\partial \mathbf{v}} \tag{15.40b,c}$$

The viscous secant matrix $\mathbf{C}_s$ has been used extensively in explicit time-stepping schemes (Zienkiewicz 1984).

Due to the simplicity of the three-node linear element the components of the tangent matrix can be found explicitly. From (15.19), (15.23) and (15.25) the membrane stress resultant can be written in terms of the nodal velocities as

$$\boldsymbol{\sigma}_m = \mu h \mathbf{D}\mathbf{B}\mathbf{v}^e \tag{15.41}$$

The discretized element internal nodal forces $\mathbf{T}^e$ can be written explicitly from (15.28) as

$$\mathbf{T}^e = \mathbf{B}^{\mathrm{T}} \boldsymbol{\sigma}_m A^e \tag{15.42}$$

from which the global viscous secant matrix (15.40a) is easily constructed as

$$\mathbf{C}_s^e = \mu h\, \mathbf{B}^{\mathrm{T}}\mathbf{D}\mathbf{B}\, A^e \tag{15.43}$$

To enable the viscous tangent matrix component $\mathbf{C}_\mu^e$ to be found it is necessary to rewrite (15.40c) as

$$\mathbf{C}_\mu^e = \frac{\partial \mathbf{T}^e}{\partial \mu}\frac{\partial \mu}{\partial \mathbf{v}^e} = \frac{\partial \mathbf{T}^e}{\partial \mu}\frac{\partial \mu}{\partial \dot{\bar{\epsilon}}}\frac{\partial \dot{\bar{\epsilon}}}{\partial \dot{\boldsymbol{\epsilon}}}\frac{\partial \dot{\boldsymbol{\epsilon}}}{\partial \mathbf{v}^e} \tag{15.44}$$

Eqs (15.43), (15.16), (15.32) and (15.25) respectively yield

$$\frac{\partial \mathbf{T}^e}{\partial \mu} = \mathbf{B}^{\mathrm{T}}\frac{\boldsymbol{\sigma}_m}{\mu}A^e, \quad \frac{\partial \mu}{\partial \dot{\bar{\epsilon}}} = \frac{\mu(m-1)}{\dot{\bar{\epsilon}}} \tag{15.45a,b}$$

$$\frac{\partial \dot{\bar{\epsilon}}}{\partial \dot{\boldsymbol{\epsilon}}} = \frac{\boldsymbol{\sigma}_m^{\mathrm{T}}}{3\mu h \dot{\bar{\epsilon}}}, \quad \frac{\partial \dot{\boldsymbol{\epsilon}}}{\partial \mathbf{v}^e} = \mathbf{B} \tag{15.46a,b}$$

from which

$$\mathbf{C}_\mu^e = \frac{(m-1)}{3\mu h \dot{\bar{\epsilon}}^2}\mathbf{B}^{\mathrm{T}}\hat{\boldsymbol{\sigma}}_m \mathbf{B}\, A^e \tag{15.47}$$

where the stress resultant matrix $\hat{\boldsymbol{\sigma}}_m = \boldsymbol{\sigma}_m\boldsymbol{\sigma}_m^{\mathrm{T}}$.

The initial stress matrix $\mathbf{K}_\sigma$ arises in all geometrically nonlinear problems. It describes the variation of the internal forces with changes of nodal position at constant membrane stress, constant local shape function derivatives and constant volume. From (15.28) the global nodal internal forces are

$$\mathbf{T}_a^e = \mathbf{B}_a^{\mathrm{T}}\boldsymbol{\sigma}_m A^e = \mathbf{R}^{\mathrm{T}}\begin{bmatrix} \dfrac{\partial N_a}{\partial x_1} & 0 & \dfrac{\partial N_a}{\partial x_2} \\ 0 & \dfrac{\partial N_a}{\partial x_2} & \dfrac{\partial N_a}{\partial x_1} \end{bmatrix}\boldsymbol{\sigma}_m A^e \tag{15.48}$$

where the local/global transformation matrix (15.4) is

$$\mathbf{R}^{\mathrm{T}} = \frac{\partial \mathbf{x}}{\partial \boldsymbol{x}}, \quad \frac{\partial \mathrm{x}_i}{\partial x_j} = \sum_{b=1}^{3}\frac{\partial N_b}{\partial x_j}\mathrm{x}_b^i \qquad \begin{matrix} i = 1,2,3 \\ j = 1,2 \end{matrix} \tag{15.49}$$

The internal force for the ith degree of freedom of node a can be concisely written as

$$T_a^i = \frac{\partial \mathrm{x}_i}{\partial \boldsymbol{x}}\bar{\boldsymbol{\sigma}}_m\left(\frac{\partial N_a}{\partial \boldsymbol{x}}\right)^{\mathrm{T}} A^e \qquad \begin{matrix} i,a = 1,2,3 \\ \boldsymbol{x} = [x_1, x_2]^{\mathrm{T}} \end{matrix} \tag{15.50}$$

where the stress resultant matrix

$$\bar{\boldsymbol{\sigma}}_m = \begin{bmatrix} N_{zz} & N_{xy} \\ N_{xy} & N_{yy} \end{bmatrix} \tag{15.51}$$

To obtain the components of the initial stress matrix it is assumed that, although the nodal positions x_b^j are allowed to vary, the stresses $\bar{\boldsymbol{\sigma}}_m$ and the coefficients of $\mathbf{G}$ (15.26) remain constant thus giving

$$K_{\sigma,ab}^{ij} = \left.\frac{\partial T_a^i}{\partial x_b^j}\right|_{\mathbf{G},\,\boldsymbol{\sigma}_m = \text{constant}} = \delta_{ij}\frac{\partial N_b}{\partial \boldsymbol{x}}\bar{\boldsymbol{\sigma}}_m\left(\frac{\partial N_a}{\partial \boldsymbol{x}}\right)^{\mathrm{T}} A^e \tag{15.52}$$

Upon expansion the initial stress matrix can be written in a clearer, but computationally more expensive form, using (15.26) as

$$\mathbf{K}_\sigma = \mathbf{G}^{\mathrm{T}}\boldsymbol{\sigma}_I\mathbf{G}\,A^e \tag{15.53}$$

where the stress resultant matrix

$$\boldsymbol{\sigma}_I = \begin{bmatrix} N_{xx}\mathbf{I}_{3\times 3} & N_{xy}\mathbf{I}_{3\times 3} \\ N_{xy}\mathbf{I}_{3\times 3} & N_{yy}\mathbf{I}_{3\times 3} \end{bmatrix} \tag{15.54}$$

Finally in this section the velocity and geometry dependent gradient vectors $\mathbf{L}_v$ and $\mathbf{L}_x$ are formulated. From (15.36b),

$$\mathbf{L}_v = \frac{\partial L}{\partial \mathbf{v}^e} = \frac{\partial \dot{\bar{\epsilon}}}{\partial \mathbf{v}^e} = \frac{\partial \dot{\bar{\epsilon}}}{\partial \dot{\boldsymbol{\epsilon}}}\frac{\partial \dot{\boldsymbol{\epsilon}}}{\partial \mathbf{v}^e} \tag{15.55}$$

which is now easily evaluated using (15.46) as

$$\mathbf{L}_v = \frac{\boldsymbol{\sigma}_m^{\mathrm{T}}}{3\mu\dot{\bar{\epsilon}}}\mathbf{B} \tag{15.56}$$

From (15.36a)

$$\mathbf{L}_x = \frac{\partial L}{\partial \mathbf{x}^e} = \frac{\partial \dot{\bar{\epsilon}}}{\partial \mathbf{x}^e} = \frac{\partial \dot{\bar{\epsilon}}}{\partial \dot{\boldsymbol{\epsilon}}}\frac{\partial \dot{\boldsymbol{\epsilon}}}{\partial \mathbf{x}^e} \tag{15.57}$$

The final term in this expression, to be denoted as $\mathbf{B}_x$, depends upon the variation of the local velocity gradients with respect to changes in the element nodal positions $\mathbf{x}^e$. It can be shown, that for the global ith degree of freedom of node a,

$$\frac{\partial}{\partial \mathbf{x}_i^a}\left(\frac{\partial \boldsymbol{v}}{\partial \boldsymbol{x}}\right)^{\mathrm{T}} = \left(\frac{\partial v_i}{\partial \boldsymbol{x}}\right)^{\mathrm{T}}\left(\frac{\partial N_a}{\partial \boldsymbol{x}}\right) \qquad \begin{matrix} i,a = 1,2,3 \\ \boldsymbol{x} = [x_1, x_2]^{\mathrm{T}} \end{matrix} \tag{15.58}$$

where the *local* velocity $\boldsymbol{v} = [v_1, v_2]^{\mathrm{T}}$.

All the terms in $\mathbf{B}_x$ are now available and the gradient $\mathbf{L}_x$ becomes

$$\mathbf{L}_x = \frac{\boldsymbol{\sigma}_m^{\mathrm{T}}}{3\mu\dot{\bar{\epsilon}}}\mathbf{B}_x, \quad \mathbf{B}_x = \mathbf{A}_Q\mathbf{G} \tag{15.59a,b}$$

where $\mathbf{A}_Q$ is a matrix of derivatives of the global velocities with respect to the local coordinate axes which, using (15.26) becomes

$$\mathbf{A}_Q = \begin{bmatrix} \boldsymbol{\theta}_1^{\mathrm{T}} & \mathbf{0} \\ \mathbf{0} & \boldsymbol{\theta}_2^{\mathrm{T}} \\ \boldsymbol{\theta}_2^{\mathrm{T}} & \boldsymbol{\theta}_1^{\mathrm{T}} \end{bmatrix} \tag{15.59c}$$

Having found the variation of the internal nodal forces and the strain-rate constraint with respect to the velocity and geometry changes the scene is now set for an examination of the time-stepping scheme and the nonlinear iteration technique which forms the solution procedure.

15.4 Solution Procedures

15.4.1 Time Integration Schemes

By now it is clear that the governing equilibrium equations constitute a system of nonlinear differential equations for the time history of the nodal positions. To obtain the shape history the velocities must be integrated numerically using a suitable time integration or time-stepping scheme. The relative merits of explicit and implicit schemes were discussed in Sect. 15.1.2.2 where it was explained that it is essential to use an implicit scheme. Introducing an implicit scheme enables the unknown geometry to be expressed in terms of the unknown sheet velocities and the differential equations are transformed into a set of nonlinear algebraic equations for the nodal velocities. These equations have to be solved at a sequence of time steps with the integration scheme providing the geometry update (see Fig. 15.9).

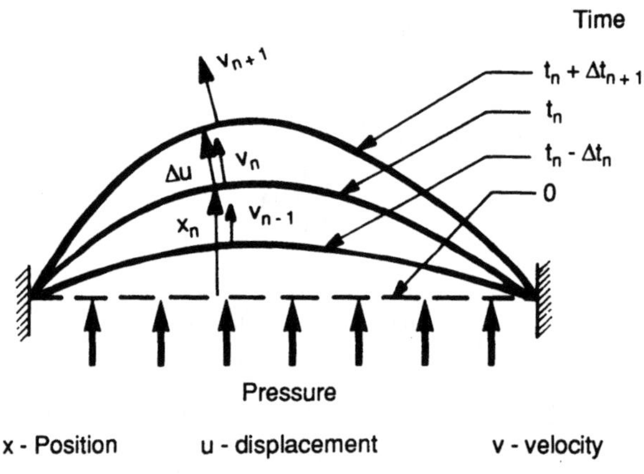

Fig. 15.9. Implicit time integration scheme – solution sequence.

In general the implicit time integration gives the nodal positions $\mathbf{x}_{n+1}$ at time t_{n+1} in terms of previous known positions and unknown velocities $\mathbf{v}_{n+1}$ as

$$\mathbf{x}_{n+1} = \mathbf{I}(\mathbf{v}_{n+1}) \tag{15.60a}$$

where

$$\mathbf{I}(\mathbf{v}_{n+1}) = \begin{cases} \mathbf{x}_n + \frac{1}{2}\Delta t_{n+1}\,\mathbf{v}_n + \frac{1}{2}\Delta t_{n+1}\,\mathbf{v}_{n+1} & \text{(Trapezoidal rule)} \\ \mathbf{x}_n + \Delta t_{n+1}\,\mathbf{v}_{n+1} & \text{(Backward Euler)} \\ \mathbf{x}_n + (1-\gamma)\dfrac{\mathbf{x}_n - \mathbf{x}_{n-1}}{\Delta t_n}\Delta t_{n+1} + \gamma \Delta t_{n+1}\,\mathbf{v}_{n+1} & \text{(Two steps back.-diff.)} \end{cases} \tag{15.60b}$$

and the parameter γ for the two steps backward-difference scheme is derived from a quadratic variation of $\mathbf{x}$ with time to give

$$\gamma = \frac{\Delta t_{n+1} + \Delta t_n}{2\Delta t_{n+1} + \Delta t_n} \tag{15.60c}$$

Experience has shown that the two steps backward-difference scheme is currently the most reliable (Bonet et al. 1988c). Combining an integration scheme with the equilibrium equations (15.30) and the strain rate constraint equation (15.31) gives a system of equations solely in terms of the unknown velocities $\mathbf{v}_{n+1}$ at time t_{n+1} as

$$\mathbf{T}(\mathbf{I}(\mathbf{v}_{n+1}), \mathbf{v}_{n+1}) - p(t)\,\mathbf{F}(\mathbf{I}(\mathbf{v}_{n+1})) = \mathbf{0} \tag{15.61a}$$

$$L(\mathbf{I}(\mathbf{v}_{n+1}), \mathbf{v}_{n+1}) - \dot{\bar{\epsilon}}_0 = 0 \tag{15.61b}$$

15.4.2 Predictor–Corrector Procedure

For the trapezoidal and backward Euler schemes the prediction of pressure, velocity and geometry are

$$p^0_{n+1} = p_n \tag{15.62a}$$

$$\mathbf{v}^0_{n+1} = \mathbf{v}_n \tag{15.62b}$$

$$\mathbf{x}^0_{n+1} = \mathbf{x}_n + \Delta t_{n+1}\,\mathbf{v}_n \tag{15.62c}$$

where the superscripts refer to nonlinear iterations and subscripts to time steps. If the two steps backward-difference is used then additional information from time step t_{n-1} is available to enhance the accuracy of the predictor which, using a quadratic extrapolation for $\mathbf{v}^0_{n+1}$ and $\mathbf{x}^0_{n+1}$ gives

$$\mathbf{v}^0_{n+1} = \eta\mathbf{v}_n + (1-\eta)\frac{\mathbf{x}_n - \mathbf{x}_{n-1}}{\Delta t_n}; \qquad \eta = \frac{2\Delta t_{n+1} + \Delta t_n}{\Delta t_n} \tag{15.63a}$$

$$\mathbf{x}^0_{n+1} = \mathbf{x}_n + \Delta t_{n+1}\,\mathbf{v}_n + \Delta t^2_{n+1}\frac{\mathbf{v}_n - (\mathbf{x}_n - \mathbf{x}_{n-1})/\Delta t_n}{\Delta t_n} \tag{15.63b}$$

These predicted values are unlikely to satisfy the governing equations and will have to be corrected using the Newton–Raphson iteration given by (15.33) and (15.34), where the increments in displacement $\Delta\mathbf{x}$ and velocity $\Delta\mathbf{v}$ are related via (15.60) as

$$\Delta\mathbf{x} = \frac{\partial\mathbf{x}}{\partial\mathbf{v}}\Delta\mathbf{v} = \theta\Delta\mathbf{v}, \qquad \theta = \begin{cases} \tfrac{1}{2}\Delta t_{n+1} & \text{(Trapezoidal rule)} \\ \Delta t_{n+1} & \text{(Backward Euler)} \\ \gamma\Delta t_{n+1} & \text{(Two steps back.-diff.)} \end{cases} \tag{15.64}$$

The Newton–Raphson procedure is now expressed as

$$\mathbf{KT}(\mathbf{v}^k_{n+1})\,\Delta\mathbf{v}^k_{n+1} = \mathbf{R}(\mathbf{v}^k_{n+1}) + \Delta p^k_{n+1}\,\mathbf{F}(\mathbf{v}^k_{n+1}) \tag{15.65a}$$

$$\mathbf{L}(\mathbf{v}^k_{n+1})^{\mathrm{T}}\,\Delta\mathbf{v}^k_{n+1} = s(\mathbf{v}^k_{n+1}) \tag{15.65b}$$

where the tangent matrix

$$\mathbf{KT}\,(\mathbf{v}_{n+1}^k) \approx \mathbf{C}_s\,(\mathbf{v}_{n+1}^k) + \mathbf{C}_\mu\,(\mathbf{v}_{n+1}^k) + \theta\mathbf{K}_\sigma\,(\mathbf{v}_{n+1}^k) \tag{15.65c}$$

and the strain-rate constraint gradient vector

$$\mathbf{L}(\mathbf{v}_{n+1}^k) = \mathbf{L}_v + \theta\mathbf{L}_x \tag{15.65d}$$

The solution of system (15.65a,b) can be efficiently achieved in two stages as follows: from (15.65a) the increment of velocity emerges as

$$\Delta\mathbf{v}_{n+1}^k = \mathbf{KT}(\mathbf{v}_{n+1}^k)^{-1}\,[\mathbf{R}(\mathbf{v}_{n+1}^k) + \Delta p_{n+1}^k\,\mathbf{F}(\mathbf{v}_{n+1}^k)] \tag{15.66}$$

which, upon substitution into Eq. (15.65b), gives the pressure increment as

$$\Delta p_{n+1}^k = \frac{s(\mathbf{v}_{n+1}^k) - \mathbf{L}(\mathbf{v}_{n+1}^k)^{\mathrm{T}}\,\mathbf{KT}\,(\mathbf{v}_{n+1}^k)^{-1}\,\mathbf{R}(\mathbf{v}_{n+1}^k)}{\mathbf{L}(\mathbf{v}_{n+1}^k)^{\mathrm{T}}\,\mathbf{KT}(\mathbf{v}_{n+1}^k)^{-1}\,\mathbf{F}(\mathbf{v}_{n+1}^k)} \tag{15.67}$$

Finally, Eqs (15.66) and (15.67) enable the pressure, velocities and nodal positions to be updated as

$$\mathbf{v}_{n+1}^{k+1} = \mathbf{v}_{n+1}^k + \Delta\mathbf{v}_{n+1}^k \tag{15.68a}$$

$$\mathbf{x}_{n+1}^{k+1} = \mathbf{I}(\mathbf{v}_{n+1}^{k+1}) \tag{15.68b}$$

$$p_{n+1}^{k+1} = p_{n+1}^k + \Delta p_{n+1}^k \tag{15.68c}$$

The corrector stage continues until convergence is obtained and then the process returns to the predictor stage for the next time step.

15.4.3 Starting Procedure

At time $t = 0$ the sheet is flat and the nodal velocities and element stresses are zero; consequently the tangent stiffness cannot be evaluated. If an artificial viscosity were introduced as a starting "ploy" the tangent stiffness matrix would be singular since the membrane sheet would, in the absence of bending, be a mechanism. To overcome the starting problem an artificial initial geometry is found by introducing a fictitious constant uniform tensile stress σ_0 which enables the initial stress component of **KT** to be found, thereby endowing the membrane with an out-of-plane stiffness. By constraining the displacements to be in the x_3 direction only, a system of linear equations is obtained from which an initial geometry and thence initial velocity can be found (Wood et al. 1990).

15.5 Applications

15.5.1 Introduction

In this section two examples are presented to demonstrate the capabilities of the numerical procedures. The truncated cone problem is included to show a correlation with experimental data and the deep square box illustrates a more complex case. Further examples are given in Wood et al. (1989).

15.5.2 Forming of a Truncated Cone

The first example concerns the superplastic forming of a circular sheet, initially 1 mm thick with an initial grain size of 8 μm, into a truncated conical die. The finite-element mesh used to describe the die is shown in Fig. 15.10. Fig. 15.11 shows the initial quarter mesh of 648 elements together with the final configuration which was reached in 74 min forming time.

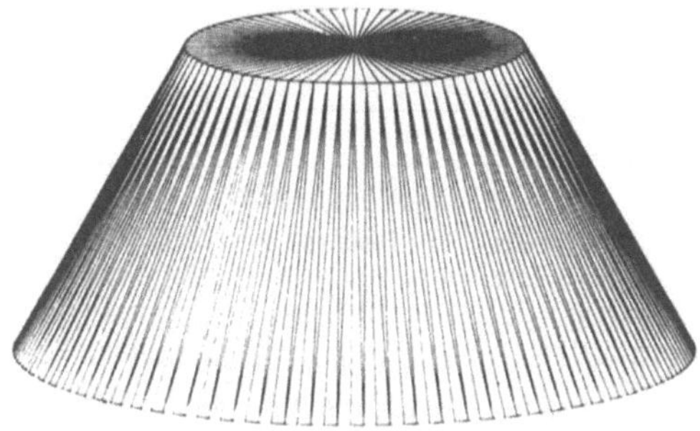

Fig. 15.10. Truncated cone: die mesh.

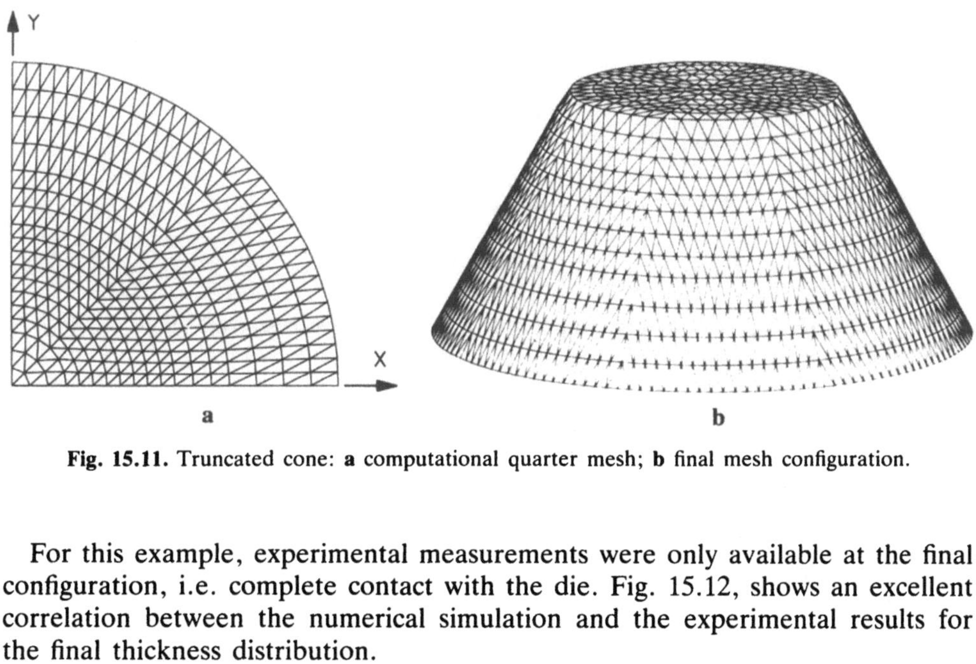

Fig. 15.11. Truncated cone: **a** computational quarter mesh; **b** final mesh configuration.

For this example, experimental measurements were only available at the final configuration, i.e. complete contact with the die. Fig. 15.12, shows an excellent correlation between the numerical simulation and the experimental results for the final thickness distribution.

15.5.3 Forming of a Deep Square Box

This example illustrates the numerical simulation of the superplastic forming of a deep square box with rounded corners. The die dimensions are defined in Fig. 15.13 and Fig. 15.14 depicts the finite-element representation of a half die. The material is titanium (Ti-6Al-4V) and an initial grain size of 10 μm was arbitrarily chosen. Given the symmetry of the problem, a quarter mesh comprising 1152

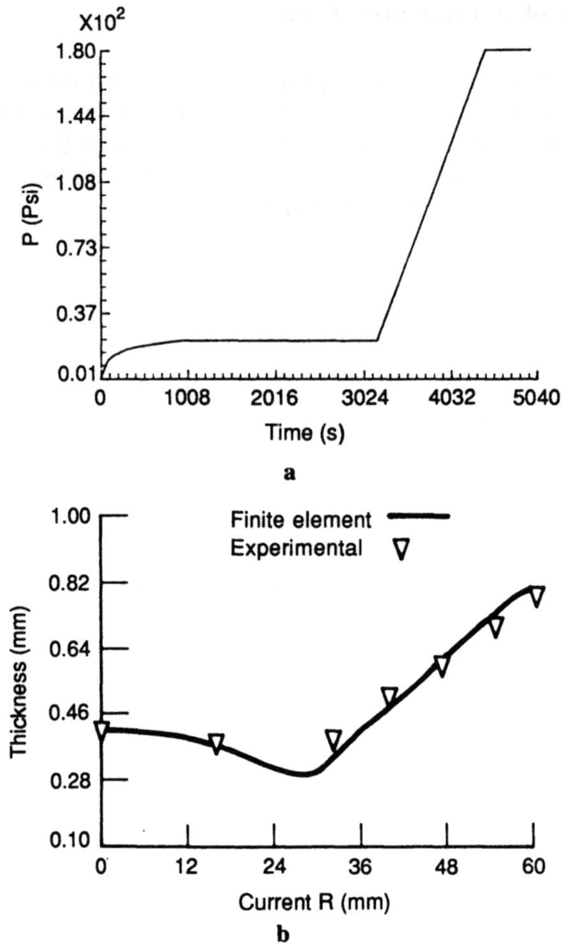

Fig. 15.12. Truncated cone: **a** pressure cycle; **b** final thickness distribution.

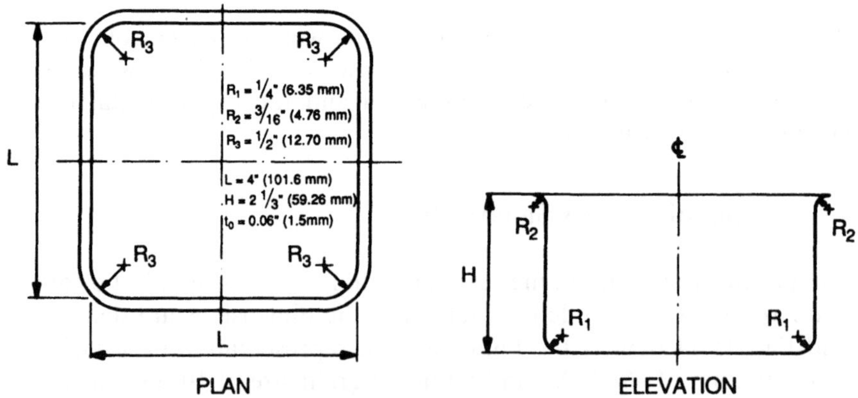

Fig. 15.13. Square box: dimensions.

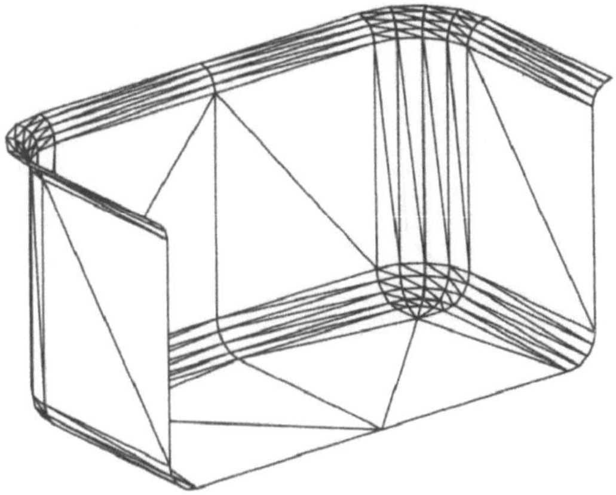

Fig. 15.14. Square box: die mesh.

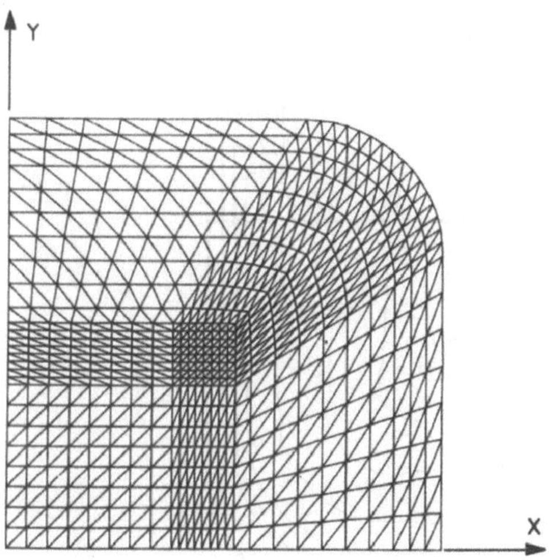

Fig. 15.15. Square box: computational mesh.

membrane elements (see Fig. 15.15) was employed to model the forming sheet. The initial thickness of the sheet was 1.5 mm.

Imposing a constant maximum equivalent strain rate of $3.0\times10^{-4}\,s^{-1}$, the pressure cycle shown in Fig. 15.16 was obtained. After an initial increase in pressure, a decrease is observed before the sheet reaches the bottom of the box. This is followed by an increase in pressure as the sheet forms into the corners of the box. This type of pressure cycle is characteristic of the forming of deep components. Additionally, Fig. 15.17 shows the thickness distribution at the end of the forming plotted at the initial configuration. Finally, deformed finite-element meshes at various times are shown in Fig. 15.18. The problem was solved in 181 time steps and 403 Newton–Raphson iterations and took 3 h 45 min on a VAX 750.

15.6 Conclusions

The numerical simulation of the superplastic forming of a thin sheet using the finite-element method is clearly a viable proposition. A simple membrane approximation appears to be adequate to model the sheet and an implicit time-

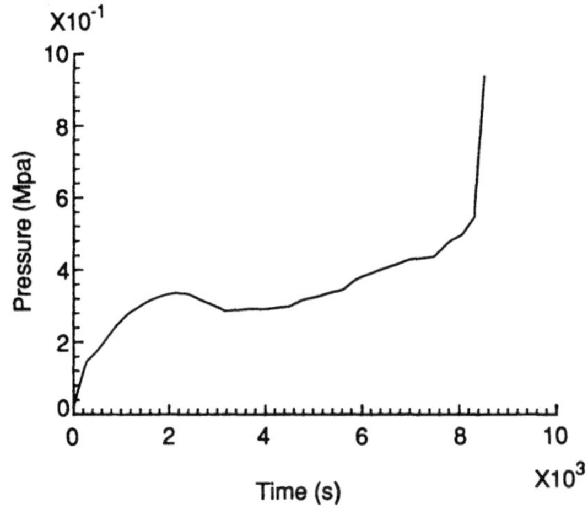

Fig. 15.16. Square box: computed pressure.

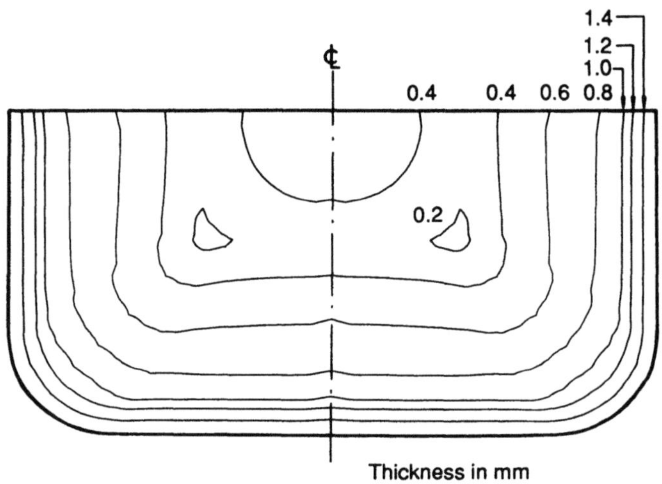

Fig. 15.17. Square box: final thickness distribution.

stepping scheme transpires to be efficient in addition to having the advantage of introducing an initial stress effect. Computational times are not unreasonable considering the complexity of the problem, but for realistic simulations supercomputing will become a necessity, especially if diffusion bonding and friction are taken into account.

Nevertheless at present, SPF simulation together with pressure cycle evaluation is a useful facility for the engineer, hopefully enabling manufacturing processes to be devised more effectively and innovative designs to be investigated without the penalty of high physical experimentation costs.

Acknowledgements The majority of the development work reported herein was funded by British Aerospace Commercial Aircraft Ltd and carried out by the Institute for Numerical Methods in Engineering in the Civil Engineering Department at Swansea University, UK. The authors gratefully acknowledge the

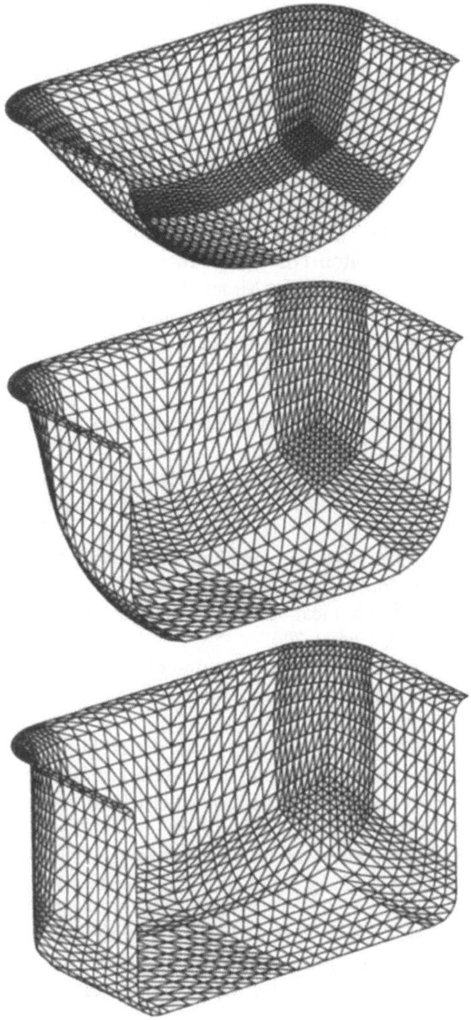

Fig. 15.18. Square box: deformed meshes at 27, 78 and 142 min.

permission given by BAe. to quote the experimental results for the truncated cone.

References

AGARD (1987) Superplasticity. Advisory Group for Aerospace Research and Development, NATO, London (AGARD-LS-154)

Argyris JH, St Doltsinis J (1984) A primer on superplasticity in natural formulation. Comput Meth Appl Mech Engng 30:83–132

Bellet M (1988) Modélisation numérique du formage superplastic de tôles. PhD Thesis, L'Ecole Nationale Superieure des Mines de Paris (in French)

Bellet M, Chenot JL (1986) Numerical modelling of 3-dimensional superplastic forming of titanium sheet. In: Proceedings of Titanium 86. Titanium Development Association, vol II, pp 1175–1184

Bellet M, Massoni E, Chenot JL (1987) A viscoplastic membrane formulation for 3-dimensional analysis of superplastic forming of thin sheet. In: Proceedings of the international conference on computational plasticity models, software and applications, part 2. Pineridge Press, Swansea, pp 917–926

Bonet J (1989) Finite element analysis of thin sheet superplastic forming. Department of Civil Engineering, University College of Swansea, C/Ph/128/1989

Bonet J, Wood RD (1987) Solution procedures for the finite element analysis of superplastic forming of thin sheet. In: Proceedings of the international conference on computational plasticity models, software and applications, part 2. Pineridge Press, Swansea, pp 927–939

Bonet J, Wood RD, Zienkiewicz OC (1988a) Finite element analysis of thin sheet superplastic forming. In: Mathematical models for metals and materials applications. Institute of Metals, London, pp 2.1–2.7

Bonet J, Wood RD, Zienkiewicz OC (1988b) Finite element modelling of the superplastic forming of thin sheet. In: Hamilton CH, Paton NE (eds) Proceedings of the international conference on superplasticity and superplastic forming. The Minerals, Metals and Materials Society, USA, p 706

Bonet J, Wood RD, Zienkiewicz OC (1988c) Time stepping schemes for the numerical analysis of superplastic forming of thin sheet. In: Chenot JL, Oñate E (eds) Modelling of Metal Forming Processes, Sophia Antipolis, (Euromech Colloquium no 233) Kluwer Academic Publishers, pp 179–186

Bonet J, Wargadipura AHS, Wood RD (1989) A pressure cycle control algorithm for superplastic forming. Comm Appl Num Meth 5(2):121–128

Chandrasekaran N, Goforth RE, Haisler WE (1985) Finite element formulation of superplastic metal forming processes. American Society for Metals, Metals/materials technology series, Materials week, Paper 8511–004, pp 1–8

Cormeau A, Cormeau I, Roose J (1984) Numerical simulation of glass blowing. In: Pittman JFT et al. (eds) Numerical analysis of forming processes. John Wiley

Ghosh AK, Hamilton CH (1979) Mechanical behaviour and hardening characteristics of a superplastic Ti-6Al-4V alloy. Metall Trans A 10:699–706

Hamilton CH (1991) Simulation of static and deformation-enhanced grain growth effects on superplastic ductility. Department of Materials Science and Engineering, Washington State University (to be published)

Hamilton CH, Paton NE (1988) Superplasticity and superplastic forming, Proceedings of an International Conference. The Minerals, Metals and Materials Society, USA

Hori S, Tokizane M, Furushiro N (eds) (1991) Superplasticity in advanced materials, Osaka (ICSAM-91), The Japan Society for Research on Superplasticity

Kobayashi S, Kim JH (1978) Deformation analysis of axisymmetric sheet metal forming processes by the rigid-plastic finite element method. In: Koinstinen et al. (eds) Mechanics of sheet metal forming. Plenum Press, New York

Massoni E, Bellet M, Abouaf A, Chenot JL (1987) Elastoplastic analysis of thin plates by the finite element method in deep drawing process. Proceedings of the international conference on computational plasticity models, software and applications, part 2, Pineridge Press, Swansea, pp 901–916

Massoni E, Bellet M, Chenot JL (1988) Thin sheet forming numerical analysis with a membrane approach. In: Chenot JL, Oñate E (eds) Modelling of metal forming processes. (Euromech Colloquium no 233) Sophia Antipolis. Kluwer Academic Publishers, pp 187–195

Nakazawa S, Pittman JFT, Zienkiewicz OC (1982) Numerical solution of flow and heat transfer in polymer melts. In: Gallagher et al. (eds) Finite elements in fluids vol. 4, John Wiley, Chap. 13

Oñate E, Zienkiewicz OC (1983) A viscous shell formulation for the analysis of thin sheet metal forming. Int J Mech Sci 25(5):305–335

Padmanabham KA, Davies GJ (1980) Superplasticity. Springer-Verlag, Berlin Heidelberg New York

Rebelo N, Wertheimer TB (1988) Finite element simulation of superplastic forming, MARC users conference. MARC Analysis Research Corporation, USA, pp 81–87

St Doltsinis J, Longinsland J, Nolting S (1987) Some developments in the numerical simulation of metal forming processes. Engng Comput 4:266–280

Taylor RL (1985) Solution of linear equations by a profile solver. Engng Comput 2:344–350

Wang N-M (1978) Elasto-viscoplastic analysis of simple stretch forming problems. In: Koinstinen et al. (eds) Mechanics of sheet metal forming. Plenum Press, New York

Wang N-M (1984) A rigid-plastic rate sensitive finite element method for modelling sheet metal forming processes. In: Pittman JFT et al. (eds) Numerical analyses of forming processes. John Wiley, Chap. 4

Wennerström H, Samuelsson A, Mattiasson K (1984) Finite element method for sheet metal stretching. In: Pittman JFT et al. (eds) Numerical analysis of forming processes. John Wiley, Chap. 14

Wood, RD, Mattiasson K, Honnor ME, Zienkiewicz OC (1986) Viscous flow and solid mechanics approaches to the analysis of thin sheet forming. In: Wang N-M, Tang SC (eds) Computer modelling of sheet metal forming processes: theory, verification and application. Metallurgical Society of AIME pp 121–142

Wood RD, Bonet J, Wargadipura AHS (1989) Simulation of the superplastic forming of thin sheet components using the finite element method. In: Thompson et al. (eds) Proceedings NUMIFORM89, numerical methods in industrial forming processes. Balkema, pp 85–93

Wood RD, Bonet J, Wargadipura AHS (1990) Simulation of the superplastic forming of thin sheet components using the finite element method. Int J Num Meth Engng 30:1719–1737

Zhang WC, Wood RD, Zienkiewicz OC (1986) Superplastic forming analysis using a finite element viscous flow formulation. In: Aluminium Technology '86. The Institute of Metals, London, Book 4, Session C, pp 111.1–111.6

Zienkiewicz OC (1984) Flow formulation for the numerical solution of forming processes. In: Pittman JFT et al. (eds) Numerical analysis of forming processes. John Wiley, Chap. 1

Zienkiewicz OC, Jain PC, Oñate E (1978a) Flow of solids during extrusion: some aspects of numerical solutions. Int J Solids Struct 14:15–38

Zienkiewicz OC, Oñate E, Heinrich JC (1978b) Plastic flow in metal forming: 1. Coupled thermal, 2. thin sheet forming. In: Applications of numerical methods to forming processes, proceedings winter annual meeting of ASME, AMD-28, pp 107–120

16 **Fracture in Forming Processes**

S.E. Clift

16.1 Introduction

The occurrence of ductile fracture is the limiting factor in many metal-forming operations. The ability to predict ductile fracture would allow early modification of a production process and so reduce the risk of failure occurring, with obvious financial savings.

Over the years, many different types of ductile fracture have been identified in industrial metal-forming operations (Johnson and Mamalis 1977). Edge cracking in rolling and chevron cracking in extrusion are oft-quoted examples. However, no general theoretical means of predicting the occurrence of this type of failure has been advanced, and the successful avoidance of fracture has been largely a matter of empirical practice.

Various metal-forming processes have now been examined in the laboratory and criteria developed to model their fracture behaviour. These criteria are generally process, geometry and material specific. Although a considerable advance in terms of understanding over the traditional empirical approach, they have proved to be of limited value in commercial manufacture.

A useful advance here would be one unified fracture criterion that could be applied with relative ease to a wide range of forming processes. It is to be expected that such a criterion would be reliant upon reasonably accurate continuum values of stress and strain, and it is here that the potential usefulness of the finite-element method becomes apparent.

This chapter reviews research work which has been performed to assess the applicability of published fracture criteria to a range of metal-forming operations and describes the application of the finite-element method in the prediction of ductile fracture initiation in metal forming.

The different forms of these published fracture criteria fall into the two broad categories of "models of void coalescence and growth" and "empirical and semi-empirical models". These are briefly presented in Sect. 16.2, and studies of their use together with the finite-element method presented in Sect. 16.3. Finally, the arguments are summarized in Sect. 16.4 and their current limitations are highlighted.

16.2 Criteria for Ductile Fracture

16.2.1 Models of Void Coalescence and Growth

Ductile fracture has been widely studied as a process of void nucleation, growth and subsequent coalescence. Voids may be formed at inclusions and second phase particles (as shown by, for example, Argon et al. (1975) and Goods and Brown (1979)), and may occur by cracking of the particles or debonding at the particle–matrix interface. This process of void creation and growth leading to ductile fracture may be interpreted as one of progressive damage: a characteristic

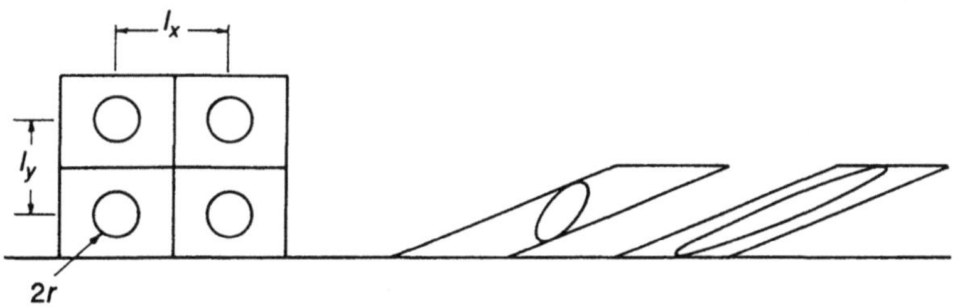

Fig. 16.1. McClintock model of void growth in a shear stress field. (After Atkins and Mai 1985.)

amount of damage may be accumulated over the plastic strain path until a critical value associated with fracture is reached. This critical value is dependent upon the material, its heat treatment and such parameters as the hardness, quantity and distribution of inclusions and second phase particles. It is independent of the geometry of the workpiece and its loading mode.

Within this framework, McClintock et al. (1966) derived approximate equations for the deformation of holes in shear bands. In a two-dimensional model large numbers of equispaced, equal-sized, cylindrical holes were assumed to be located uniformly throughout the material. The material was assumed to be divided into a number of regular elements, each containing a single centrally located hole (Fig. 16.1). Since the spacings between the holes were considerably greater than the hole diameter over much of the strain to fracture, the motion of each hole was calculated as if it lay in a semi-infinite body. The calculations predict that the holes will eventually meet the boundaries of their elements, at which point fracture is assumed to occur. From this, two modes of fracture were predicted, depending on which boundary of the element the edge of the hole touched. The following hole growth damage functions were derived to quantify the disorder:

1. For hole growth in the longitudinal direction (L) of the shear band

$$D_L = \ln F_L = \ln(1 + \gamma_{xy}^2) + \frac{\gamma_{xy}}{2(1-n)} \sinh(1-n)\frac{\sigma_H}{\tau_{xy}} \tag{16.1}$$

2. For hole growth in the thickness direction (T) of the shear band

$$D_T = \ln F_T = \frac{\gamma_{xy}}{2(1-n)} \sinh(1-n)\frac{\sigma_H}{\tau_{xy}} \tag{16.2}$$

where D is the hole growth damage parameter, F is the hole growth factor, γ is the shear strain, σ_H is the hydrostatic stress, τ is the shear stress and n is the work-hardening exponent. The hole growth damage parameter D, is required to reach a critical value for fracture to occur. In this way, ductile fracture is treated as an accumulation of damage, and is dependent on the entire history of stress and strain. Two simple experiments were performed by McClintock et al. to validate the theory, using plasticine as a model material. As a test of hole growth in pure shear a plasticine cylinder with a small transverse circular hole was twisted. The experimental result quoted for the measured shear strain and orientation of the hole at closure agreed well with that predicted by theory.

The second test consisted of pulling a grooved plasticine bar, as shown in Fig. 16.2. The results agreed well with theory except for very large holes which were

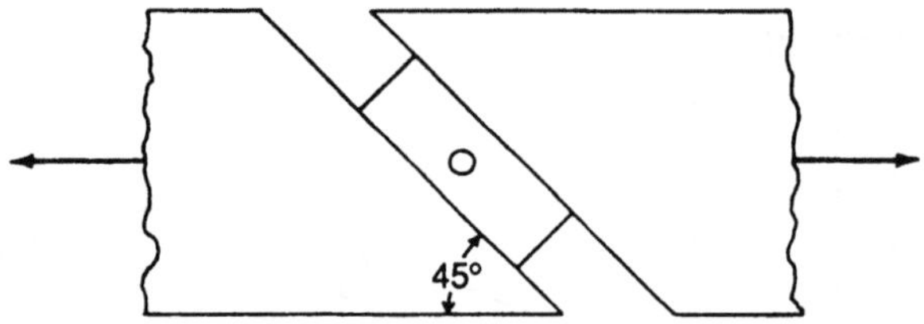

Fig. 16.2. Grooved plasticine test specimen for 45° band in tension. (After McClintock et al. 1966.)

found to grow more rapidly than predicted. This may be due to the breakdown of one of the assumptions on which the model is based: In the derivation of the hole growth functions it is assumed that the hole diameter is small when compared with the interhole spacing. However, it may be that the hole diameter becomes too large in this test for this assumption to remain valid.

This model of void growth under shear loading was extended by McClintock (1968) to apply to the growth and coalescence of pre-existing holes in plastically deforming materials for various ratios of the principal components of stress. Referring to Fig. 16.3, the following hole growth damage function was derived:

$$\ln (l_y/2r) = \int_0^{\bar{\epsilon}_f} \left[\frac{\sqrt{3}}{2(1-n)} \sinh \left(\frac{\sqrt{3}(1-n)}{2} \frac{(\sigma_1 + \sigma_2)}{\bar{\sigma}} \right) + \frac{3}{4} \frac{(\sigma_1 - \sigma_2)}{\bar{\sigma}} \right] d\bar{\epsilon} \quad (16.3)$$

where $\bar{\epsilon}_f$ is the generalized plastic strain at fracture, $\bar{\sigma}$ is the generalized stress and σ_1, σ_2 are the principal stress components. There are two other equations for the coalescence of holes with their axes parallel to the y and z axes, obtained by cyclic interchange of the subscripts.

A similar approach to void growth and coalescence was taken by Rice and Tracey (1969). They considered the growth of a spherical void in a general remote

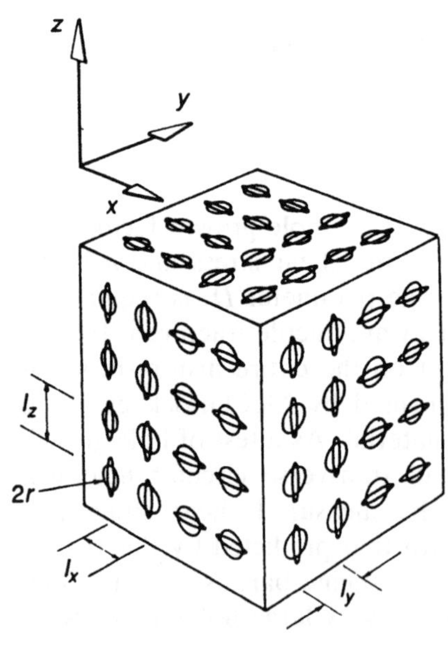

Fig. 16.3. McClintock model of void growth in a normal stress field. (After Atkins and Mai 1985.)

stress field. The results suggest a rapidly decreasing fracture ductility with increasing hydrostatic tension.

Oyane and co-workers (Oyane 1972, Oyane et al. 1980) formulated criteria for the ductile fracture of pore-free and pore-containing materials. It was postulated that when a critical material-dependent volumetric strain is reached the material fractures. This criterion was expressed in the following form for initially pore-free materials:

$$\int_0^{\bar{\epsilon}_f} \left(1 + \frac{\sigma_H}{A\bar{\sigma}}\right) d\bar{\epsilon} = G \qquad (16.4)$$

where A and G are material constants and G is given by

$$G = \int_0^{\epsilon_{vf}} (f^2 \rho^{2n-1})\, d\epsilon_v \qquad (16.5)$$

where ϵ_{vf} is the volumetric strain at fracture, ρ is the density and f is a function of the relative density (ratio of apparent density of porous material to the density of its pore-free matrix). The density dependence of both these functions was slightly modified in the form of this fracture criterion modelling the behaviour of initially pore-containing materials. A method for estimating the values of the constants required using compression test results was provided.

In addition to many studies of bulk forming, the void growth and coalescence model has also been applied to the fracture of sheet materials. Ghosh (1976) proposed a fracture criterion based on the shear joining of voids in the stretch forming of sheet materials (Fig. 16.4). This was given by

$$k_f = \left(\frac{\sigma_1 - \sigma_3}{2}\right) \sigma_H \qquad (16.6)$$

The constant contains the information regarding the role of inclusions in the material, but was determined without detailed metallographic analysis. Biaxial fracture limits produced using this criterion are in reasonable agreement for brass, aluminium and some steels.

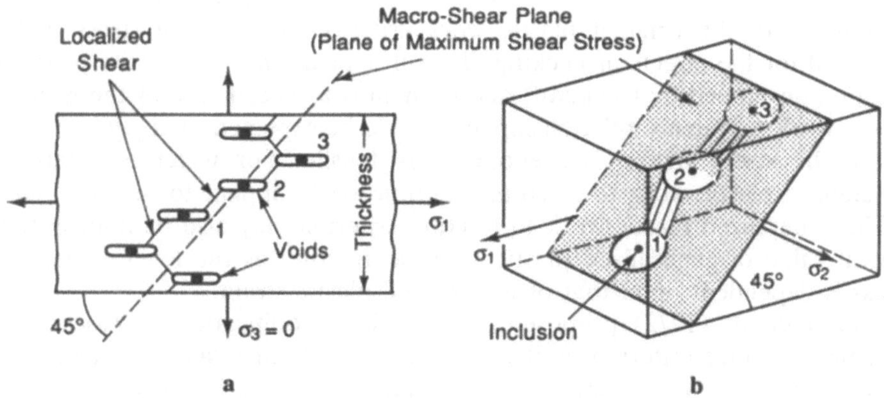

Fig. 16.4. Shear joining of voids. (After Ghosh 1976.)

16.2.2 Empirical and Semi-empirical Models

This section reviews some of the empirical and semi-empirical models that have been developed to model ductile fracture. Like the void growth criteria, these functions are accumulated over the plastic strain path until a critical value associated with fracture is reached.

In 1950, Freudenthal postulated that strain energy to fracture is the critical parameter. Mathematically, this can be expressed as

$$\int_0^{\bar{\epsilon}_f} \bar{\sigma}\, d\bar{\epsilon} = C1 \tag{16.7}$$

i.e. a critical value of generalized plastic work per unit volume at fracture.

Gillemot (1976) used this criterion in an investigation of brittle fracture in 0.35% carbon steel. In tension, compression, and low cycle fatigue tests, a crack appeared after a characteristic amount of energy had been absorbed.

The validity of this generalized plastic work criterion was also assessed by Cockcroft and Latham (1966, 1968). They used tensile test specimens of EN2A mild steel which show a well-defined neck at fracture. However, a correction to their experimental results to compensate for the triaxial nature of the stress system at fracture was not made. Such corrections have been presented in the literature (Bridgman 1944, Davidenkov and Spiridova 1946). The measured axial stress was therefore assumed to be that which led to separation of the material under uniaxial conditions. Cockcroft and Latham chose to modify the work criterion explicitly for the effect of the peak tensile stress in the necked region, rather than implicitly by using the corrected value of axial stress. Thus the modified equation assumed the following form:

$$\int_0^{\bar{\epsilon}_f} \sigma_1\, d\bar{\epsilon} = C2 \tag{16.8}$$

This criterion was validated using further experimental work in which the tensile test specimens were remachined after appropriate amounts of deformation to avoid the development of a significant neck. From these experiments Cockcroft and Latham concluded that the integral of tensile plastic work per unit volume must reach a critical, material dependent, value for fracture to occur. Following the way in which this modified criterion has been developed, it would be expected to agree with the original plastic work criterion if the axial stress had been corrected for the effects of necking. Thus it appears that the tensile plastic work criterion developed by Cockcroft and Latham is a special case of the generalized plastic work to fracture relationship and would be expected to apply in operations where the stress system is predominantly tensile. For wider application, the generalized plastic work to fracture criterion would appear to be more suitable.

The generalized plastic work to fracture criterion may also be interpreted as a critical value of generalized plastic strain at fracture as the level of generalized stress is dependent upon that of generalized plastic strain.

Brozzo et al. (1972) proposed an empirical modification to Cockcroft and Latham's fracture criterion as it predicted generalized strains at fracture which were too low when compared with experimental results for sheet forming. This led to the following criterion:

$$\int_0^{\bar{\epsilon}_f} \frac{2\sigma_1}{3(\sigma_1 - \sigma_H)} d\bar{\epsilon} = C3 \tag{16.9}$$

This integral must reach a critical value, in this case dimensionless, for fracture to occur. This critical value was estimated from the results of a tensile test on a notched specimen. The criterion was used to predict fracture strains in bulge tests on three low-carbon deep-drawing steels. Good correlation between theory and experiment was found if the material was assumed to be equally ductile in all directions in the plane.

Oh and Kobayashi (1976) and Oh et al. (1979) used a modification to Cockcroft and Latham's criterion in a form originally suggested by Latham and Cockcroft (1966). They considered the ratio of the maximum tensile stress to the generalized stress is the important parameter in ductile fracture, expressed mathematically as

$$\int_0^{\bar{\epsilon}_f} \frac{\sigma_1}{\bar{\sigma}} d\bar{\epsilon} = C4 \tag{16.10}$$

Osakada and Mori (1978) developed the following fracture criterion which includes an explicit dependence upon generalized plastic strain:

$$\int_0^{\bar{\epsilon}_f} \langle a + \bar{\epsilon} + b\,\sigma_m \rangle\, d\bar{\epsilon} = C5 \tag{16.11}$$

where σ_m is the mean stress, a and b are material constants and $C5$ is the critical value accumulated over the strain path to fracture. The influence of hydrostatic stress is such that a compressive value will increase the value of plastic strain to fracture.

Norris et al. (1978) developed the following fracture criterion:

$$\int_0^{\bar{\epsilon}_f} \frac{1}{(1 - c\,\sigma_H)} d\bar{\epsilon} = C6 \tag{16.12}$$

This was arrived at using both experimental results and results of a finite-difference computer program for several tensile test geometries including a simple axisymmetric tensile test and a Charpy V-notch specimen.

Atkins (1981) modified the criterion of Norris et al. to agree with experimental results obtained from sheet metal working. This led to the following criterion:

$$\int_0^{\bar{\epsilon}_f} \frac{\left(1 + \frac{1}{2L}\right)}{(1 - c\,\sigma_H)} d\bar{\epsilon} = C7 \tag{16.13}$$

where L is the strain ratio $d\epsilon_1/d\epsilon_2$. The modification was made as it was found in some tests that at the instant of fracture the current hydrostatic stress state was most tensile for the greatest strains to fracture. This is in contradiction to the established inverse relationship between hydrostatic stress and strain to fracture (see, for example, Hancock and Mackenzie (1976) and Goods and Brown (1979)). Therefore, in order to produce a correct prediction a path-dependency term was included in the damage integral.

There are, therefore, a significant number of fracture criteria that have been developed in a largely empirical manner. The criteria are based upon the accumulation of continuum values of stress and strain and can therefore be predicted from finite-element analysis without a highly detailed knowledge of the microstructure of the material. The critical value of each of these criteria at fracture does, however, need to be determined experimentally. This could be found from, for example, a simple tensile test. This value is then available to be compared with finite-element results from more complex loadings and geometries. Fracture would then be predicted to occur when the accumulated value equalled that obtained for the tensile test.

The next section of this chapter reviews experimental and finite-element studies of fracture in specific metalforming operations, first plane-strain side pressing, followed by simple upsetting and finally axisymmetric extrusion.

16.3 Predictions of Ductile Fracture Using the Finite-Element Method

16.3.1 Plane-Strain Side Pressing

In an experimental study, Jain and Kobayashi (1970) examined the side pressing of 7075 aluminium alloy rods. The behaviour of specimens of initially circular cross-section and specimens with machined flats of the initial geometry shown in Fig. 16.5 was studied. Using high-speed photography, an attempt was made to record the initiation of fracture and subsequent crack growth. Unfortunately, the camera speed of 120 frames per second proved inadequate to record the site of fracture initiation. However, the gross fracture paths appeared qualitatively to be coincident with the lines of maximum shear stress calculated from slip-line field theory.

$2W_0$ (in.)	$2H_0$ (in.)
0	1.00
0.300	0.95
0.430	0.90
0.525	0.85
0.600	0.80
0.725	0.70

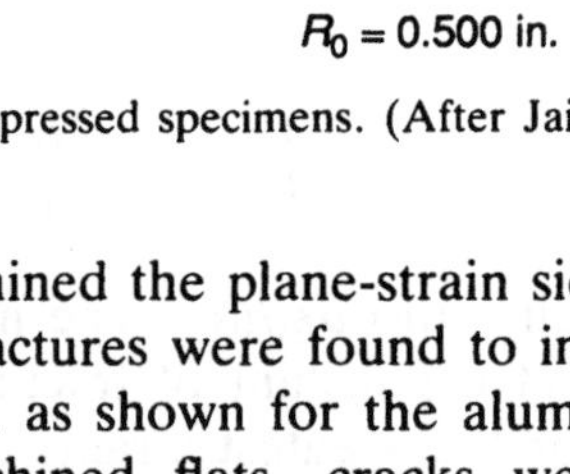

$R_0 = 0.500$ in.

Fig. 16.5. Initial geometry of plane-strain side-pressed specimens. (After Jain and Kobayashi 1970.)

Clift et al. (1985, 1987, 1990) examined the plane-strain side pressing of 7075 aluminium alloy and 60–40 brass. Fractures were found to initiate at the centre of the initially circular section billets, as shown for the aluminium alloy in Fig. 16.6. For the geometries with machined flats, cracks were found to have propagated from the surfaces of the billet in contact with the platens (Fig. 16.7). Using elastic-plastic finite-element solutions, predictions from a total of nine continuum fracture criteria were compared with experiment. From these results it was found that the criterion of a critical value of generalized plastic work per unit volume was successful in predicting both the site of fracture and the level of deformation at fracture for all the shapes studied.

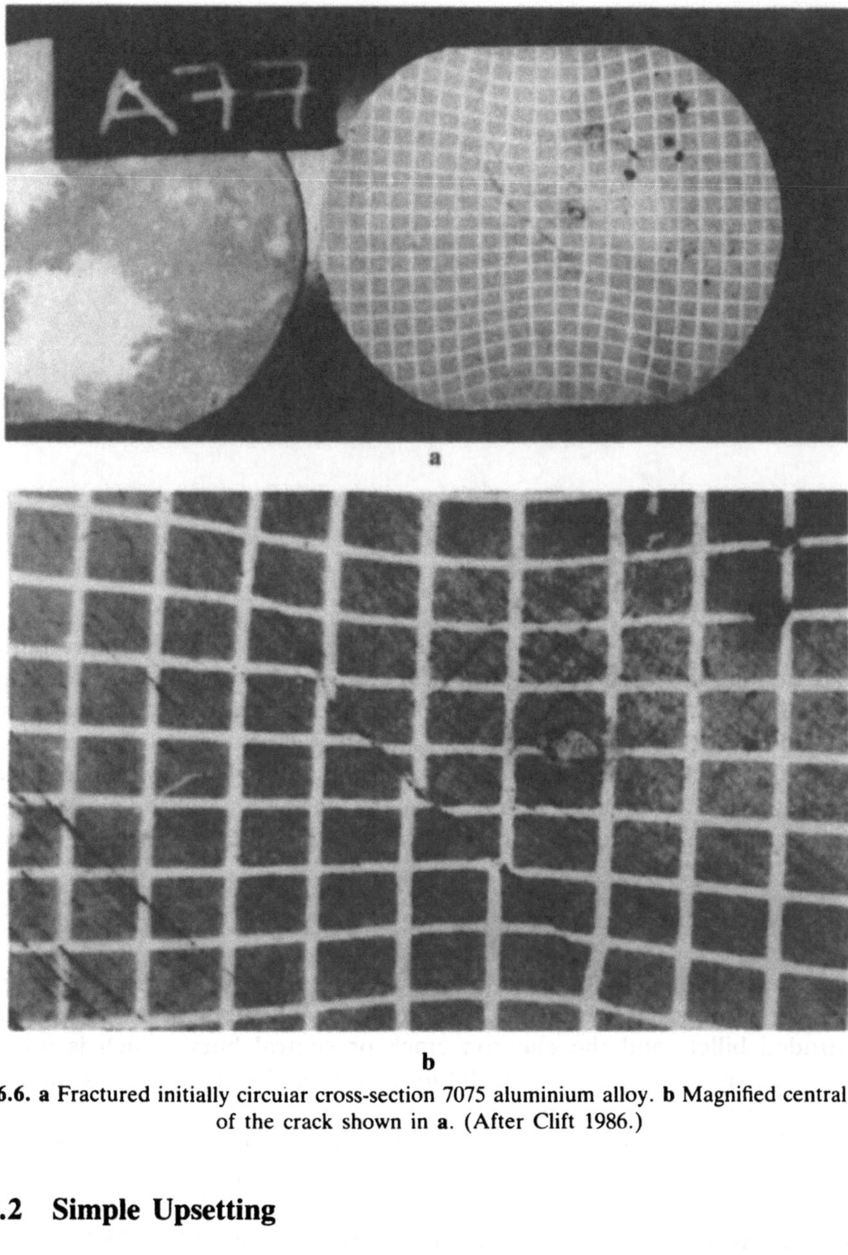

a

b

Fig. 16.6. **a** Fractured initially circular cross-section 7075 aluminium alloy. **b** Magnified central region of the crack shown in **a**. (After Clift 1986.)

16.3.2 Simple Upsetting

Using a finite-element program to examine the stress and strain behaviour of preselected elements, Sowerby et al. (1985) investigated the prediction of damage accumulation during upsetting tests based on McClintock's model. The accumulated damage was found to be greatest near the surface of the model. The experimental fracture sites were not explicitly stated but would be expected to be surface fractures (Kuhn and Lee 1971) and therefore agree with the predicted site. No predictions of the level of the deformation at fracture were reported.

Clift et al. (1990) examined the simple upsetting of 60–40 brass and 7075 aluminium alloy. As with their study of plane-strain side pressing, the continuum values of stress and strain produced by an elastic-plastic finite-element analysis

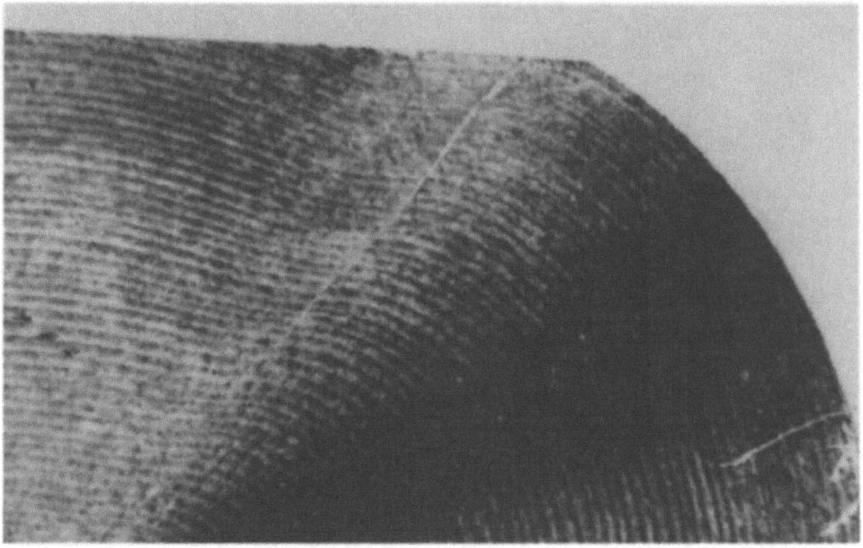

Fig. 16.7. Fracture in 7075 aluminium alloy with initial $H/W = 2.03$. (After Clift 1986.)

were used, together with a range of fracture criteria, to predict fracture initiation sites and amount of deformation at fracture. Here it was found that the McClintock criterion of 1968, the criterion of Brozzo and the criterion of a critical value of generalized plastic work per unit volume at fracture were all successful in predicting the surface fracture observed experimentally. The level of deformation at fracture was only compared with experiment for the generalized work criterion, but did not give the agreement previously found for plane-strain side pressing.

16.3.3 Fracture in Extrusion

Fracture in extrusion may be broadly classified into two categories; the so-called "fir tree" or "Christmas tree" effect where cracks are present on the surface of the extruded billet, and the chevron crack or central burst which is produced within the metal during extrusion. Johnson and Kudo (1962) give examples of these two general types of defect.

An insight into the nature of the fracture process in extrusion was first sought using the finite-element method by Iwata et al. (1972) and Lee et al. (1973). Both studies identified tensile stresses along the free surface of the workpiece extending from the die exit, as shown from Iwata's study in Fig. 16.8. Lee et al. additionally predicted tensile stresses in the central region of the workpiece also near the die exit.

Oh et al. (1979) also examined fractures in extrusions using finite-element generated stress and strain values. Although the authors reported that their results were inconclusive, Kobayashi and co-workers were the first research group to use the output of a finite-element analysis in ductile fracture prediction and thereby demonstrate the potential of such an approach.

Both Dung (1988) and Clift et al. (1990) have successfully predicted the fir tree cracking in brass and aluminium alloy previously found experimentally by Clift (1986). Dung used a McClintock formulation to extend the success previously

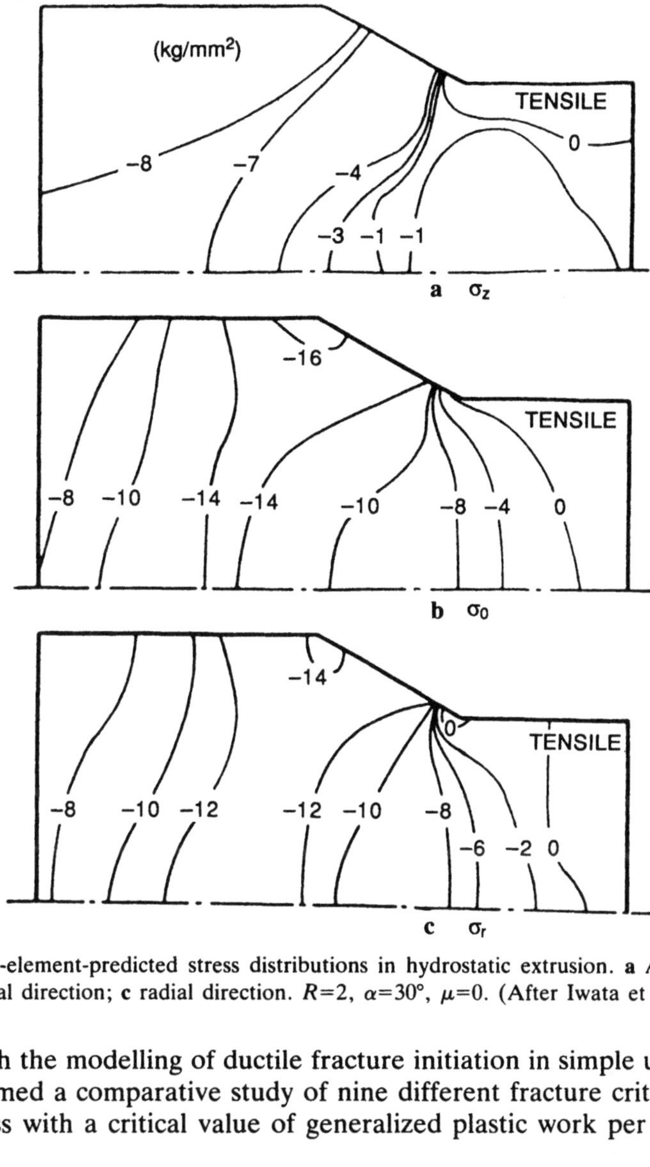

Fig. 16.8. Finite-element-predicted stress distributions in hydrostatic extrusion. **a** Axial direction; **b** tangential direction; **c** radial direction. $R=2$, $\alpha=30°$, $\mu=0$. (After Iwata et al. 1972.)

obtained with the modelling of ductile fracture initiation in simple upsetting. Clift et al. performed a comparative study of nine different fracture criteria and again found success with a critical value of generalized plastic work per unit volume.

16.4 Conclusions

A large number of different investigations have now been performed using the finite-element method in predictions of ductile fracture initiation.

Dung and co-workers have obtained successful predictions of ductile fracture initiation from McClintock's criteria. However, the most successful fracture criterion reported in the literature is that of generalized plastic work per unit volume. Studies of a range of geometries performed by Clift and co-workers have shown that the site of fracture could only be predicted correctly using this generalized plastic work criterion, as shown in Table 16.1. However, there were clear discrepancies on the prediction of level of deformation at fracture in some of the geometries studied. An obvious omission from this criterion is the effect

Table 16.1. Summary of fracture initiation site predictions; correct predictions indicated

Model	McClintock (1968)	McClintock (1966)	Ghosh (1976)	Oyane (1972)	Cockroft and Latham (1966)	Brozzo et al. (1972)	Work criterion	Norris et al. (1978)	Atkins (1981)
Side pressing									
Brass circular					×	×	×		
7075 circular					×	×	×		
7075 H/W=1.33							×		×
7075 H/W=2.03								×	×
Upsetting									
H/D=0.5		×	×	×	×	×	×	×	
H/D=1.0	×	×	×	×		×	×		
H/D=1.5	×	×		×	×	×	×		
H/D=2.0	×	×	×			×	×		
Strip									
Compression			×	×	×	×	×	×	×
Milled	×		×	×	×	×	×	×	×
Forged							×		
Extrusion									
7075	×			×	×	×	×		×
Brass	×			×	×	×	×		×

of hydrostatic stress which may account for some of the differences observed. The criterion of Osakada and Mori (1978) incorporates the effect of both plastic strain and hydrostatic stress and is worthy of further investigation.

References

Argon AG, Im J, Safoglu R (1975) Cavity formation from inclusions in ductile fracture. Met Trans 6A:825–837

Atkins AG (1981) Possible explanation for the unexpected departure in hydrostatic tension-fracture strain relations. Metals Sci (February)

Atkins AG, Mai Y-W (1985) Elastic and plastic fracture. Ellis Horwood, Chichester

Bridgman PW (1944) The stress distribution at the neck of a tension specimen. Trans ASM 32:553–574

Brozzo P, DeLuca B, Rendina R (1972) A new method for the prediction of formability limits in metal sheets. In: Sheet metal forming and formability, proceedings of the 7th biennial conference of the International Deep Drawing Research Group

Clift SE (1986) Identification of defect locations in forged products using the finite-element method PhD thesis, University of Birmingham

Clift SE, Hartley P, Sturgess CEN, Rowe GW (1985) Fracture initiation in plane-strain forging. Proceedings 25th machine tool design and research conference, University of Birmingham, pp 413–420

Clift SE, Hartley, P, Sturgess CEN, Rowe GW (1987) Further studies on fracture initiation in plane-strain forging. In: Tooth and Spence (eds) Applied solid mechanics II, Elsevier, pp 89–100

Clift SE, Hartley P, Sturgess CEN, Rowe GW (1990) Fracture prediction in plastic deformation processes. Int J Mech Sci 32:1–17

Cockcroft MG, Latham DJ (1966) A simple criterion of fracture for ductile materials. National Engineering Laboratory report number 240

Cockcroft MG, Latham DJ (1968) Ductility and the workability of metals. J Inst Metals 96:33–39

Davidenkov NN, Spiridova NI (1946) Mechanical methods of testing: Analysis of the state of stress in the neck of a tension test specimen. Proc Am Soc Test Mater 46:1157–1158

Dung NL (1988) Identification of defect locations in metalforming using a personal-computer-oriented finite-element method. In: Chenot JL, Oñate E (eds) Modelling of metalforming processes. Kluwer Academic, pp 245–252

Freudenthal AM (1950) The inelastic behaviour of solids, John Wiley

Ghosh AK (1976) A criterion for ductile fracture in sheets under biaxial loading. Met Trans 7A:523–533

Gillemot LF (1976) Criterion of crack initiation and spreading. Eng Fract Mech 8:239–253

Goods SH, Brown LM (1979) The nucleation of cavities by plastic deformation. Acta Metall 27:1–15

Hancock JW, Mackenzie AC (1976) On the mechanisms of ductile fracture in high strength steels subjected to multi-axial stress states. J Mech Phys Solids 24:147

Iwata K, Osakada K, Fujino S (1972) Analysis of hydrostatic extrusion by the finite-element method. Trans ASME J Eng Ind 94:697–703

Jain SC, Kobayashi S (1970) Deformation and fracture of an aluminium alloy in plane-strain side-pressing. In: Proceedings of the 11th machine tool design and research conference, University of Birmingham

Johnson W, Kudo H (1962) The mechanics of metal extrusion. Manchester University Press

Johnson W, Mamalis AG (1977) Defects in metalworking processes. In: Tobias SA, Koenigsberger F (eds) Proceedings of the 17th machine tool design and research conference, MacMillan Press, pp 607–621

Kuhn HA, Lee PW (1971) Strain instability and fracture at the surface of upset cylinders. Met Trans 2:3197–3202

Latham DJ, Cockcroft MG (1966) The effect of the stress system on the workability of metals. National Engineering Laboratory Report number 216

Lee CH, Iwasaki H, Kobayashi S (1973) Calculation of residual stresses in plastic deformation processes. Trans ASME J Eng Ind 95:283–291

McClintock FA (1968) A criterion for ductile fracture by the growth of holes. Trans ASME J Appl Mech 35:363–371

McClintock FA, Kaplan SM, Berg CA (1966) Ductile fracture by hole growth in shear bands. Int J Fract Mech 2:614–627

Norris DM, Reaugh JE, Moran B, Quinnones DF (1978) A plastic strain mean-stress criterion for ductile fracture. J Eng Mat Tech 100:279–286
Oh SI, Chen CC, Kobayashi S (1979) Ductile fracture in axisymmetric extrusion and drawing. Part 2, workability in extrusion and drawing. Trans ASME 101:36
Oh SI, Kobayashi S (1976) Theories of flow and fracture in metalforming processes. Part 2: A theory on ductile fracture in metalworking processes. AFML-TR-76-61
Osakada K, Mori K (1978) Prediction of ductile fracture in cold forging. Ann CIRP 27:135–139
Oyane M (1972) Criteria of ductile fracture strain. Bull JSME 15:1507–1513
Oyane M, Sato T, Okimoto K, Shima S (1980) Criteria for ductile fracture and their applications. J Mech Work Tech 4:65–81
Rice JR, Tracey DM (1969) On the ductile enlargement of voids in triaxial stress fields. J Mech Phys Solids 17:201–217
Sowerby R, Chandrasekaran N, Dung N, Mahrenholtz O (1985) The prediction of damage accumulation during upsetting tests based on McClintock's model. VDI-Forsch Ingenieurwesen 51:147–150

Part IV
Computer Integrated Systems

17 Integration of Numerical Modelling and CAD/CAM Systems

R. Kopp and G.H. Arfmann

17.1 Introduction

Product and process development play an important role in highly industrialized countries. Two important development tools are testing and computer simulation. The importance of computer simulation and optimization is growing as computing speeds and user-friendliness increase. High-performance information systems for companies in the production sector can be achieved by linking simulation and optimization programs to CAD/CAM systems. Within the context of a general concept for the computerized planning and simulation of metal-forming processes, the following section discusses a number of problems involved in interlinking CAE systems (and thus numerical modelling, simulation and optimization techniques) with CAD/CAM systems in an integrated data-processing network.

The most important computer-aided activities in the production engineering sector are:

- CAD Computer-aided design (computer-aided drafting)
- CAP Computer-aided planning
- CAM Computer-aided manufacturing
- CAQ Computer-aided quality control
- CAS Computer-aided service
- CAE Computer-aided engineering

Integration of these activities results in the computer integrated manufacturing (CIM) concept.

This chapter deals primarily with the integration of the CAE and CAD/CAM fields. Current planning, simulation and optimization potentials are indicated in the case of the metal-forming process. Methods for postprocessing the resulting data via interfaces in an integrated system using CAD/CAM are then illustrated.

17.2 Typical Tasks

In many areas of metal-forming technology, the engineer is confronted by the task of manufacturing a finished or semifinished product from a blank. Figs 17.1 and 17.2 show two typical examples. During the process, both the shape and the properties (e.g. strength) of the product are altered. The limits within which metal-forming processes may take place are increasingly restricted by geometrical and quality tolerances. Especially in multistage processes, this entails a cost-intensive development effort to convert the initial concept into a finished product. As a means of minimizing costly trial-and-error procedures, a variety of numerical simulation and optimization techniques transferring optimized data via interfaces to computer-aided manufacturing are available. This concept may involve a machine tool capable of producing the desired effective surface with the aid of

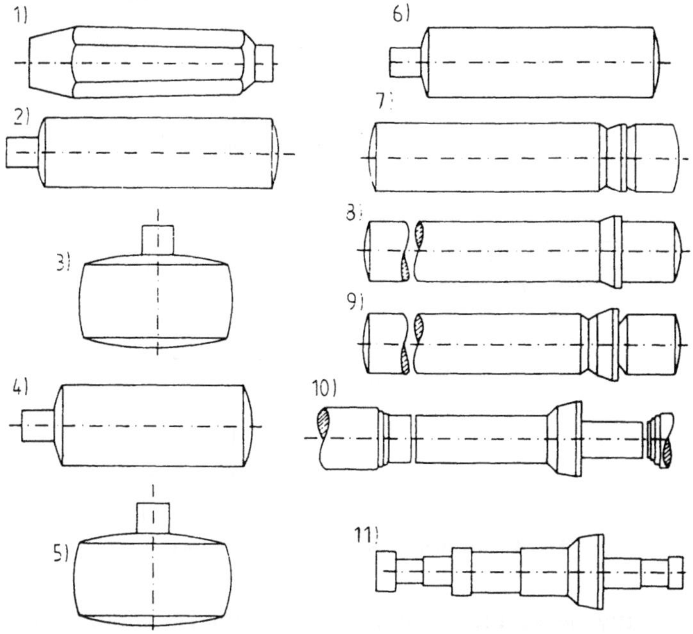

Fig. 17.1. Forming stages in the forging of a turbine rotor.

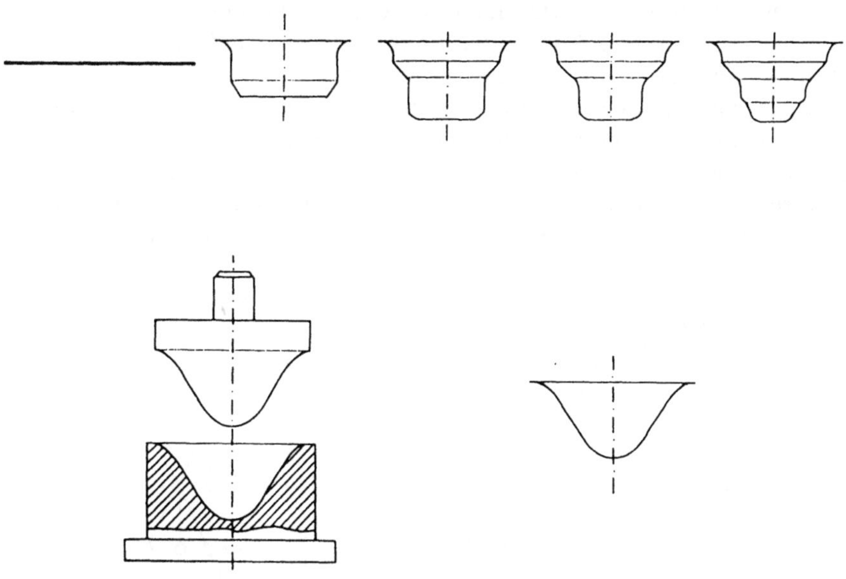

Fig. 17.2. Stages in the deep drawing of a reflector.

the transferred data, or a flexible metal-forming machine using the optimized data as an input to control the die position or speed.

An engineer designing forming processes generally proceeds according to the routine indicated in Fig. 17.3. The main activities are embraced by planning, including simulation, evaluation and optimization. The cycle can be run a number of times to arrive at a final process design or operating schedule. The chief task is therefore to proceed from State A to State B in one or a number of stages.

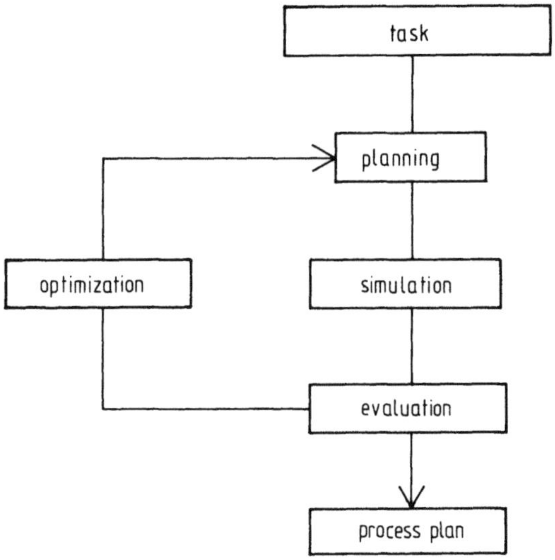

Fig. 17.3. Engineer's planning of a process.

There will generally be several combinations of intermediate stages capable of leading to the desired result (Fig. 17.4). The optimum combination of substeps has to be identified, subject to various constraints (the smallest possible number of energy-optimized steps, attainment of defined properties, etc.). The basic prerequisite is appropriate knowledge of the process, and particularly the most accurate possible models for process simulation.

The corresponding target variables are determined by a large number of variables, generally with complex mutual interdependences. In a forming process, the workpiece, die and machine have to be described. The associated descriptions of the various subsystems die, workpiece and machines, require systems of differential equations of exceptional mathematical complexity. It is therefore

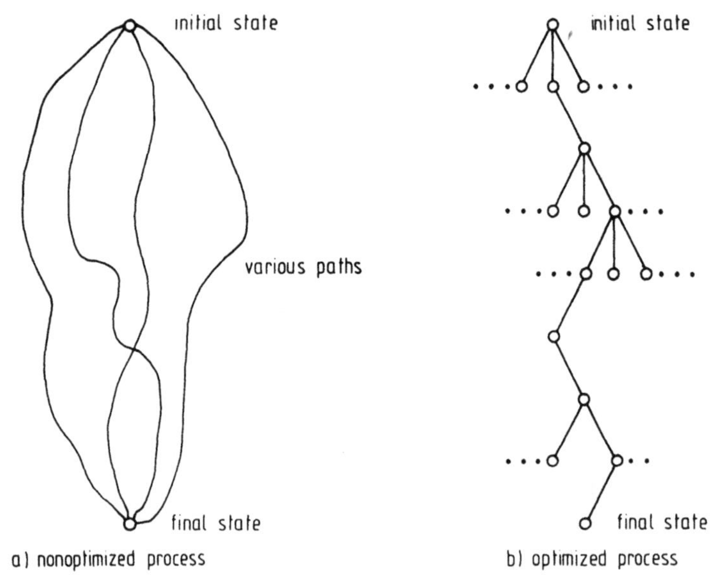

Fig. 17.4. Process steps.

advisable to use adapted models appropriate to the desired target variables which are to be described. A multilevel approach is indicated.

17.3 Multilevel Approach

In a multilevel approach, the process is considered in various degrees of detail, depending on requirements. In the following section, this concept is exemplified in the case of the workpiece.

Fig. 17.5 subdivides the descriptive variables of the workpiece into three different levels:

- Integral or global variables (Level 1)
- Local, continuum mechanics variables (Level 2)
- Local, metal physics variables (Level 3)

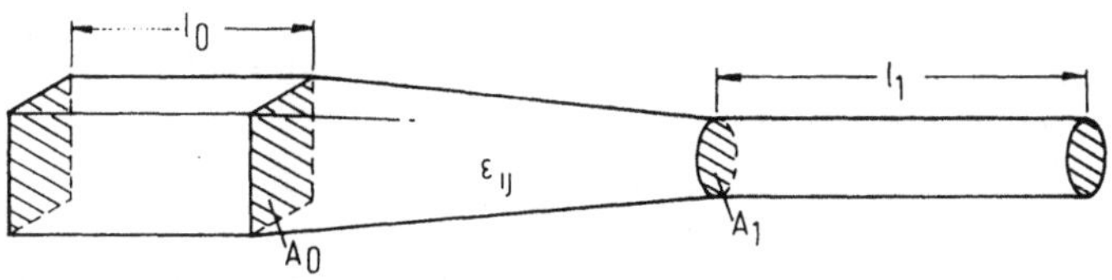

level	initial state	process	final state
1	h_0, b_0, l_0, r_0 ϑ_{m0}, v_{m0} k_{f0}	h, b, l, r ϑ_m, v_m k_{fm}	h_1, b_1, l_1, r_1 ϑ_{m1}, v_{m1} k_{f1}
2	σ_{eig_0}, ε_{ij_0}, ϑ_0	$\varepsilon_{ij}, \sigma_{ij}, \dot{\varepsilon}_{ij}, \vartheta$ $\mu = f(p, \vartheta \ldots)$	σ_{eig_1}, ε_{ij_1}, ϑ_1
3	texture grain boundaries roughness anisotropy σ_{eig_0}	texture grain boundaries roughness anisotropy σ_{ij}	texture grain boundaries roughness anisotropy σ_{eig_1}

Fig. 17.5. Multilevel simulation. h: height; b: width; l: length; r: radius; ϑ: temperature; v: velocity; k_f: yield stress; σ: stress; ϵ: strain; $\dot{\epsilon}$: strain rate; μ: friction; p: pressure.

In a system free of contradictions, defined target variables at each level must constitute the integral of the corresponding target variables at the "next-finest" level. Die pressure at Level 1 must, for example, represent the mean value of the normal stresses in the effective die line. This entails continuity between the solutions used at the different levels of the process description. Fig. 17.6 indicates the interrelationships between the variables at the three levels. When a die penetrates a workpiece, a deformation stress k_w = force/area is created. This deformation stress represents stress variable σ^I at the first level and is then the mean value of the normal stress distribution σ^{II} at the second level. Given sufficiently accurate measurement techniques, the procedure would be to determine in accordance with microdeformations (grain boundary slip, dislocation movement, etc.) the mean stress curves σ^{III} for a single grain resulting from the actual curve σ^{IV}. The mean value of the σ^{III} or σ^{IV} curve is then represented by the σ^{II} curve.

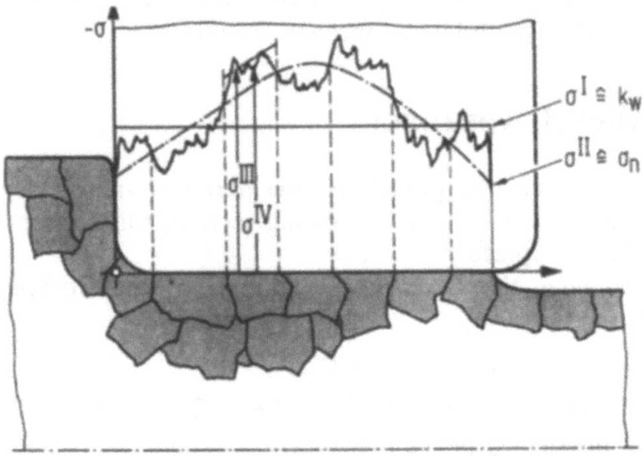

Fig. 17.6. Stress distribution at the three different levels.

Models matched to this degree of detail at the different levels are required in order to calculate the process variables.

Coordination and processing of a program package run within this structure are centrally controlled (see Fig. 17.7).

Interface problems and strict separation between dialogue and autonomous background programs are of central importance in the creation of program packages within this structure. A wide range of program structure, interface and data organization definitions are required, as illustrated by the following examples at different levels of detail.

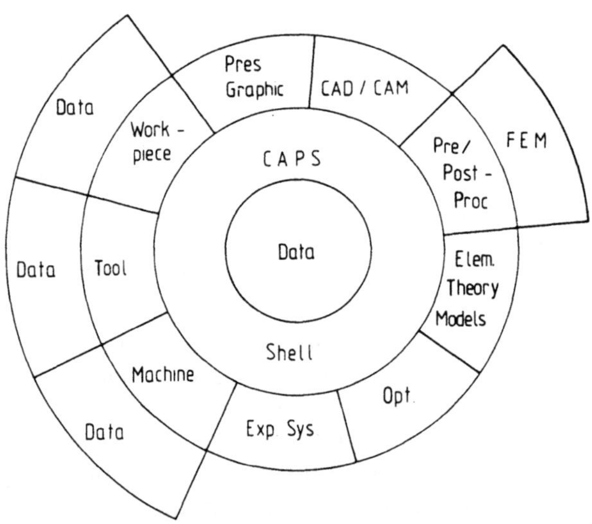

Fig. 17.7. CAPS structure.

17.3.1 Simulation Model at Level 1

The creation of program packages within the CAPS structure may be exemplified by a pass design system for flat and bar steel rolling cases.

The first-level simulation system clearly illustrates the principles on which program packages within the structure are based. The program package comprises five components:

- Description of rolling lines (machines)
- Materials data bank (workpiece)
- Calibration (tool design)
- Simulation models for flat and bar steel (element theory models)
- Graphic processing of geometries and results including variant design of common pass shapes (presentation graphics, CAD/CAM)

Fig. 17.8 shows the program package structure.

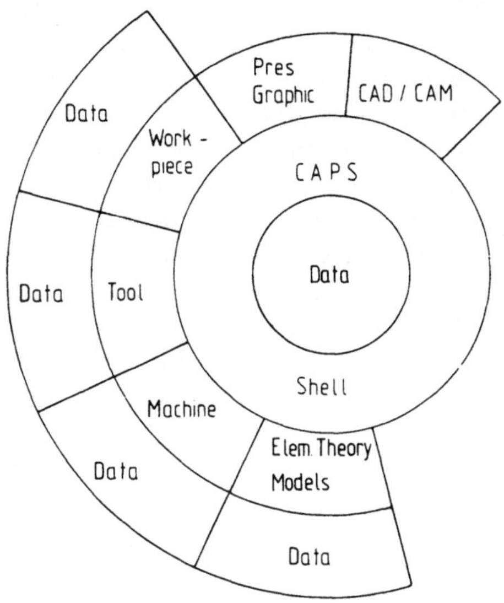

Fig. 17.8. CAPS–Walzen system structure.

The *rolling-line description component* contains the parameters of a rolling plant. Data describing the number of stands in a rolling line, the stand intervals, the performance data of the individual stands (maximum forces, maximum torques, maximum power) and data on stand frame spring and type are stored in this component. Data for various rolling plants can be stored ready for retrieval when a simulation is invoked.

The *material description component* supplies the system with material data from a materials data bank containing, for example, the stress–strain curves of the materials. The required data are made accessible to the simulation program by selecting the material.

The *pass component* generates and stores the data for a particular pass sequence, i.e. it stores the number of roll passes, the type of pass and the corresponding geometrical data. These are later accessed by the simulation program.

The *simulation component* contains algorithms for temperature computations, rolling case statics, rolling case kinematics, rolling power, frame spring, etc. Since many of these variables are mutually interdependent, the calculation is performed iteratively for each case. Some of the technological modules included in the system are:

- Temperature
- Yield stress
- Material flow
- Spread
- Material throughput
- Required force, torque, power
- Geometry (form categories)
- Drafting
- Material database
- Presentation graphics

These are essentially modules based on algorithms already discussed in previous studies (Kopp and Mauk 1982, 1983, Kopp et al. 1984).

Appropriate *utilities* enable the user to display the calculated data following completion of the simulation. The display may be in the form of tables relating to a particular roll pass, or in the form of graphics showing the entire rolling line (e.g. temperature distribution, force distribution).

The CAPS rolling system represents the realization of a system for the metal forming process "rolling" within the CAPS structure, i.e. there is a strict separation between program functions and data flow in the system. Each of the main components uses or modifies assigned data and makes these available in the system via a classified structure (e.g. common blocks in Fortran). The data structure in the system is subdivided accordingly into:

- Plant-related data
- Die-related data
- Workpiece-related data
- Simulation results

Data are further classified as purely geometrical or technological information.

The system described and realized for rolling can be realized for any other metal-forming process. The sole requirements are that the rolling-line description component and the pass component must be adapted to suit the corresponding metal-forming plant and strategy with their characteristic rules and laws for die and process design, and that appropriate data must be supplied.

The following section describes further examples integrating finite-element methods (FEM).

17.3.2 Simulation Models at Levels 2 and 3

Integration of an FEM program in the CAPS program environment enables second-level data to be generated and coupled with the top level.

Figs 17.9 and 17.10 show results in this sector. Passes with defined local minimum strains in the rolling stock are to be designed. The stress distribution is also to be determined. The first step is to generate a rough pass using the CAPS rolling program at Level 1. The resulting geometry data are then interfaced to a preprocessor (St Gilges, personal communication, IBF 1989) and supplied to the FEM program. The FEM program calculates the individual passes. Material

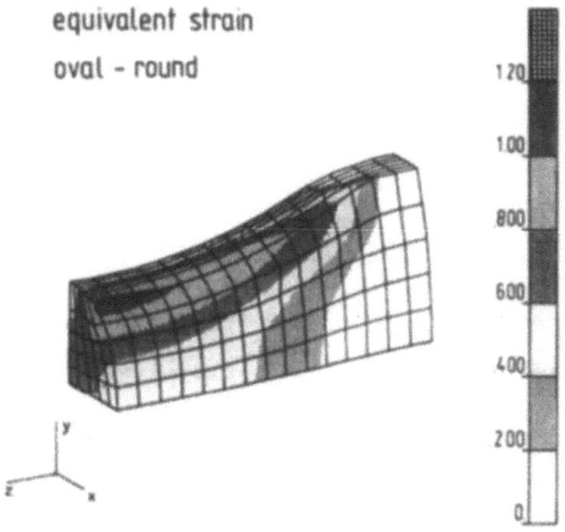

Fig. 17.9. FEM: oval–round pass.

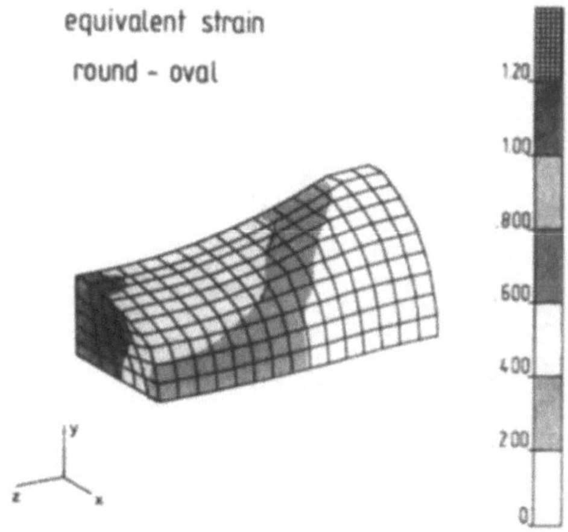

Fig. 17.10. FEM: round–oval pass.

flow events are analysed with the aid of a rigid-plastic static FEM calculation. Appropriate graphic postprocessing (PDA 1987) provides material flow data for the pass designer, enabling him to evaluate second-level variables when appraising the pass he is designing. Figs 17.9 and 17.10 show the oval–round–oval rolling case and the resulting equivalent strain.

The integral of the stress σ_y over the xz plane within the roll/roll-slit contact area yields values of 2630 kN or 2332 kN for the rolling cases in Figs 17.9 and 17.10 respectively. The equivalent values calculated with the CAPS rolling program package are 2440 kN or 2284 kN (Figs 17.11 and 17.12), so that the

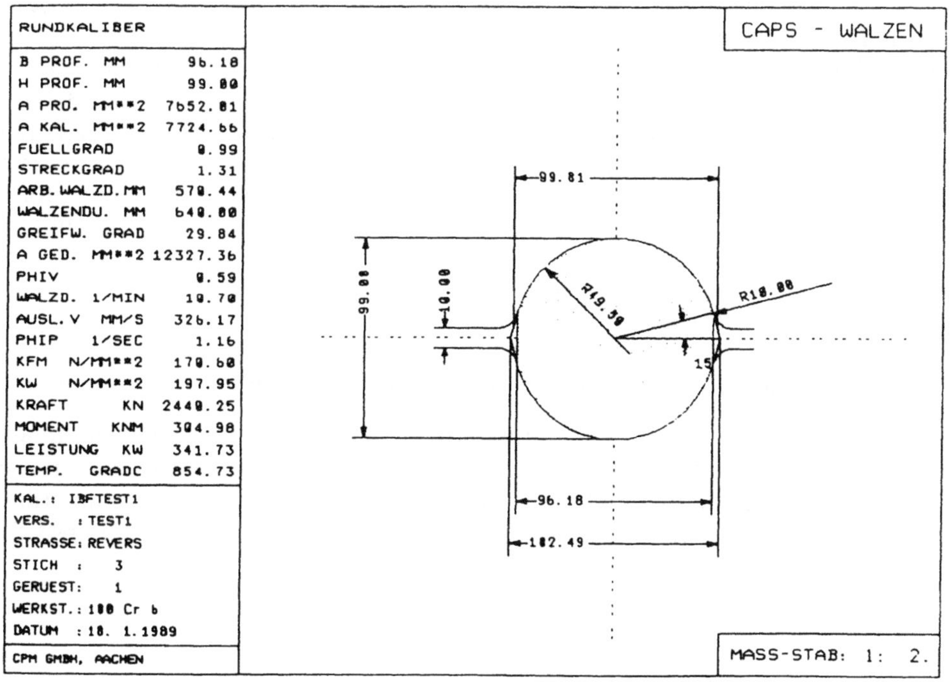

Fig. 17.11. CAPS: oval–round pass.

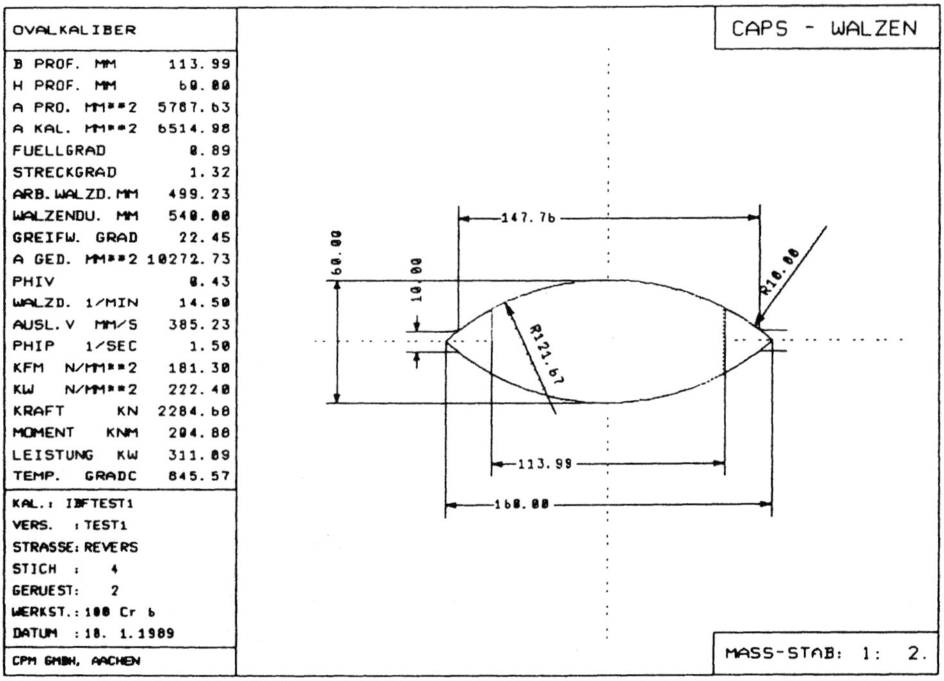

Fig. 17.12. Round–oval pass.

integration condition between two levels and two different calculation principles (elementary theory and FEM) may be regarded as a good approximation.

Elementary theory can no longer be used to simulate complex three-dimensional material flow phenomena. Here FEM is also suitable as a simulation tool at Level 1, as indicated below.

Fig. 17.13 analyses the influence of the number of elements on the accuracy of the target variable force for the case of a cylinder upsetting process. It is evident that even with four elements the upsetting force up to the specific height change of 50% can already be calculated with great accuracy. This analysis is required for each forming case, in order to provide a reliable concept for a multistage, matched FEM simulation technique. The concept is based on the following constraints:

- Level 1: FEM program, number of elements: $n_{\min}$
- Level 2: FEM program, number of elements: n_{opt}
- Level 3: FEM program, including material models, number of elements: n_{opt}

At Level 3, a knowledge of material properties such as recrystallization and recovery is required in addition to optimum use of FEM (n_{opt}). An example of this type of material model and the coupling of Level 3 with Level 2 is described in de Souza (1989). The resulting equation for the critical strain required for dynamic recrystallization of a steel is

$$\epsilon_1 = 0.8\,\epsilon_m \tag{17.1}$$

$$\epsilon_m = 3.47 \times 10^{-3}\,(\alpha\gamma)^{0.35}\,z^{0.11} \tag{17.2}$$

where

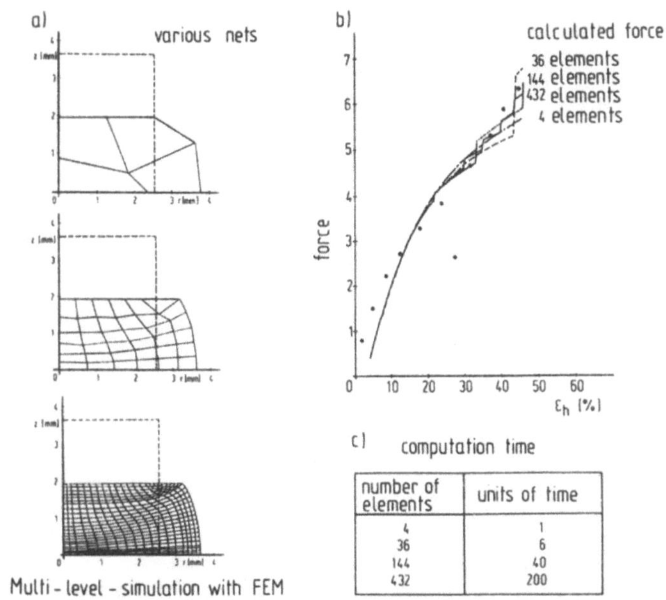

Fig. 17.13. Multi-level: simulation with FEM.

$$z = \dot{\epsilon} \exp\left(Q_w/(RT)\right) \tag{17.3}$$

Using a suitable thermal–mechanically coupled FEM program, information on local strain can be derived from Level 2. This presupposes that the yield stress k_f should be known as a function of ϵ, $\dot{\epsilon}$, ϑ, $d\gamma$ and t_p.

Given this information on local strains, it is possible to predict the location, time and extent of dynamic recrystallization. Fig. 17.14 shows an example.

17.3.3 Optimization Techniques

Optimization plays an important role alongside simulation. The problems in metal-forming technology described at the beginning of this chapter constitute a multiparameter optimization task with restrictions and subconstraints. Two groups of optimization methods or strategies are available for the solution of this task: algorithmic optimization methods and expert systems. *Hill-climbing methods* occupy an important position in the first group. Together with a process model using (for example) FEM, these can be used to optimize the process (Fig. 17.15). Details of this procedure are given in Becker and Kopp (1989).

Figs 17.16 and 17.17 show the results of such an optimization. The task consists of varying the preform of a disc in such a way that following the drop-forging operation in the centre of the disc the same minimum equivalent strain of $\epsilon = 0.4$ is attained at all points along the line AB. After roughly 100 variations, the optimization model attained at least a suboptimum (Fig. 17.17).

A significant aspect in this case is again the realization of suitable interfaces between, for example, the material data bank, the optimization program, FEM, pre- and postprocessing and CAD.

An important step in the planning and development of metal-forming processes is, however, the expert knowledge contributed by the engineer during the design

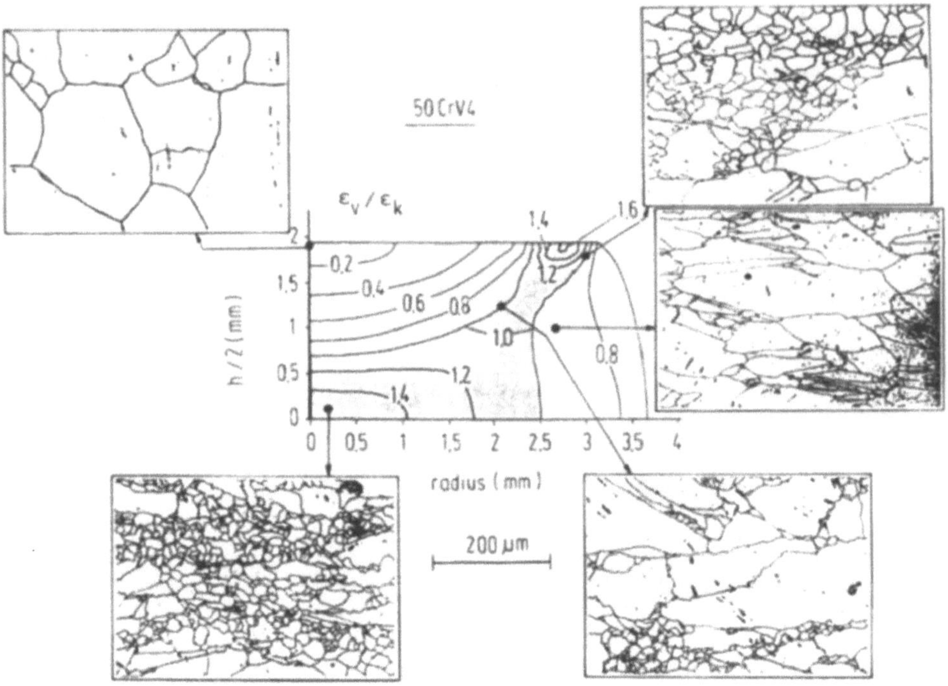

Fig. 17.14. Dynamic recrystallization.

of the process. This includes empirical values for the determination of forming geometries and knowledge of past failure cases which exclude particular, only theoretically conceivable process steps. Knowledge of material failure, e.g. susceptibility to cracking, special forming properties of materials and methods for preventing process-induced errors, also belongs in this category.

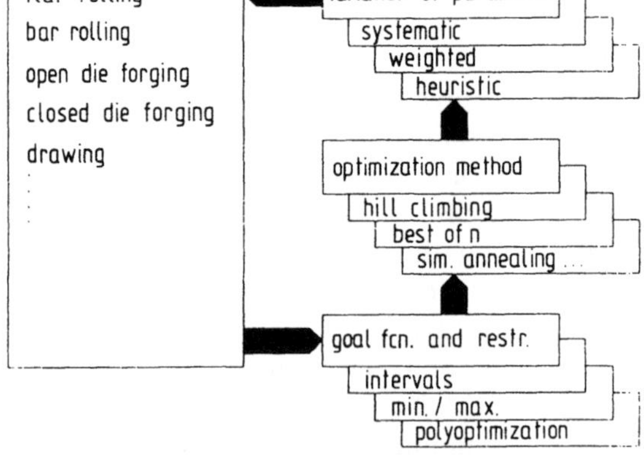

Fig. 17.15. Tools for optimizing metal-forming processes.

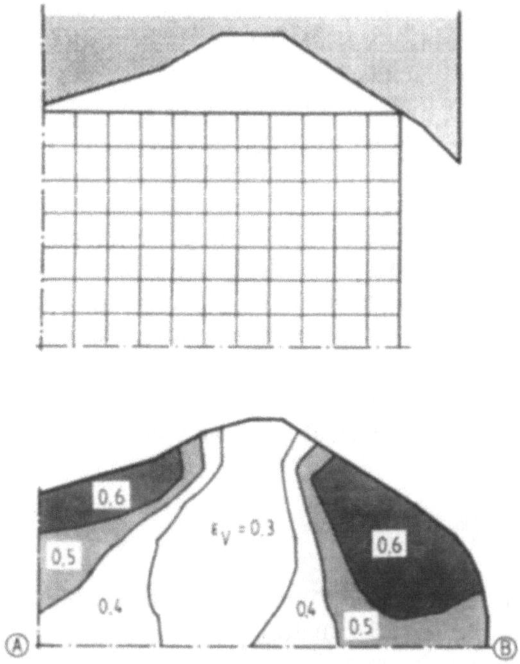

Fig. 17.16. Axisymmetric closed-die forging: preform and equivalent strain distribution.

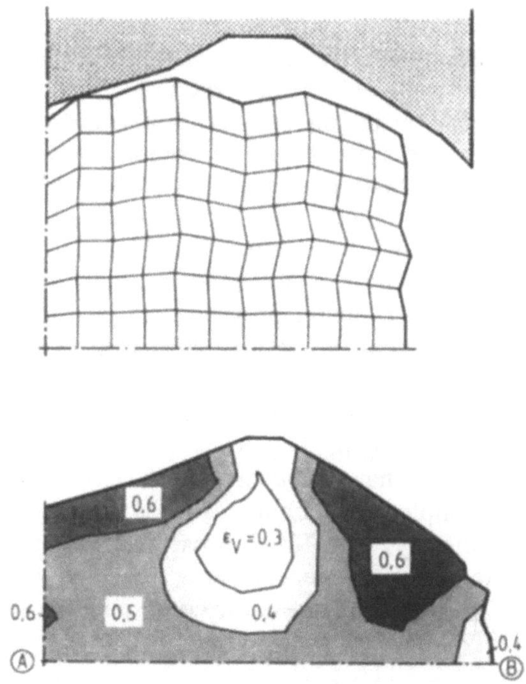

Fig. 17.17. Preform and equivalent strain distribution after optimization with the greedy method.

The expert is thus in a position to impose sensible restrictions on the theoretically available range of solutions. Application of his knowledge to computer-aided process design can greatly enhance its efficiency. The expert's knowledge is also of decisive importance in the field of evaluation and correction.

The opportunity of securing knowledge, classifying empirical methods and, not least, combining the knowledge of a number of experts opens up an interesting potential in this field.

One of the principal problems is to integrate *expert systems* into existing structures, e.g. the CAPS structure described above. In Arfmann (1990), a number of expert systems for the planning of forming processes are described.

Graphic data processing in the work presented here was by means of the PATRAN pre-/postprocessing program, the CATIA CAD/CAM system and a GKS program. Integration of these systems into the CAPS structure is dependent on the realization of a wide range of interfaces discussed in the next section.

17.4 Interface Problems

This section presents examples of interface problems encountered in realizing program packages in the CAPS structure.

Fundamentally, it may be stated that apart from standards describing pure geometry data such as VDA and IGES (Spur and Krause 1984, Brauner et al. 1981), or GKS (GSS 1989), no standardized interfaces are available.

Since the transfer of geometry data alone is insufficient within complex systems in the CAPS structure, these standards can at best represent useful aids.

It should rather be assumed that the creation of complex systems integrating a wide variety of programs demands development of a specific data structure containing not only geometry data but other information relating to, for example, materials science or technology. The corresponding interfaces for the individual programs should then be matched to this structure. Examples are given below.

The interlinking of FEM programs provides a particularly clear illustration of the interface problem. These programs all require the same type of input (geometry, materials data, constraints etc.), but each FEM program needs them in a different form. For this reason, access to these data for the individual FEM programs should be via an integral structure. For the realization of systems within the CAPS structure, these data are available in an internal data structure, as noted above. Appropriate conversion programs access these data in the CAPS data structure and present them in a form which can be read by the system in question (Fig. 17.18).

In the same way, data from the CAPS system are transferred to CAD/CAM systems or pre-/postprocessors (Fig. 17.19).

In turn, pre-/postprocessors have their own interfaces to various FEM programs. Fig. 17.20 shows, for example, the coupling of FEM programs with the PATRAN pre-/postprocessor. Internally, PATRAN possesses a so-called neutral system from which input data can be generated and to which resulting data can be transferred (PDA 1987). Fig. 17.21 shows the corresponding specific file structure.

The CATIA CAD/CAM system can be linked to other systems via a Fortran interface available in the CATIA package Figs 17.22 to 17.24 show examples of the appropriate Fortran interface specifications defining an axial system, a rotating

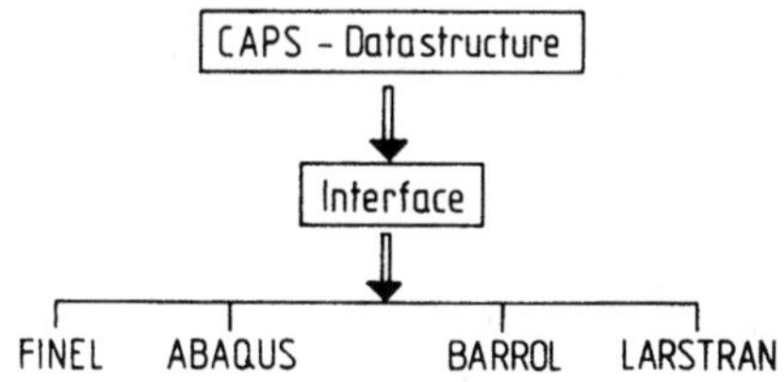

Fig. 17.18. CAPS interface: FEM.

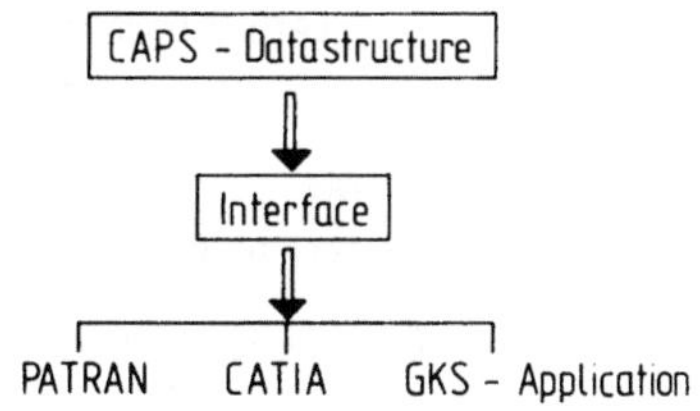

Fig. 17.19. CAPS interface: graphics.

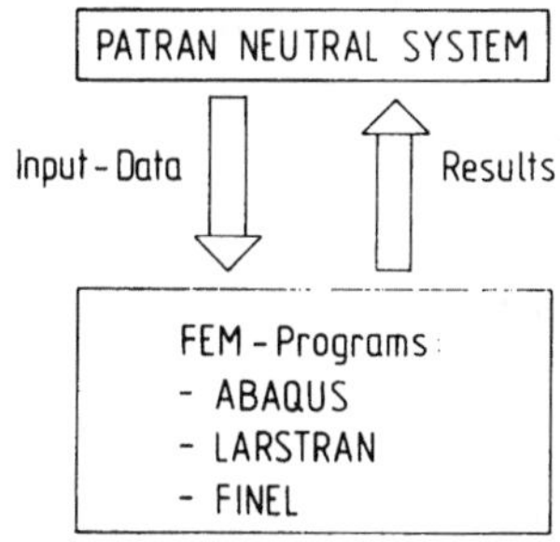

Fig. 17.20. PATRAN neutral system.

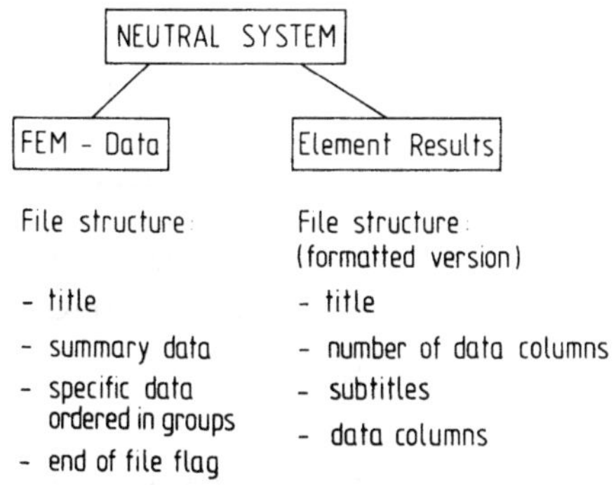

Fig. 17.21. File structure of neutral system.

surface and a line. There a similar standardized interfaces for linking-in applications realized with GKS (GSS 1989).

One of the principal problems of interface definition lies in the processing of complex geometries. Currently, there are no adequate definitions, especially in the three-dimensional sector. Only attempts at data exchange at a geometrically simple level can therefore be realized.

Since the metal-forming calculations in FEM programs can generally be carried out with linear elements, interface definitions can be confined to simple geometries (lines, points etc.). The complex structures are created and sophisticated within the CAD/CAM system, using the wide range of options available (Dassault 1986).

Interfaces permitting, for example, the transfer of material data must be realized in a similar way. Fig. 17.25 shows the structure of a material data management system with interfaces to simulation programs and test evaluation systems (CPM

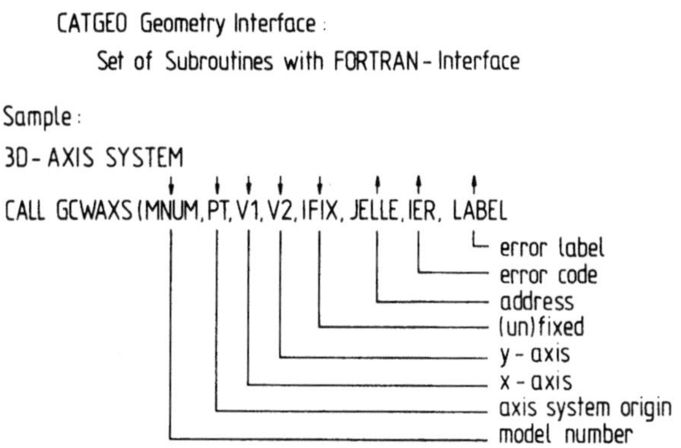

Fig. 17.22. Interface 3D-axis: axis system.

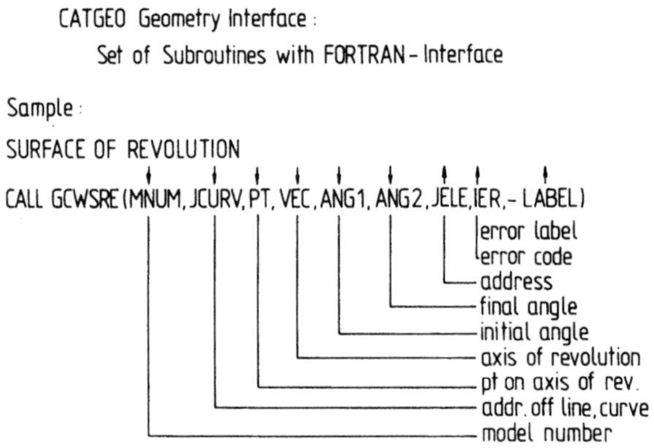

Fig. 17.23. Interface: surface of revolution.

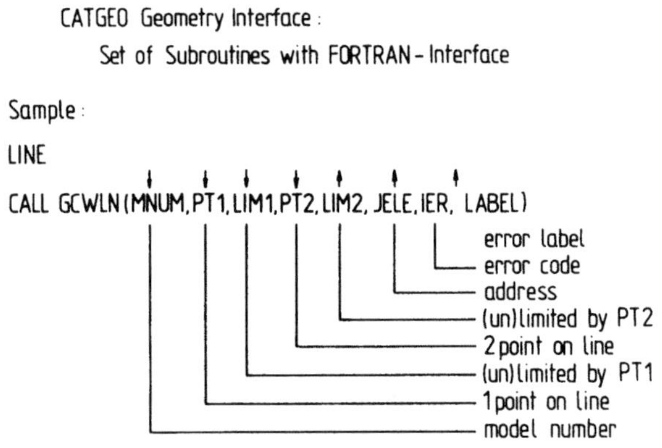

Fig. 17.24. Interface: line.

1989a). This system enables data to be handled at all stages from acquisition via management to use in simulation programs.

Standards are currently being developed for material data management systems of this kind, so that here again no adequate universal interfaces are available.

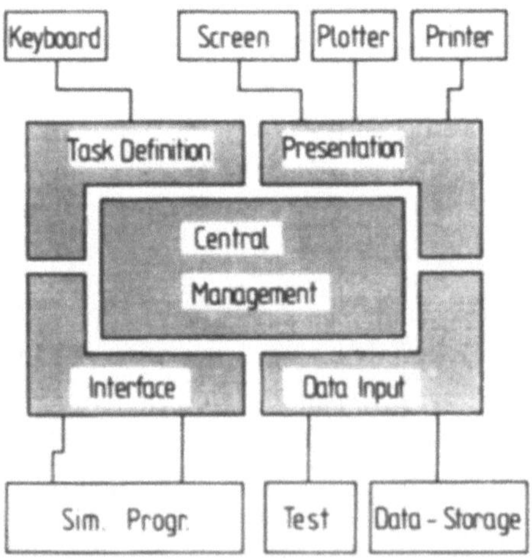

Fig. 17.25. Material data management system.

An interface specially suited to metal-forming technology has therefore been defined within the CAPS structure. This enables stress–strain curves and material parameters such as specific heat, density, etc. to be processed. The data are managed in the system and accessed via a Fortran interface of the CAPS data structure. In addition, corresponding interfaces for technology data have been defined and programmed in the CAPS structure.

Fig. 17.26 shows the example of the data structure contained in the CAPS rolling package referred to above (CPM 1989b). It is defined in common blocks (Fortran) and classified according to program needs into plant data, material data, pass data, simulation result data and control data.

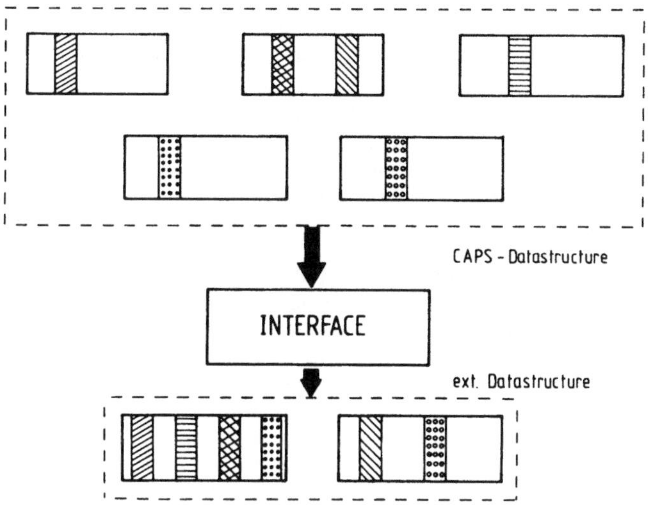

Fig. 17.26. CAPS interface: external program.

Interface programs permit the integration of FEM and CAD systems referred to above. Data from the CAPS structure are converted for the other structure in each case. In this way FEM and CAD techniques can be integrated in metal-forming program packages.

17.5 Summary

Modern production technology can no longer operate effectively without the use of the computer. The combination of production engineering with information technology has opened up new potentials for cost reduction and improved quality. The closed information flow from planning via design to production and quality assurance has acquired stategic importance.

In the individual stages of production, highly effective computer programs are already available and in use. New program packages, models, etc. are constantly being developed. This chapter has discussed the interlinking of these modules to provide a closed information flow. An adapted, multistage simulation and optimization process has been analysed, with special attention to the interlinking of elementary models, FEM programs and expert systems. Interface problems with CAD/CAM programs have also been discussed. Ways have been indicated in which different program modules from different sources can be integrated in order to provide a continuously extendable optimum program system for the CAD/CAE/CAM sectors.

References

Arfmann GH (1990) Rechnereinsatz in der Umformtechnik unter Einbindung von Expertensystemen. Dissertation RWTH Aachen

Becker M, Kopp R (1989) A new approach to optimization of metalforming processes. Numiform '89, Balkema, Rotterdam, pp 107–113

Brauner K et al. (1981) IGES. American National Standards Institute, New York

CPM (1989a) CAPS-Mat, Benutzerhandbuch. CPM GmbH, Aachen

CPM (1989b) CAPS-Walzen, Benutzerhandbuch. CPM GmbH, Aachen

Dassault (1986) CATIA, Benutzerhandbuch. Dassault Systems, Suresnes

de Souza MM (1989) Simulation und Planung des thermomechanischen Walzens von Profilen durch Verwendung von Umformmodellen und Werkstoffmodellen. Dissertation, RWTH Aachen

GSS (1989) GSS-GKS, Manual. GSS Inc, Portland, Oregon

Kopp R, Mauk P-J (1982) Stofflußuntersuchungen beim Walzen von I-Trägern. Abschlussbericht zum Forschungsvorhaben FAA-Nr 668, RWTH Aachen

Kopp R, Mauk P-J (1983) Rechnerunterstütztes Kalibrieren. Abschlussbericht zum Forschungsvorhaben FA 8646, RWTH Aachen

Kopp R, Helsper T, Kallabis H-P, Osterburg H (1984) Modelle zur Berechnung des Stoffflusses für Stabstahl und Formstahlkaliberreihen zur Weiterentwicklung der Kalibrierungstechnik. Abschlussbericht zum BMFT-Forschungsvorhaben S 234, RWTH Aachen

PDA (1987) PATRAN – manual. PDA Engineering, München

Spur, G, Krause F-L (1984) CAD-Technik. Carl Hauser Verlag, München Wien

18 Application of Knowledge-Based Techniques

T. Altan and R.A. Miller

18.1 Introduction

In a deformation process, dies are used to transform a given material with a simple shape into a useful part. The resulting part has a complex geometry with well-defined shape, size, tolerances, appearance and properties (Altan et al. 1983). The desired part geometry is stored in the dies and imparted onto the material with pressure through the material–die interface.

The physical phenomena describing a deformation process are difficult to express with quantitative relationships. The material flow, the heat generation and transfer during the process and the relationships between the process variables and product properties are difficult to predict. As a result, the design of a deformation process and the associated dies is more an art than a science. Ideally, the engineering design of a deformation process consists of: (1) establishing the kinematic relationships (shape, velocities, strain rates, strains) between the incoming material and the shaped product, i.e. predicting material flow, (2) establishing the limits of producibility, i.e. determining whether it is possible to deform the part without surface or internal defects, and (3) predicting the forces and stresses necessary to execute the deformation so that the tooling and equipment can be designed and selected.

In most practical forming applications, the shape of the final product is too complicated to be formed in one forming operation. As a result, intermediate deformation, or preforming steps must be used to transform the incoming simple geometry into a final complex shape. In most cases the die and process designers are forced to use trial and error before they can achieve a satisfactory process sequence, i.e. number and geometry of intermediate preforming steps.

In establishing the sequences in practical deformation processing, it is appropriate to consider the differences between sheet and massive forming. In both cases, the surfaces of the deforming material and the dies are in contact, and friction between them has a major influence on the process. However, other differences in load, stresses, and springback influence design and tolerances (Altan 1987).

In massive forming, the input material is in billet, rod or slab form, and a considerable increase in the surface to volume ratio and changes in cross-sections take place. The portion of the workpiece undergoing permanent plastic deformation is generally much larger than the portion undergoing elastic deformation. As a result, elastic recovery after deformation is usually negligible but the deformation stresses are much higher, three to ten times, than those observed in sheet forming.

In sheet forming, a sheet blank is plastically deformed into a three-dimensional object without significant changes in sheet thickness. In some cases, the magnitudes of permanent plastic and recoverable elastic deformations are comparable; therefore, elastic recovery or springback may be significant. Thus, it is appropriate to consider the modelling of sheet and massive deformation processes separately.

18.2 Application of Numerical Modelling and Knowledge-Based Systems

With the widely accepted use of computer-aided design/manufacturing (CAD/CAM) die manufacturing has been extensively computerized. The present trends now are to (1) improve die and process design methods by modelling and optimizing various deformation processes, and (2) establish an integrated design and manufacturing activity that utilizes not only modelling techniques but also knowledge-based systems (KBS) for preliminary die design and selection of initial process conditions.

KBS may be defined as a computer-based system that uses an explicit representation of knowledge, facts and reasoning techniques to solve problems that normally require the abilities and skill of human expertise. A typical KBS consists of

1. A knowledge acquisition facility, that receives and collects rules and facts from human experts in a specific area of expertise, or domain. Such facility may also include a "rule editor", allowing the improvement of existing and addition of new rules by the user.
2. Data bases to store facts and rules usually with interfaces to other data base systems.
3. An inference engine that determines how to apply the knowledge rules to solve a problem.
4. A user interface that allows the non-expert user to utilize the KBS for solving specific problems.

In order to make a KBS suitable for a task, it is desirable that:

- The KBS will save time and money or will perform the task in a better way than it is done conventionally.
- Experts are available to gather the knowledge for building the KBS and nonexperts are available who require this expertise.
- The time required to perform the task is neither too trivial nor too involved.

Some obvious benefits obtained from using a KBS include:

- The knowledge of the expert is captured and is available to users when the expert is not available.
- The system is thorough and systematic; it is easy to add new knowledge as it becomes available.
- The KBS can be used as a learning tool; a large number of people may benefit from the expert's knowledge.
- A well-built KBS helps to clarify the knowledge in the field because knowledge must be precisely characterized to be entered into the KBS.

The application of numerical modelling and KBS in material deformation processes has the following objectives:

1. Assist die design and process development, by
   - Investigating and assuring die fill
   - Preventing flow-induced defects such as laps and cold-shuts

- Avoiding exceeding forming limits so that internal and external cracks are prevented
- Predicting and controlling temperatures so that part properties, friction conditions and die life may be controlled

2. Improve part quality and reduce costs associated with deformation processes, by
   - Predicting and improving grain flow and microstructure
   - Reducing die try-out and lead times
   - Reducing rejects and improving material yield

As seen in Fig. 18.1, these objectives are best achieved by carrying out the following steps:

1. Establish a preliminary die design and initial selection of process parameters by an experienced die designer or process engineer, or a knowledge-based system (KBS), sometimes called an expert system.
2. Verify initial design and process conditions via numerical modelling.
3. Modify die design and process variables, as needed, based on the results of modelling.
4. Complete die design and manufacture the dies.
5. Conduct die and process try-outs on production equipment.
6. Make modifications on dies and process conditions, if necessary (hopefully, and if modelling is accurate, at this stage very little, if any, modifications will be needed).

As seen in Fig. 18.1, excellent applications for KBS in deformation processing exist in (1) designing for producibility by a given metal deformation process, (2) designing the initial design of dies and process sequences, and (3) establishing the initial process conditions such as die and material temperatures and deformation rates. In addition, KBS can be quite useful in increasing the efficiency of the finite-element method (FEM)-based simulation of metal flow. In this application, KBS can assist in selecting initial conditions, establishing the initial mesh on the deforming material, and, most importantly, in developing a procedure for automatic remeshing during simulation.

The combined application of KBS and FEM has been recently introduced to practice in the UK (Dean et al. 1990). This "intelligent knowledge-based system" is used for forging die design and has been incorporated into a commercial CAD system.

18.3 Design of Parts and Process Sequences in Cold Forging

Cold forging is a type of deformation process wherein material at room temperature is forced to flow plastically under compressive force into a variety of shapes (Altan et al. 1983). These shapes are usually axisymmetric with relatively small nonsymmetric features, and, unlike impression die forging, the process does not generate flash. The terms "cold forging"and "cold extrusion" are often used interchangeably and refer to well-known forming operations such as extrusion, upsetting or heading, coining, ironing and swaging (Lange 1985).

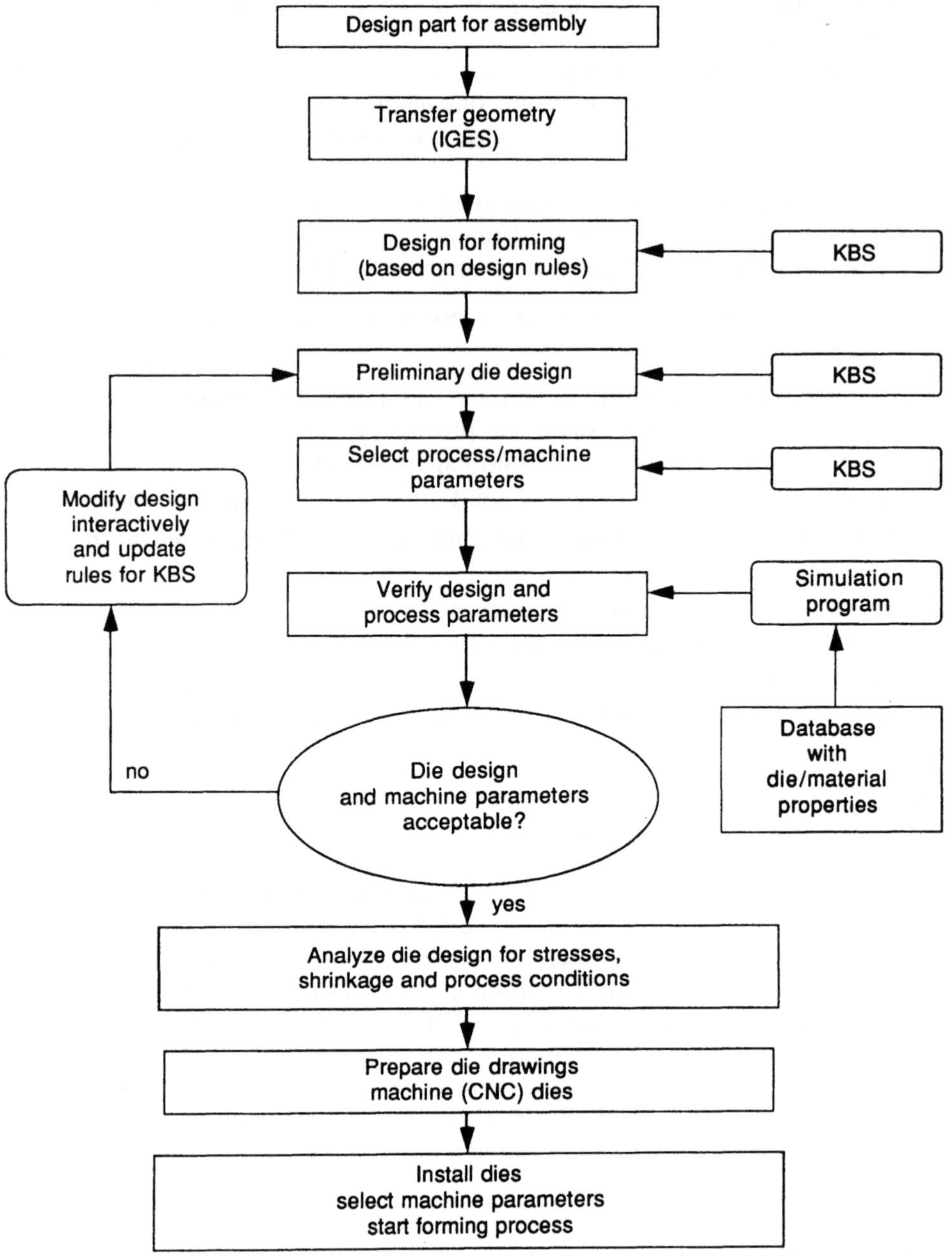

Fig. 18.1. Application of numerical modelling and KBS in material deformation processes.

Cold forging requires several preforming operations to transform the initial simple billet geometry into a more complex product geometry without any surface or internal defects, Fig. 18.2 (Sagamuller 1968). Process design for cold forging includes: (a) the design of the cold-forged part, (b) establishing the cold-forming sequence, and (c) the design of the dies for each station. Establishing the forming sequence is more of an art than a science, involving creativity, intuition and experience. Therefore, it is desirable to develop an integrated process planning

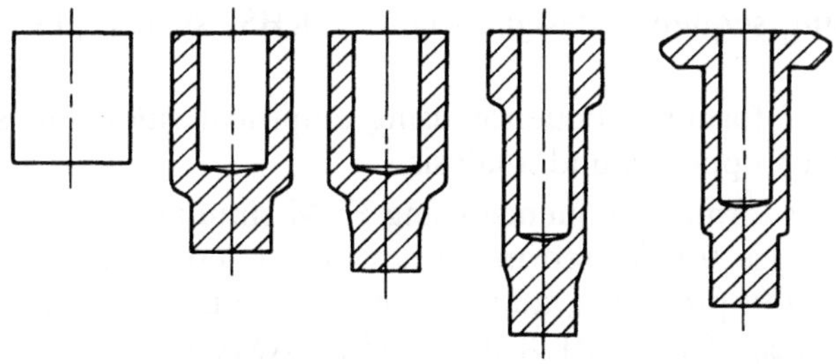

Fig. 18.2. Schematic illustration of forming sequences in cold forging of a gear blank (Sagemuller 1968). Left to right: Sheared blank, simultaneous forward rod and backward cup extrusion, forward extrusion, backward cup extrusion, simultaneous upsetting of flange and coining of shoulder.

system, using KBS methodology, as illustrated in Fig. 18.3. This system consists of the following major building blocks (Sevenler et al. 1989):

1. Final assembly-ready product design, where cold-forging process characteristics should be considered.
2. Cold-forged product design, where a KBS, such as PREFORMEX, may be used.

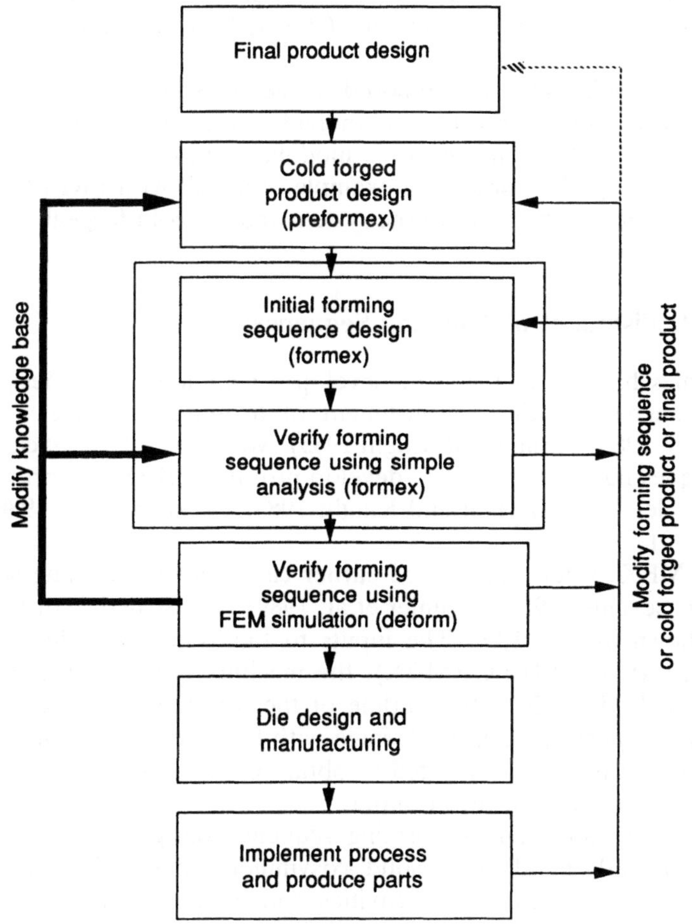

Fig. 18.3. Integrated process planning system.

3. Initial forming sequence design, where a KBS, such as FORMEX may be used.
4. Verification of forming sequence using empirical rules and simple analyses such as slab or upper bound analysis.
5. Verification of forming sequence using FEM simulations, to (a) optimize the forming sequence for desired metal flow and microstructure, (b) predict and avoid formability problems, and determine if annealing is necessary, and (c) modify the rules of FORMEX and PREFORMEX.
6. Die design and manufacture, using commercially available CAD/CAM systems.
7. Verify process in actual operation, hopefully with minimum amount of corrections.

18.3.1 Design for Cold Forging

A prototype KBS was developed to assist die designers in modifying the final part geometry to design the cold-forgeable part (Ganapathy et al. 1988). The structure of the system, called PREFORMEX, is shown in Fig. 18.4. The input to the system is the assembly-ready part. PREFORMEX is interfaced through Initial Graphics Exchange Specifications (IGES) files to commercial CAD/CAM systems.

The PREFORMEX shell is composed of the rules used for the control of the overall system. It hands control to the individual modules in the sequence shown in Fig. 18.4. The rest of the system is modular, and each module accomplishes a specified task. Fig. 18.5 shows an example obtained with PREFORMEX. The assembly-ready part is the input to the system and the cold-forgeable part drawing is the output.

18.3.2 Establishment of Forming Sequences

Several studies have been made to develop a computer-aided design system to assist die designers in the establishment of forming sequences (Noack 1973, Badaway et al. 1985, Bariani and Knight 1985). More recently, artificial intelligence (AI) techniques have been used to develop prototype KBS for forming sequence design (Sevenler 1986, Bariani and Knight 1988, Osakada et al. 1988, Azushima and Kim 1990).

The FORMEX system mentioned earlier can establish forming sequences for cold forging (Sevenler 1986, Sevenler et al. 1989). The structure of the FORMEX system is shown in Fig. 18.6. The inputs to the system are the cold-forgeable part geometry (from PREFORMEX), the machine type and the material type. The output of FORMEX is the outline of the forming sequence as well as the operations to be performed in each stage with the relevant dimensions. Machine and material data files are consulted to obtain the specifications for the machine and material type specified in the input.

Fig. 18.7 shows an example forming sequence designed by FORMEX. The system can currently handle only solid axisymmetric parts. Work is underway to modify the system for parts with cavities and hollow parts. FORMEX is also being modified so that it can evaluate existing sequences for any violation of known cold forging guidelines. When this is completed, the user will be able to

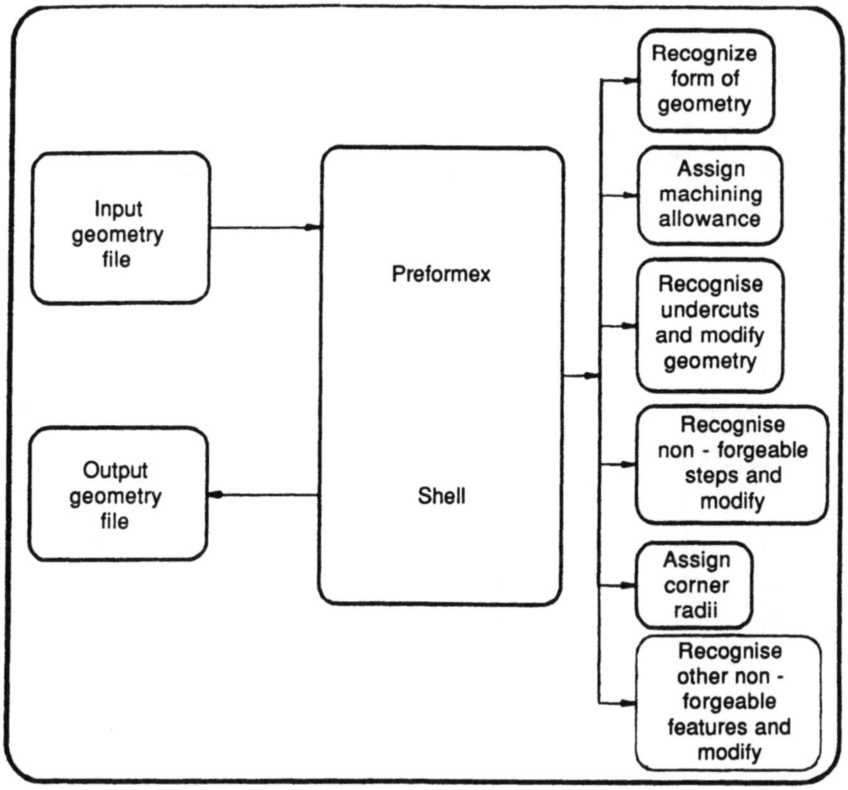

Fig. 18.4. Structure of the PREFORMEX system (Ganapathy et al. 1988).

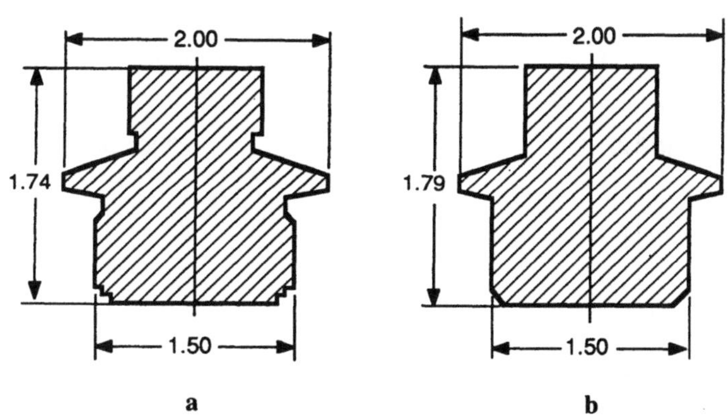

Fig. 18.5. PREFORMEX example: **a** input of assembly-ready part; **b** cold-forgeable part drawing output.

enter existing sequences or his own guesses to the program to evaluate the practicality of these designs.

18.4 Application of KBS in Hot Forging

Similar to cold forging, KBS may also be used in hot-forging process design, specifically to carry out the following functions: (1) the conversion of the machined

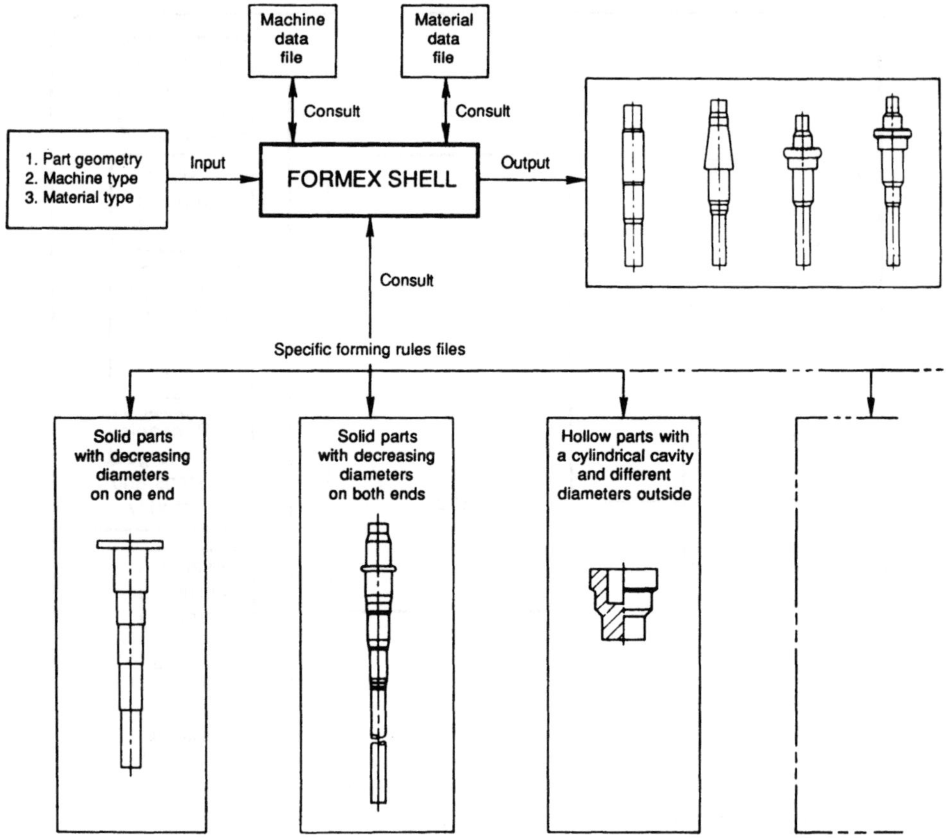

Fig. 18.6. Structure of the FORMEX system (Sevenler 1986).

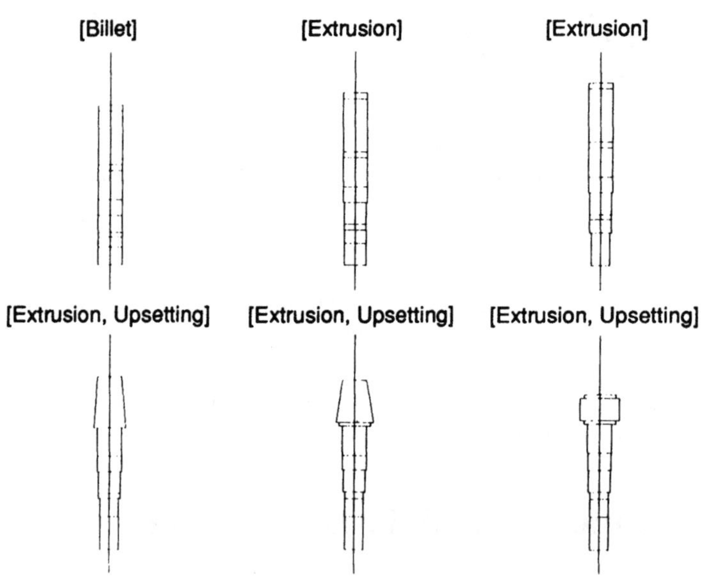

Fig. 18.7. Example forming sequence designed by FORMEX.

part geometry into a forging design geometry using guidelines associated with forging design and process limitations, and (2) the design of blocker or preform dies, including the calculation of stock volume with flash allowance and the estimation of blocker and preblocker die geometries. In addition, in hot forging the selection of process conditions such as die and billet temperatures, ram or deformation speed, and lubrication conditions can also be accomplished by using a KBS.

18.4.1 Design of Finish Forgings

Given the geometry and material of assembly-ready parts and the forging process conditions, a forging designer uses experience-based know-how to generate an optimal forging part geometry that can be produced by the intended forging and subsequent machining operations. The designer evaluates and modifies key design features, seen in Fig. 18.8 (Tang et al. 1988). In general these features include: parting line, forging plane, finish allowances, web thickness, rib width, draft angle, corner radii and fillet radii. To reduce the complex forging design problem to more manageable subprograms, the designer usually decomposes the forging design task into subtasks, each corresponding to a key design feature. This design procedure is mostly experience based and is affected by the following process variables: forging type (blocker, conventional, precision), forging material, forging geometry (complexity, thickness of ribs and webs, etc.), forging machines (hammer, mechanical or hydraulic press), forging and die temperatures, and production quantity.

When using a knowledge-based approach to automate the forging design process, the representation and manipulation of the forging geometry and forging

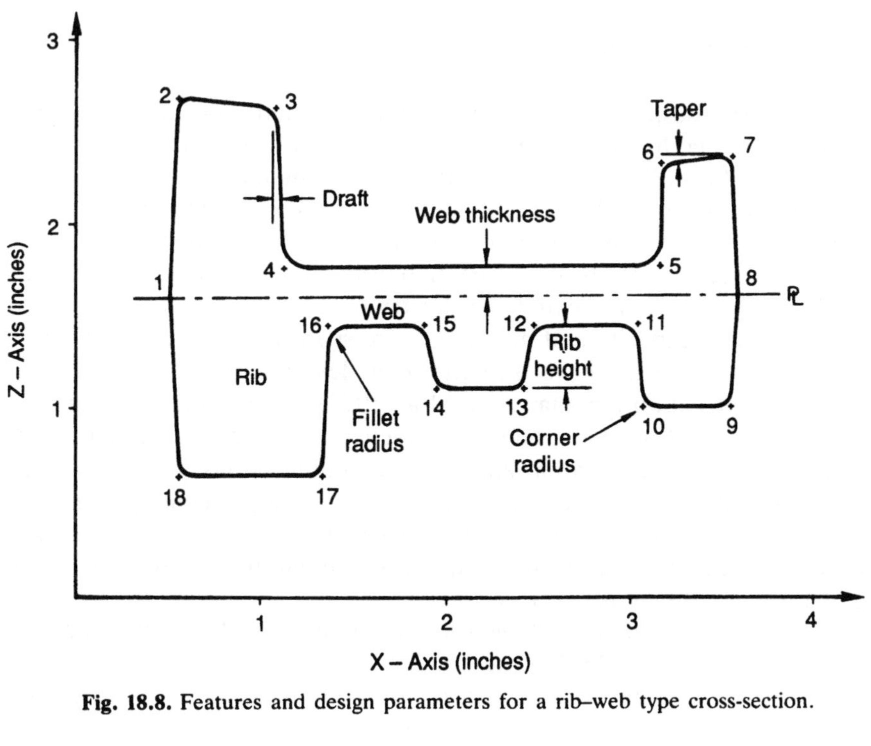

Fig. 18.8. Features and design parameters for a rib–web type cross-section.

design guidelines are essential. In the development of a KBS called an automatic forging design (AFD) program, a mixed represention scheme that combines the advantages of rule and frame-based schemes was used. The AFD system is capable of designing forgings in two dimensions, i.e., by considering sections of a given part. AFD consists of a forging design data base and four main modules, namely: AFD control, AFD input, AFD design and AFD output modules. The forging design data base contains forging design specifications such as forging part geometries, process conditions, and special forging design considerations. It is used as a working memory and its information can be accessed by all AFD modules, which use a combination of knowledge-based and algorithmic techniques. Fig. 18.9 illustrates two example forging section designs, generated by AFD, for two different materials, using the same assembly-ready part geometry. The forging had a plan view area of 106 in^2 and was to be manufactured in a production lot size of 200 pieces in a hydraulic press.

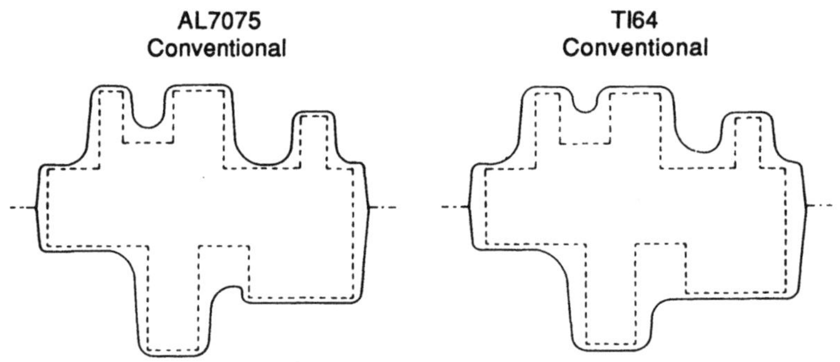

Fig. 18.9. Example forging sections generated by AFD for two different materials.

18.4.2 Design of Preforms or Blockers

In hot forging the design of intermediate stages, called blockers or preforms is of critical importance for the success of the forging process, particularly in closed die forging with flash. This design activity is predominantly manual because the determination of proper blocker configuration is a very difficult task, an art requiring skills achieved only through years of experience. Most of the time, the designers are forced to use trial and error before they can achieve a satisfactory blocker design. As a result, extensive die try-outs are necessary and expensive productive machine capacity and man-hours are spent in developing blocker dies.

Some prototype systems have already been developed for CAD of blockers (Yu and Dean 1985, Subramanian and Altan 1978) using interactive graphics. In designing forgings and blockers, common practice is to consider the critical cross-sections of a part in order to approximate, in two dimensions, the complex three-dimensional metal flow present in a practical forging. In principle, given a forging cross-section, Fig. 18.8, designing a blocker can be considered as modifications of appropriate design features or parameters, similar to the approach used in design of forgings, discussed in Sect. 18.4.1.

By changing the procedure followed for modifying the design parameters of a forging cross-section, Fig. 18.8, different blocker designs can be generated. To ensure that these designs are practical, the modification procedures should consider the effect of various process variables, such as: finished part geometry, material characteristics (forgeability and flow stress), forging equipment to be used (type, speed, and load/energy capacity), forging tolerances (closeness to machined shape), production lot size, and forging temperatures (of dies as well as forgings). Blocker design is a highly knowledge-intensive activity, suggesting the usefulness of the KBS approach. Using KBS techniques, a prototype computer program, called BID (blocker initial design) was developed. It is planned that the predictions made by BID must be further evaluated by FEM simulations, as illustrated in Fig. 18.1.

Being a KBS, the program BID separates the knowledge base and the mechanism for using this knowledge, i.e. the inference engine. The knowledge base is structured in the form of "IF condition THEN action" instructions, or rules. Such a clean separation of blocker design knowledge from the rest of the computer system simplifies the evolutionary growth for the knowledge base. Reasoning over knowledge contained in this KBS system is accomplished using symbolic instead of numerical processing. BID is implemented using LISP and Knowledge Craft, an expert system shell. In addition to the symbolic processing, FORTRAN is used for numeric computations. An example of blocker geometry, designed automatically with the program BID for the forging section seen in Fig. 18.8, is given in Fig. 18.10.

One of the most significant aspects of developing a KBS for blocker design is the knowledge acquisition. Since very little blocker design information is available in the literature, information gathering relied heavily on practising die designers, and the literature could provide only a general background as a starting point.

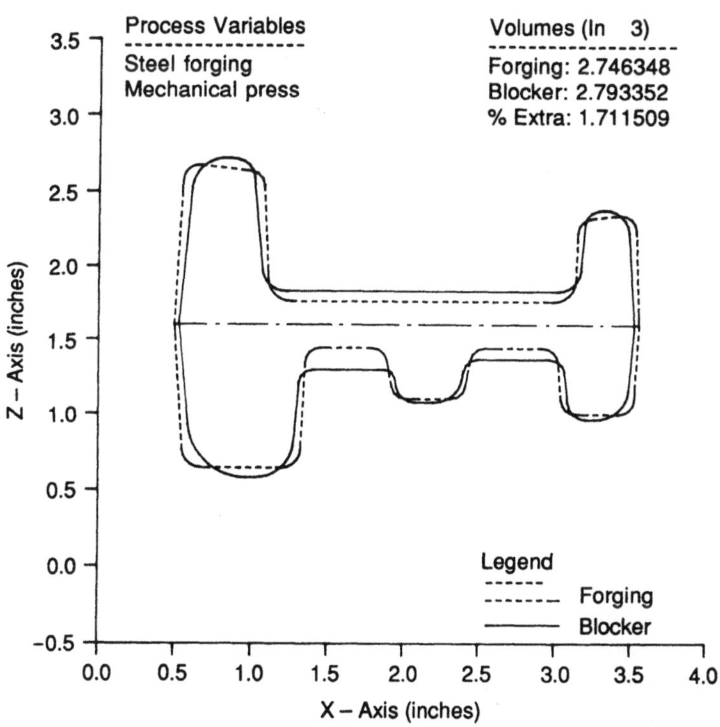

Fig. 18.10. Blocker design for the forging cross-section shown in Fig. 18.8 assuming a steel forging and a mechanical press.

Accordingly, blocker design knowledge was compiled by (1) reviewing the literature, (2) mailing a questionnaire on blocker design to various forging companies, (3) discussing the topic with designers from a few companies and (4) obtaining assistance from two designers on a consultancy basis (Vemuri et al. 1989).

18.5 Applications in Sheet-Metal Forming

The state of the art in die design in sheet forming is very similar to that in cold and hot forging. After many years of practice, the sheet-forming technology today still remains more an art than a science. However, CAD/CAM techniques are being applied by various companies to improve and partially automate the die design and manufacturing functions (Sfiligoj 1983, Murakami et al. 1980, Okamoto et al. 1988). Attempts are also being made to computerize process design in sheet forming by using both KBS as well as finite-element- or finite-difference-based systems. As in many applications in manufacturing, process planning and design of progressive dies in sheet forming are important fucntions that can benefit from the application of KBS. The prototype systems developed so far are very similar, in principle, to the systems developed for process sequence design for cold forging, discussed in Sect. 18.4.2.

Automatic generation of multitechnology process outlines (AGMPO) is an experimental system for developing a multitechnology process outline for the manufacture of axisymmetric, monotonic cups of nominal uniform wall thickness that are deep drawn and machined (Eshel et al. 1986). The input to the system is a CAD representation of the cross-section of the final object geometry and the output from the system is the process outline to manufacture the part. The system, in the present form, can handle only monotonic geometries. In other words, it cannot handle geometries requiring reverse operations. The system relies only on experience-based die-design guidelines for process sequence design.

Tisza's (1986) work describes a modular CAD/CAM system for developing a deep-drawing process sequence and designing tools for the manufacture of sheet-metal components having axisymmetric and rectangular cross-sestions. It consists primarily of a geometric description module for reading in the object geometry, a blank module for determining the optimum shape, size, and nesting of blanks, a technological design module for designing a process sequence based on empirical rules and technological parameters, a tool design module for designing the tools and selecting a tool of standard size, and an NC/CNC postprocessor module for preparing control programs for NC/CNC manufacturing of tool elements.

A hybrid computer-aided engineering (CAE) system for designing the process sequence for the manufacture of axisymmetric sheet metal components is under development. The principle of this system is very similar to the general approach depicted in Fig. 18.1. The input to the CAE system is the final-sheet metal object geometry to be manufactured, and the output is the recommended process sequence with the geometries of the intermediate forming steps. Initial effort is concentrated on axisymmetric parts because of the large range of applicability and the ease of development (Sitaraman et al. 1989).

Two main components of the hybrid CAE system are a KBS or expert system module (symbolic module) and a process-modelling analysis module (numeric module). The KBS module and the analysis module are separate and stand-alone. The user would first run the KBS module to design an initial process sequence. This module contains experience-based die design guidelines and empirical formulas. The user would then run the analysis module having the geometry output by the KBS as the input to the analysis module and determine the stress and strain distributions. If the predicted strain distribution is within agreeable limits (as suggested by a certain failure-prediction criterion for the material), the process sequence needs no further modification. Otherwise, the analysis results are used to improve the sequence design as well as the rules used by the KBS.

Ultimately, the KBS and analysis modules will be integrated to produce a CAE system in which the KBS will communicate with the analysis module to test and rectify the generated process sequence. Potential sources, used to obtain knowledge for constructing the KBS module include: (1) technical literature, (2) industrial brochures and nonproprietary industry publications, and (3) industrial visits and discussions with industry experts.

The KBS module of the hybrid system, described here has been used to design forming sequences, for example industrial multistage forming sequences. In all cases, the predicted designs were either the same or very close to those used in industrial practice. An example KBS design for a flanged shell with 0.065 in wall thickness is seen in Fig. 18.11.

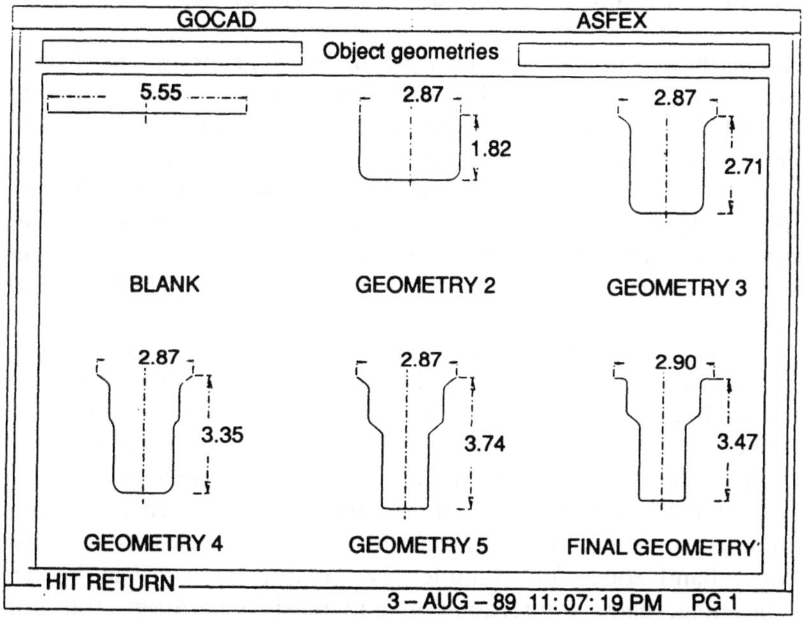

Fig. 18.11. Process sequence for a flanged shell with two diameters – KBS (Sitaraman et al. 1989).

18.6 Applications in FEM Modelling of Forming Processes

18.6.1 Mesh Generation

When using FEM-based programs to analyse metal-forming problems, the complex geometry needs to be divided into a finite number of simple geometric elements.

The collection of the elements, i.e. the mesh, is generated using a mapping method. That is, the user divides the geometry into four-sided patches and assigns desired density distributions for the patches, using interactive graphics. A mesh generator will then generate the FEM mesh on each patch (Wu et al. 1990). This procedure is not only time consuming but also error prone. In addition, the mesh density to be assigned by the user usually requires prior knowledge of the solution behaviour or extensive user experience.

To achieve good results from FEM with a reasonable amount of computational effort, the mesh density should be large in regions where large gradients of the computed values, e.g. temperature, stress, strain, etc. are expected. Thus a KBS could be potentially useful in the selection of mesh density for specific problems since it may allow the user to benefit from the past experience obtained in running FEM programs for a variety of geometries and boundary conditions. To determine the mesh density automatically, several adaptive and KBS systems have been suggested (Rank and Babuska 1986, Zienkiewicz and Zhu 1987). Starting with a simple initial mesh, the adaptive method refines the mesh based on estimated local error that occurs in the previous solution. The adaptive method has been successfully applied to various structural and fluid dynamics problems where the geometry and boundary conditions do not change during computations.

18.6.2 Remeshing and Automated Mesh Generation

In simulating the forming processes with an FEM program, the FEM mesh is updated at every time step by adding the displacements to the coordinates of the nodes. Therefore, the mesh may degenerate after several steps depending upon the die geometry. At that time, another FEM mesh within the updated deformed geometry must be generated in order to continue the simulation. An example is given in Fig. 18.12 (Wu and Oh 1989, personal communication). For this particular simulation, nine remeshings were required. It is obvious that an automated procedure for remeshing would greatly facilitate the use of FEM for simulation of large deformation processes.

Remeshing is also required when a large mesh element does not have the degree of freedom necessary to describe the details of material flow expected because of die boundary conditions. An example is given in Fig. 18.13. The large element near the flash area must be divided into many small elements for a realistic simulation of the metal flow around the flash area. Similar conditions may exist within the material, in deformation regions where the strains, strain rates or temperatures exhibit large gradients, such as the flow conditions present near a shear band zone, for example. Two major tasks are to be performed during remeshing, namely interpolation and mesh generation. Interpolation is necessary when material properties are history dependent. The process variables such as strain and temperature need to be interpolated from the old mesh into the new mesh. This procedure is well established.

The automation of mesh generation, especially the automatic generation of an "optimum mesh" for an arbitrarily shaped part, is no trivial task. The input into such an automatic system must be: (1) geometry of the blocker or preform and (2) process related information, including die geometry and kinematic conditions in the process. The optimum mesh must, on one hand, be capable of describing

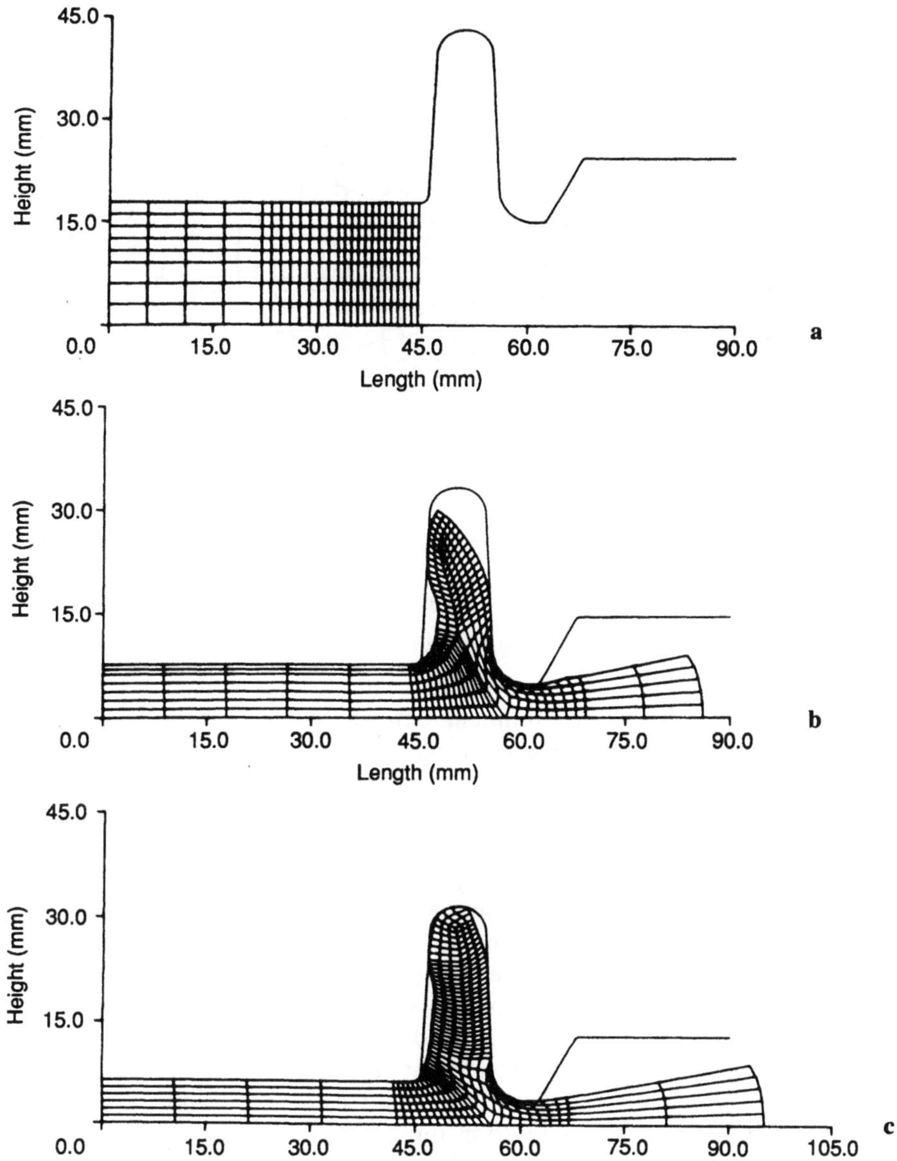

Fig. 18.12. Material flow during rib–web forging predicted by ALPID: **a** undeformed; **b** die stroke = 0.1 mm; **c** die stroke = 11.5 mm (Wu and Oh 1989, personal communication).

the local deformations adequately while, on the other hand, it must maintain the total mesh density to a minimum to be computationally efficient. Thus, the higher density elements/nodes must be generated in the area of high gradients of calculated values (displacements, velocities, temperatures, strain rates) while a lower density of elements/nodes should be generated in the area of low gradients.

In general, error estimation may be the best approach to determine the mesh density, since error distributions indicate where elements should be refined and where elements should be unrefined, regardless of the type of the problem.

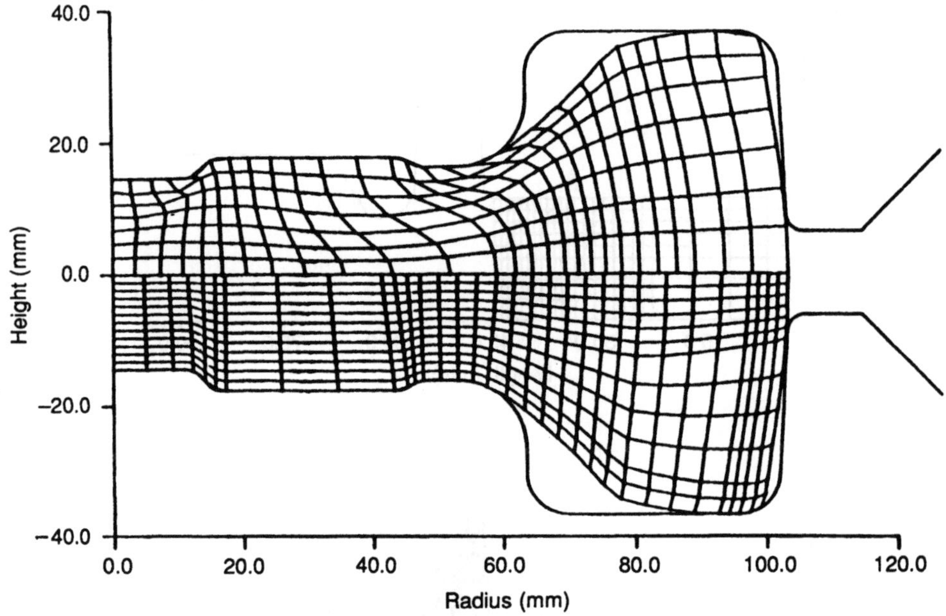

Fig. 18.13. Example of remeshing required because of die boundary conditions.

However, in simulating large nonlinear deformations, as in metal forming, this approach cannot be used because of the very large number of computations required. The alternative may be to utilize KBS for estimating a reasonable density distribution, as it has been done in computational fluid dynamics (Andrews 1988). In metal forming, this approach would require the consideration of (1) current deformation mode, which may be determined by using the correlations between the die geometry, die motion, and the deforming material, (2) the future deformation mode, where the user must estimate which free material surface will be contacted by the die so that mesh density may be increased in that area to improve accuracy, or where deformation is simple so that mesh density may be reduced to save computing time. In addition, in estimating the mesh density distribution, it is necessary to consider at least the formation of flow-induced defects, internally as shear bands and externally as laps or folds. This requirement also makes automatic mesh generation quite complex and could possibly benefit from the use of a KBS.

Research on automatic meshing and remeshing is being conducted in various laboratories around the world and it may be expected that, within a few years, automatic mesh generation and remeshing for metal-forming simulation will be a practical reality.

References

Altan T (1987) Design and manufacture of dies and molds. Ann CIRP 36:455

Altan T, Oh SI, Gegel HL (1983) Metal forming: fundamentals and applications. American Society for Metals, Metals Park, Ohio

Andrews AE (1988) A knowledge-based approach to automate flow field zoning for computational fluid mechanics. PhD dissertation, Department of Mechanical Engineering, Stanford University

Azushima A, Kim M (1990) A PC based expert system for forming sequence design. Ann CIRP 1:39(1)
Badaway AA et al. (1985) Computer-aided design of multistage forging operations for round parts. J Mech Work Technol 13:259–274
Bariani P, Knight WA (1985) Computer aided cold forging process design, determination of machine setting conditions. Ann CIRP 34/1/1985:245–252
Bariani P, Knight WA (1988) Computer aided cold forging process design, a knowledge-based system approach to forming sequence generation. Ann CIRP 1:243
Dean TA, Pillinger I, Hartley P, Sturgess CEN (1990) Research and industrial application of CAD/CAM and CAE in the British forging industry. Metallurgia 57:32–36
Eshel G, Barash MM, Johnson W (1986) Rule based modeling for planning axisymmetric deep drawing. J Mech Work Technol 14(1):1–115
Ganapathy R, Miller RA, Altan T (1988) A knowledge-based systems approach to design parts for cold forging. NSF Research Center for Net Shape Manufacturing, report no ERC/NSM-88-44, The Ohio State University, Columbus, Ohio
Lange K (1985) Handbook of metalforming. McGraw-Hill
Murakami H et al. (1980) A CAD system for progressive dies. In: Proceedings of 21st machine tool design and research conference, pp 587–591
Noack P (1973) Computer-aided determination of operation sequence and costs in cold forging of rotational-symmetric workpieces. SME Technical Paper, MF73-141
Okamoto I et al. (1988) Computer aided design and evaluation system for stamping dies in Toyota, advances and trends in automotive sheet steel stamping. SAE, Inc., Warrendale, Pennsylvania, pp 79–89
Osakada K et al. (1988) Application of AI technique to process planning of cold forging. Ann CIRP: 239
Rank E, Babuska I (1986) An expert system for the optimal mesh design in the HP-version of the finite element method. Technical note BN-1053, Institute for Physical Science and Technology, University of Maryland, College Park, Maryland
Sagamuller Fr (1968) Cold impact extrusion of large formed parts. Wire (no 95):2
Sevenler K (1986) Knowledge-based systems approach to forming sequence design for cold forging. PhD dissertation, The Ohio State University, Columbus, Ohio
Sevenler K et al. (1989) Progressive die design for cold forging, ERC for net shape manufacturing. Report no ERC/NSM-B-89-16, The Ohio State University, Columbus, Ohio
Sfiligoj M (1983) Sheet metal system built. AMM November:10
Sitaraman S et al. (1989) Process sequence design for multi-stage forming of axisymmetric sheet metal parts. ERC for net shape manufacturing, report no ERC/NSM-S-89-4, The Ohio State University, Columbus, Ohio
Subramanian TL, Altan T (1978) Application of computer aided techniques to precision closed-die forging. Ann CIRP 27 (1):123
Tang I et al. (1988) A knowledge-based approach to automated forging design. J Mater Shap Technol 6(1):7
Tisza M (1986) A CAD/CAM system for deep-drawing processes. Proceedings of 2nd ICTP conference, Stuttgart, pp 145–152
Vemuri KR et al. (1989) BID: a knowledge-based system to automate blocker design. Int J Mach Tools Manuf 28(4):505
Wu WT et al. (1990) Automated mesh generation for forming simulation-I, submitted to ASME international conference on computers in engineering, Boston, Massachusetts
Yu GB, Dean TA (1985) A practical computer aided approach to mould design for axisymmetric forging cavities. Int J Mach Tools Manuf 25(1):1
Zienkiewicz OC, Zhu JZ (1987) A simple error estimator and adaptive procedure for practical engineering analysis. Int J Num Meth Engng 24:337

Subject Index

MIX
Papier aus verantwortungsvollen Quellen
Paper from responsible sources
FSC® C105338

If you have any concerns about our products,
you can contact us on
ProductSafety@springernature.com

In case Publisher is established outside the EU,
the EU authorized representative is:
Springer Nature Customer Service Center GmbH
Europaplatz 3, 69115 Heidelberg, Germany

Printed by Libri Plureos GmbH
in Hamburg, Germany